W0225813

HIGHLIGHTS OF ASTRONOMY

INTERNATIONAL ASTRONOMICAL UNION
UNION ASTRONOMIQUE INTERNATIONALE

HIGHLIGHTS OF ASTRONOMY

AS PRESENTED AT THE XIIIth GENERAL ASSEMBLY
OF THE I.A.U.
1967

EDITED BY
LUBOŠ PEREK
(General Secretary of the Union)

D. REIDEL PUBLISHING COMPANY

DORDRECHT-HOLLAND

1968

Published on behalf of
the International Astronomical Union
by
D. Reidel Publishing Company, Dordrecht, Holland

ISBN-13: 978-94-010-3419-7   e-ISBN-13: 978-94-010-3417-3
DOI: 10.1007/978-94-010-3417-3

© *International Astronomical Union, 1968*
Softcover reprint of the hardcover 1st edition 1968

No part of this book may be reproduced in any form, by print, photoprint,
microfilm, or any other means, without permission from the publisher

PREFACE

The XIIIth General Assembly of the International Astronomical Union, held in Prague from 22 to 31 August, 1967, brought together more than 1800 active astronomers from 40 countries. The agenda of the Assembly ranged from administrative questions to highly intricate scientific problems. Thirty-eight Presidents of the IAU Commissions contributed by their reports to a detailed survey of Astronomy for the period 1964–67. These reports were published in the *Transactions* of the IAU, Volume XIIIA. Transactions volume XIIIB bring the results of the administrative meetings and of the meetings of Commissions held during the General Assembly.

The present volume is devoted to the most important scientific results of the General Assembly as presented in the *Invited Discourses*, *Joint Discussions*, and at *Special Meetings*.

It is an established policy of the Union to invite prominent astronomers to hold, during the General Assembly, formal discourses on topics of major importance due to recent significant developments. These Invited Discourses cover, as a rule, broad fields of astronomical interest and are meant to give basic information also to scientists from other branches. Subjects which call for a more detailed technical approach are dealt with in the Joint Discussions or at Special Meetings. Here, invited speakers present authoritative surveys complemented by communications on current or envisaged projects.

The proceedings of two other meetings, on Moon Probes and on Coordination of Solar Observations Made at Ground-based Observatories and with Space Vehicles, are of a sufficiently wide interest to warrant the inclusion into this volume.

The proceedings published in the present volume are certainly the Highlights of Astronomy 1967, but only the highlights. It should be emphasized that the 140 Commission meetings held during the General Assembly brought a wealth of highly important scientific results which, however, could not be included in this volume. These results can be found in astronomical journals and in the publications of observatories.

L. PEREK

December 1967

TABLE OF CONTENTS

SPECIAL MEETINGS

A. *Lunar Probes*

B. *Coordination of Solar Observations made at
Ground-Based Observatories and with Space Vehicles*

INVITED DISCOURSES

EXPLORING THE MOON

Invited Discourse A given on August 23, 1967 in the Lucerna Hall

A. A. MIHAJLOV

(Main Astronomical Observatory, Pulkovo, Leningrad, U.S.S.R.)

The exploration of this heavenly body has many different aspects. First of all there is the problem of the Moon's motion, which is one of the most complicated problems of celestial mechanics; then of its figure, which is a joint question of geodesy, astrometry and also of mechanics; then of the Moon's rotation, depending on the orbital motion, figure and distribution of mass; then comes the study of the structure and properties of the lunar surface and of the physical conditions on it, which was formerly restricted to collecting and interpreting photometric, colorimetric and radiometric data; the topographic features and different formations of the lunar surface have attracted many enthusiastic observers, and the explanation of the genesis of these formations pertains to the domain of geophysics and geology.

At last, or to say better at first, comes the most baffling question of all – that of the Moon's origin. Is the Moon a former planet, captured by the Earth, or is the Earth its mother, which gave birth to such an enigmatic child, or are the Earth and Moon twin sisters – we are still quite uncertain and very much in the dark, notwithstanding many ingenious theories advanced during more than a century.

There are still more problems connected with the Moon, such as its influence on the Earth, but they necessarily remain outside the scope of our discourse.

The most spectacular achievements of the latest years accomplished with the help of astronautics have greatly furthered our knowledge of the Moon, and we are sure that many even more brilliant results will shortly follow. Perhaps it is expected that my discourse will be restricted to these newest scientific and engineering conquests, but in order to recall them we must first pay a tribute to those branches of science that have made it possible to explore the Moon from a close vicinity and even reach its surface and probe its structure.

Thus first of all we must turn our attention to the motion of the Moon. Owing to the very strong attraction by the Sun and also to the flattening of the Earth, the Moon's movement around the Earth is extremely complex, pertaining to the most difficult problems of celestial mechanics. Only few mathematicians of the highest rank, beginning with Newton and Euler, tackled this problem, but it is astounding with what perfection and precision it was ultimately solved on the basis of the gravitational theory alone.

Perek (ed.), Highlights of Astronomy, 3–11. © I.A.U.

Quoting the great American astronomer Simon Newcomb on this subject "the men who have done it are in intellect the select few of the human race – an aristocracy ranking above all others in the scale of being. The astronomical ephemeris is the last practical outcome of their productive genius."

Let us consider for one moment this question. The Moon's ephemeris gives its position in advance, say for a score of years to within nearly 1 sec of arc, corresponding to about 2 km in its orbit. During 20 years it would travel some 700 million km and yet its position could be predicted with such a precision, amounting to a few units of 10^{-9}. The gravitational theory alone is sufficient to represent the motion of the Moon much more precisely than by the rotation of the Earth around its axis, now serving on this account to study the variations in the length of the day.

Now comes the still very difficult problem of the axial rotation of the Moon, closely connected to the Moon's figure, or more accurately speaking depending on the principal moments of inertia of the lunar globe. If this rotation were strictly uniform, it would have presented no difficulty producing the relatively simple effect of optical libration. The difficulties arise from the physical libration, consisting of very small pendulum-like oscillations excited by the attraction of the Earth on the Moon's bulge in the direction of the Earth. It is mainly the forced physical libration, induced and maintained by the deviation of the Moon's bulge from the exact direction towards the Earth. The amount of this libration depends upon the so-called mechanical ellipticity of the Moon

$$f = \frac{B(C-B)}{A(C-A)},$$

denoting the excess of the flattening of the Moon's meridian directed towards the Earth over the flattening of the meridian with longitude $\pm 90°$.

Now the mathematical treatment of this problem shows that there is a critical value $f = 0.662$ when the denominator of one of the terms in the expression for the forced physical libration tends towards zero and the libration itself increases indefinitely. It seems, however, that this is only a result of the mathematical treatment of the phenomenon, not inherent to its physical nature. Professor Koziel of Cracow has recently shown that there is no discontinuity in the Moon's rotation for this value of f. The observations also give for f a value quite distinct from the critical one, but it is difficult to decide on which side it is located. There are two values of f, namely 0.73 and 0.60, around which the different determinations are grouped. Jeffreys in an important paper gives $f = 0.639 \pm 0.014$. Not going into details, we can sum up saying that the difference between the semi-axes of the Moon's ellipsoid directed towards the Earth and the Moon's poles lies between 1100 and 650 m, the third axis at right angles to the line of sight being from 140 to 280 m longer than the polar one.

The other dynamical method of determining the figure of the Moon is much simpler. The mean polar flattening is derived from the angular speed of rotation and the elongation towards the Earth is found from the force of the Earth's gravitational pull. Owing to the very slow axial rotation of the Moon, and consequently the smallness of centrifugal force, the polar flattening $(b-c)/b$ is very small, confined between the limits $\frac{5}{4}q$ and $\frac{1}{2}q$, q being the ratio of centrifugal force to the force of gravity. The upper limit corresponding to a homogeneous body, that in the case of the Moon is highly probable. In this case $b-c$ is only 17 m or from 8 to 16 times smaller than as derived from the physical libration.

The tidal force produced by the Earth's gravitational attraction at the present mean distance of the Moon can create a bulge $a-c=65$ m high, also some 10–17 times less than found previously. As the figure of the Moon deviates from a sphere much more than these amounts, it was supposed that the Moon acquired its present figure a long time ago, when it was much nearer to the Earth, its rotation was much faster and the tidal force much stronger. As the centrifugal force increases as the square of angular velocity, a flattening 16 times greater corresponds to a rotation 4 times faster than at present or to one revolution in 7 days. The tidal action is inversely proportional to the third power of the distance and the bulge mentioned corresponds to a distance 2·5 times nearer to the Earth, when according to Kepler's third law the orbital revolution was again 4 times shorter. Thus, the present shape of the Moon could have been acquired at a time when its distance from the Earth was about 150 000 km, and owing to the Moon's rigidity this figure was preserved till the present time. Thus the large discrepancy of the Moon's figure as determined from physical libration and the forces to which the Moon is at present subjected can be explained by the hypothesis of a fossilized Moon that has kept its figure from bygone ages.

The third most direct though not easiest method is by optical observations. We have here two different sides of the problem – the determination of the figure of the Moon's limb or marginal zone and of the general shape of the Moon's globe, in a first approximation as an ellipsoid with three unequal axes. It would seem that it is very simple to determine the figure of the limb as projected on the background of the sky, but unfortunately we can usually see only the illuminated half of the same and it is very difficult to connect it with the other half, which remains hidden in darkness. Even the observations of occultations of stars do not remove this difficulty, as the errors of observation of ingress and egress are systematically different.

The figure of the visible limb and its irregularities have a direct influence on the determination of the Moon's position either with meridian instruments or photographically as well as by occultations of stars. Detailed charts of the profile of the Moon's marginal zone were prepared by Hayn in Leipzig, Weimer in Paris, Nefediev in Kazan, and Watts in Washington. These mappings meet the common difficulty that the entire limb can never be seen, as at the time of perfect full Moon a lunar eclipse occurs. Approximate to this is a penumbral eclipse when photographic observations allow a sufficiently exact determination of the figure of the entire limb. The most propitious method consists in photographing annular solar eclipses when the whole limb is projected on the solar disk. Unfortunately little attention was formerly given to such observations, and only in the last years this method was applied in praxis. Good results were obtained by an Anglo-American expedition which observed the eclipses of July 31, 1962 and January 26, 1963, in Africa. The fear that the two halves of the limb have different centres proved to be unfounded, but the best approximation was given by an ellipse with a flattening of about 1/1000 and direction of the major axis inclined by 35° to the polar axis of the Moon. A very close result was obtained also by Potter at Pulkovo from observations during the penumbral lunar eclipse of September 27, 1958. The semi-axes of such an ellipse differ by some 1·5–2 km, which is in strong disaccord with the dynamical results. It is possible that this discrepancy can be at least partly explained by the following considerations.

First it must be pointed out that the observations of two or three eclipses are insufficient for a determination of the whole profile of the marginal zone, as each eclipse can give the figure for only one phase of libration; meanwhile Yakovkin has shown that the curvature of the Southern part of the Moon's limb depends on the libration in latitude. But still more important is the real difference between the geometric and mechanical figures of the Moon. The first pertains to the physical surface somewhat distorted by a curious phenomenon, while the second applies to an equipotential surface of gravity. The distortion arises from the fact that an elevation near the limb is seen as projected on the sky at different phases of libration, whereas a valley can be noticed only if directed straight towards the Earth and even then only if it is of sufficient length. Thus a mountainous region is seen permanently and therefore exaggerated, but depressions are noticeable only if they are of very large extent, such as Oceanus Procellarum. The contour charts of the marginal zone show this very clearly. It is even probable that the elevations are isostatically compensated, in which case they will have very little connection with the mechanical figure of the Moon.

It is still more difficult to determine by optical means the general figure of the Moon's visible hemisphere. Heliometric and photographic observations give only the relative positions of points on the lunar surface in projection on the sky. The libration produces a stereoscopic effect, which can be used for determining the third coordinate of the measured points not too near the limb. This effect is rather small, moreover the changes of illumination create systematic errors so that the best measurements still have a poor precision.

Extensive measurements were made by Hartwig, Franz, Hayn in Germany, Sounder in England, Baldwin in the U.S.A., a group of observers at the Engelhardt observatory near Kazan in Russia, and recently at the observatory near Kiev. The measures of 150 points by Franz had been newly reduced by Schrutka-Rechtenstamm in Vienna. Professor Hopmann has also made valuable contributions to this question. New photographs were obtained for this purpose at the Pic-du-Midi observatory in France and taken with the U.S. Navy astrometric reflector at Flagstaff. Meyer and Ruffin have determined the selenocentric coordinates of 196 points to be used in the construction of large-scale lunar maps with contour lines of the Aeronautical and Information Center at St. Louis. Notwithstanding all pains the precision of the absolute elevations of these points or the distance from the Moon's centre of mass is not as high as desired, the probable error often exceeding 1 km, which gives hardly any reliable data on the general figure of the Moon. Moreover, the measured points are not without a statistical bias, representing chiefly peaks or other elevated features which are not typical for a figure approximating an equipotential surface. However, Goudas used them for an elaborate study of the Moon's figure by means of development into spherical functions.

The standard value of the mean radius of the Moon's globe is 1738·0 km. In order to determine the force of gravity on its surface, which is an important physical characteristic, we must know the Moon's mass as expressed in units of the mass of the Earth. Till recent times there was a perceptible uncertainty in this important constant, ranging from Newcomb's value of 1/81·58 to Spencer Jones' 1/81·27. This uncertainty arose chiefly from the difficulty of locating by observations the position of the barycentrum of the Earth–Moon system inside the Earth's body. At present the perturbations of artificial satellites give a much more precise method, which allowed the IAU at its Hamburg meeting in 1964 to adopt, among other astronomical constants, the value 1/81·30 with a possible error less than 0·01%. This gives the Moon's mean density equal to 3·343 g/cm^3, the mean acceleration of gravity 162·314 cm/sec^2, and the velocity of escape 2375 m/sec. Thus the force of gravity on the Moon is just 6 times less than on the Earth, and with the small velocity of escape the Moon could not keep an atmosphere, especially as the temperature on the illuminated side rises up to $+120\,°C$ when the kinetic velocity of gaseous molecules is too great for a permanent detention.

The visible side of the Moon amounting to 59% of its entire surface has been studied in great detail. Several enthusiastic observers spent many years of intensive labours in measuring the relative positions of reference points and drawing detailed charts. Such were the works of Mädler, Lohrman, Schmidt, Fauth, Neison, and others. Beginning from the end of the last century, photography was applied to this task with great success. The beautiful atlas of the Paris observatory prepared by Loewy and Puiseux must be mentioned. The observatories of Lick, Yerkes, Mount Wilson, Pic-du-Midi and the 200-inch reflector of Mount Palomar contributed large series of photographs, some of which were used by Kuiper in his lunar atlases. On the base of photography a chart of the visible side of the Moon is prepared by the U.S. Air Force on the scale 1 : 1 000 000 with a successful attempt of drawing contour lines with an interval of 300 m.

This is our knowledge of the topography of the visible side of the Moon. Its reverse side, comprising 41% of its total surface, has never been seen by a human eye. The more interesting and important are the first photographs transmitted to us by Luna-3 in 1959. Although at that time the television technique was not of the present high standard and the images transmitted from a distance of hundreds of thousand kilometres were not as distinct as at the second subsequent experiment in 1965, still an atlas of the Moon's far side could be compiled showing many details. In one important instance the reverse side differed from the visible one – in the nearly total absence of the so-called maria, i.e. extensive dark depressions which occupy an area of about 40% of the visible side. On the far side only one conspicuous rather small round sea was discovered – the sea of Moscow. In this connection I recall a most remarkable prediction made by Professor Franz of the Breslau observatory in his admirable booklet *Der Mond* published in 1906: "Auf der Rückseite des Mondes hinter seinem Nordostrand ein ausgedehntes helles, kraterreiches Hochland ohne Meere jenseits des Nordrandes des Gürtels der Meere liegt", or in English: "On the reverse side of the Moon behind the Northeastern limb lies an extended, bright, crater-rich highland without seas beyond the northern rim of the belt of seas."

The excellent photographs taken in 1965 by Zond-3 filled up the remaining part of the reverse side and confirmed the absence of extended maria. The far side proved to be very mountainous, covered by innumerable craters of different sizes, some of them forming regular, nearly rectilinear chains, consisting sometimes of a score of craters from very small ones and up to 20–30 km in diameter and many hundreds of kilometres long. In such numbers and extension we do not know similar formations on the visible side of the Moon.

The latest photographs transmitted by the American Orbiters showing many minute details, also contain many crater chains and no maria. Instead they show depressions called 'thalassoids', differing from maria in being not dark, but strewn with many bright craters, in this respect resembling the so-called continents.

We come now to the problem of the origin of the lunar relief. There are two different theories: one of exogenous or external origin through meteoric impacts, and the other owing to endogenous or volcanic action. There are very forcible adherents to either of these theories to the total exclusion of the opposite opinion. It seems to me, however, that the face of the Moon was formed by the action of both of these factors. First of all let us consider the formation of maria. It is quite improbable that such vast areas, being depressions thousands of kilometres in extent, originated from meteoritic impacts. If all craters were of meteoritic origin, we could expect a much more even distribution over the whole surface of the Moon with perhaps only a small preponderance on its Western side, which meets the meteorites with a slightly greater velocity. On the other hand, the polar regions of the Moon have slightly less chances of being hit by meteorites, although an excess of craters is observed near the Moon's Southern pole, which is again in conflict with the impact theory.

We could expect some indications from a statistical study of lunar craters. If, as seems probable to me, a part of chiefly large craters, more irregularly distributed, are of volcanic origin, whereas most of the smaller craters were caused by the fall of meteorites, we could expect some difference in distribution or in some other instance of these two kinds of craters. However, the recent statistical evaluation by Cross of the distribution of craters as counted on Ranger pictures, has shown the absence of any difference between the vast range of sizes from 1 m to 70 km, which, according to the author proves their meteoritic origin. It seems, however, that if the mixture of craters of these two different origins is the same for the whole range of sizes, there would be only an influence on the dispersion, but no difference in the law of distribution.

On the other hand, Fielder and Marcus found a very pronounced clustering of craters and also formations of crater chains, which can be explained only by an internal volcanic origin. This is confirmed by the chains revealed by Zond-3 and the Orbiters, which can be explained by the existence of rifts or fissures in the lunar crust through which gases or molten lava were erupted. The well-known observation of the Pulkovo astronomer N. Kozyrev, who obtained a spectrogram showing the escape of gases containing carbon from the central peak of Alphonsus, proves that vulcanism is not yet dead.

It is impossible to discard the idea of vulcanism when inspecting some of the pictures of the reverse side of the Moon. On the map drawn from the photographs of Luna-3 the crater Tsiolkovsky is clearly seen. On the image transmitted by Orbiter-I this crater has a level dark floor with a composite central elevation. It looks very like a caldera on Java, where the central mountain is surrounded by a lake of dark solidified lava and the inner wall has a complex structure. The perspective view taken with a telephoto lens by Orbiter-II closely resembles a landscape of a mountainous region on the Earth, the only difference being a black sky and the absence of aerial haze.

On the other hand, there are many superimposed craters, e.g. when a smaller crater is located on the rim of a larger one, or a crater interrupts a regular mountain ridge, giving the impression of an external influence. Therefore it seems very likely that both factors played a substantial role in the formation of the Moon's macrorelief, although we are yet unable to assign to which factor is due every individual feature. It is possible and even probable that in some cases a meteoric impact acted as a trigger weakening the Moon's crust and facilitating at the particular place the eruption of lava.

Passing to the microstructure of the lunar surface we must recall that quite a short time ago it was generally assumed that the Moon was covered with a deep layer of dust formed in bygone ages through innumerable meteoric impacts. This was emphasized by the belief that the very strong monthly variations of temperature activated the disrupture of rocks on the surface. A dust layer explained also the exceedingly small thermal conductivity and heat capacity of the outer stratum as revealed by the very rapid cooling of the surface during lunar eclipses, when the temperature falls from over $+100\,°C$ to about $-60\,°C$ at the beginning of total eclipse and after its ending again quickly attains the previous high level.

However, it was soon found that fine dust in a high vacuum especially exposed to great temperature fluctuations would coagulate forming grains of millimetre size without losing its low conductivity. Such a grainy surface is capable to bear a pressure of about 0·5–1 kg per 1 cm² in conditions of lunar gravity, still having a small density owing to its porosity. Photometric and polarimetric observations are in accord with such a structure.

Volcanic tuff or porous lava are good approximations to such material, which by its dark colour corresponds to the small albedo of the Moon. As radio-astronomical observations had shown, below this light and thermo-insulating layer lies a more dense rocky substance which during a lunation undergoes a much smaller variation of temperature. In many separate places were discovered 'hot spots' that remain much longer warm during eclipses or at the beginning of night. Many of them coincide with the floors of craters. The explanation seems to be that at these places the insulating layer is much thinner or is even absent, so that we receive the radiation from the deeper, warmer and more slowly cooling stratum. However, such a simple explanation is not sufficient to account for all observed peculiarities connected with the origin and subsequent history of these hot spots.

We have now touched some problems which have obtained a spectacular development owing to the astounding achievements of modern cosmonautics, especially of the soft landing of automatic stations on the Moon, which transmitted to us close panoramic images of the lunar surface. This was first accomplished by Luna-9 on February 3, 1966. After this followed the American Surveyor-I and the Russian Luna-13. The great difficulty of such an operation is obvious from the following considerations. The parabolic velocity on the Moon's surface is 2375 m/sec, but in order of passing the neutral point of attraction of the Earth and a passable time of transition the final velocity must be somewhat greater, about 2600 m/sec. For a soft landing to prevent damage to the spacecraft the velocity of landing must not exceed some 2–3 m/sec, and this must be achieved by reactionary breaking at the right moment with a precision of 0·1% in velocity and its direction. The landing place was every time chosen on level ground of maria in order to avoid steep inclines or deep valleys and also as a reconnaissance of a possible landing place for manned spacecraft.

The panoramas obtained confirmed the absence of dust, the grainy or porous structure of the outer surface, sufficiently resistant to bear the weight of the station itself and, in the future, to enable a cosmonaut to walk on the Moon without crushing the surface or sinking into the dust. A very interesting feature are stones or clumps apparently lying loosely upon the grainy surface of the Moon. They seem to be debris ejected by volcanic eruption or meteoritic impact some distance away.

Very important was the launching of satellites around the Moon. The symposium of the IAU on Astronomical Constants held in Paris in spring of 1963 adopted the following resolution: "We direct the attention of the authorities concerned to the importance of creating artificial satellites of the Moon, very desirably observable by various means, to obtain reliable information concerning the Moon's gravitational field and geometric form. We further recommend that specialists in celestial mechanics consider and specify the best values for the orbital elements of such satellites in order to obtain data for the Moon's gravitational field with the least number of satellites."

Less than 3 years had passed since this symposium, when the first lunar satellite was launched in April 1966 in the U.S.S.R. American Orbiters followed and this kind of research is continuing.

The symposium of the IAU restricted itself to problems directly connected with astronomical constants. However, satellites of the Moon give besides much important information on magnetic and radiational properties of the Moon, and also yield detailed pictures of both sides of the Moon. It is evident that before sending a cosmonaut to the Moon we must study its physical and chemical properties and get as much general information as possible in order to prevent him from many surprises and enable him the fullest utilisation of its natural resources. Otherwise it would be impossible to ensure his return to the Earth or to obtain the vast complex of information for which such a most difficult experiment would be undertaken.

I could only briefly touch some aspects of the investigation of our natural satellite, that seemed to me of main importance. Lunar research is carried on and developed in the U.S.S.R., the U.S.A., and some other countries at such a rate, so much information is received and accumulated, mostly concordant but sometimes disagreeing, that it is difficult to keep pace with it.

From the time when I finished the compilation of this discourse some further progress has been made; it is even possible that during the present General Assembly something new will be learned about the nature of the Moon. I am sure that at the meetings of the corresponding commissions, first of all of Commissions 16 and 17, we shall hear some more specialized and more profound discussions. At present I had to restrict myself to such aspects that seemed to me of general importance.

The unprecedented successes of the last years in the study of the nature and properties of the Moon give us the assurance that in the nearest future we shall witness new and possibly quite unexpected achievements in this field. We have every reason to look forward full of optimism and not follow the French philosopher Auguste Comte, who proclaimed in 1830 in his *Cours de philosophie positive* about the heavenly bodies that "nous ne saurions jamais étudier par aucun moyen leur composition chimique ou leur structure minéralogique...".

Well, the first – the chemical composition – is being investigated well nigh for a century, after the invention of spectral analysis. The second – the mineralogical composition of the Moon – we are on the threshold of investigating in our earthly laboratories when we will receive samples of rocks from the Moon or in lunar laboratories, first automatic ones and subsequently by human investigators transported by spacecraft to the Moon.

We can be confident that we are on the eve of still more brilliant and spectacular achievements in this field of science and technology.

COUCHES EXTÉRIEURES ET STRUCTURE INTERNE
DES ÉTOILES

Invited Discourse B given on August 25, 1967 in the Lucerna Hall

P. LEDOUX

(Institut d'Astrophysique, Université de Liège, Belgique)

Monsieur le Président, Mesdames, Mesdemoiselles, Messieurs,

C'est un honneur redoutable que d'être invité à prendre la parole devant cette imposante assemblée, d'autant plus redoutable dans ce cas, qu'en l'acceptant j'ai assumé deux tâches parallèles: d'une part, celle de renforcer quelque peu, et espérons-le honorablement, la présence d'une des deux langues officielles de l'Union et, d'autre part, celle de vous présenter un problème qui relève certes du domaine de mes intèrêts mais qui aurait pu trouver parmi vous des porte-paroles sans aucun doute plus attitrés et plus compétents.

Le problème des interactions entre la structure interne des étoiles et les propriétés de leurs couches extérieures présente une grande variété d'aspects dont certains, par leur complexité, peuvent échapper longtemps encore à tout effort d'élucidation. C'est d'autre part un problème fondamental car, à part des renseignements précieux sur la masse et la luminosité totales des étoiles et, dans quelques cas assez rares, sur leurs rayons et sur leur condensation centrale et, peut-être à l'avenir grâce aux neutrinos, quelques renseignements directs sur les réactions nucléaires qui se déroulent dans le soleil, toutes nos connaissances sur la constitution interne des étoiles dérivent, en fin de compte, de la possibilité d'extrapoler, vers l'intérieur, les conditions directement observées dans les couches atmosphériques les plus extérieures dont la masse ne représente cependant qu'une fraction infime de la masse totale.

Dans le passé, cette dernière circonstance a souvent été invoquée pour justifier l'usage dans la théorie de la structure interne des étoiles de conditions aux limites fort sommaires à la surface. Les développements ultérieurs que nous nous proposons de passer en revue ici nous incite à présent à plus de prudence, mais on peut néanmoins se demander si toutes les propriétés de ces couches ou certains détails de leur structure sont réellement significatifs pour l'intérieur de l'étoile. L'agitation générale à la surface du soleil, la granulation, les ondes plus régulières qu'on y discerne à plus grande échelle, les champs magnétiques irréguliers et leurs variations au cours du temps, la rotation différentielle, la présence d'une chromosphère et d'une couronne, le vent solaire lui-même sont-ils des phénomènes d'origine purement superficielle sans consé-

Perek (ed.), Highlights of Astronomy, 12–32. © I.A.U.

quences pour la constitution des couches internes? Sont-ils au contraire le reflet de certaines conditions réalisées en profondeur? Les mêmes questions peuvent se poser à propos des champs magnétiques stellaires, de certains types de variabilités erratiques ou régulières, de la distribution des vitesses de rotation ou, encore, à propos des anomalies d'abondances ou même des abondances moyennes observées dans les atmosphères. Et dans les circonstances de l'astrophysique, il est bien difficile d'y répondre sinon par l'élaboration patiente de toutes les possibilités théoriques que la présence de ces facteurs suggère, l'élimination progressive d'un certain nombre d'entre elles et la mise en évidence de l'importance de quelques autres soit par des déductions logiques basées sur une bonne connaissance de la physique stellaire soit par la découverte de corrélations entre la présence et l'intensité de ces facteurs et d'autres données significatives telles que l'âge des étoiles ou leur appartenance à différents systèmes ou sous-systèmes stellaires. Mais d'ailleurs dans la plupart des cas, nous en sommes réduits à une information fort limitée concernant essentiellement la température et la gravité dans les couches superficielles et la luminosité totale de l'étoile. Historiquement, le problème de la structure interne a été formulé tout naturellement en termes des trois paramètres correspondants, la masse totale M, le rayon R, et la luminosité L.

Même ainsi, la nature du problème doit nous inciter à quelque prudence. En effet, supposons que la composition chimique est connue soit qu'elle soit supposée uniforme et identique à celle de l'atmosphère soit que des calculs préliminaires liés à l'étude de l'évolution stellaire aient fourni sa variation avec la profondeur. Supposons également que nous disposons de toutes les lois physiques appropriées: équations d'état très générales, opacité, taux de génération d'énergie ainsi que de critères locaux permettant de juger de l'importance relative des différents modes de transport d'énergie. Dans ce cas, la construction d'un modèle stellaire instantané en équilibre hydrostatique ou quasi-statique, pour une masse donnée M, revient à chercher la (ou éventuellement les) solution d'un système différentiel du quatrième ordre dépendant de R et L comme paramètres et qui satisfait (ou satisfont) à deux conditions au centre et à deux conditions à la surface. Comme divers auteurs l'ont fait remarquer (1), il s'agit là, en réalité, d'un problème aux limites fortement non linéaire et, à ma connaissance, la question de l'existence et de l'unicité de la solution n'a pas reçu jusqu'ici de réponse satisfaisante d'un point de vue mathématique strict. Cependant, l'expérience considérable acquise dans le domaine des modèles stellaires, grâce notamment aux grands ordinateurs électroniques, montre qu'en pratique la solution existe et qu'elle est presque toujours unique sauf dans de rares cas comme celui, p. ex., des modèles possédant un noyau isotherme ou quelque autre type de noyau très condensé de masse relative donnée. Mais dans ces cas, sans doute, une seule des solutions possibles est-elle réellement stable et peut-être même le calcul continu de l'évolution stellaire nous conduit-il automatiquement et univoquement à celle-ci bien qu'à l'occasion, une telle situation puisse soulever des problèmes délicats là notamment où deux telles séries stable et instable se rejoignent ou se croisent.

Une autre caractéristique de ces problèmes aux limites est la sensitivité souvent considérable des solutions aux conditions exactes imposées aux extrémités de l'intervalle d'intégration. Ceci est déjà apparent dans les modèles les plus simples définis par une relation polytropique $P = K\rho^{\gamma}$ ou, paramétriquement, $P = P_c\theta^{n+1}$, $\rho = \rho_c\theta^n$. Dans ces modèles, la surface est déterminée par la condition $\theta = 0$ (c.-à-d., physiquement $P = \rho = T = 0$) pour une valeur de la variable radiale d'Emden disons ξ_s qui peut être prise égale à l'unité. Les solutions construites à partir de ce point sont de différents types : pour une valeur bien déterminée $(d\theta/d\xi)_{s,E}$ de la dérivée en $\xi_s = 1$, la solution, dite d'Emden, croît uniformément pour atteindre le centre $\xi = 0$ avec une valeur finie et un coefficient angulaire nul ; pour des valeurs initiales de $(d\theta/d\xi)_s$ plus petites en valeurs absolues, les solutions tendent vers l'infini au centre et sont dites solutions M ou 'centrally condensed', tandis que pour des valeurs initiales plus grandes que $(d\theta/d\xi)_{s,E}$ les solutions tendent à s'annuler de nouveau avant d'atteindre le centre et sont dites solutions F ou du type 'collapsed'.

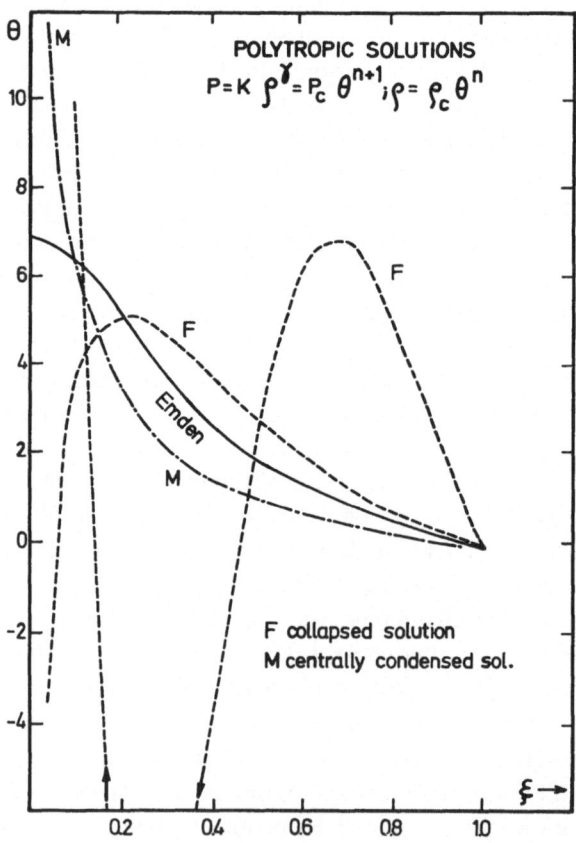

FIG. 1. *Illustration des différents types de solutions polytropiques.*

L'existence des solutions des types F et M était déjà connue d'Emden mais leur discussion doit beaucoup (2) à l'intérêt que Milne (3) y prit. A cette époque déjà lointaine où bien des données physiques faisaient défaut, la considération de ces solutions et particulièrement des solutions 'centrally condensed' ainsi que de conditions à la surface plus précises tenant compte de la température finie qui y règne lui paraissait justifier une remise en question générale de vues courantes sur la structure interne des étoiles. Bien que ce point de vue extrême se soit vite révélé injustifié (4), il n'en reste pas moins que l'assertion de Milne concernant l'impossibilité de déterminer univoquement la composition chimique profonde à partir de M, L et R ou des relations qui les lient a reçu de nos jours une confirmation au moins partielle. De même, certains résultats formels issus des recherches qu'il incita ont pavé la voie à des developpements importants.

À ce dernier point de vue, nous sommes particulièrement intéressés ici par une des conditions de jonction dans les configurations composites, à savoir qu'une solution extérieure d'indice polytropique plus petit que celui du noyau central ne peut être réunie à celui-ci que si elle est du type 'centrally condensed'. De plus, c'est la valeur de la constante K dans la relation $P = K\rho^\gamma$ qui permet de distinguer entre elles les solutions extérieures de ce type. Or la valeur de K sera déterminée par les valeurs de P et ρ obtenues par intégration à partir de la mince frange atmosphérique en équilibre radiatif au travers d'une zone de transition sous-jacente plus ou moins profonde à la base de laquelle la solution polytropique ou pseudo-polytropique envisagée devient valable. Dans ces circonstances, la structure interne profonde peut être fortement affectée par les conditions dans les couches extérieures et en particulier par l'ionisation de l'hydrogène.

En fait, dès 1930, Unsöld (5) fit remarquer que cette ionisation peut réduire suffisamment les rapports généralisés Γ des chaleurs spécifiques pour créer une instabilité convective violente dans les couches subphotosphériques des étoiles de températures effectives pas trop élevées, disons inférieures à 10000°. Dans le cas du soleil, il trouva que la convection ne pouvait manquer de se développer à partir d'une profondeur optique relativement faible de l'ordre de $\tau = 3$ et il n'hésita pas à y rattacher l'origine de la granulation observée aux niveaux photosphériques.

Au début, ce fut sans doute cette dernière possibilité qui retint l'attention et qui consacra l'intérêt de la découverte de Unsöld. Dans la suite, l'importance de cette zone ne fit que croître au fur et à mesure que la prédominance de l'hydrogène dans la composition chimique des étoiles s'affirmait et que le calcul précis de l'opacité aux basses températures ($T < 10^4\,°K$) due à l'hydrogène et son ion négatif (6) révélait que son accroissement rapide avec la température hâtait encore l'apparition à faible profondeur de l'instabilité convective. D'autre part, les travaux de Biermann (7) en particulier attirèrent bientôt l'attention sur le fait que la zone de convection de l'hydrogène pouvait s'étendre bien plus profondément qu'on ne l'avait prévu initialement. En effet, les premières discussions reposaient sur la considération d'enveloppes en équilibre purement radiatif dans lesquelles l'instabilité convective due à l'ionisation de l'hydrogène était limitée à une zone relativement étroite et, malgré la violence de cette instabilité, on admettait généralement en première approximation que la convection ne s'étendrait guère qu'à cette zone. Mais en réalité, la convection en s'établissant réduit le gradient de température, là particulièrement où les Γ sont petits, ce qui tend déjà à étendre vers le bas la région de l'ionisation critique. De plus, les températures relativement basses qu'elle associe ainsi à des pressions ou densités données, même bien au-delà de la zone d'ionisation de l'hydrogène, réduisent très fortement la contribution du flux radiatif au transport total d'énergie si bien que la convection persiste jusqu'à des profondeurs considérables (**8**). En réalité, dans ces régions où les Γ reprennent rapidement leurs valeurs normales, la convection est si efficace que la distribution des conditions physiques y est pratiquement isentropique et correspond, dans le cas où la pression de radiation est négligeable, à une des solutions polytropiques $n = \frac{3}{2}$, $P = K\rho^{5/3} = QT^{5/2}$.

Malheureusement, il n'en est pas de même dans la partie supérieure de la zone convective de l'hydrogène. Dans cette région, l'efficacité de la convection est beaucoup plus faible par suite de la capacité calorifique réduite, de la transparence relative de la matière et de la limite supérieure relativement basse que la vitesse du son y impose aux vitesses de convection. Les gradients de température et de densité qui s'y établissent à l'équilibre dépendent des valeurs relatives du flux radiatif et du flux convectif et ce dernier devrait donc pouvoir être évalué, comme le premier, directement en termes des conditions physiques. Malheureusement, nous n'avons guère à notre disposition qu'une théorie de la convection turbulente dans laquelle le libre parcours moyen l_t reste un paramètre relativement arbitraire. Certes, il peut être associé à quelque sous-multiple de l'épaisseur totale de la zone convective ou à la hauteur d'échelle des variations de la pression ou de la densité. Mais il dépend ainsi des grandeurs mêmes qu'on se propose de déterminer et l'utilisation d'un processus d'itération ne garantit même pas, dans tous les cas, des résultats exempts de toute contradiction interne. Afin de pallier à ces difficultés, différentes suggestions ont été avancées récemment tendant, p. ex., à lier le libre parcours moyen l_t à une hauteur d'échelle moyenne plutôt qu'à sa valeur locale (**9**) ou même à se passer de l'usage, au moins explicite, de cette notion (**10**). Mais, dans les deux cas, les avantages réels paraissent plutôt minimes. D'autre part, il est peut-être un peu surprenant que, dans les développements en série des variables d'état par lesquels l_t s'introduit dans la théorie, seuls les termes linéaires aient été retenus jusqu'ici alors qu'en tenant compte au moins des termes en l_t^2 on pourrait s'attendre à améliorer les cas les plus difficiles où, justement, les conditions varient très rapidement avec la profondeur.

De plus, les effets de la dissipation par conduction radiative, par conduction et friction turbulente aux échelles intermédiaires et par viscosité à l'échelle la plus petite compliquent encore le problème. Ceci d'ailleurs soulève la question du spectre de la convection turbulente et de l'utilisation des théories semi-heuristiques appropriés, en combinaison, peut-être, avec les résultats de l'étude des perturbations linéaires non-radiales de la couche (11). Cette analyse linéaire étendue aux facteurs non-conservatifs du problème pourrait peut-être fournir les dimensions et le temps de vie des perturbations les plus favorisées (12) fixant ainsi l'ordre de grandeur de l_t dans la théorie élémentaire ou permettant, dans les théories plus raffinées, une normalisation non-totalement arbitraire du spectre.

En tout cas, l'insuffisance de nos théories actuelles de la convection est un handicap sérieux dans ce problème.

Dans ce qui précède, suivant en cela le développement historique de la question, nous avons mis l'accent sur les effets de l'ionisation de l'hydrogène mais il est clair que l'ionisation d'un autre élément abondant comme l'hélium peut entraîner également des effets importants qui peuvent d'ailleurs se manifester dans des classes d'étoiles distinctes et différer quelque peu p. ex. pour les étoiles de population I et II.

Peut-être n'avons-nous pas non plus assez insisté sur l'importance des tables détaillées d'opacité, pour les températures et densités intermédiaires, mises à notre disposition par quelques chercheurs (13) passés maîtres en physique atomique et disposant de puissants ordinateurs électroniques. Sans leurs résultats, bien des progrès substantiels dans la physique difficile des couches de transition entre l'atmosphère et l'intérieur stellaire proprement dit n'auraient pas été possibles.

Enfin, une théorie satisfaisante des couches atmosphériques en équilibre radiatif est également nécessaire afin de nous fournir les conditions régnant au dessus de la couche d'ionisation de l'hydrogène. Ceci requiert au moins une bonne connaissance de l'opacité dans les couches très extérieures ce qui pose parfois des problèmes bien difficiles surtout aux très basses températures. Mais il est d'ailleurs d'autres effets (14) qui peuvent être importants dans certains cas et qui restent bien difficiles à évaluer : effet de serre, anisotropie de la radiation, courbure des atmosphères étendues, convection pénétrative, modification des conditions à la base de l'atmosphère qui, strictement, rend caduques (15) les solutions radiatives pour atmosphères semi-infinies généralement utilisées.

Comme d'habitude, dans les problèmes complexes de ce genre, les progrès sont issus d'attaques incomplètes certes mais efficaces comme celle de Vitense qui, dès 1953 (16) permit, dans le cas du soleil, d'encadrer la solution dans des limites raisonnables. Améliorée en 1958 (17), cette méthode a servi de modèle à presque toutes les investigations subséquentes. De plus, c'est en grande partie son emploi critique et ses généralisations qui ont dégagé peu à peu les différents aspects du problème, y compris ceux qui, jusqu'à présent, restent bien difficiles ou impossibles à traiter.

Malgré ces imperfections et l'indétermination qui pèse sur les solutions du fait de l'arbitraire qui subsiste dans le choix du libre parcours moyen l_t, une réelle moisson de résultats importants et d'implications intéressantes a déjà été récoltée, même en se bornant à la séquence principale. Si nous passons aux étoiles plus chaudes que le soleil, la zone d'ionisation de l'hydrogène se déplace vers la surface en s'amincissant et la convection qui en résulte devient de moins en moins efficace, la radiation devant assurer le transport d'une fraction de plus en plus grande de l'énergie. Ses effets sur la structure interne profonde s'atténuent donc progressivement pour disparaître pratiquement au-dessus d'une température effective de l'ordre de 8000 à 10000°K dans les dernières classes spectrales de type A. Le rôle des ionisations de l'hélium dépend naturellement de l'abondance adoptée pour cet élément et, dans la population de type I, elles peuvent renforcer quelque peu les effets de l'ionisation de l'hydrogène mais leur influence semble cependant s'atténuer très rapidement dès que la température effective approche une dizaine de milliers de degrés (**18**).

Si, comme il paraît vraisemblable, une partie de l'activité solaire et la présence d'une chromosphère et d'une couronne sont liées étroitement à l'existence de la zone convective de l'hydrogène, on peut prévoir que des phénomènes du même genre se rencontreront sans doute jusque dans les premiers types spectraux F. Schatzman (**19**)

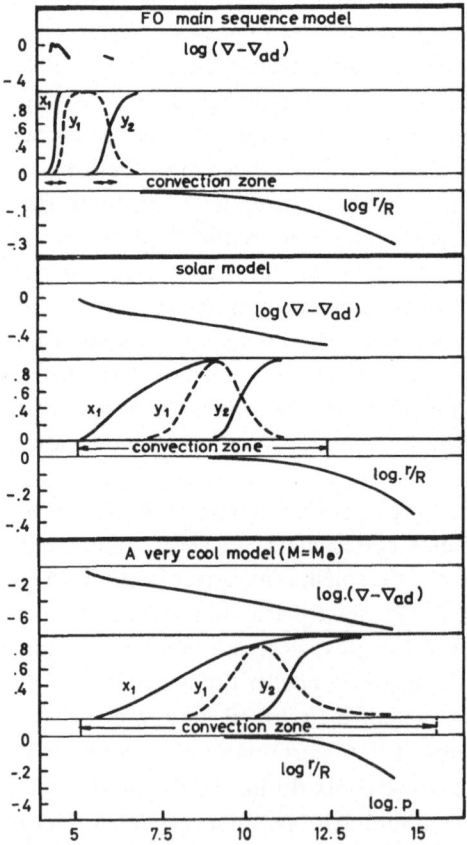

FIG. 2.   *Importance relative des zones de convection de l'hydrogène et de l'hélium pour différents modèles (tirée de Baker et Temesvary,* (**18**)*).*

considère d'ailleurs que, dans les classes spectrales plus froides, cette activité en présence de champs magnétiques peut conduire, au cours des dernières phases de la contraction gravifique, à une dissipation de moment cinétique capable d'expliquer la variation rapide de la vitesse de rotation qui se produit justement dans cette classe F. Peut-être, les nouvelles possibilités d'observation dans l'UV et dans le domaine des rayons X ouvertes par l'avènement des satellites artificiels pourront-elles un jour fournir des données décisives sur cette question?

D'autre part, si nous descendons le long de la séquence principale vers les étoiles moins chaudes que le soleil, l'influence de la zone d'ionisation de l'hydrogène, renforcée dans les couches très extérieures par les effets d'opacité et par la dissociation de la molécule H_2, s'accentue et la convection tend à s'étendre de plus en plus profondément dans l'étoile. En même temps, son efficacité s'accroît dans les couches extérieures et tend à y établir des conditions isentropiques pratiquement jusqu'au point, bien au-dessus de la zone d'ionisation de l'hydrogène, où le critère de Schwarzschild est violé. Ainsi la région difficile à gradient superadiabatique qui compliquait fortement le problème dans les étoiles plus massives tend à s'évanouir ici. Dans ce cas, une simple zone à entropie constante peut servir à joindre, au travers des zones d'ionisation de l'hydrogène et de l'hélium, l'atmosphère radiative à la région convective profonde qui, pour ces faibles masses, peut être représentée par une solution polytropique $P = K\rho^{5/3}$ ou $P = QT^{5/2}$ où K et Q ou encore E^* reliée à Q par

$$E = 4\pi G^{3/2} \left(\frac{\mu}{R}\right)^{5/2} R^{3/2} M^{1/2} Q$$

sont fixées par les valeurs que prennent les variables d'état là où l'ionisation de H et He devient complète.

Comme il s'agit de réunir éventuellement ces enveloppes convectives à un noyau radiatif dont l'index polytropique effectif sera certainement plus grand que $\frac{3}{2}$, seules, comme nous l'avons rappelé plus haut, des solutions polytropiques d'indice $n = \frac{3}{2}$ du type 'centrally condensed' peuvent convenir pour représenter ces enveloppes. Elles correspondent à des valeurs de E inférieures à la valeur $E = 45\cdot5$ qui caractérise la solution d'Emden. Elles doivent être réunies aux solutions radiatives issues du centre au point où, le long de ces dernières, l'index polytropique effectif devient égal à $\frac{3}{2}$. Sur la Figure 3, utilisée par Osterbrock (20) pour construire des modèles des étoiles de faible masse Castor C et α Cen B ($0\cdot65 \leqslant M/M_\odot < 1$), le lieu de ces points est indiqué et la jonction avec la solution 'centrally condensed' fixée par la valeur de E se fait facilement par interpolation. Osterbrock établit ainsi que, pour les étoiles

---

* E est lié au paramètre classique

$$\omega_n = - \xi_s^{\frac{n+1}{n-1}} \left(\frac{d\theta_n}{d\xi}\right)_s$$

défini par Chandrasekhar (I.S.S.S., Chap. IV, Eqs. 73, 337, 477) par la relation
$$E = (\omega_{3/2})^{1/2} (\tfrac{5}{2})^{3/2} \text{ si } n = \tfrac{3}{2}.$$

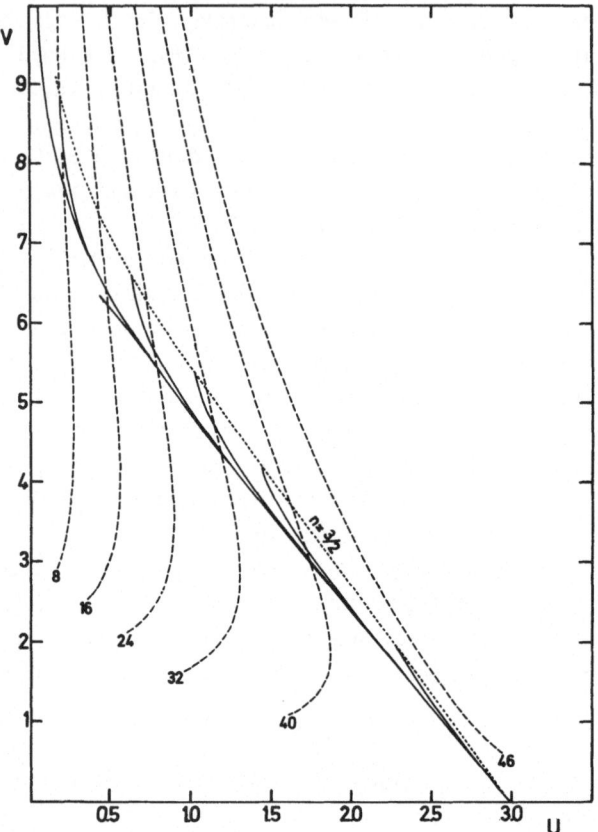

FIG. 3. *Figure empruntée à l'article d'Osterbrock* (**20**), *et illustrant le problème de la jonction entre enveloppes convectives (courbes en trait interrompu correspondant à diverses valeurs de E) et noyaux radiatifs (courbes en trait plein). La courbe pointillée représente le lieu des points où le long des solutions radiatives n devient égal à ½.*

considérées, l'enveloppe convective s'étend très profondément couvrant, pour $M = 0 \cdot 65 M_\odot$, des fractions du rayon et de la masse égales respectivement à $0 \cdot 33$ et $0 \cdot 11$.

Ces résultats furent étendus successivement à des masses de plus en plus faibles par Limber (**21**), qui prend en considération les effets de la dégénérescence partielle des électrons et par Hayashi et Nakano (**22**), qui, en plus, tiennent compte de la molécule H_2 et de sa dissociation. Ces études révélèrent que la convection s'étendait effectivement jusqu'au centre dans les masses plus petites ou égales à $0 \cdot 27 \ M_\odot$, conclusion qui fut confirmée récemment par une étude plus détaillée encore due à Gabriel (**23**). De plus, ce dernier travail montre que les modèles correspondants sont vibrationnellement instables sous l'action des réactions de la chaine p–p et des effets de la convection dans les zones extérieures d'ionisation, instabilité qui doit d'ailleurs s'étendre quelque peu au-dessus de la masse limite $0 \cdot 27 \ M_\odot$.

Bien que cette instabilité ne soit pas très violente, elle devrait pouvoir se manifester, le temps d'amplification de l'ordre de 10^8 années restant court vis-à-vis de la durée de vie de 10^{11} à 10^{12} années de ces étoiles. Devons-nous dès lors, nous attendre à y découvrir quelque forme de variabilité très rapide avec des périodes de l'ordre d'une vingtaine de minutes? Ou avons-nous laissé échapper quelque facteur capable de rétablir la stabilité au prix sans doute du rejet des modèles?

Quoiqu'il en soit, comme nous venons de le rappeler, l'influence des couches extérieures s'est révélée capitale pour l'interprétation des étoiles de la séquence principale des types les plus froids au type F. Mais, c'est sans doute dans l'étude de l'évolution stellaire que l'importance des conditions superficielles et des zones de transition extérieures a reçu la consécration la plus spectaculaire.

Dès que les principales réactions nucléaires d'intérêt stellaire eurent été isolées par Bethe et Von Weiszäcker en 1939, l'interprétation des étoiles de la séquence principale parût immédiatement assurée au moins en première approximation. Par contre, la structure des étoiles géantes et la signification des branches correspondantes dans les diagrammes de Hertzsprung-Russell des amas restaient mystérieuses (24). Un pas décisif fut accompli par Sandage et Schwarzschild (25) quand ils établirent l'existence de modèles à noyau d'helium pseudo-isotherme en contraction entouré d'une enveloppe en expansion permettant d'assurer la continuité de l'évolution au-delà du modèle critique de Schönberg-Chandrasekhar. Ce résultat fut d'ailleurs bientôt étendu (26) aux masses assez faibles pour que le gaz d'électron dégénère partiellement dans les régions centrales. Dans tous les cas, le chemin évolutif dans le diagramme de Hertzsprung-Russell tourne brusquement vers la droite par suite de l'accroissement rapide du rayon global quand la masse du noyau d'hélium atteint quelque 10% de la masse totale. Le coude caractéristique des amas anciens trouve ainsi une interprétation immédiate et sa position permet une estimation directe de leur âge. Mais l'évolution ultérieure sur la base de mêmes modèles à enveloppes purement radiatives dans lesquelles T s'annule à la surface révélait bientôt une nouvelle difficulté. En effet, le point représentatif du modèle continuait à s'écarter vers la droite presque horizontalement au lieu de remonter vers le coin supérieur droit du diagramme après un court palier comme les données d'observation l'exigeaient.

En 1955, Hoyle et Schwarzschild (27) confrontés par cette difficulté remarquèrent que, quelque part le long de cette évolution, la température effective devient si basse que des conditions aux limites réalistes avec une température de surface finie requièrent, comme dans les modèles d'Osterbrock, l'existence d'une zone convective extérieure qui devient rapidement importante. En tenant compte de ceci, ils purent montrer que, pour des valeurs acceptables de la constante E, le chemin d'évolution se redresse vers le haut en accord raisonnable avec la branche observée des géantes rouges. Les nombreuses investigations ultérieures (28) adaptées aux divers perfectionnements auxquels nous avons déjà fait allusion confirmèrent entièrement ces résultats bien que l'indétermination qui pèse toujours sur le libre parcours moyen turbulent entraîne inévitablement une certaine marge d'incertitude sur la position de la branche théorique (voir (**14**)).

À l'autre extrémité de l'évolution stellaire, les travaux de Hayashi allaient bientôt dégager une autre conséquence fort inattendue et tout aussi importante puisqu'elle allait révolutionner nos vues sur les phases de contraction gravifique précédant l'arrivée sur la séquence principale. Au cours d'une investigation sur l'interprétation de la branche des géantes rouges (**29**), il remarque que, pour les configurations où la pression de radiation est négligeable, il ne peut pas exister de modèles en équilibre pour une valeur du paramètre E supérieure à la valeur 45·5 associée à la solution d'Emden ce qui n'est rien d'autre que la proposition déjà rappelée qui limite les solutions acceptables pour l'enveloppe convective au type 'centrally condensed'. Comme la discussion n'est guère affectée qualitativement par la pression de radiation, nous continuerons ici à l'ignorer. Ainsi, pour une masse et une composition chimique données, on peut, par une théorie appropriée des couches photosphériques et de la zone de transition, exprimer le rapport $(P/T^{5/2}) = Q$ à la limite inférieure de cette region, là où commence la zone convective polytropique, en termes de la luminosité L et de la température effective T_e. Réintroduisant cette expression de Q dans la définition de E qui prend sa valeur critique $E = 45·5$, on obtient une équation qui détermine dans le diagramme de Hertzsprung-Russell une courbe à droite de laquelle aucun modèle d'équilibre ne peut exister. La position de cette courbe qui est presque verticale dépend quelque peu de la composition chimique et glisse légèrement vers la gauche quand la masse augmente mais au total ces déplacements latéraux ne couvrent guère au plus que quelques centaines de degrés en température effective.

Hayashi n'hésita pas à tirer les conséquences logiques (**30**) qui découlaient de l'existence de cette zone interdite et qui impliquaient notamment le rejet des chemins évolutifs correspondant à la contraction gravifique de modèles radiatifs (**31**) qui traversaient justement cette région et qui avaient été acceptés jusque là. En réalité, toute configuration momentanément dans cette partie doit évoluer très rapidement vers un point de la courbe $E = 45·5$ où elle peut se stabiliser en réajustant sa structure à celle d'un modèle entièrement convectif. Elle descend ensuite par contraction gravifique le long de cette courbe, restant complètement convective jusqu'au moment où la diminution de l'opacité dans les régions centrales y favorise l'apparition d'un noyau radiatif qui dévie le chemin d'évolution vers des luminosités et des températures effectives correspondant à des valeurs de E inférieures à sa valeur critique et le rapproche du chemin d'évolution radiative de Henyey *et al.* avec lequel il coïncide pratiquement quand les réactions nucléaires commencent à être efficaces (pour illustration, voir p. ex. (**32**)).

Etant donné l'importance de ces conclusions, elles ont été soumises à une critique aiguë (**33**) qui n'a pu que les confirmer au moins qualitativement car une certaine marge d'indétermination subsiste par suite notamment des incertitudes de la théorie de la convection (voir p. ex. (**14**)). Les conséquences en sont multiples et nous forcent en particulier à nous pencher sérieusement sur les toutes premières phases de la contraction capables d'amener l'étoile dans la région de hautes luminosités caractérisant l'évolution à la Hayashi. Et d'abord, par suite des hautes luminosités, les phases convectives de contraction gravifique sont parcourues beaucoup plus rapidement que dans le cas radiatif, surtout pour les faibles masses, ce qui a des incidences considérables sur l'interprétation des amas très jeunes tels que NGC 2264 et sur les quantités d'éléments légers tels que le lithium et le béryllium qui peuvent subsister à la fin de ces phases de contraction convective au cours desquelles ils ont été brassés à grande profondeur jusqu'à des températures élevées.

L'évolution extrèmement rapide qui amène les étoiles sur la séquence convective de contraction gravifique n'a guère fait l'objet encore que d'investigations très approchées (34) mais elle doit être déterminée essentiellement par les processus de refroidissement disponibles aux très basses densités et températures des condensations interstellaires (35) et par les instabilités dynamiques (34, 36) correspondant à l'ionisation initiale de l'hydrogène et de l'hélium et à la dissociation de la molécule H_2.

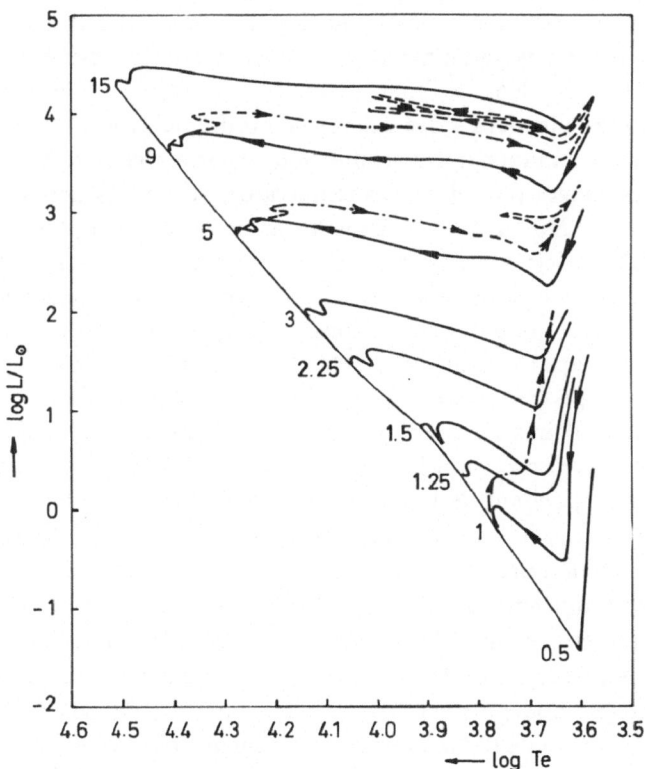

FIG. 4. *Evolution schématique dans le diagramme de Hertzsprung-Russell pour quelques valeurs de la masse. Les phases de contraction sont représentées par les courbes en trait plein, les phases postérieures au rebroussement sur la séquence principale, en trait d'axe.*

Pourtant comme Schatzman (37) le fait remarquer, cette instabilité pourrait être réduite fortement ou même supprimée si la condensation initiale était le siège d'une turbulence violente telle que la pression de turbulence cesse d'être négligeable vis-à-vis de la pression gazeuse.

D'autre part, jusqu'ici, on n'a guère fait que soulever la question de l'influence possible d'une rotation ou de champs magnétiques originels qui peuvent sans doute jouer un rôle important à la fois dans la détermination des chemins évolutifs et dans la distribution finale du moment angulaire ou du flux magnétique.

Il est cependant fort douteux que ces facteurs puissent modifier appréciablement les limites de la zone interdite à la droite du diagramme de Hertzsprung-Russell. Ainsi, en pratique, aucun point représentatif d'étoiles ne devrait être observé, dans cette région, qu'il s'agisse d'étoiles fortement évoluées atteignant le sommet de la branche des géantes rouges, car leur évolution vers la droite est limitée par la même courbe critique, ou d'étoiles dans les premiers stades de la contraction gravifique car les phases d'évolution dynamique à droite de la limite doivent être si rapides que les chances de les observer sont extrêmement faibles (voir Figure 4).

Naturellement ceci entraîne la possibilité d'un test précis et quelque peu redoutable de la théorie. Jusque très récemment, les meilleurs calculs, résumés par Auman et Bodenheimer (38), situaient la limite extrême de l'extension des modèles vers les basses températures effectives à des valeurs de l'ordre de 3100 à 3200°K. Bien qu'en gros l'accord avec les branches des géantes soit remarquable (29, 39), il existe cependant des géantes rouges possédant des températures effectives nettement inférieures de l'ordre, p. ex., de 2200 à 2500°K dans le cas de o Ceti. De plus, dans le cas des étoiles infrarouges observées ces dernières années (40), les estimations de température effective sont tombées encore bien plus bas, aux environs de 800 à 1000°K quoique Johnson (41) considère que des corrections de rougissement capables de ramener la température beaucoup plus haut, vers 3000°K, ne sont pas exclues.

D'autre part, le spectre de ces étoiles froides révèle dans plusieurs cas des bandes d'absorption moléculaires intenses dues à H_2, et Auman et Bodenheimer (38) ont cherché à évaluer avec précision les effets de l'opacité due à la vapeur d'eau. Mais l'abaissement correspondant de la limite des températures effectives reste très faible de l'ordre de 100°K si les autres paramètres plus ou moins indéterminés dans la théorie de la convection prennent les valeurs habituelles. Toutefois toute modification de ces paramètres qui tend à réduire l'efficacité de la convection dans la zone de transition et à pousser les modèles vers les basses températures voit ses effets largement amplifiés par l'opacité de H_2 qui grandit très rapidement quand la température diminue. Auman et Bodenheimer trouvent ainsi qu'une réduction de l_t/H par un faible facteur de l'ordre de 0·7 combinée à une augmentation des effets de la friction turbulente et de la dissipation par conduction radiative peuvent, en présence de vapeur d'eau, abaisser la température effective limite aux environs de 2600°K. Toutefois, il s'agit là d'une limite extrême pour des modèles entièrement homogènes et convectifs dont les températures centrales ne dépassent pas $2 \cdot 2 \times 10^5 \,°K$ et il n'est pas sûr qu'une étoile issue de la séquence principale puisse jamais l'atteindre. Ainsi, la question de l'accord entre la théorie et l'observation reste malgré tout assez critique et, s'il s'avérait que les étoiles infrarouges ont des températures nettement inférieures à $2000 \,°K$, la théorie serait confrontée avec un problème bien ardu.

Jusqu'ici, nous n'avons considéré que la question des modèles en équilibre hydrostatique ou quasi-statique et nous n'avons rien dit de l'influence des conditions aux limites et des zones d'ionisation de l'hydrogène et de l'hélium sur la stabilité globale des étoiles si ce n'est de rappeler l'instabilité dynamique qui caractérise les toutes premières phases de contraction d'une proto-étoile et l'instabilité vibrationnelle qui affecte les modèles entièrement convectifs d'étoiles de masses très faibles ($M \leqslant 0 \cdot 27 M_\odot$). Mais cette même instabilité vibrationnelle est d'une importance fondamentale dans l'interprétation des étoiles variables intrinsèques et, là aussi, il semble bien établi à présent que les zones extérieures, sièges de l'ionisation de l'hydrogène et de l'hélium, jouent un rôle capital. C'est Eddington (42) qui attira l'attention sur ce point dans les deux derniers articles qu'il consacra, en 1941–42, aux céphéides, ces mêmes étoiles qui avaient amorcé son intérêt dans le problème de la structure stellaire un quart de siècle plus tôt. Son argument principal reposait sur l'accroissement considérable de la capacité calorifique de ces couches par suite de l'ionisation en cours de l'hydrogène ce qui permettait d'y accumuler de l'énergie à hautes températures, durant la compression et de la libérer à températures plus basses, lors de l'expansion. Elles constituaient ainsi une véritable machine thermique capable de fournir de l'énergie mécanique à l'oscillation compensant l'énergie dissipée par la conduction radiative. Toutefois, il apparut dans la suite qu'en l'absence d'un apport supplémentaire d'énergie dans les couches profondes (43), ce mécanisme était insuffisant et Zhevakin (44), entretemps, avait eu le grand mérite d'attirer l'attention sur l'importance, dans ce cas, des ionisations de l'hélium, particulièrement de la seconde, et des effets d'opacité. Il put montrer que ces facteurs suffisaient pour provoquer une instabilité vibrationnelle appréciable pourvu que le transfert radiatif prédomine encore dans ces régions malgré la convection qui résulte des basses valeurs des Γ. Dans la suite, la prise en considération de coefficients d'opacité précis et de modèles détaillés des couches extérieures renforça encore cette source d'instabilité et les discussions détaillées de Baker et Kippenhahn (45) de J.P. Cox (46) et de Hoffmeister (47) confirmèrent définitivement l'effet global déstabilisant des couches extérieures dans les céphéides quoique les rôles respectifs des différentes zones d'ionisation et des effets de capacité calorifique ou d'opacité soient souvent si intimement mêlés qu'il n'est pas toujours très aisé de les distinguer. Ceci est vrai également de la question délicate des déphasages typiques entre les variations de luminosité et de rayon (46).

Enfin, les remarquables résultats de Christy (48) déduits de l'intégration directe des équations non-linéaires du problème, confirmés dans certains aspects par les investigations de J.P. Cox et A.N. Cox (49) et de leurs collaborateurs ainsi que par Alyoshin

(50) montrent que ce type d'instabilité peut effectivement conduire, sous l'action des effets non-linéaires, à des oscillations finies dont les amplitudes et l'anharmonicité sont en bon accord avec l'observation.

De plus, le mécanisme déstabilisant ne semble effectif que dans une mince bande du diagramme de Hertzsprung-Russell en accord satisfaisant avec la localisation des céphéides et partant fournit également une explication réelle de la relation Période-Luminosité.

D'autre part, ces résultats combinés à la théorie de l'évolution qui montre qu'une même étoile peut traverser plusieurs fois dans un sens et dans l'autre la bande d'instabilité, ouvrent d'intéressantes possibilités d'interpréter les variations de périodes parfois observées et suggèrent toute une série de tests statistiques.

La plus grande faiblesse de la théorie tant du point de vue linéaire que non-linéaire réside, comme ses protagonistes eux-mêmes l'ont souvent remarqué, dans les difficultés très sérieuses qui s'opposent à présent à une évaluation réaliste des effets mécaniques et thermiques de la convection dans les couches extérieures au cours de la pulsation. D'autre part, les conditions aux limites à la 'surface' de l'étoile restent assez délicates à formuler d'une façon entièrement satisfaisante en particulier si l'étoile est entourée d'une couronne à température élevée (51). Enfin, le passage du cas linéaire au cas non-linéaire soulève des problèmes mathématiques et physiques difficiles et parfois intrigants. Ainsi, les calculs non-linéaires semblent stabiliser assez rapidement l'oscillation autour d'un des modes linéaires, parfois le mode fondamental, parfois le premier mode alors que l'instabilité linéaire affecte à peu près également ces deux modes et peut même, dans certains cas, s'étendre à des modes plus élevés, tous ces modes ayant d'ailleurs des périodes, en général, strictement incommensurables.

De toute façon, on peut certainement conclure que la prise en considération de conditions de surface plus réalistes et de leurs implications pour les zones de transition quasi-convectives liées à l'ionisation de l'hydrogène et de l'hélium a contribué puissamment, au cours des 10 ou 12 dernières années, à la solution de problèmes relevant typiquement du domaine de la structure interne des étoiles: interprétation de la branche des géantes rouges, modifications profondes de nos idées sur l'évolution par contraction gravifique, existence d'une région interdite dans le diagramme de Hertzsprung-Russell aux températures effectives plus basses que quelque 2500°K, progrès marquants dans l'interprétation des variables intrinsèques des types céphéides, RR Lyra et peut-être des céphéides naines.

D'autre part, j'ai à peine fait allusion aux interactions plus délicates entre ces zones d'ionisations quasi-convectives et l'atmosphère qui, p. ex. dans le soleil, sont sans doute responsables de la grande variété de phénomènes transitoires ou semi-périodiques observés à sa surface aussi bien que de l'échauffement des couches chromosphériques et de la couronne. Tout en m'écartant du sujet même de cette conférence et en l'allongeant indûment, la discussion de ces différents aspects m'eût aussi rapidement entraîné dans un domaine qui sort largement de ma compétence. Cependant, la découverte par Leighton (52) de la nette périodicité ($\simeq$ 300 sec) sur quelques cycles successifs de champs de vitesses à petites échelles ($<$5000 km) à la surface du soleil a renouvelé l'intérêt (53) dans l'étude des ondes et des oscillations atmosphériques du soleil. La plupart de ces investigations se rapportent soit aux modes propres d'oscillations des couches supérieures convectivement stables ou à la propagation, dans ces couches, d'ondes excitées par la convection turbulente sous-jacente soit par pénétration, soit par effets gravifiques, soit encore par des variations aléatoires des conditions aux limites à la surface de séparation entre les deux zones. Maintenant que les familles possibles de modes d'oscillations non-radiales dans un milieu contenant une zone superadiabatique sont mieux connues (54), il me paraîtrait intéressant de réétudier le problème global des oscillations des couches extérieures du soleil couvrant à la fois la zone atmosphérique stable et la zone d'ionisation de l'hydrogène vers le sommet de laquelle, malgré l'établissement de la convection, de forts gradients superadiabatiques peuvent subsister.

Dans le cas des champs périodiques de vitesse de Leighton, étant donné la rapidité des oscillations observées, il s'agit vraisemblablement de modes acoustiques ou modes-p mais leur excitation pourrait être liée à l'existence de ces gradients super-adiabatiques (55) et les possibilités de couplage avec les modes de gravité, ou modes-g, stables et instables devraient être prises en considération.

Notons aussi qu'en présence de rotation ou de champs magnétiques, certains des modes-g instables et sans doute particulièrement ceux de grandes longueurs d'ondes horizontales, les autres se confondant avec la convection turbulente proprement dite, peuvent devenir 'overstable' (56) et acquérir des fréquences réelles (57, 58) liées à la vitesse angulaire de rotation ou à l'intensité du champ magnétique. Les périodes correspondantes seraient vraisemblablement très longues mais elles pourraient être significatives pour d'autres aspects des phénomènes solaires.

Poussant les choses plus loin, on peut se demander si, à part la convection profonde ordinaire, toute possibilité de connexion directe entre l'atmosphère superficielle et les couches beaucoup plus profondes est totalement exclue. D'habitude, du moins dans les modèles possédant des zones radiatives intermédiaires étendues, on considère que l'intérieur profond n'influence guère l'atmosphère que par le champ gravifique qu'il y établit et par la quantité totale d'énergie qu'il dégage. Celle-ci, sous forme de radiation, ne progresse d'ailleurs que très lentement vers la surface et en prenant dans chaque couche des propriétés d'équilibre qui effacent totalement les caractéristiques des couches plus profondes. Mais, p. ex., la théorie des taches solaires et de leur cycle, proposée par Alfvén (59) il y a quelques années ainsi que la version de Walén (60) faisaient appel à une telle interaction profonde liée à la propagation jusqu'à la photo-sphère d'ondes hydromagnétiques nées à la surface du noyau convectif supposé, à l'époque, exister dans les régions centrales du soleil. La diminution de densité accom-pagnant la concentration du champ magnétique dans un tube ou dans un globule stabilisé par ce champ même et en équilibre de pression totale avec le milieu ambiant fait apparaître, comme l'ont signalé Parker (61) et Jensen (62), une force d'Archimède qui pousse ce globule ou ce tube vers la surface. Récemment, cette possibilité a été discutée (63) du point de vue du transport des propriétés de composition chimique profonde vers la surface.

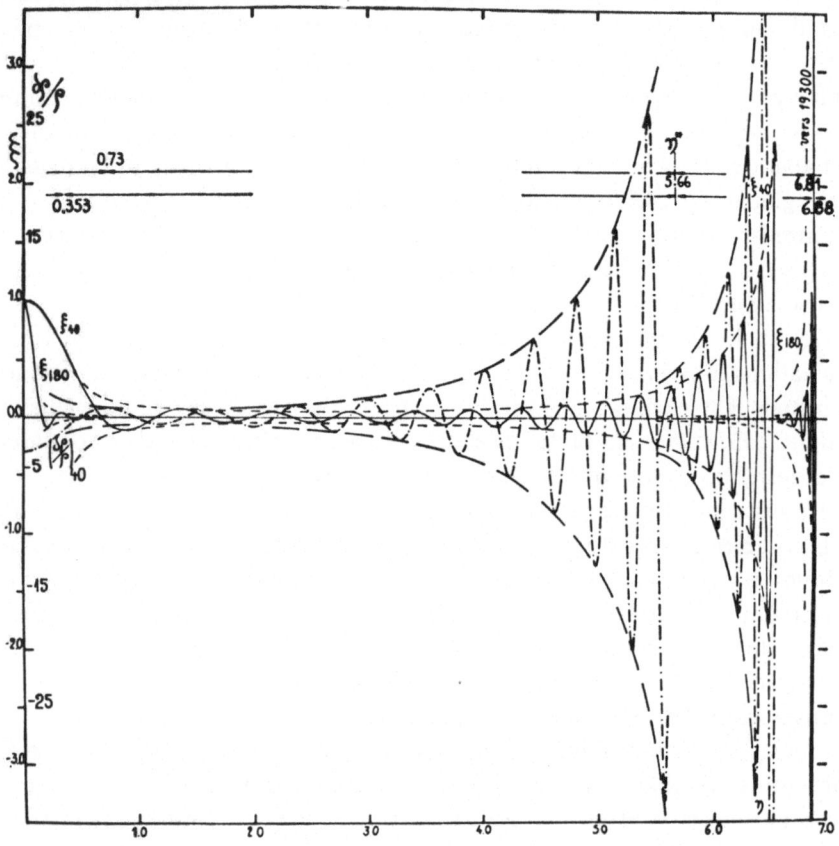

FIG. 5.   *Comportement de l'amplitude relative $\xi = \delta r / r$ (trait plein) et de $\delta p / p$ (trait pointillé) pour des modes élevés de pulsation radiale.*

Le comportement de l'amplitude des modes très élevés d'oscillations radiales possède un aspect assez curieux (64) qui suggère également quelque possibilité d'inter-action. En effet, cette amplitude commence par décroître très rapidement à partir du centre pour prendre des valeurs très faibles partout sauf près de la surface où elle peut atteindre des valeurs extrêmement grandes. Dans la région intermédiaire étendue, l'amplitude est si petite que la dissipation radiative ou visqueuse doit être très faible et ainsi il se peut que la plus grande partie de l'énergie gagnée par l'onde très près du centre par suite des réactions nucléaires ne soit dissipée qu'aux environs de la surface. Bien que nous ne disposions pas, dans le cas des oscillations non-radiales, d'une théorie asymptotique adéquate, on peut néanmoins s'attendre à un comportement quelque peu analogue au moins pour les modes-p. Dans ce cas, quels peuvent être, p. ex., les effets de ces modes-p certainement excités dans les noyaux convectifs de toute une classe d'étoiles?

Enfin, au cours des 10 à 15 dernières années, une autre notion nouvelle, celle de 'vent solaire' ou plus généralement de 'vent stellaire' est venue préciser le comportement des couches les plus extérieures des étoiles. Suggérée d'abord par Biermann (65) sur la base d'observations cométaires, cette notion s'est vue, sur cette courte période, dotée d'un statut théorique satisfaisant à beaucoup de points de vue grâce surtout aux travaux de Parker (66), puis définitivement confirmée, dans le cas du soleil, par les mesures effectuées à bord des satellites artificiels. Tout récemment, les fluctuations très rapides (< 1 sec) dans le flux des sources-radio de très petits diamètres angulaires (67) y ont révélé une structure surprenante caractérisée par des irrégularités de très petites dimensions (≃ 200 km) qui en constitue peut-être une autre caractéristique significative.

Si, dans le cas du soleil, la perte de masse correspondante est si faible qu'on puisse la rejeter immédiatement en tant que facteur susceptible d'influencer sa structure interne ou son évolution, il n'en est sans doute pas de même pour d'autres étoiles notamment les supergéantes (68) et peut-être même les géantes rouges qui ont longuement retenu notre attention ici. A ce point de vue, la théorie des vents stellaires ne sera vraiment satisfaisante qu'au moment où elle couvrira adéquatement son origine aussi bien que sa description cinématique et reliera son intensité aux autres propriétés de l'étoile. Il s'agit là, certainement d'un problème très difficile lié à la théorie détaillée de l'évaporation compte tenu de l'anisotropie des fonctions de distribution dans les couches très extérieures et aux conséquences, au moins dans certains cas, de l'absorption sélective d'ions accélérés et de l'activité acoustique engendrée par la convection turbulente dans les couches sub-photosphériques.

Quoiqu'il en soit, tous ces derniers points que nous n'avons pu qu'effleurer très superficiellement suggèrent une foule de problèmes dont la solution, sans aucun doute, tissera des liens de plus en plus étroits entre les questions de structure interne et le comportement des couches les plus extérieures, nous livrant enfin une vue globale de l'étoile y compris ses interactions avec le mileu interstellaire dans lequel elle baigne.

Bibliographie générale

Hayashi, C. (1966)   *A. Rev. Astron. Astrophys.*, **4**, 171.
Hayashi, C., Hashi, R., Sugimoto, D. (1962)   Evolution of the Stars, *Progr. Theor. Phys. Suppl.*, **22**.
Pecker, J.-C. (1960)   9e Coll. lntern. Astrophys. Liège 1959, *Mém. Soc. Roy. Sci. Liège*, 5e sér., **3**, 343.

Références

1. Voir p. ex.: Odgers, G.J. (1957)   *Publ. Dom. Astrophys. Obs., Victoria*, **10**, 393.
2. Voir en particulier Fowler, R.H. (1930)   *Mon. Not. R. astr. Soc.*, **91**, 63. – Strömgren, B., *ibid.*, 466. – Cowling, T.G., *ibid.*, 472. – Hopf, E., *ibid.*, 653.
3. Milne, E.A. (1930)   *Mon. Not. R. astr. Soc.*, **91**, 4; (1932) *ibid.*, **92**, 610; (1945) *ibid.*, **105**, 146; (1931) *Z. Astrophys.*, **3**, 253.
4. Russell, H.N. (1931)   *Mon. Not. R. astr. Soc.*, **91**, 952. – Rosseland, S. (1932)   *Z. Astrophys.*, **4**, 255. – Cowling, T.G. (1932)   *Z. Astrophys.*, **4**, 331.
5. Unsöld, A. (1930)   *Z. Astrophys.*, **1**, 138.
6. Chandrasekhar, S., Münch, G. (1946)   *Astrophys. J.*, **104**, 446. – Unsöld, A. (1955)   *Physik der Sternatmosphären*, 2ème éd., Springer-Verlag, Berlin.
7. Bierman, L. (1937)   *Astronom. N.*, **264**, 361 et 395.
8. Schwarzschild, M. (1958)   *Structure and Evolution of the Stars*, Princeton Univ. Press, § 11.
9. Ezer, D., Stein, R.F., Cameron, A.G.W. (1963)   *Astron. J.*, **68**, 278. – Hofmeister, E., Weigert, A. (1964)   *Z. Astrophys.*, **59**, 119.
10. Faulkner, J., Griffiths, K., Hoyle, F. (1964)   *Mon. Not. R. astr. Soc.*, **129**, 363. Voir aussi: Ruben, G.V. (1963)   *Astr. Zu.*, **40**, 855.
11. Ledoux, P., Schwarzschild, M., Spiegel, E. (1961)   *Astrophys. J.*, **133**, 184.
12. Böhm, K.H., Richter, E. (1959)   *Z. Astrophys.*, **48**, 231. – Böhm, K.H. (1963)   *Astrophys. J.*, **137**, 881. – Spiegel, E. (1964)   *Astrophys. J.*, **139**, 959.
13. Cox, A.N. (1965)   *Stars and Stellar Systems*, **8**, Ed. by L.H. Aller and D.B. McLaughlin, Univ. of Chicago Press, Chicago, p. 195; (1964) *J. Quantit. Spectrosc. Radiat. Transfer*, **4**, 737. – Cox, A.N., Stewart, J.H., Eilers, D.D. (1965)   *Astrophys. J. Suppl.* **11**, 1. – Cox, A.N., Stewart, J.H., *ibid.*, 22. – Mayer, H.L. (1964)   *J. Quantit. Spectrosc. Radiat. Transfer*, **4**, 585.
14. Voir p. ex.: Henyey, L.G., Vardya, M.S., Bodenheimer, P. (1965)   *Astrophys. J.*, **142**, 841.
15. Sack, N. (1966)   *Ann. Astrophys.*, **29**, 633.
16. Vitense, E. (1953)   *Z. Astrophys.*, **32**, 135.
17. Böhm-Vitense, E. (1958)   *Z. Astrophys.*, **46**, 108.
18. Voir p. ex.: Baker, N.H., Temesvary, S. (1966)   Tables of Convective Stellar Envelope Models, Inst. Space Studies, NASA, New York.
19. Schatzman, E. (1962)   *Ann. Astrophys.*, **25**, 18.
20. Osterbrock, D.E. (1953)   *Astrophys. J.*, **118**, 529.
21. Limber, D.N. (1958)   *Astrophys. J.*, **127**, 387.
22. Hayashi, C., Nakano, T. (1963)   *Progr. Theor. Phys.*, **30**, 460.
23. Gabriel, M. (1966)   Coll. on Late-Type Stars, Trieste, p. 347.
24. Gamow, G., Keller, G. (1945)   *Rev. Mod. Phys.*, **17**, 125. – Hayashi, C. (1947)   *Progr. Theor. Phys.*, **2**, 127. – Hayashi, C. (1949)   *Phys. Rev.*, **75**, 1619. – Gardiner, J.G. (1951)   *Mon. Not. R. astr. Soc.*, **111**, 102. – Dumézil-Curien, P. (1954)   *Ann. Astrophys.*, **17**, 197.
25. Sandage, A.R., Schwarzschild, M. (1952)   *Astrophys. J.*, **116**, 463.
26. Schwarzschild, M., Rabinowitz, I., Härm, R. (1953)   **118**, 326.
27. Hoyle, F., Schwarzschild, M. (1955)   *Astrophys. J. Sup.* **2**, 1.
28. Haselgrove, C.B., Hoyle, F. (1956)   *Mon. Not. R. astr. Soc.*, **116**, 527; (1958) *ibid.*, **118**, 519; (1959) *ibid.*, **119**, 112. – Kippenhahn, R., Temesvary, S., Biermann, L. (1958)   *Z. Astrophys.*, **46**, 257. – Simoda, M., Obi, S. (1958)   *Publ. Astr. Soc. Japan*, **10**, 26. – Simoda, M. (1961)   *Publ. Astr. Soc. Japan*, **13**, 424. – Schwarzschild, M., Selberg, H. (1962)   *Astrophys. J.*, **136**, 150. – Swamy Krishna, K.S., Kushwaha, R.S. (1962)   *Astrophys. J.*, **135**, 802. – Iben, I., Jr. (1967)   *Astrophys. J.*, **147**, 624.
29. Hayashi, C., Hoshi, R. (1961)   *Publ. Astr. Soc. Japan*, **13**, 442.
30. Hayashi, C. (1961)   *Publ. Astr. Soc. Japan*, **13**, 450.
31. Henyey, L.G., Le Levier, R., Levée, R.D. (1955)   *Publ. Astr. Soc. Pacific*, **67**, 154. Cf. also Brownlee, R.R., Cox, A.N. (1961)   *Sky and Telescope*, **21**, 252.
32. Hayashi, M. (1965)   *Publ. Astr. Soc. Japan*, **17**, 177. – Ezer, D. (1966)   Coll. on Late-Type Stars, Trieste, p. 357.

33. Faulkner, F., Griffiths, K., Hoyle, F. (1963)   *Mon. Not. R. astr. Soc.*, **126**, 1; (1965) *ibid.*, **129**, 363. – Ezer, D., Cameron, A.G.W. (1963)   *Icarus*, **1**, 422; (1965) *Canadian J. Phys.*, **43**, 1497. – Weymann, R., Moore, E. (1963)   *Astrophys. J.*, **137**, 522.
34. Cameron, A.G.W. (1963)   *Icarus*, **1**, 13. – Hayashi, C., Nakano, T. (1965)   *Progr. Theor. Phys.*, **34**, 754. – Upton, E.R.L., Little, S.J., Dworetsky, M.M. (1967)   *Astrophys. J.* (sous presse); cf. also *ibid.*, (1967) *Astron. J., *, **72**, 308.
35. Gaustad, J.E. (1963)   *Astrophys.J.*, **138**, 1050. – Gould, R.J. (1964)   *Astrophys.J.*, **140**, 638.
36. Ledoux, P. (1958)   *Hdb. Physik*, Ed. by S. Flügge, Springer-Verlag, Berlin, t. LI, p. 660.
37. Schatzman E. (1967)   *Ann. Astrophys.* **30**, 963.
38· Auman, J.R., Bodenheimer. (1967)   *Astrophys. J.*, **149**, 641.
39. Barbaro, G., Dallaporta, N., Nobili, L. (1966)   Coll. on Late-Type Stars, Trieste, p. 368.
40. Neugebauer, G., Martz, D.E., Leighton, R.B. (1965)   *Astrophys. J.*, **142**, 399. – Johnson, H.L., Mendoza, E.E., Wisniewski, W.Z. (1965)   *Astrophys. J.*, **142**, 1249. – Johnson, H.L. (1966)   *A. Rev. Astron. Astrophys.*, **4**, 193.
41. Johnson, H.L. (1966)   121st meeting A.A.S., Hampton, Va, March.
42. Eddington, A.S. (1941)   *Mon. Not. R. astr. Soc.*, **101**, 182; (1942) *ibid.*, **102**, 154.
43. Cox, J.P. (1955)   *Astrophys. J.*, **122**, 286.
44. Zhevakin, C.A. (1952)   *Astr. Zu.*, **29**, 37; (1954) *ibid.*, **31**, 335; (1955) *ibid.*, **32**, 124 etc. Cf. *A. Rev. Astron. Astrophys.*, **1**, 1963, 367.
45. Baker, N., Kippenhahn, R. (1962)   *Z. Astrophys.*, **54**, 114; (1965) *Astrophys. J.*, **142**, 868.
46. Cox, J.P. (1959)   *Astrophys. J.*, **130**, 296; (1963) *ibid.*, **138**, 487.
47. Hofmeister, E. (1965)   Thesis (München); (1965) *Mitt. Astr. Ges.*, **19**, 90; (1965) *Proc. I.A.U. Coll.*, *Bamberg*, p. 224 (*Kl. Veröff. Remeis-Ternw. Bamberg*, **4**, no. 40).
48. Christy, R.F. (1966)   *Astrophys. J.*, **144**, 108; (1966)   *A. Rev. Astron. Astrophys.* **4**, 353; (1967)   *I.A.U. Symp.* no. 28; *Aerodynamic Phenomena in Stellar Atmospheres*, Nice 1965, p. 105.
49. Cox, J.P., Cox, A.N., Olsen, K.H., King, D.S., Eilers, D.D. (1966)   *Astrophys. J.*, **144**, 1038. – King, D.S., Cox, J.P., Eilers, D.D. (1966)   *Astrophys. J.*, **144**, 1069.
50. Alyoshin, V.I. (1964)   *Astr. Zu.*, **41**, 201; (1966) *ibid.* (sous presse).
51. Voir p. ex.: Unno, W. (1965)   *Publ. Astr. Soc. Japan*, **17**, 205. – Simon, R. (1964)   *Astrophys. Norv.*, **9**, 113.
52. Leighton, R.B. (1961)   *Nuovo Cimento, Suppl.* **22**, 321. – Leighton, R.B., Noyes, R.W., Simon, G.W. (1962)   *Astrophys. J.*, **135**, 474. – Evans, J., Michard, R. (1962)   *Astrophys. J.*, **136**, 493. – Leighton, R.B. (1963)   *A. Rev. Astron. Astrophys.*, **1**, 13.
53. Whitney, C.A. (1958)   *Smithsonian Contr. Astrophys.*, **2**, 365; (1963) *Astrophys. J.*, **138**, 537. – Kahn, F.D. (1961)   *Astrophys. J.*, **134**, 343; (1962) *ibid.*, **135**, 547. – Schmidt, H.V., Zirker, J.B. (1963)   *Astrophys.J.*, **138**, 1310. – Bahng, J., Schwarzschild, M. (1963)   *Astrophys. J.*, **137**, 901. – Schatzman, E. (1964)   *Astrophys. Norv.*, **9**, 283. – Moore, D.W., Spiegel, E.A. (1964)   *Astrophys.J.*, **139**, 48. – Orrall, F.Q. (1965)   *Astrophys. J.*, **141**, 1131. – Uchida, Y. (1965)   *Astrophys. J.*, **142**, 335. – Souffrin, P. (1966)   *Ann. Astrophys.*, **29**, 55. – Kato, S. (1966)   *Astrophys. J.*, **143**, 893; *ibid.*, **144**, 326.
54. Ledoux, P., Smeyers, P. (1966)   *C.R. Acad. Sci., Paris*, **262**, 841. – Smeyers, P. (1966)   *Ann. Astrophys.*, **29**, 539.
55. Spiegel, E. (1964)   *Astrophys. J.*, **139**, 959.
56. Chandrasekhar, S. (1961)   *Hydrodynamic and Hydromagnetic Stability*. Clarendon Press, Oxford, Chaps. III and IV.
57. Voir p. ex.: Ledoux, P. (1951)   *Astrophys.J.*, **114**, 373. – Ledoux, P., Simon, R. (1957)   *Ann. Astrophys.*, **20**, 185. – Ledoux, P. (1967), *The Magnetic and Related Stars*. AAS-NASA Symposium, Greenbelt (U.S.A.), 1965, Ed. by R.C. Cameron, Mono Book Corporation, Baltimore, p. 65.
58. Cretin, M., Tassoul, J.L. (1965)   *Ann. Astrophys.*, **28**, 982.
59. Alfvén, H. (1945)   *Mon. Not. R. astron. Soc.*, **105**, 3 and 382.
60. Walén, C. (1944)   *Ark. Nat. Astr. Fys.*, **30A**, no. 15; (1949) *On the Vibratory Rotation of the Sun*. Henrik Lindstähls Bokhandel, Stockholm.

61. Parker, E. N. (1955)   *Astrophys. J.*, **121**, 491.

62. Jensen, E. (1955)   *Ann. Astrophys.*, **18**, 127; (1957) *Astrophys. Norv.*, **5**, 289.

63. Gurm, H. S., Wentzel, D. G. (1967)   *Astrophys. J.*, **149**, 139.

64. Ledoux, P. (1963)   *Bull. Acad. R. Belgique*, Cl. Sci., 5e série, **49**, 286.

65. Biermann, L. (1951)   *Z. Astrophys.*, **29**, 274; (1957) *Observatory*, **77**, 109.

66. Parker, E. N. (1958)   *Astrophys. J.*, **128**, 664; (1960) *ibid.* **132**, 821; (1964) *ibid.*, **139**, 72, 93 and 690; (1965) *ibid.*, **141**, 322.

67. Hewish A., Scott, F. P., Wills, D. (1964)   *Nature*, **203**, 1214. – Little, L. T., Hewish, A. (1966) *Mon. Not. R. astron. Soc.*, **134**, 221. – Salpeter, E. E. (1967)   *Astrophys. J.*, **147**, 433.

68. Deutsch, A. J. (1956)   *Astrophys. J.*, **123**, 210. Voir aussi: Weymann, R. (1963)   *A. Rev. Astron. Astrophys.*, **1**, 97.

RADIO GALAXIES AND QUASARS, I

Invited Discourse C given on August 28, 1967 in the Lucerna Hall

M. RYLE

(University of Cambridge, England)

If we observe the sky with a radio telescope operating at metre wavelengths, we find first a continuous background mainly due to the Milky Way – which we observe undimmed by the obscuration which spoils our optical view. In addition we find compact sources, a few minutes of arc or less in extent. About 8000 of these compact sources have now been discovered using large instruments, but only a few hundred of the most intense have been studied in any detail; a few are within our Galaxy and represent the emission from the remnants of supernova explosions, but many of them are found to be associated with faint galaxies. From their distances we can conclude that their radio emission is very great, in some cases more than a million times greater

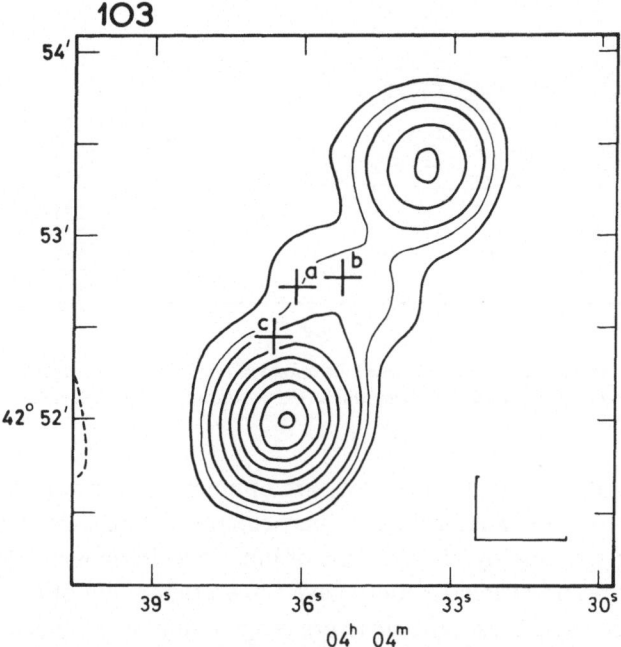

FIG. 1.   *3C 103 consists of two compact components separated by 88″ arc; the cross (a) represents the position of a 16ᵐ D galaxy with which the radio source is probably related.*

Perek (ed.), Highlights of Astronomy, 33–44. © I.U.A.

than that from our own Galaxy or the Andromeda nebula. These powerful sources
are known as 'radio galaxies'.

Observations with instruments of high resolving power show that about 60% have
a double structure, with radio emission from two components, one on each side of the
related galaxy; the two components are frequently of unequal intensity, and they may
be located at unequal distances from the galaxy.

Their appearance and other evidence, which will be discussed by Allan Sandage,
suggest that the radio-emitting material has been ejected by an explosion within the
galaxy.

The maps shown in Figures 1–4 were made with the one-mile telescope at Cambridge – which has a resolution of about 25 sec of arc when operating at a wavelength
of 21 cm.

Some sources show a more complex structure, in which a number of components
can be distinguished, sometimes linked with bridges of emission (Figures 2, 3, 4).
Such sources might be the result of a series of lesser explosions.

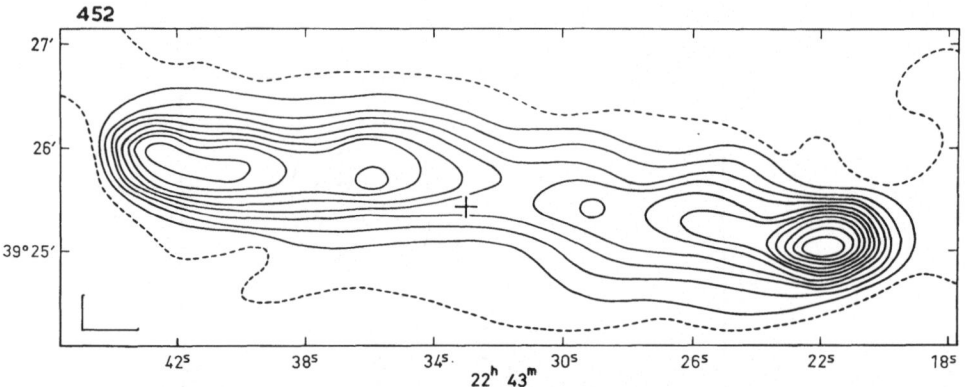

Fig. 2.   *3C 452 contains a number of peaks in a long ridge of emission, suggesting that more than
one release of energy has occurred. The cross marks the position of a 16ᵐ ED galaxy, which may be the
source of the energy.*

In addition to these sources, some 35% are of much smaller angular size; observations with long-baseline interferometers, particularly that between Jodrell Bank and
Malvern in England, and by observations of lunar occultations notably at Parkes in
Australia and Arecibo in Puerto Rico, have shown that many of them are less than
1″ arc in extent; in some cases the structure is again double, with components ∼0·1″
arc in diameter, separated by a few seconds of arc.

Many of these compact sources have been found to be associated with very small
optical objects which look like stars, and they have become known as quasi-stellar
sources (QSS). Dr. Sandage will describe how their optical spectra contain lines which
can only be interpreted if very large red-shifts are adopted. The origin of these red-

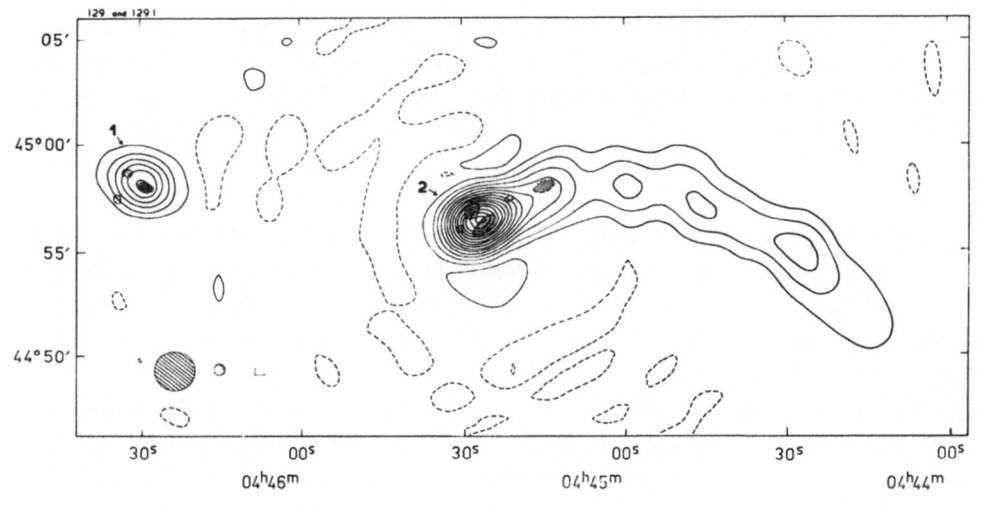

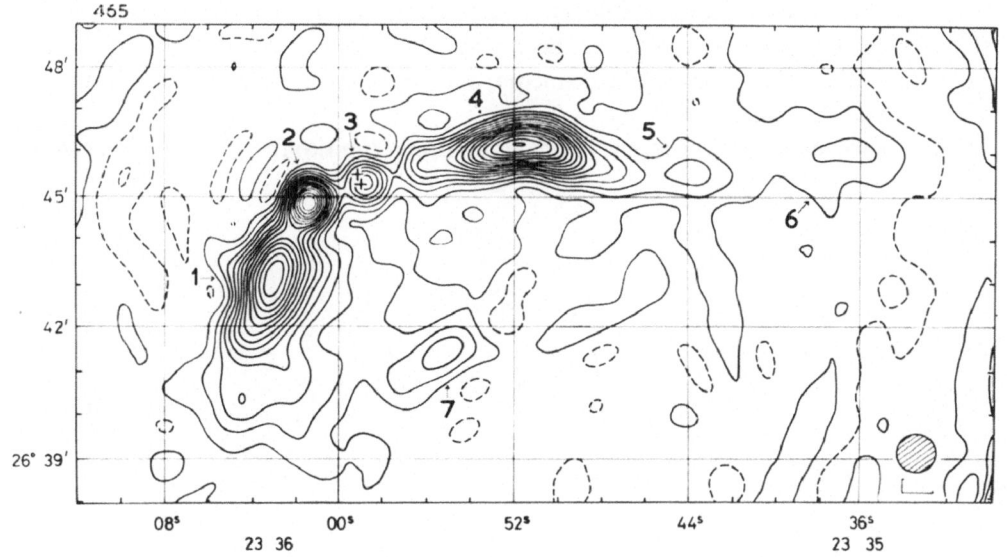

shifts is still controversial; the most obvious interpretation is that they are Doppler shifts due to the general expansion of the Universe, and that the QSS are at great distances. In order to avoid the necessity for objects having such very great radio, optical and UV emission, and to explain some of the peculiarities of the observed red-shifts, some astronomers believe that they are comparatively local – though still extra-galactic. In this case some other explanation must be sought for the red-shifts. Allan Sandage will discuss these two alternatives in detail, but I shall be presenting evidence which supports the former interpretation, that they are at great distances.

Observationally we can distinguish QSS from radio galaxies in a number of ways, although it is very difficult to draw a sharp dividing line; many sources show characteristics belonging to both classes of object.

(a) QSS show fluctuations of radio intensity with periods of ~ 1 sec, due to diffraction in the interplanetary medium; this effect, which is analogous to optical 'seeing' only occurs for sources having an angular diameter $< 1''$ arc, and is proving valuable as an additional method of investigating their angular structure in the range $1''$ to $0''.01$ arc.

Here we should note that 3C 295, which is associated with a massive galaxy, and has a compact double structure, also scintillates, revealing the presence of at least one very small component. On the other hand, 3C 47 and 263, which are both identified with optical QSS, are extensive double radio sources. 3C 225 and 267 are also extensive double sources, but each contains one very compact component as revealed by interplanetary scintillation; neither has a conclusive optical identification.

(b) Observations at short wavelengths (< 10 cm) have revealed longer-term variations, of some months, which cannot be explained in terms of diffraction in the interplanetary, interstellar or intergalactic media, and which must be attributed to intrinsic variations of the source itself.

The existence of such variations indicates an upper limit to the *physical* dimensions

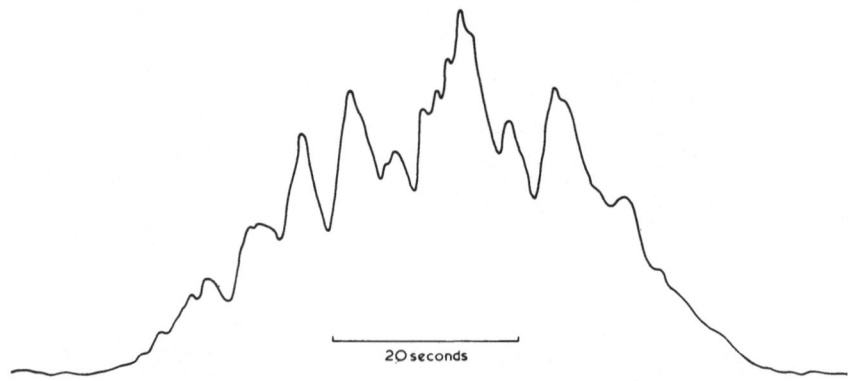

20 seconds

FIG. 5. *Record showing the fluctuations in the intensity of 3C 48 due to diffraction in the interplanetary medium. The observations were made with a transit instrument.*

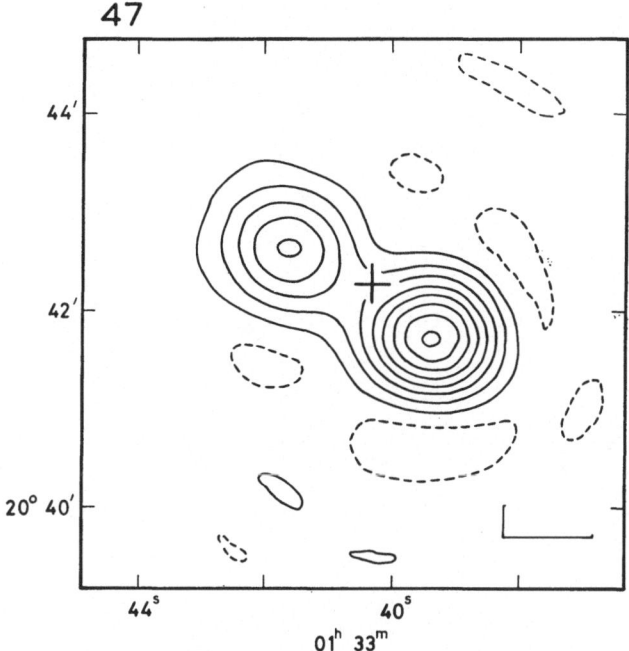

FIG. 6.   *3C 47 consists of two components either side of an optical QSS; if the red-shift of 0·425 is interpreted as of cosmological origin their physical separation is 208 kpc, a figure similar to that of many radio galaxies.*

of the source of a few light months. There are in fact even more rapid variations in the optical emission, as Dr. Sandage will discuss.

Here again, it has been found that the radio source associated with the galaxy NGC 1275 also shows fluctuations at high frequencies, and shows that very compact radio components certainly exist within some galaxies.

(c) Figure 8 shows a third way of distinguishing QSS by their radio spectra; most radio galaxies have a spectrum which may be described by an exponential, but QSS normally show a marked cut-off at low frequencies, as shown in Figure 8.

In some cases this cut-off occurs at quite a high frequency, and it is these sources which are found to be the most compact as shown by their variability; this result suggests that the cut-off should be attributed to self-absorption effects within the source itself, and also indicates why variability is not observed at low frequencies.

The physical explanation for this cut-off is now reasonably well understood. The only satisfactory mechanism which has been suggested to account for the generation of radio waves in both QSS and radio galaxies is the synchrotron process, in which electrons of high energy are accelerated in a magnetic field. This mechanism not only provides a simple explanation for the observed spectra – including the low-frequency cut-off in the most compact sources – but also predicts the presence of linear polariza-

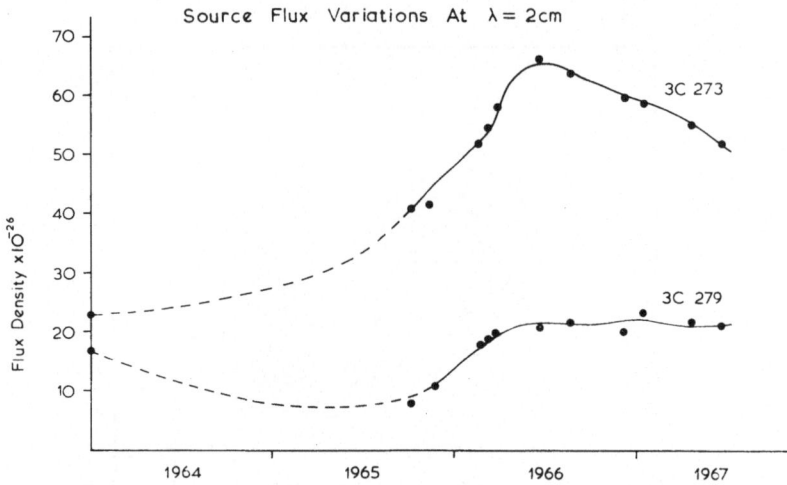

FIG. 7. *Observations made at Greenbank of the emission from 3C 273 and 279 at a wavelength of 2 cm show marked variations of intensity.*

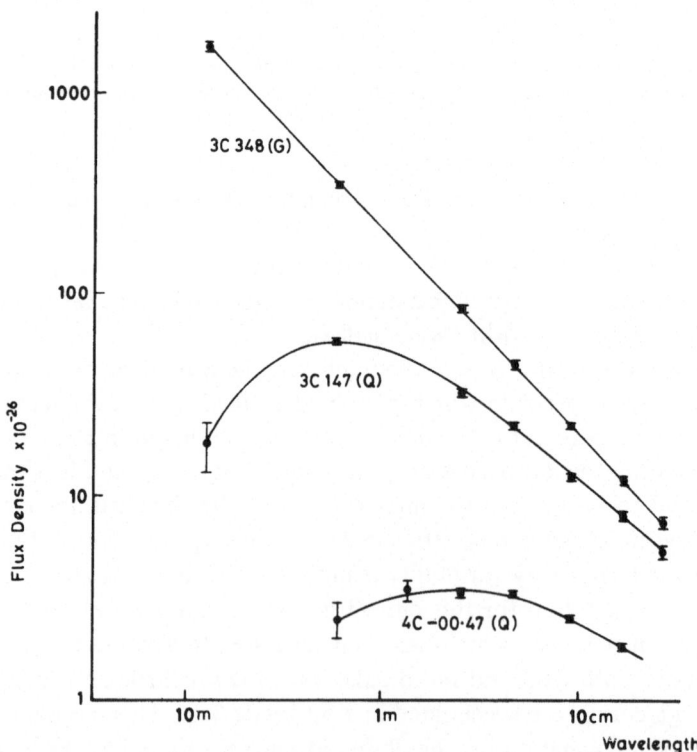

FIG. 8. *The radio spectra of a radio galaxy (3C 348) and two QSS (3C 147 and 4C-00·47).*

tion in the emission from sources where the geometry of the magnetic field is simple; the observation of such polarization in many sources has provided further confirmation of the synchrotron mechanism.

For sources with a low-frequency cut-off whose angular extent and distance are known, we can derive the magnetic field and the number and energy distribution of the high-energy electrons, so that the total energy required to account for the radiation may be found. In other cases where no cut-off in the observed spectrum occurs it is still possible to establish a *lower limit* to the energy in the particles and magnetic field. Estimates of the rate at which the energy will be lost may also be made so that the life time of different types of source may be estimated.

The energy which is necessary to account for the most powerful radio galaxies – which must presumably be released within the parent galaxy – is of the order of 10^{61} ergs, equivalent to the conversion to helium of a mass of hydrogen of about $10^9 \, M_\odot$. It is significant that if the red-shift of the QSS is interpreted as of cosmological origin, the emission from the most powerful of these sources also implies energies of the same order.

Even in the extensive radio galaxies the lifetime of the electrons is only about a million years, and the large separation of the emitting regions from the parent galaxy (of 100 kpc or more) implies that they must have been ejected from the galaxy with velocities comparable with the velocity of light.

These results suggest that QSS and radio galaxies may simply be different stages in the evolution of the same class of source. It seems certain that radio galaxies must have their origin within the related galaxy and that the source must therefore have passed through a more compact stage; can we indeed release 10^{61} ergs within a time $< 10^6$ years in a galaxy and *not* observe it by optical or radio means? It is equally certain that the energy density in QSS is so high that they must expand very rapidly; where are the sources resulting from such an expansion if they are not the radio galaxies?

It is suggestive that the recent high-resolution observations(Jodrell Bank–Malvern) of a number of the most compact QSS known, also reveal a double structure, so that if we adopt the cosmological interpretation of their red-shifts, two-component sources are now known which cover a range of separations from 1–450 kpc.

A number of double sources from the 3C catalogue are indicated in Figure 9, together with the physical size and intrinsic radio power derived on the assumption that the red-shift is cosmological. The nature of the related optical object is also shown.

If the red-shifts of the QSS are *not* cosmological, the similarity in the derived powers would be a remarkable coincidence. The impossibility of making any clear division between QSS and radio galaxies suggests strongly that the two classes of source are related.

In order to visualize how QSS and radio galaxies might evolve, a simple model may be constructed which supposes that a large amount of energy ($\sim 10^{61}$ ergs) is suddenly released at the centre of a galaxy, and that this gives rise to the ejection of two plasma

Sources 3C   48 147 295 191 254 405   79 263   47 348 219

Separation   1·2 2·2 16   17   56   78 147 179 208 260 263

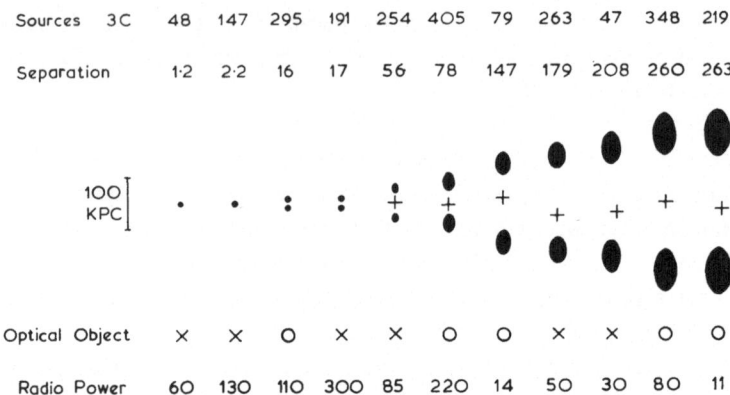

100
KPC

Optical Object   ×   ×   O   ×   ×   O   O   ×   ×   O   O

Radio Power   60 130 110 300   85 220   14   50   30   80   11

FIG. 9. *The physical extent (kpc) and radio power (in units of 10^{25} watts ster^{-1} Hz^{-1}) of a number of double radio sources, together with the nature of the related optical object ($+$ QSS; $\circ$ Galaxy).*

clouds travelling out in opposite directions with an initial velocity $\sim c$. Except in the case where the ejection has occurred along a line perpendicular to the line of sight the components will be observed at significantly different ages because of their relativistic velocities, and they will also not be separated by equal angles from the galaxy.

By using the *observed* angular separations of the components from the optical object in a number of actual sources, the velocity of ejection and ages of the individual components may be found, so that for each source the variation of emission with time may be compared on the basis of the model.

Less accurate data may be obtained in a similar way for the more compact sources. It is seen that the observations of a large number of sources are in reasonable agreement with a model in which the emission at short wavelengths remains constant for $\sim 10^5$ years and then rapidly decreases to a small value after $\sim 3 \cdot 10^6$ years. If a similar curve is drawn for the emission at long wavelengths the effects of self-absorption may be seen to persist until $\sim 3 \cdot 10^4$ years.

These results suggest that *all* powerful extra-galactic radio sources may belong to the same class and that the wide spread in the properties is mainly due to their different ages. Sources showing the characteristics of both QSS and radio galaxies occur within, and only within, the age limits $3 \cdot 10^4$–$3 \cdot 10^5$ years. The rapid decrease in emissivity after $\sim 10^5$ years corresponds approximately to the age at which the plasma clouds pass out of the galaxy, and agrees with the expected decrease of emission from a plasma cloud expanding into a vacuum.

Many suggestions have been made to account for the large amount of energy needed, including the possibility of a chain reaction of supernovae, mechanisms occurring during star formation, the annihilation of matter and anti-matter, and the gravitational collapse of a supermassive star. Some support for the first of these hypotheses has been provided by some recent observations of the Seyfert galaxy NGC 1275, which gives the impression of a QSS in slow motion.

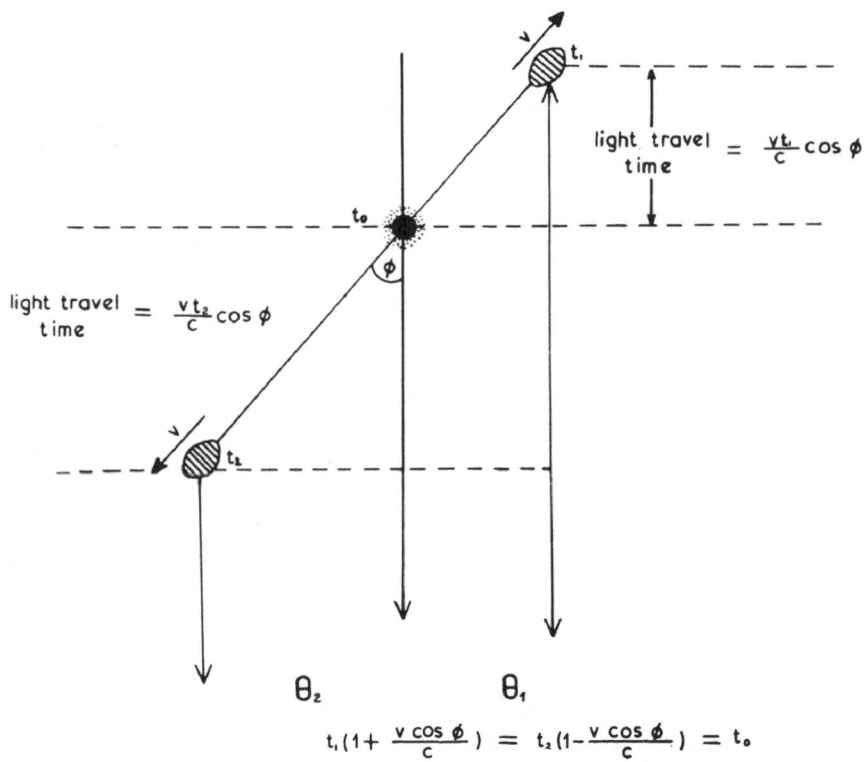

FIG. 10. *The ejection of two plasma clouds in opposite directions from a galaxy at relativistic velocities will give rise to a double source whose components will be observed at different proper ages.*

The nucleus of the galaxy contains a compact radio source, variable in periods of a few years, with a low-frequency cut-off of 5000 MHz; it is thus similar, though of much less power, to some of the most compact QSS. This very compact source could be accounted for by a large supernova event comparable with that which has produced the intense galactic radio source Cassiopeia A. The radio emission from the nucleus as a whole also shows a low-frequency cut-off (400 MHz) and could be maintained by the occurrence within the nucleus of one such supernova every few years. The optical evidence supports the idea that there has been a succession of supernovae within the nucleus.

Recent observations have revealed radio emission from more extensive components, including the distorted halos of two neighbouring galaxies within the Perseus Cluster which can only be accounted for in terms of particles ejected from NGC 1275. A rate of production of particles at least as great as that occurring at the present time must therefore have persisted for 5–10 million years, a result confirmed by the observation by Margaret Burbidge of gas streams in the outer parts of NGC 1275. The total energy released is therefore comparable with that of the most powerful sources known,

M. R Y L E

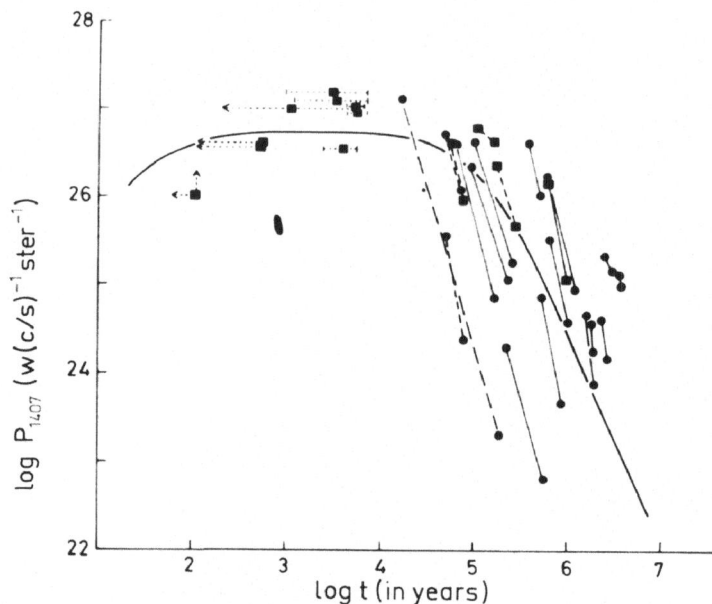

FIG. 11.   *The emission from each component of a number of double sources plotted against the component age, as derived on the basis of the simple source model.*

but it has been released over a period of several million years. The possibility of such a supernova process occurring very rapidly, perhaps by a chain reaction as suggested by Geoffrey Burbidge, which would allow all the energy to be released within the light travel time across the nucleus, might therefore provide the energy source for QSS and radio galaxies.

Whatever their detailed nature, the existence of intrinsically very powerful radio sources is of great importance in testing cosmological models; with large radio telescopes they can be detected at great distances and hence should provide information on the structure of the Universe at earlier epochs than can be reached with optical telescopes.

The most important of these cosmological tests are of three kinds:

(i) Determination of the numbers of sources in different ranges of flux density;

(ii) Examination of the isotropy of the sources;

(iii) Determination of the contribution to the background radiation of sources below the limit of detection.

The results obtained at Cambridge in the 4C survey and at Parkes have shown a distribution of sources which is isotropic at flux densities which correspond to values of the red-shift z of at least 1, but their distribution is not uniform in depth; with decreasing flux density the number of sources initially increases much faster than expected for a uniform population, suggesting that at earlier epochs either the number or the intrinsic power of radio sources was greater than at present. This increase has been attributed to QSS alone but there are important observational selection effects which make it less likely that faint radio galaxies will be identified than QSS; the greater intrinsic optical luminosity of QSS and the difficulty of defining the position of a galaxy associated with an extensive double source are particularly important. It seems probable that all powerful radio sources exhibit the effect, and that it can only be explained in terms of a cosmological model which permits evolutionary changes.

The greatest excess of sources appears to be due to sources having red-shifts $z \sim 2$–3.

New observations with the one-mile telescope at Cambridge have allowed the detection of sources some 100 times fainter than the limit of the 4C and Parkes surveys. The results are shown in Figure 12, in which the number of sources (N) observed in unit solid angle having a flux density greater than S, is compared with N_0, the number expected on a static Euclidean universe. The variation of N/N_0 with S for an Einstein-de Sitter model is also shown.

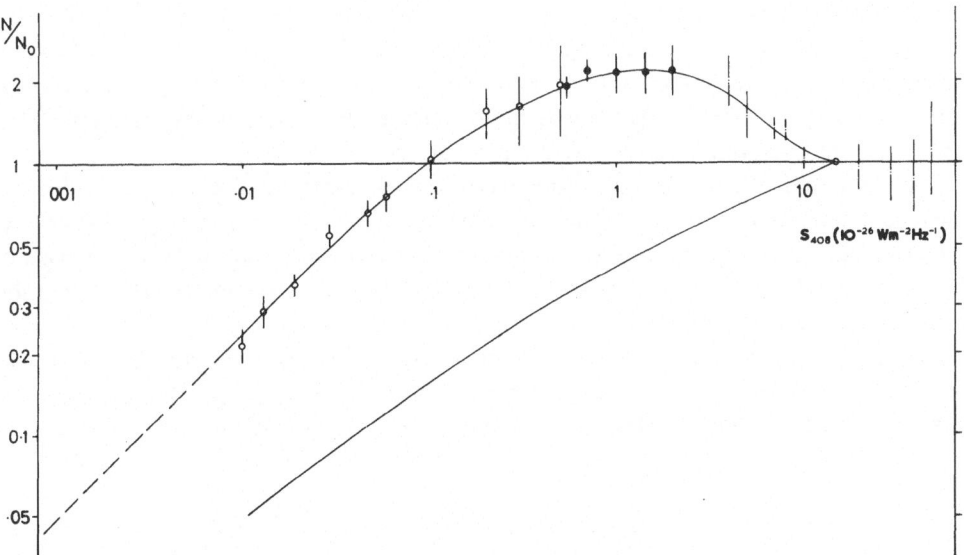

FIG. 12.   *Plot of the ratio of the number of sources (N) per unit solid angle having a flux density greater than S, to the number (N₀) which would occur in a static Euclidean universe. The lower line represents the curve for an Einstein-De Sitter model. – The extrapolated dotted curve would account for effectively the whole of the integrated extragalactic emission.*

These results show that at still smaller flux densities, the excess of sources which is so marked at $S \sim 10^{-26}\ w \cdot m^{-2}\ Hz^{-1}$ is followed by a remarkable convergence in the number of sources. This result indicates that a cut-off in the number of sources must occur at red-shifts $z \sim 3$. Since any cut-off will be blurred by the spread in the intrinsic radio luminosity, the relatively rapid convergence observed implies that this cut-off must be very sudden.

The existence of this cut-off is confirmed by considering the integrated emission from extra-galactic sources; the contribution from the sources which have now been observed individually is about half that of the total integrated emission, and it is evident that most of the remainder must be associated with sources of smaller intrinsic power at smaller red-shifts.

These observations provide strong additional evidence against any explanation of the source counts in terms of a local non-uniformity in the distribution of radio galaxies or QSS; the observed isotropy implies that we would have to be situated very close to the centre of such a local system, and since the observed sources account for most of the extra-galactic background emission, there can be few other such systems within $z \sim 1$. We can therefore only accept a local origin if we suppose that we are situated in a specially favoured place in the Universe – a situation which has been distasteful to astronomers since the time of Copernicus.

It appears that the source counts and the integrated extra-galactic radio emission reveal important evolutionary effects associated with the expansion of the Universe; prior to some epoch corresponding to a red-shift $z \sim 3$ (which may be related to the formation of galaxies) radio sources apparently did not exist. Subsequently, galaxy formation may have led to the birth of radio sources which were either more powerful or more numerous than they are at the present epoch.

Entirely independent evidence for an evolutionary cosmology has been provided by the recent discovery of isotropic microwave background radiation having a blackbody spectrum; the only explanation which has been proposed to account for this emission is that it represents the fossil radiation from the 'fireball' associated with the highly condensed initial stages in evolutionary cosmologies.

I have tried to summarize the radio evidence on radio galaxies and QSS, but much of our information on these strange objects has of course been provided by optical observations. Dr. Allan Sandage will continue with this side of the story.

RADIO GALAXIES AND QUASARS, II

Invited Discourse C given on August 28, 1967 in the Lucerna Hall

DR. A. SANDAGE

*(Mount Wilson and Palomar Observatories, Carnegie Institution of Washington,
California Institute of Technology, Pasadena, Calif., U.S.A.)*

1. The Present Status

It is my pleasant duty and high privilege to continue the story of radio galaxies and quasars begun by Professor Ryle. Because the intenseness of activity has been so great, and the advance of knowledge so remarkably rapid in this field, it is surprising to recall that only 9 quasars were known at the time of the Hamburg meeting of the Union just 3 years ago. Several hundred are now known, and techniques exist for the discovery of thousands more at will. At the time of Hamburg, less than a year had passed since M. Schmidt's discovery of the red-shift in 3C 273. Hoffleit and Smith, and Matthews and myself had only just established the optical variation of 3C 273 and 3C 48, respectively. There was no knowledge of how extensive the phenomenon of optical variability was, nor of the exceedingly short time-scales of the variation. Ideas of the number of quasars, radio and non-radio, were not yet formed, and the feature of radio doubling had not yet been established as a general characteristic.

Now, 3 years later, the whole of the systematic identification of the 3C R catalogue has nearly been achieved, red-shifts exist for over 100 quasars, optical variation has been found to be a general phenomenon, the existence of a class of weakly emitting or radio-quiet quasars established, and patterns of similarity between QSS and radio galaxies have emerged.

The initial thrust of this advance was made possible by the final solution to the optical identification problem – a problem which was so discouraging in the 1950's because of the major failure to continue the initial success in identification of such objects as Tau A, Cen A, Cyg A, and Virgo A, each of which had early been associated with very peculiar optical objects. However, once radio-position accuracies became known to $\pm 15''$ in both coordinates, as are now routinely available, optical identification proved to be rapid and relatively easy. This fundamental advance, bordering on the spectacular, was due to ingenious new techniques invented by the radio astronomers.

As the identification work proceeded, primarily by the Cambridge group in England, by Mills and Bolton in Australia, by the workers in Nancy, and by such men as Matthews and Minkowski in California, the pattern clearly emerged that there were only two types of optical objects associated with radio sources in high galactic latitudes. Galaxies accounted for about 70% of the identifications, and the starlike quasars made up the remaining 30%.

Most of the galaxies are ellipticals, normal in appearance, often in clusters, and then, invariably among the several brightest cluster members. The quasars are isolated, not obviously in clusters, and have an image diameter below the optical resolution of present techniques ($\sim \frac{1}{2}''$).

Professor Ryle has emphasized that the available radio data show rather clearly that explosions have taken place in radio sources, be they galaxies or quasars, sending clouds of plasma away from the central body. Although the evidence is not as abundant in optical astronomy, explosive events have, in fact, been directly observed in a few of the nearby optical objects. That the phenomenon of explosion is involved at all is positively proved by the Crab Nebula (3C 144), which was observed to explode by the Chinese and perhaps by the Arizona American Indians (Miller, 1955) in 1054 A.D.

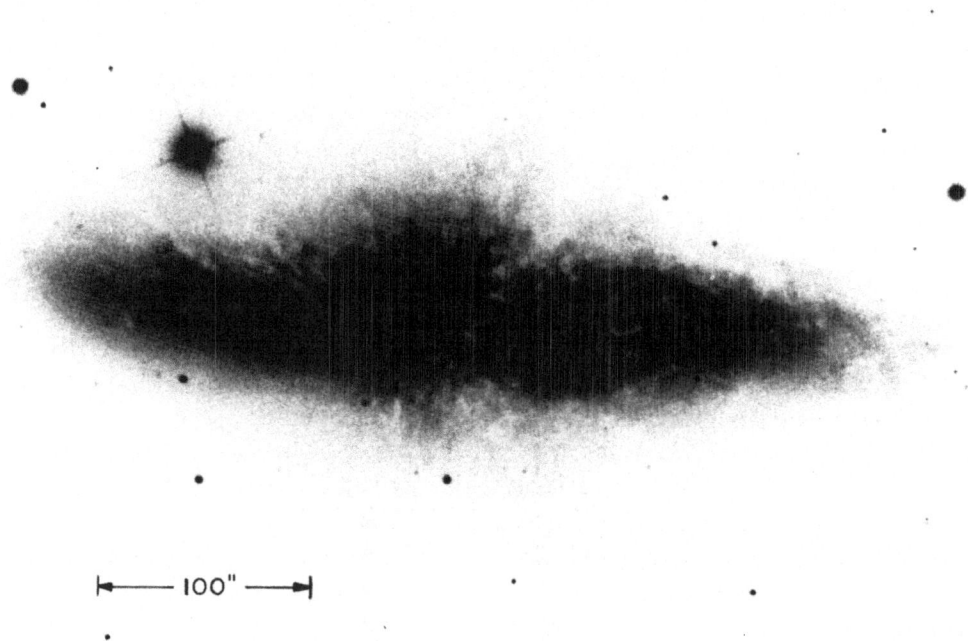

$\longmapsto$ 100" $\longmapsto$

FIG. 1.   *Photograph in blue light of the exploding galaxy M82 (3C 231). A chaotic pattern of dust exists across the face, and suggestions of thin filamentary structures near the direction of the minor axis are present. Picture from a plate of 30-min exposure made with the 200-inch Hale reflector on Eastman 103aO emulsion.*

Although the Crab is in our own Galaxy, the physics involved in its radio emission undoubtedly has relevance to the larger-scale problem of exploding galaxies themselves.

The most direct evidence for galactic explosions is provided by M 82 (Lynds and Sandage, 1963; Sandage and Miller, 1964) and by NGC 1275 (Burbidge and Burbidge, 1965), where filamentary structures in motion have been observed by a combination of direct photography and spectroscopy. Figures 1, 2, and 3 show the change of character of the optical image of M 82 between a short-exposure blue plate ($\lambda\lambda$ 3900–5000 Å), an Hα plate (λ 6563 $\pm$ 40 Å), and a composite print made by superposition of three blue photographs showing the extended outer filaments. The structures of Figure 3 are very highly polarized and are believed to be caused by optical synchrotron emission. Spectroscopic studies show that the Hα structures are expanding with a linear law and that all the gaseous matter was probably ejected from the central regions of M 82 about 2 million years ago (Lynds and Sandage, 1963; Burbidge *et al.*, 1964).

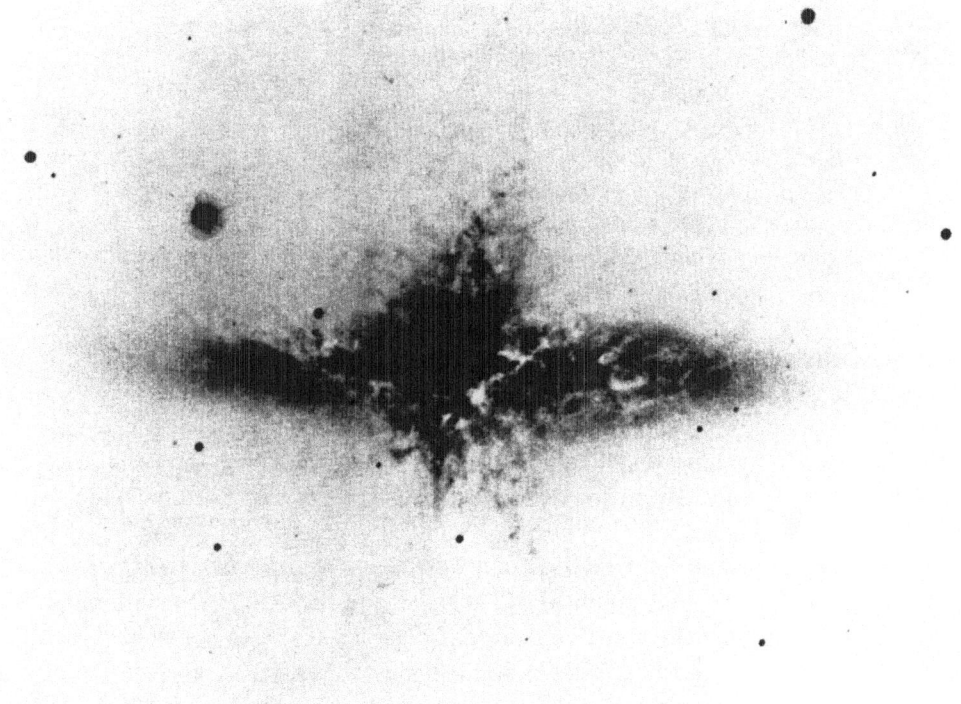

FIG. 2.   *M 82 in the nearly monochromatic light of Hα, reproduced to the same scale as Figure 1. The plate was made with the Hale reflector on Eastman 103aE emulsion behind an 80 Å total half-width interference filter tuned 15 Å redward of Hα. The filaments along the minor axis are visible in the line radiation of Hα.*

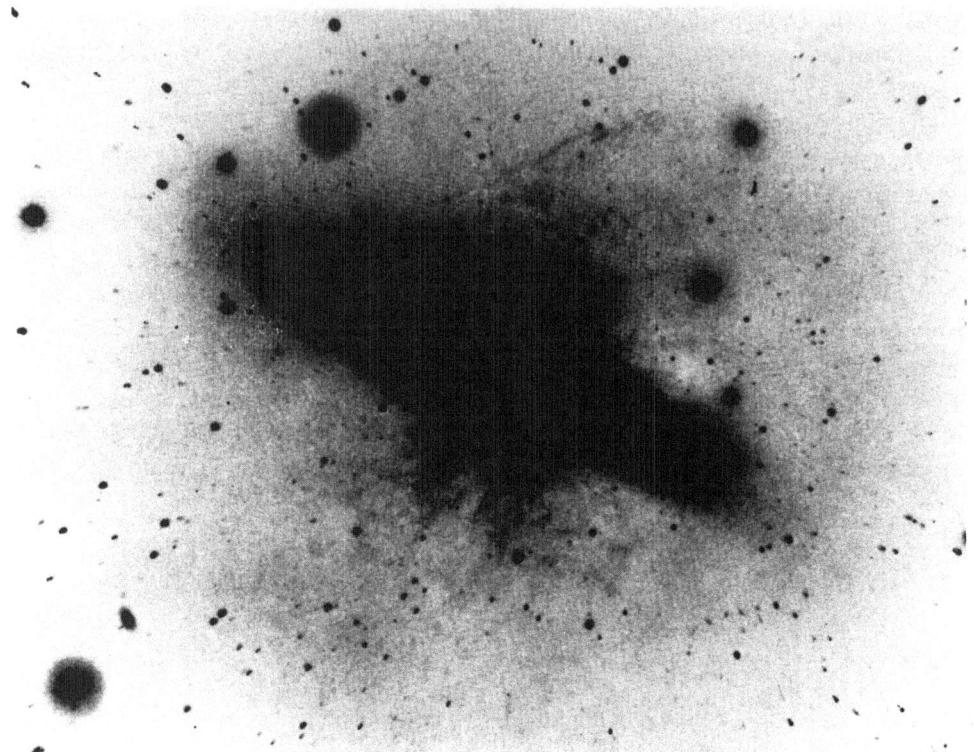

FIG. 3.   *Composite print of M 82 made from three 103aO plates. An extensive set of outer blue filaments is present which emit continuous radiation. The light from these filaments is highly polarized and is believed to be due to optical synchrotron emission (Sandage and Miller, 1964).*

The optical forms of most radio galaxies are not as spectacularly peculiar as those for M 82, M 87, NGC 5128, and NGC 1275. The majority of the radio galaxies with radio powers greater than 10^{40} ergs/sec (in a bandpass from 10^7 to 10^{11} Hz) are normal elliptical systems. This fact, established rather late in the identification program, was not fully foreseen because of the obvious peculiarities of the objects initially identified in the 1950's.

The quasars exhibit an entirely different optical appearance from radio galaxies. Figure 4 is a 200-inch photograph of 3C 273 showing the diffraction pattern of the central image and the well-known optical jet.

Although known since 1960, QSS are still a mystery. Schmidt's identification of the Lyman-α hydrogen line in 3 C9 at an observed wavelength of $\lambda = 3666$ Å, giving $\Delta\lambda/\lambda_0 = 2 \cdot 012$, suggested that quasars partake of the general expansion of the universe. No other explanation for such large apparent velocities had heretofore been successful in a scientific sense, and the evidence is clear and abundant that the red-shifts for normal galaxies are strictly correlated with distance.

Quasars immediately captured the imagination because, if red-shifts are cosmological, we look back in time some 80% of the way to the creation event in a Friedmann-type universe for redshifts of $\Delta\lambda/\lambda_0 = 2$. A very early sample of matter is then available, from which deductions can possibly be made of the evolution of the world.

However, if we put the QSS at the Hubble distance, they become the most powerful radiators known. The brightest would emit 10^{47} ergs/sec over all wavelengths from the radio to the optical, or more than 10^{60} ergs in a lifetime of 10^6 years. And the energy input must be very much more because the conversion processes from injection to radiation are not 100% efficient. If 10^{62} ergs are involved, assuming 1% efficiency, then 10^{10} solar masses must be converted from hydrogen to helium for the most intense QSS.

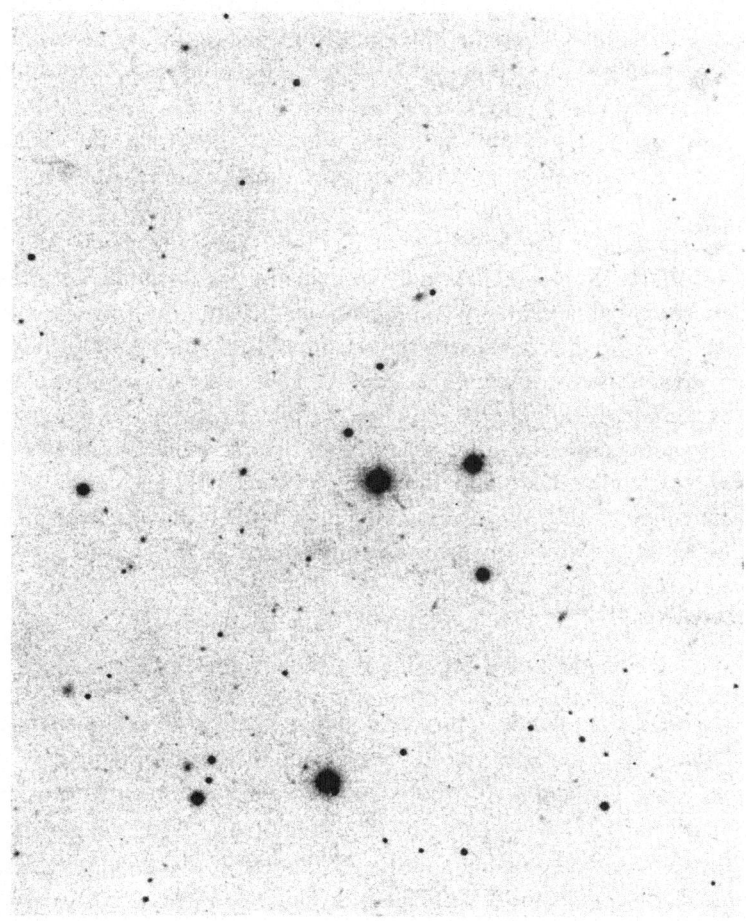

FIG. 4. *Photograph of the field of 3C 273 from a plate taken with the Hale reflector on Eastman 103aD emulsion of 50-min exposure time, in February 1962. The jet and the diffraction spikes on the central optical image are clearly visible.*

This problem, combined with the consequences of the rapid light variations, has caused some astronomers to doubt the cosmological interpretation of the red-shifts. There is no general agreement. The current debate is like that of the 1920's concerning galaxies as island universes. The subject is the same – that of distance; only the objects are different. In its original version, the debate centered around the distances to galaxies. Today, it is the QSS. Are they local or at the Hubble distance? What is needed now, as then, is a crucial experiment. As echoed in every issue of the current journals, it is clear that either no crucial experiment has yet been performed or, if it has, it has not been so recognized; or, if recognized by some, it is not acknowledged by all.

All versions of all theories can, of course, not be right. Although the evidence seems quite sufficient to eliminate some of the current suggestions, there is still no over-whelming proof of the type presented by Hubble in 1924 on the nature of the galaxies.

However, the evidence just presented by Ryle on the continuity of radio properties between galaxies and quasars, together with other evidence reviewed in later sections here, seem in my opinion, to establish the cosmological interpretation of the redshifts. Conversely, the difficulties in the other interpretations seem, at the moment, para-mount.

Questions concerning quasars are of two types. By far the easiest to answer are those of traditional astronomy. These may be called, perhaps unfairly, the superficial problems concerned with such items as the surface distribution over the plane of the sky, the number of QSS, the prevalence and nature of the optical and radio variations, and, most important, their distance. The second class concerns the physics of the situation. What is the energy source? What triggered the explosion? By what mechanism are the electrons accelerated? Can we understand the ontogenetic nature of the event? It is not likely that complete answers to these questions will be found until the astronomical problems are solved. We address ourselves in the following sections to certain questions of the first set, however bringing in problems of physics where answers are available and germane.

2. Identification and Surface Distribution of QSS

Extensive optical identification programs have fairly well exhausted the 3C R catalogue to the limit of the Palomar 48-inch Schmidt telescope. This work is due to many people at many places, in many countries, and the cordial and enthusiastic international cooperation of radio and optical astronomers clearly marks the result. Following the early classical work of Bolton and Stanley, Mills, Baade and Minkowski, and F. C. Smith, the most important modern names connected with the work are Matthews, Véron, Fomalont, Wyndham, Longair, Parker, Wills, and Dewhirst.

There are 328 radio sources in the 3C R. The number of confirmed QSS with redshifts and/or UBV photoelectric photometry is 44. There are a few additional possibilities which have not yet been completely verified. Of the 328 sources, 229 are in latitudes greater than $\pm 15°$. Of these, 41 are confirmed QSS, 38 are empty fields to the limit of the Palomar Schmidt, and 150 are either positive or possible radio galaxies. 27% of all possible identifications now made in the 3C R are QSS. The total number of QSS positively identified, either by spectra or by photometry, in the 3C, the 4C, the Mills, Slee, Hill, and the Parkes catalogs is at least 150 as of May 1967, and the number can apparently be increased at will, especially among the radio-quiet quasars. To July 20, 1967, red-shifts of 103 quasars had been determined, primarily by four extremely active groups of observers which include Schmidt at Mount Wilson and Palomar, Lynds at Kitt Peak, E. M. Burbidge and Kinman at Lick, and Ford and Rubin of the Carnegie Institution.

Figure 5 shows an equal-area projection in galactic coordinates of all known quasars which had been verified to April 1967. There are 134 objects plotted. Dots are QSS from the 3C R, triangles from the Parkes, squares from the 4C that have been identified by Scheuer and Wills (1966) as blue stellar objects from the Haro-Luyten catalogue, and crosses are radio-quiet quasars found in a special study by Luyten and myself (1967) in the PHL field at 1^h36^m, $+6°$. The only unbiased sample are the black dots, because identification programs are not yet statistically complete for the Parkes and the 4C catalogs.

The distribution appears to be isotropic in the area searched north of $\delta = 0°$ to

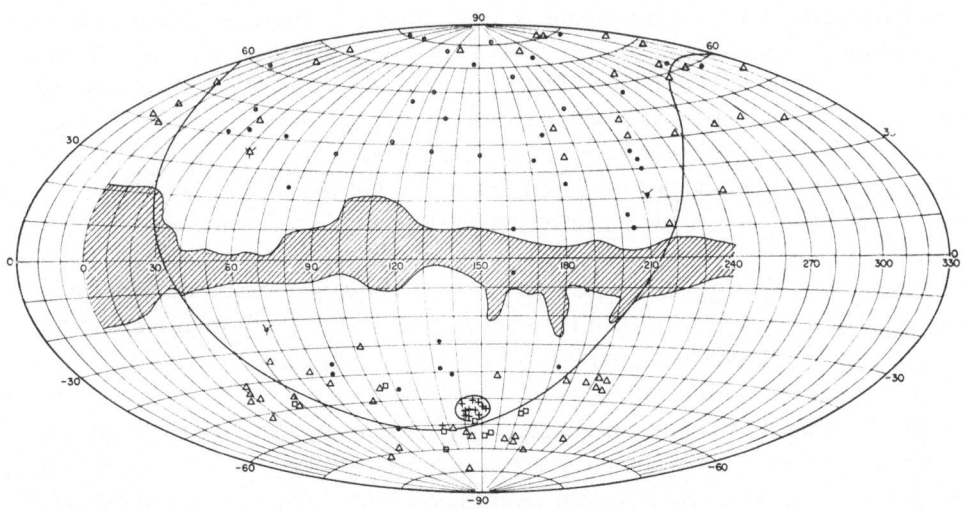

FIG. 5.   *The distribution in galactic coordinates of quasars known to May 1967. The zone of avoidance in the galactic plane is shown, together with the great circle of zero declination, North of which the 3C R radio catalogue is complete to 9 flux units. Dots represent radio quasars identified from the 3C R, triangles are from the Parkes catalogue, squares from the 4C for objects which overlap the PHL catalogue, and crosses are radio-quiet quasars found in a 40□° search area at 1^h36^m, $+6°$. Spikes represent quasars which violate the suggested anisotropy of Strittmatter et al. (1966).*

within the accuracy attainable with the small number of points. There is a general avoidance of the galactic plane, and a relative concentration toward the poles with a roughly cosecant distribution. The crucial conclusion on isotropy can be improved by noting that the Cambridge observers have analyzed segments of the 4C catalog and found no evidence of anisotropy to the accuracy of their statistics, which is better than 1%. If 30% of the 4C class are QSS, then isotropy is assured to better than 3%. This high degree is important for the question of local versus cosmological distances for certain versions of the local hypothesis.

3. Selected Physical Properties of QSS

The problem of optical identification of quasars divides into two parts. Those QSS which are strong radio emitters can be, and have been, found by inspection of photographs at the already known radio position on the plane of the sky. However, it now appears that most of the quasars are weak radio emitters and must be found by optical methods, for which knowledge of the optical properties is important. These properties include the nature of the line spectra, the peculiar energy distribution of the optical continuum, and the prevalence of optical variation.

A. OPTICAL LINE SPECTRA

Most quasars show broad emission lines of both forbidden and permitted transitions. Many lines of the abundant elements are present, such as the Balmer and Lyman series of hydrogen, lines of helium, and the ground-state transitions of carbon, nitrogen, silicon, and magnesium. Forbidden lines of oxygen and neon are often strong.

All emission lines show the same red-shift to within narrow limits. This appears to be a strong argument against the interpretation of the red-shift as due to an intense gravitational field of a single, compact, massive object. The forbidden lines must be formed in regions of low pressure so as to prevent collisional de-excitation of the metastable levels. Such regions, for a single compact body, will almost certainly have a lower gravitational potential than the region of permitted line formation, and hence a smaller red-shift, which is contrary to the well-established fact.

To circumvent this difficulty, Fowler and Hoyle proposed an *ad hoc* gravitational model where the emission lines originate inside a strong gravitational shell, formed by a uniform clustering of, say, neutron stars. It is here necessary that the gravitational potential-well be exceedingly uniform over the region of line formation to explain the data on the constancy of red-shift from line to line. Furthermore, there is no explanation on this model of the range of redshifts among the quasars as a class, spreading from $z = 0.158$ for 3C 273 to larger than 2 for objects such as 3C 9. It is now known (Schmidt, 1968) that the number of quasars at a given redshift, z, in a given red-shift range, Δz, is just that required to explain the $\log N$, $\log S$ curve for radio galaxies on the cosmological hypothesis. This crucial fact must be taken as coincidence on the gravitational-shell model, but, in fact, finds ready explanation on the cosmological hypothesis because radio galaxies are known to be at the Hubble distance (Section 6). This, and the arguments of Greenstein and Schmidt (1964) seem persuasive against the gravitational interpretation.

The emission lines of quasars are usually very broad – often with a 50 Å half-width. Interpreted as a Doppler motion, internal velocities of the order of several thousand km/sec are indicated. To stabilize such internal motions in a body of say 0·01 pc radius requires a mass of at least $10^7 M_\odot$, as shown by Setti and Woltjer (1966).

QSS may, of course, not be stable. Evidence comes from the several sources that have absorption lines. Of the 103 QSS with known redshifts, at least 20 have absorption lines, and in many of these cases the absorptions are displaced blueward of the corresponding emission lines, as in P Cyg stars. The simplest explanation is that of an expanding cool envelope with apparent subsequent mass loss. Setti and Woltjer have concluded from the available observations that a minimum initial mass on any assumption of distance is $10^5 M_\odot$ per average QSS – a number which is important in the later discussion of the local vs. the cosmological interpretation of the red-shift.

The most remarkable absorption-line object is PHL 5200 (4C: 05·93) which Lynds (1967) found to have wide, deep absorption bands starting at the emission edges of the lines of C IV (1550), Si IV (1403), and N V (1243) and extending blueward. These bands are from 100 to 150 Å wide, indicating a continuum of expansion velocities reaching values of about 10000 km/sec. All other sources studied so far are considerably less extreme.

B. OPTICAL CONTINUUM, COLORS, ENERGY DISTRIBUTION

The principal distinguishing feature of the observed radiation of QSS is the abnormal ultraviolet flux relative to ordinary stars. This ultraviolet excess was used as a discovery technique from 1964 to 1965 before precise radio positions were available.

Figure 6 shows a conventional two-color diagram for about 100 QSS with colors known to January 1967. N-type galaxies, as defined by Matthews et al. (1966), are shown as triangles. Fourteen radio-quiet quasars (QSO) are shown as crosses.

Half of the plotted QSS fall above the black-body line and half below in the region of white dwarfs, U Gem stars, and old novae. Although the distribution looks scattered and random, such is not the case. The red-shifts in various ranges are preferentially concentrated in discrete regions of the diagram, as shown in Figure 7. 82% of the plotted QSS obey this red-shift-area relation. The phenomenon was originally discovered by Kardeshev and Kolmberg from only 16 objects known in March 1966. McCrea and Barnes independently found the same thing in a different representation. Many authors, including Lynds, Schmidt, Strittmatter and G. R. Burbidge, Grewing, Lari and Setti, and myself, have shown that Figure 3 results from features of the energy distribution being shifted into and out of the broad-band U, B, V filters as the red-

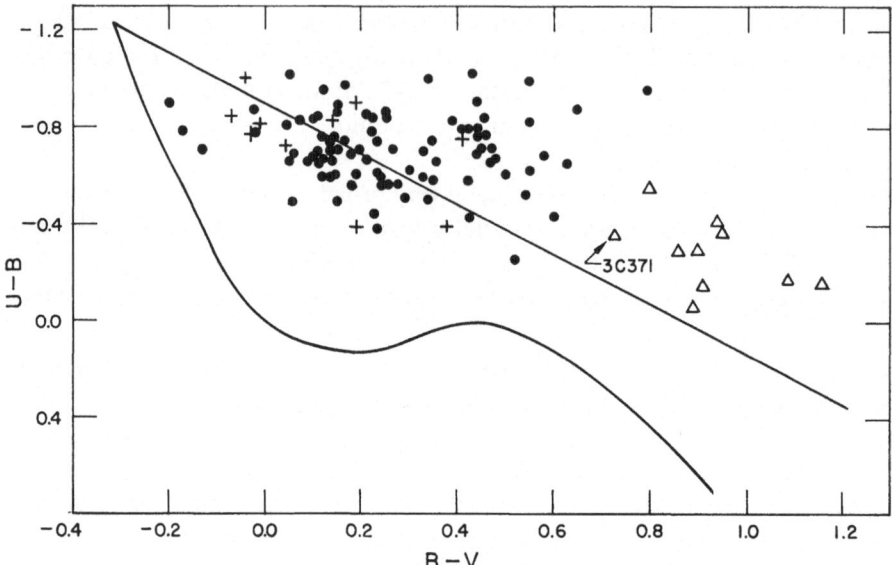

FIG. 6. *The two-color diagram for radio quasars (dots), radio-quiet quasars (crosses), and N-type galaxies (triangles). The black-body line and the normal stellar relations are shown for comparison.*

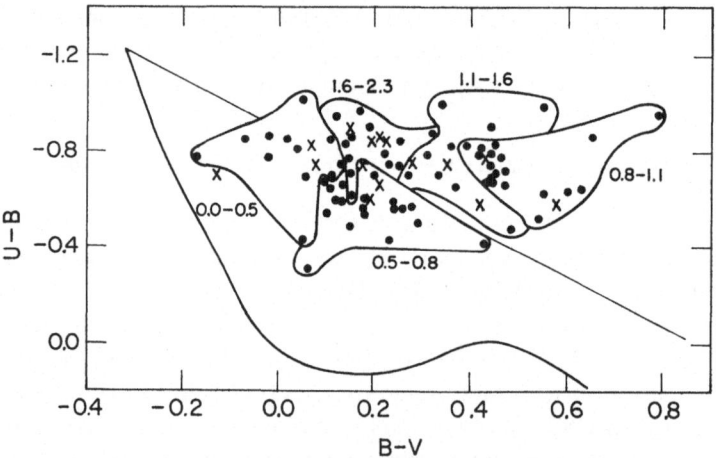

FIG. 7. *The two-color diagram for radio quasars of known red-shift. Dots are objects which obey the red-shift-area assignment; crosses are quasars which violate the area designation.*

shifts increase. These features are probably lines superposed on a continuum energy distribution which is statistically similar from quasar to quasar.

It is possible to construct a composite $I(\lambda)$ curve for an 'average' QSS from these results (*Astrophys. J.*, **146**, 13, 1966). Figure 8 shows a first attempt for the interval $\lambda_0 = 1000$ Å to $\lambda_0 = 5000$ Å. Schmidt, Oke, and Wampler have each also contributed to this problem.

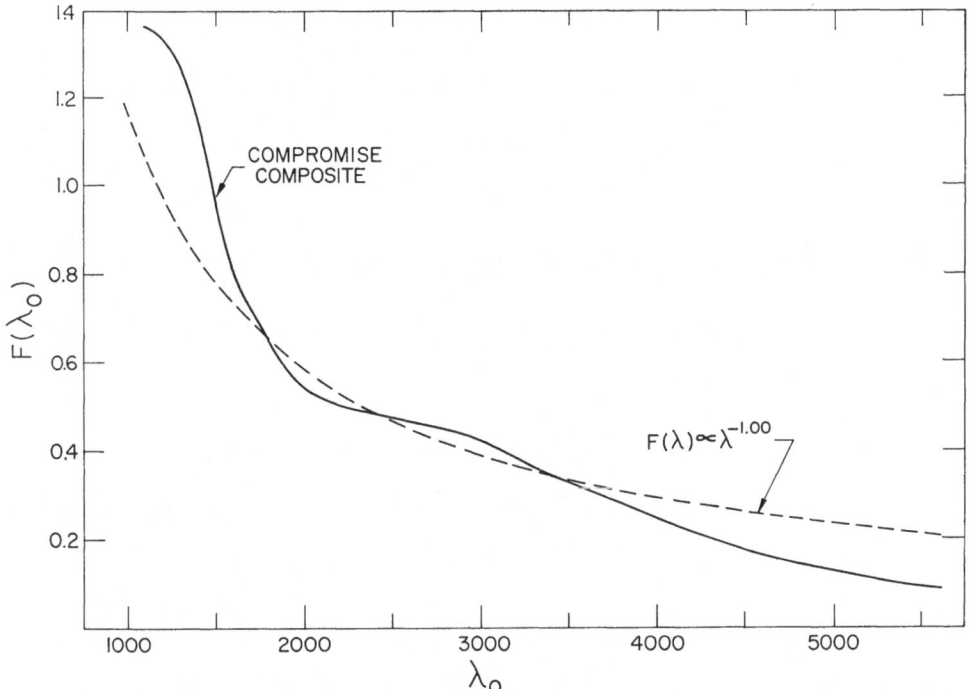

FIG. 8.   *The energy distribution per unit wavelength in the rest frame for an average quasar in the optical and near X-ray region. The dashed curve is a power spectrum, $F(v) \propto v^{-1}$, shown for comparison. The compromise composite represents the true continuum plus line radiation, smoothed by 1000-Å band-pass filters characteristic of the UBV system.*

The origin of the continuum radiation is still not known with certainty. In the beginning (1963) it seemed possible that the optical flux could be due to the synchrotron process and would connect up simply with the radio spectrum. The best early case was 3C 48, where it was shown (*Astrophys. J.*, **138**, 30, 1963) that the measured optical flux from $v = 3 \times 10^{14}$ Hz to 8.3×10^{14} Hz ($\lambda = 10000$ Å to $\lambda = 3600$ Å) agreed with a theoretical synchrotron spectrum (with parameters fitted to the radio data) to better than ± 0.1 magnitude, as illustrated in Figures 9 and 10.

We know now that the situation is much more complex. The results of Low and Johnson (1965) and of Epstein (1965) for 3C 273 show that a simple extrapolation is not possible. Furthermore, in many well-documented cases, such as 3C 9, 3C 186, 3C 191, 3C 204 and 3C 208, the optical points in the $\log F(v)$, $\log v$ diagram lie as much as a factor of 100 brighter than the expected synchrotron spectrum, as extended by theory from radio frequencies using a monotonic electron-energy distribution. If we are to maintain that the optical radiation is synchrotron for these cases, then at least a two-regime electron-energy distribution is needed.

The crucial observation showing that the radio and optical radiation are not simply

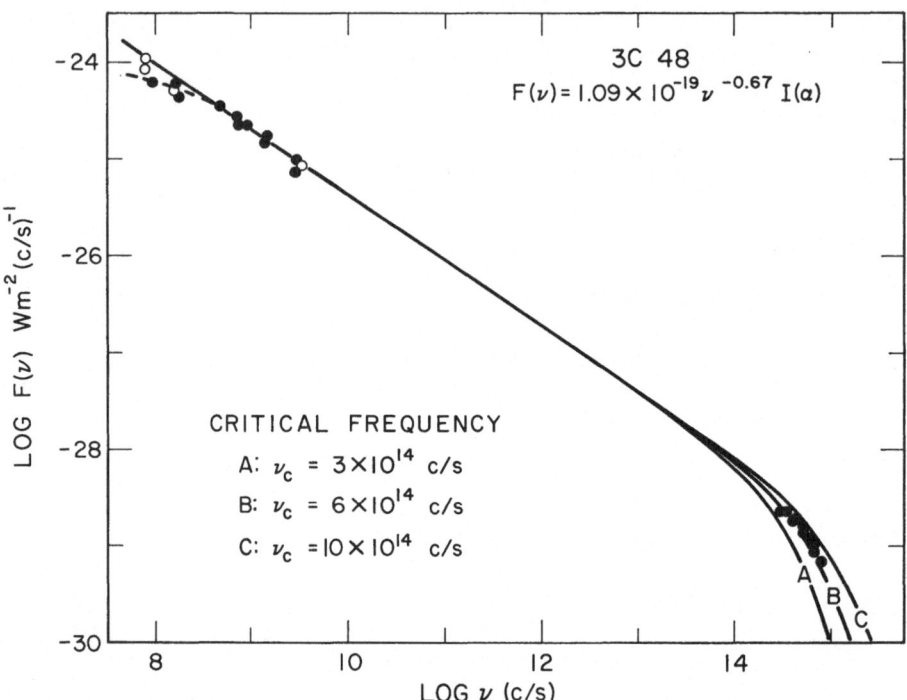

FIG. 9. *The energy spectrum, per unit frequency interval, of 3C 48 from the radio to the optical region. Three theoretical synchrotron curves are shown for various choices of the critical frequency at the high-energy end.*

connected in some QSS concerns the radio source MSH 14-121 (Véron, 1965), which is a radio double separated by 20″, and where a central blue optical object exists which is the optical quasar. The clear physical separation between the regions of optical and radio emission proves that the radio and optical emission must be decoupled here and must, therefore, result from different electron distributions. As Professor Ryle points out, many other QSS are radio doubles and are, therefore, similar to MSH 14-121.

However, the strongest clue that the optical radiation may after all be of synchrotron origin comes from data on optical variation and polarization.

C. OPTICAL VARIATION

Nearly every adequately observed QSS shows optical variations. The data for 3C 273 and for 3C 48 are so well known that they need not be here reviewed (Smith and Hoffleit, 1965; Matthews and Sandage, 1963).

The largest outburst so far observed occurred in 3C 446, where an increase in intensity by a factor of 20 was found in June 1966. The broad-band UBV colors changed between pre- and post-outbursts, and the two emission lines of CIV (1550) and CIII

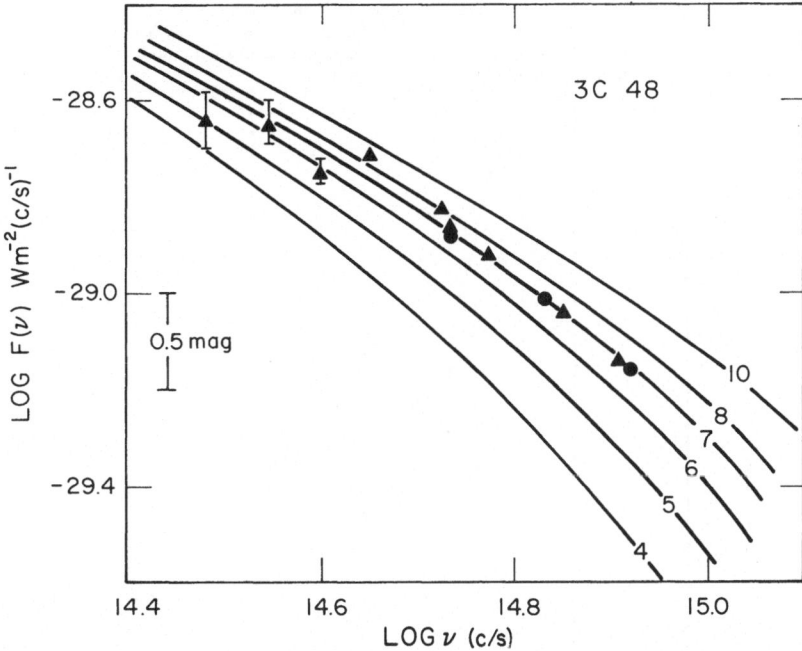

FIG. 10.   *Detail of Figure 9 in the optical region, with curves of critical frequency marked. Discussion is from Astrophys. J., 138, 1963, 30. The closed circles are from UBV data. The triangles are from measurements of W.A. Baum.*

(1909), which were bright and easily visible on spectrograms by Schmidt before outburst, all but disappeared at maximum light. Westphal, Strittmatter, and Sandage showed that the absolute intensity of the lines remained constant and were swamped by the increased continuum during outburst – results that were confirmed and improved by Oke and Wampler. The fact that the lines remained constant while the continuum changed is the first direct proof of a layered quasar model where the two types of radiation are emitted in different regions of the object.

Kinman *et al.* (1966) followed the outbursts in 3C 446 and found the most remarkable light curve shown in Figure 11. The rate of dimming, in the rest frame of the quasar, was about 0·2 mag/day, with a total drop of 1·6 magnitudes in 8 days. Even greater rates were observed in 3C 345 by Goldsmith and Kinman (1965). Other active QSS include 3C 2, 3C 43, 3C 196, 3C 279, and 3C 454·3.

The connection of these results with the question of optical synchrotron emission can be made using the recent observations of Kinman and of Visvanathan that a number of active sources are highly polarized. 3C 446 has a polarization of about 30%, 3C 279 of 15%, 3C 345 of 8%, and 3C 454·3 is very high but the final value is not yet known. Kinman has presented preliminary evidence that the plane of polarization varies as the optical flux changes and, further, that the percentage polarization be-

A. SANDAGE

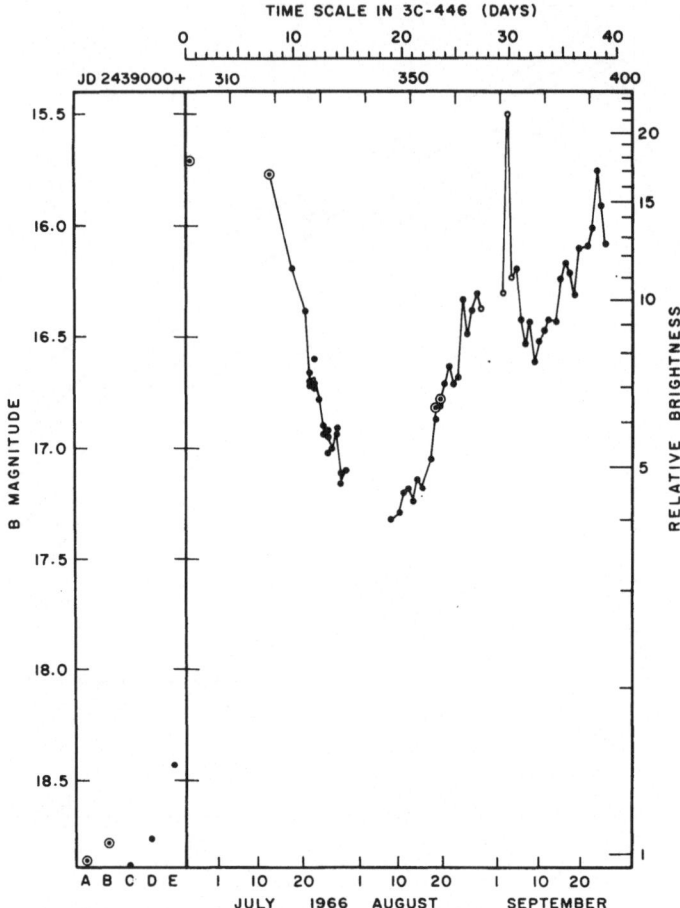

FIG. 11. *The light curve for 3C 446 according to Kinman et al. (1966). Dots are photographic determinations; circled dots are photoelectric measurements.*

comes small when the QSS becomes quiescent. The data are quite suggestive that (1) at least the variable part of the optical radiation may be due to synchrotron emission, (2) that the magnetic fields are not randomly tangled on the scale of the region emitting the variable radiation, and (3) that the aspect of the field seen either by the electrons or by the observer changes with time. The suggestion is that the outburst may be caused by beamed photons due to bursts of fresh electrons traveling along semi-regular field lines. As the aspect of the field changes, due either to rotation of the object or to different spatial regions of curved field lines that are sampled by the electron bursts as the particles travel outward, the intensity and the polarization of the radiation must change. During the quiescent phase, the net polarization may be low because of an integration over the entire body. In this regard, Kellerman's report (Prague IAU,

1967) is important in that the polarization of the radio flux from quasars is greatest during those times when the radio flux is changing most rapidly, which is analogous to the optical case.

D. OPTICAL VARIATIONS AND THE DISTANCE

In the past, the rapidity of the light variations has raised doubts about the cosmological origin of the red-shifts. Terrell has shown that the time-scale of the optical fluctuation of the order of days requires that the linear size of that part of a QSS which is outbursting must be less than a few light days across. A few years ago this was considered the death knell of cosmological red-shifts because of (a) the seemingly impossibly small angular size which was required, and (b) the apparent impossibility of maintaining synchrotron radiation for any appreciable time in such a compact region of high energy density due to the inverse Compton losses of the electrons as they collide with their own photons. This last effect was pointed out by Hoyle *et al.* (1966).

Both views are perhaps pessimistic. Radio angular-diameter measurements from the long base-line interferometer experiments of the Jodrell Bank group and the Royal Radar Establishment in England, and by the Green Bank workers in the U.S.A., together with the radio-scintillation results at Cambridge and at Arecibo, Puerto Rico, show that all active QSS have angular diameters less than $\simeq 0.02''$. These measurements are upper limits and, therefore, do not yet constitute an argument against the cosmological distances. Furthermore, it has now been shown that the Seyfert-like nuclei of N-type galaxies exhibit optical variations like those of QSS (Oke, 1967; Sandage, 1967), and these systems are known to be at the Hubble distance. This observation is perhaps the most powerful argument now available against the necessity for rapid optical variations to demand the objects be local.

The argument of energy loss by inverse Compton effect appears to be overcome by a proper arrangement of the magnetic fields and electron trajectories (Woltjer, 1966). Although the requirements for such a regular field may seem severe, a field of this type is, in fact, apparently demanded by the high degree of optical polarization observed by Kinman and Visvanathan.

4. The Number of Quasars: Radio-Quiet QSS

The bulk of the quasi-stellar objects are fainter than the flux limit of present radio surveys, and must be found by optical methods. The isolation of 'radio-quiet' quasars came from the first attempts in 1964 to identify radio quasars by the powerful two- or three-color photographic method invented by Haro, making use of the ultraviolet excess relative to normal stars which these objects possess. During a special photographic survey for radio quasars, many blue stellar objects situated far from the radio positions were found, spread over plates taken with the Mount Wilson 100-inch and the Palomar-Schmidt. After some time, it was realized that these objects were similar to those found in earlier Schmidt surveys by Haro and Luyten in 1961 in the magnitude range $15 < B < 19$.

Subsequent spectrograms proved that a number of the blue objects had intense, broad, red-shifted emission lines, together with the other attributes of normal QSS, save measurable radio flux as bright as 9 flux units. The color distribution of these radio-quiet red-shifted objects is the same as the QSS in the U–B, B–V plate (Figure 6), and quite different from the compact N-type galaxies. The objects look completely stellar on the largest available plate-scale, many vary in optical brightness, they have no measurable proper motion, and the absolute luminosities for the available sample are indistinguishable from those of radio QSS.

Figure 12 is a reproduction of parts of Palomar-Schmidt plates showing the radio QSS 3C 9 ($\Delta\lambda/\lambda_0 = 2\cdot012$), and the first radio-quiet quasar BSO1 ($\Delta\lambda/\lambda_0 = 1\cdot24$). The two objects are similar in every optical property.

Not all faint ($B > 16^m$) blue stellar objects found in high galactic latitudes are quasars. Spectra by Kinman, Greenstein, and Sandage show that nearby white dwarfs are the most frequent contaminant in any faint blue-star sample. This was predicted

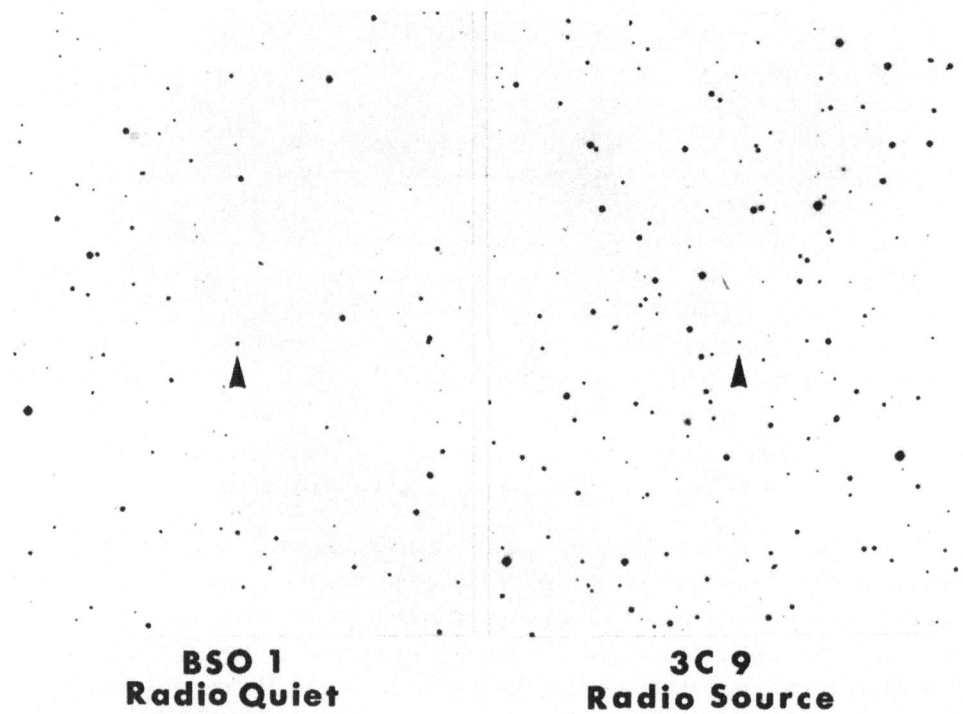

BSO 1
Radio Quiet

3C 9
Radio Source

FIG. 12.   *Comparison of photographs of the radio quasar 3C 9 and the radio-quiet quasar BSO1 made with the Palomar 48-inch Schmidt telescope.*

by Kinman (1965) in a discussion which showed that my earlier expectation of most of the Haro-Luyten objects as quasars was an overestimate by about a factor of 4. Luyten has experimentally separated some of the white dwarfs from potential QSO's by proper-motion studies. The method was particularly successful in a field at 1^h36^m, $+6°$, where we found (Sandage and Luyten, 1967) about one white dwarf per square degree to $B \simeq 18 \cdot 1$ mag., and a minimum number of $0 \cdot 3$ radio-quiet quasars per square degree to the same limit. Figure 13 shows the spectra of 3 of the 11 radio-quiet quasars found in this field.

We have estimated, on the basis of photometry of blue objects in a field studied by N. Richter near Messier 3, that between 1 and 3 QSO exist per square degree to $B \simeq 19 \cdot 7^m$. This gives a total of perhaps 10^5 such objects over the plane of the sky to a relatively bright optical limit. Similar studies by Véron and by Van den Bergh show that the number of potential blue objects to $B \simeq 20$ is greater than $10/\square°$, among which must be white dwarfs and QSO. It thus appears likely that the number of genuine quasars to $B = 20^m$ is very large indeed, as has already been shown to $B \simeq 18^m$ from available spectrographic results by Lynds, Kinman, Hiltner and his co-workers, and myself. The existence of the objects is now established, and their large number is germane to the current debate over distances of the QSS.

'Radio-quiet' means only that the object is a weak enough emitter to remain un-detected with the largest available radio telescopes. Assessment of the selection effects and of the sensitivity of present equipment shows that many 'radio-quiet' quasars can, in fact, be radio emitters at the faint end of the radio-luminosity function, such that

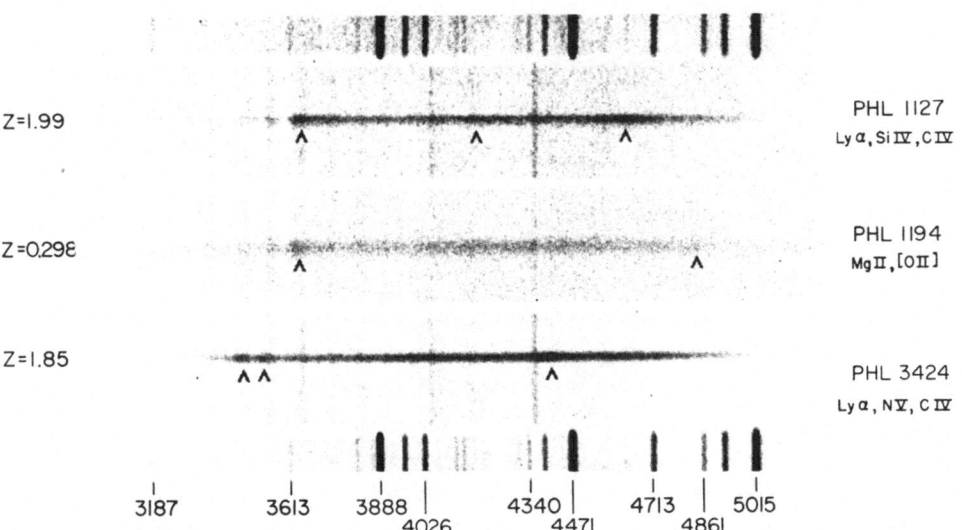

Fig. 13.   *Spectra of three radio-quiet quasars in the Haro-Luyten catalogue located in a systematic study of the area marked in Figure 5 at $1^h 36^m$, $+6°$. Reproduced from Astrophys. J., **148**, 1967, 767.*

$L_R \leqslant 10^{43}$ ergs/sec (absolute power between 10^7 and 10^{10} Hz; see Sandage and Luyten, 1967, Figure 4). However, if 10^5 quasars do, in fact, exist over the entire sky, most will be fainter than 2 f.u. at 178 Mc/sec, which is the limit of the 4C catalogue. This is because the 4C contains 10^4 sources, 30% of which may be QSS, giving only 3×10^3 objects – far short of the 10^5 estimated number.

5. Interpretation of the Red-Shift

Faced with the heretofore unknown combination of enormous total power and small linear size which result if QSS are at the Hubble distance, Terrell (1966, summary) proposed as an alternative to the cosmological explanation of the red-shifts that quasars are nearby objects that have been ejected from the center of our Galaxy in a violent event. He interpreted the red-shifts as Doppler motion with velocities reaching $0.8c$ $(\Delta\lambda/\lambda_0 = 2)$, assumed to originate in the initial acceleration processes associated with the explosion. Another version of a local hypothesis is available from Hoyle and Burbidge (1966) where ejection of QSS from a nearby radio galaxy such as M87 or NGC 5128 is contemplated.

Although each of these suggestions does have the property of reducing the total power of any given quasar by the square of the ratio of the Hubble distance to the 'local' distance, the advantage is empty because radio galaxies such as Cyg A and 3C 295 (which are at the Hubble distance) already radiate as much power in radio wavelengths as do the most powerful quasars when they are put at the Hubble distance. Thus the energy problem is not solved because it is clear that radio galaxies themselves present the same dilemma of enormous radio power. Furthermore, each local hypothesis itself runs into energetic difficulties when the problem is seen *in toto*, if there are as many quasars as the optical studies of radio-quiet QSS suggest. Setti and Woltjer have shown that a very energetic event is required to accelerate so many objects to their observed velocities. The argument is that, since a minimum mass per QSS is about 10^5 $M_\odot$, as earlier derived, the explosion in our galactic system must have had at least 10^{61} ergs, assuming 10^5 QSS and 100% efficiency of energy conversion into motion. This is considerably more than the energy of the brightest quasar on the cosmological hypothesis.

The same problem exists for the hypothesis that QSS are ejected from a single radio galaxy, such as M 87 or NGC 5128, in addition to other problems concerning the lack of anisotropy in the observed distribution, the lack of observed blueshifts (Noerdlinger *et al.*, 1966; Faulkner *et al.*, 1966), and the resulting long time-scale.

Although arguments concerning 'reasonable' limits to the energy carry little weight because the entire QSS phenomenon is itself still so manifestly strange and no rules of reasonableness are yet available, it is clear from the preceding that the proponents of local hypotheses cannot use the energy argument to support their position because they themselves are on the high side of the energy scale if all QSS come from a single parent body.

However, far more serious than the energy totals is the problem, already discussed, of the observed number of quasars at a given redshift, z, in range Δz. The characteristics of all known natural explosions is that the distribution of velocities among the given fragments is such that, some time after the explosion, most of the pieces will be in the low-velocity end, and hence close to the seat of the explosion. Only a few of the fragments will be given excessively large speeds. These, of course, will be found farthest from the explosion centre as time progresses. In no case will the fragment distribution approach that of a uniform density such that successive shells of radii r_i, each of volume $4\pi r_i^2 \, \Delta r_i$, will each have the same number of fragments. Even less likely is the case of a positive density gradient outward. And yet such is the observed situation with quasars, according to the recent work of Kafka and of M. Schmidt.

Only by making the most *ad hoc* assumptions can these observed facts be understood on a local hypothesis, whereas the explanation is natural and convincing with normal cosmological models and a reasonable variation of the frequency of production of QSS with cosmic time.

Rather than recite further difficulties of the local hypotheses, a more positive approach is to describe new observations germane to the opposite conclusion. There now exist new data which seem to support the cosmological interpretation of the redshifts. The tests all rest, ultimately, on the fact that radio galaxies are, themselves, at the Hubble distance, as will now be shown.

6. Evidence for the Cosmological Distances of Radio Galaxies and Quasars

Figure 14 shows the most recent data on the red-shift-apparent magnitude relation for the brightest galaxy in each of 41 clusters of galaxies. In 29 of the clusters, the brightest galaxy is not a radio source, while the opposite is true in the remaining 12 clusters, which include the famous cases of 3C 295, Per A (NGC 1275), and 3C 40 (NGC 545).

Plotted in Figure 14 as abscissa is the photoelectric V magnitude corrected for (1) aperture effect to an isophote of about 25^m per square second of arc, (2) K dimming due to red-shifting the energy distribution through the fixed measuring band, and (3) galactic absorption, A_V, assuming a cosecant law with a galactic half-thickness of $A_V(\tfrac{1}{2}) = 0 \cdot 18^m$. Red-shifts, corrected for galactic rotation, are plotted as ordinate.

The plotted line has not been put through the data as regards slope. It is given a slope of 5, which is the theoretical value for all cosmological models as $\Delta\lambda/\lambda_0 \to 0$, and is the exact value for all $\Delta\lambda/\lambda_0$ values in a Friedmann model with $q_0 = +1$, $\Lambda_0 = 0$, and the evolutionary correction $dL/dt = 0$. The data follow this line with remarkable fidelity. The dispersion ΔM is exceedingly small ($\sigma_M = 0 \cdot 28^m$), which shows that the brightest galaxy in a cluster defines a standard candle to within $\pm 30\%$ in absolute luminosity – a truly remarkable fact.

The important point to recall about Figure 14 is that the line represents the linear

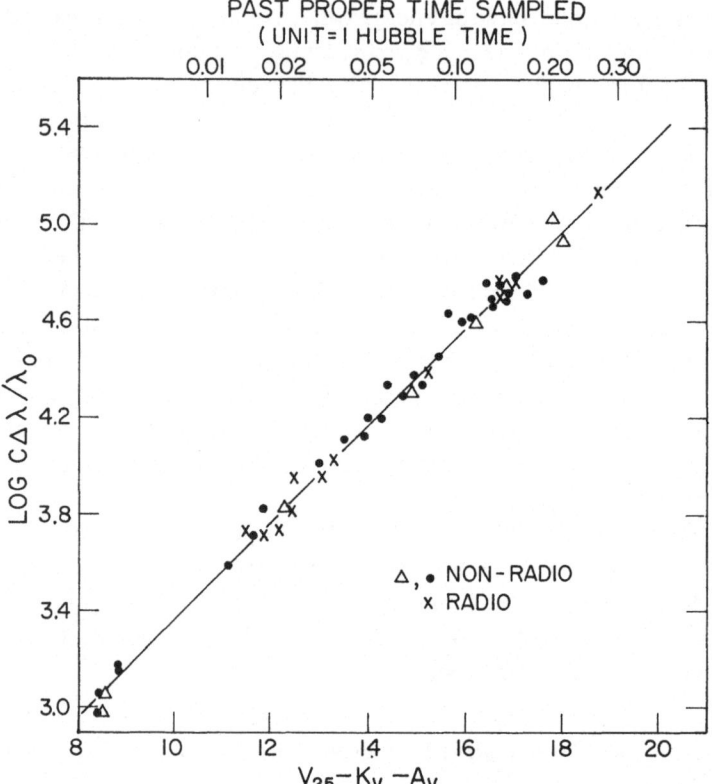

FIG. 14.   *The Hubble diagram of first-ranked cluster galaxies in 41 clusters. Dots are non-radio galaxies, crosses are radio galaxies, and open triangles are data by Baum for eight clusters, some of which are duplicated. All photometry is photo-electric and was obtained with the 200-inch Hale telescope. Correction for aperture effect, for K dimming, and for galactic absorption have been applied. The line has a slope of 5·0 and has been fitted to the data only in zero point.*

Hubble expansion law. The fact that all the galaxies obey the line proves beyond doubt that they are at their Hubble distances. The same conclusion follows for radio galaxies in general, as Figure 15 demonstrates. Here data for 65 radio galaxies are plotted. While only 30% of the galaxies are members of clusters, the conclusion is the same as for Figure 14. Again the line has the theoretical slope of 5, and it is evident that the galaxies obey linear red-shift relation. The increased scatter in Figure 15 compared with that of Figure 14 is undoubtedly due to a less homogeneous statistical sample, and therefore to an increased spread in absolute magnitude M. We are here sampling deeper into the optical luminosity function of radio galaxies for objects shown in Figure 15. Yet this diagram is still remarkably tight, proving that *a necessary condition for a galaxy to be a strong radio source is that it be in the class of the most luminous galaxies known, namely, the several brightest members of clusters.*

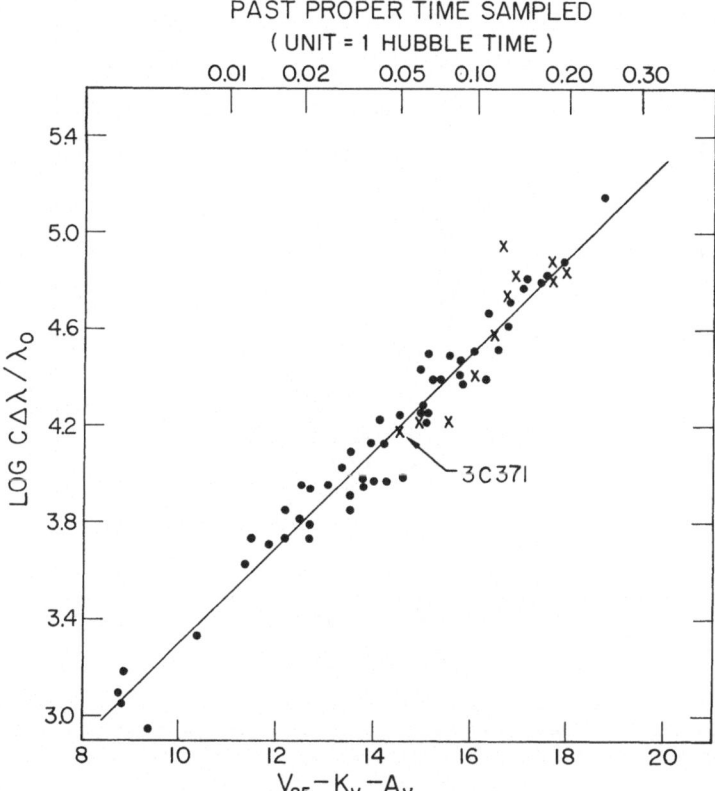

PAST PROPER TIME SAMPLED
(UNIT = 1 HUBBLE TIME)

FIG. 15. *Same as Figure 15 for all identified radio galaxies with red-shifts and photoelectric data known to August 1967. The dots are normal elliptical galaxies, crosses are N-systems. There is no evidence that the compact N-galaxies deviate from the Hubble law.*

It should here be emphasized that all of the galaxies shown in Figures 14 and 15 are elliptical systems. No spirals are amongst them. In Figure 14 this is because, as known since Hubble's work in 1936, most galaxies in the great clusters are ellipticals, almost without exception. The same is true in Figure 15, because the fact that emerged from the optical identification work was that the most powerful radio galaxies are, almost without exception, pure population-II systems.

Figures 14 and 15, taken together, show that all radio galaxies studied so far obey the Hubble law, and are therefore at their cosmological distance. The same conclusion follows from study of angular diameters, where the data for radio galaxies and first-ranked cluster members obey $\theta \propto z^{-1}$.

With radio galaxies now established to be at their Hubble distances, we can proceed to three crucial tests concerning quasars. The first is due to Heeschen (1966) and is shown in Figure 16, where the radio surface brightness, B, of those radio galaxies and QSS with resolved radio-disks is plotted against the absolute radio power, L_R,

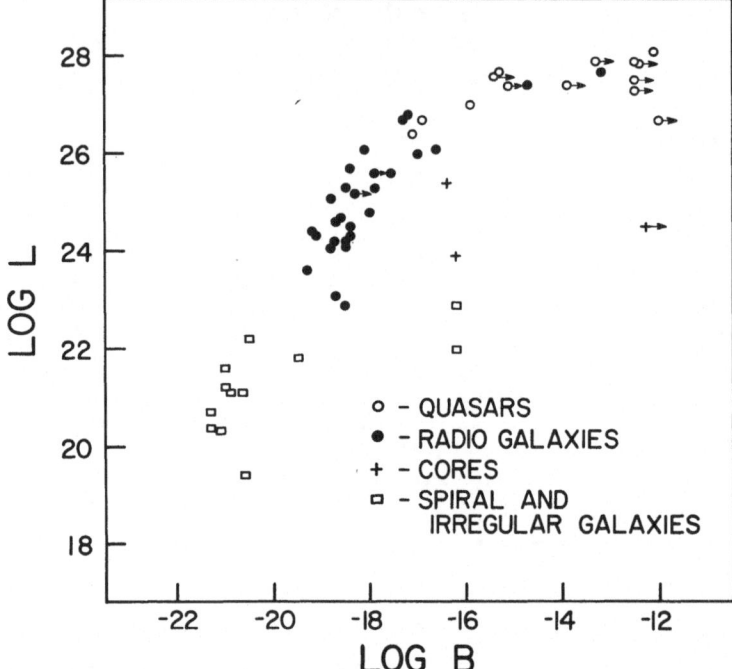

FIG. 16. *Heeschen's diagram of the correlation of surface brightness and absolute radio power for quasars and radio galaxies, calculated as if all objects are at their Hubble distances. Continuity between QSS and radio galaxies would be destroyed if the red-shifts of QSS are not cosmological.*

calculated as if all sources are at their Hubble distances. Because B is independent of distance and L_R depends on the distance squared, the observed correlation and conti-nuity would be destroyed if the QSS did not follow the Hubble law. Heeschen's result appears to be a crucial experiment which shows that quasars do indeed obey the red-shift-distance relation of the expanding universe.

A second set of data on the linear-size distribution of radio splitting for galaxies and QSS has already been mentioned by Professor Ryle. The distances between the radio components of galaxies are distributed between 1 and 450 kpc. The same range is covered by the double radio components of QSS if they are at the Hubble distance, but not if the QSS are local (Kellerman, 1966; Hazard 1967). These and newer data are particularly well shown in the diagrams discussed by Ryle in the first of these lectures.

A third indication of the similarity of quasars and the violent events of galaxies has become known only in recent weeks. The N-type galaxy, 3C 371, shown as a cross in Figure 15 at the smallest red-shift for the N systems, was discovered to vary in optical luminosity (Oke, 1967; Sandage, 1967) by one magnitude on a long time-scale, and by 15 to 20% from night to night. Subsequently, the N-type galaxies, 3C 390·3 and 3C 109, also shown as crosses in Figure 15, were similarly found to vary, as was the Zwicky blue compact galaxy I Z1727 + 50, which is of the class of radio-quiet analogs of the N-type radio galaxies. All these N galaxies obviously follow the Hubble law (Figure 15), and yet they vary in a similar fashion to QSS. They form the single counter-examples needed to prove that rapid variation of optical flux does not require that the objects need be local.

The continuity of properties between the Seyfert nuclei, the N-type galaxies, and the quasars now seems complete. They all appear to be manifestations, in different degrees, of the same type of non-thermal violent event taking place in the nuclei of galaxies. The physical cause still eludes us. The reason why only the most massive galaxies known can become strong radio sources is equally unknown. What is clear phenomenologically is that the violence of the event itself possesses a great dispersion in optical-power level, as shown in the next diagram.

Figure 17 is the Hubble diagram for radio galaxies and for QSS. The large spread in the distribution of the triangles is evident. The observed dispersion in absolute luminosity of QSS is at least 4 magnitudes. It is significant that the QSS merge into the radio galaxies in M_V. The fact that no QSS is known to exist to the right of the distribution of galaxies may be significant. Perhaps we are justified in suggesting that when the violent event giving rise to the ultraviolet, non-thermal emission is relatively mild, it cannot outshine the stellar component of an underlying galaxy, and the system would be classed as N. If the outburst is still weaker, it can only dominate the nuclear regions, and we observe a Seyfert galaxy.

Finally, it should be pointed out that the lack of a strong correlation among the triangles in Figure 17 of the red-shift and apparent magnitude surely does not indicate a failure of QSS to follow the linear Hubble law. Such a conclusion is incorrect when the sample of objects considered has a large dispersion in intrinsic luminosity. The steepening of the upper envelope of the QSS distribution with increasing red-shift is, in fact, what is expected for a sample of objects with no upper cut-off to the luminosity function as the volume of space sampled becomes larger (higher z values). This is the Scott effect (Scott, 1957). It is not justified, as some have done, to conclude that the spread of the triangles in Figure 17 implies that the QSS show no red-shift-distance effect. When the selection effects in the presence of a large intrinsic dispersion are considered, the opposite would appear to be the case.

7. Epilogue

Four years ago high hopes were expressed that quasars would play a dominating role in the solution of the geometrical cosmological problem. Although the red-shifts reach enormous values of $\Delta\lambda/\lambda_0 \simeq 2$, which means we look back 80% of the way in time to a singularity of the Friedmann universe, we cannot separate out amongst the intrinsic spread of absolute optical power the effects of deceleration which, through

A. SANDAGE

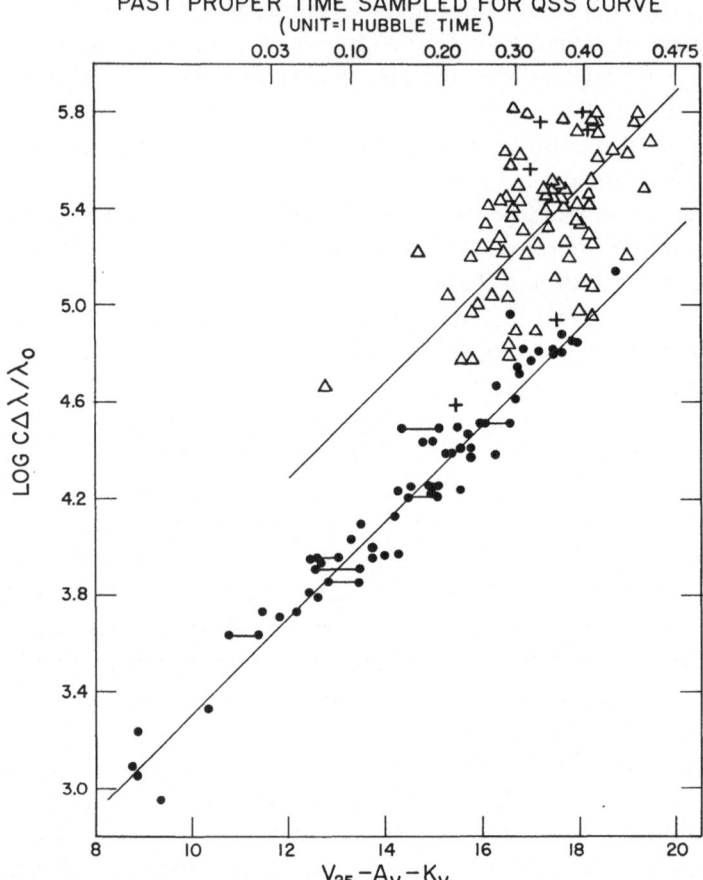

FIG. 17. *The Hubble diagram for radio galaxies (dots), radio quasars (triangles), and radio-quiet quasars (crosses). Horizontal bars connect points for dumbbell galaxies which represent the combined light of both components and the light of the brightest component. The time into the past to which we look at a given red-shift is shown along the top in units of one Hubble time, which is $H_0^{-1} \simeq 13 \times 10^9$ years with the present calibration.*

Einstein's coupling of motion and geometry, leads to a value of the spatial curvature of the world. Disappointing as this may be, quasars and radio galaxies have nonetheless provided something quite unique in the physics of astronomical structures. Instead of serving only as markers of an underlying geometry, which quasars would have done had their intrinsic luminosities shown less spread, they now provide an ontogenetic mystery whose solution may lead us even deeper into historical events of the distant past. If Schmidt's belief is true, that creation of quasars is strongly time-dependent, being more prolific near the birthday of the world, then a new dimension to cosmogony is opened up. Radio galaxies, because of their remarkable statistical stability of absolute luminosity, still appear capable to serve as the test objects – the standard candles – the fundamental probes for geometrical cosmology. High hopes, therefore, still remain that the deceleration parameter will be determined from data such as those of Figures 14 and 15. But the even more fundamental problems of the explosion physics remain.

Nowhere in this lecture has the name of Ambartsumian appeared. He foreshadowed much of what has here been said. Ten years ago he began to emphasize the role of nuclei of galaxies. Consistently, at the Solvay conferences, at the Berkeley IAU and at numerous symposia, he, almost alone at first, suggested that violent processes were indeed occurring in galactic nuclei, and that astronomers should take full heed. The fulfillment of his prophecy is only now beginning. No astronomer would today deny that mystery does indeed surround the nuclei of galaxies, and the first man to so realize the rich reward in store was Victor Ambartsumian.

The events preceding the violent release of energy in QSS, N galaxies, and Seyfert nuclei, and the reasons for the instabilities still remain a cryptic mystery. But, at least we know now somewhat better the questions to be solved – questions which drive the theoreticians to think, and the observers to the dark and quiet of their telescopes, both radio and optical, to help unravel the greatest problem known to man – the scientific story of creation revealed through the history of galactic systems and their predecessors.

References

Burbidge, E. M., Burbidge, G. R. (1965)   *Astrophys. J.*, **142**, 1351.
Burbidge, E. M., Burbidge, G. R., Rubin, V. C. (1964)   *Astrophys. J.*, **140**, 942.
Epstein, E. E. (1965)   *Astrophys. J.*, **142**, 1285.
Faulkner, J., Gunn, J. E., Peterson, B. A. (1966)   *Nature*, **211**, 502.
Goldsmith, D. W. Kinman, T. D. (1965)   *Astrophys. J.*, **142**, 1693.
Greenstein, J. L., Schmidt, M. (1964)   *Astrophys. J.*, **140**, 1.
Hazard, C. (1967)   New York Third Texas Meeting on Cosmology.
Heeschen, D. S. (1966)   *Astrophys. J.*, **146**, 517.
Hoyle, F., Burbidge, G. R. (1966)   *Astrophys. J.*, **144**, 534.
Hoyle, F., Burbidge, G. R., Sargent, W. L. W. (1966)   *Nature*, **209**, 751.
Kellerman, K. I. (1966)   *Astrophys. J.*, **146**, 621.
Kinman, T. D. (1965)   *Astrophys. J.*, **142**, 1241.
Kinman, T. D., Lamla, E., Wirtanen, C. A. (1966)   *Astrophys. J.*, **146**, 964.
Low, F. J., Johnson, H. L. (1965)   *Astrophys. J.*, **141**, 336.
Lynds, C. R. (1967)   *Astrophys. J.*, **147**, 396.
Lynds, C. R., Sandage, A. (1963)   *Astrophys. J.*, **137**, 1005.
Matthews, T. A., Sandage, A. (1963)   *Astrophys. J.*, **138**, 30.
Matthews, T. A., Morgan, W. W., Schmidt, M. (1966)   *Astrophys. J.*, **140**, 35.
Miller, W. C. (1955)   *Astr. Soc. Pacific Leaflet No. 314.*
Noerdlinger, P. D., Jokipii, J. R., Woltjer, L. (1966)   *Astrophys. J.*, **146**, 523.
Oke, J. B., (1967)   *Astrophys. J.*, **150**, L5.
Sandage, A. (1967)   *Astrophys. J.*, **150**, L9.
Sandage, A., Luyten, W. J. (1967)   *Astrophys. J.*, **148**, 767.
Sandage, A., Miller, W. C. (1964)   *Science*, **144**, 405.
Scheuer, P. A. G., Wills, D. (1966)   *Astrophys. J.*, **143**, 274.
Schmidt, M. (1968)   *Astrophys. J.*, **151**, 393.

Scott, E. L. (1957)   *Astron. J.*, **62**, 248.
Setti, G., Woltjer, L. (1966)   *Astrophys. J.*, **144**, 838.
Smith, H. J., Hoffleit, E. D. (1965)   *Gravitational Collapse and Quasi-Stellar Sources*, Chapter 16.
Strittmatter, P., Faulkner, J., Walmsley, M. (1966)   *Nature*, **212**, 1441.
Terrell, J. (1966)   *Science*, **154**, 1281.
Véron, P. (1965)   *Astrophys. J.*, **141**, 1284.
Woltjer, L. (1966)   *Astrophys. J.*, **146**, 597.

JOINT DISCUSSIONS

A. JOINT DISCUSSION OF COMMISSIONS 12, 40, 43, AND 44
NEW TECHNIQUES IN SPACE ASTRONOMY

(Friday, August 25, 1967)

Organizing Committee: L. Goldberg (Chairman), J. E. Blamont, L. Davis, G. G. Fazio, C. de Jager, R. Lüst, R. Michard, V. K. Prokof'ev, F. G. Smith
Chairman of the programme: V. K. Prokof'ev

Contents:

INTRODUCTION

V. K. Prokof'ev

Ten years ago, on October 4, 1957, Sputnik-I was launched into an orbit around the Earth by the U.S.S.R. It was the first man-made object in space. The U.S.A. launched its first satellite, Explorer-I, on January 31, 1958. The first space probe, Lunik-I, was launched on January 2, 1959, into solar orbit and became the first artificial planet. The satellites initiated the intensive development of a new region of astronomy: space astronomy. During these 10 years, our television eyes have been on the Moon and near Mars. At the present time, we know the conditions in space between the orbits of Venus and Mars. More than 100 satellites operate in space, thus yielding rich information about the phenomena in the region near the Earth and in the interplanetary medium. There are now two space stations in flight to Venus, and Lunar Orbiter-5 photographs the Moon's surface.

We meet here to discuss new astronomical techniques, the techniques of space astronomy. We must also mention here, with gratitude, the many engineers and workers whose efforts made this possible.

Noting our successes in space astronomy, we must not forget the painful losses which all humanity suffered in the course of the year 1967. The tragic loss of four astronauts, Komarov, Grissom, White, and Chaffee, during the tests of new spaceships during flight and on the ground, was marked with grief throughout the world. I invite you to stand in honor of their memory. (Pause.)

Our joint discussion is now open.

Perek (ed.), Highlights of Astronomy, 74. © I.A.U.

RECENT TECHNIQUES IN SOLAR OPTICAL ASTRONOMY
FROM SPACE

E. M. REEVES

(Harvard College Observatory, Cambridge, Mass., U.S.A.)

ABSTRACT

The article summarizes some of the recent advances in the techniques of solar space research, particularly over the last 4 years, in an attempt to review the current state of instrument technology. The present state of development in rocket and satellite vehicles for solar observations, far ultraviolet detectors, optical materials, ultraviolet reflection coatings, filters, and photographic film fogging, are among the topics described.

1. Introduction

The experimental objectives of solar space research frequently follow the general trend towards ever-increasing spectral resolution and photometric accuracy coupled, whenever possible, with spatial resolution on the disk. The experiment flux is frequently limited by the attempt to achieve these aims simultaneously, rather than limited by instrumental diffraction effects, and then one seeks ways to obtain sufficiently large collecting apertures to achieve the best compromise between spatial and spectral resolution. The advances in technology useful in this process are improvements in pointing controls, optical efficiencies, and detector efficiencies. Only rarely are solar satellite experiments limited by telemetry capabilities since these purely electronic areas are managing to keep up with or ahead of available experiments. This will shortly begin to change, however, and various forms of data compression techniques will be required for solar instrumentation. This problem becomes especially critical with the development of photoelectric image storage devices with storage capabilities for large amounts of data, as is currently the case for the simple photographic plate.

Since the last General Assembly of the IAU in 1964 there have been two launches of satellites for solar optical astronomy. These are the OSO-II and OSO-III Orbiting Solar Observatories, which were launched in February 1965 and March 1967. Figure 1 shows the general arrangement of the OSO-IV, which is a gyroscopically stabilized wheel containing a number of experiments that view the Sun once per revolution of 2 sec. Two or more experiments are carried in the sail portion, which contains the solar cells for spacecraft power and orients the instruments to the centre of the solar disk to an accuracy of approximately 20 sec of arc. Data are transmitted to the ground in real time and also recorded on magnetic tape in the satellite and played back to the ground station, on command, once per orbit.

Perek (ed.), Highlights of Astronomy, 75–93. © *I.A.U.*

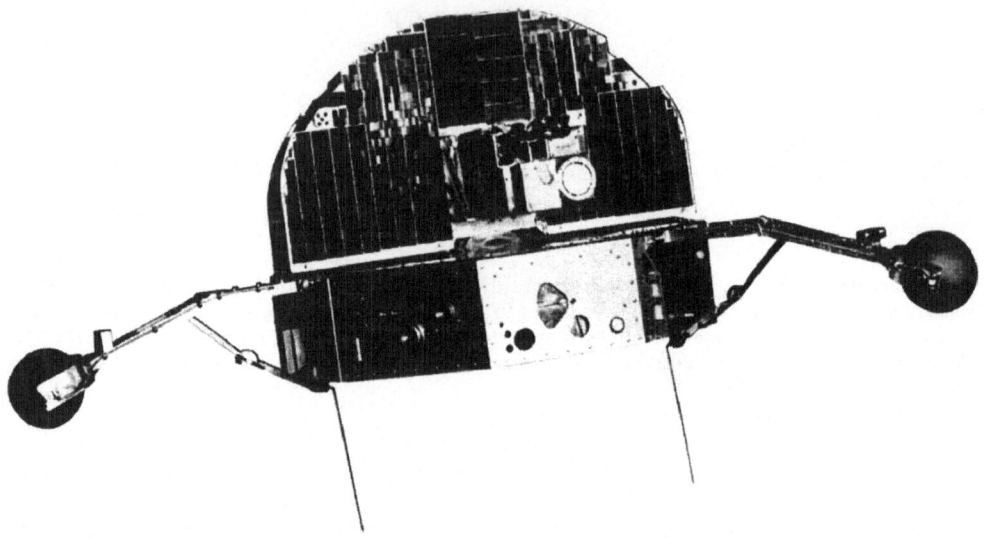

FIG. 1.   *General arrangement of an Orbiting Solar Observatory spacecraft.*

Unlike the first OSO (March 1962), which pointed the experiments at the centre of the Sun, the second OSO satellite provided a raster pattern of about 40 min of arc on the side, as indicated in Figure 2. The scientific instruments in the pointed section had a field of view of about 1 min of arc, and the raster pattern of the spacecraft was used to build up spectroheliograms with this resolution every 5 min. One of the next OSO's, namely OSO-G, will be provided with an additional small raster pattern of 7 arc minutes square which will provide 30 sec of arc spatial resolution images in 30 sec, and will also provide the first opportunity for 'offset pointing' to any location on the solar disk to permit wavelength scans from selected solar formations. These and other improvements in the capabilities of solar satellites are the cornerstones for the design of scientific instruments and provide the bases for a wider range in the type of solar optical observations which can be made.

2. Pointing Controls For Rockets

For quite a number of years, sounding rocket experimentation for the solar ultra-violet has been well served with the biaxial pointing control, which is unstabilized in roll angle about the solar vector but provides quite reasonable pointing accuracies to the Sun's centre in the range of 1–2 min of arc.

Just a few weeks before the last IAU Assembly in Hamburg, a group of British experimenters under the direction of R. Wilson successfully flew an Elliott Bros. 3-axis stabilized rocket in which the roll angle was controlled by magnetometer error signals. The system achieved better than 10 min of arc of pointing in pitch and yaw,

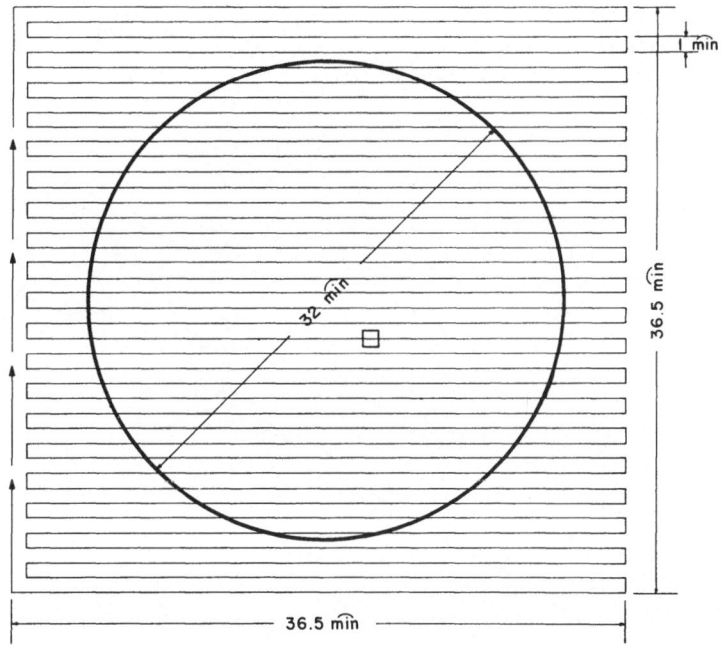

FIG. 2.   *The raster pattern of OSO spacecraft for generating spectroheliograms.*

and a roll stability of approximately 10° peak to peak. To provide greater accuracy in pitch and yaw, the experiment telescope mirror was servo-controlled with a secondary mirror and split field solar sensor, stabilizing the solar image to several seconds of arc. The optical arrangement is shown in Figure 3. A significant achievement resulted from the development of this system when Dr. Wilson's group obtained the chromospheric spectrum in April 1965 by setting the spectrograph slit about 10 sec of arc outside the visible limb with a slit width of 4 sec of arc. The experiment used Kodak Pathé SC-5 and SC-7 Schumann film and recovery techniques (Burton and Wilson, 1965; Burton *et al.*, 1967).

An interesting achievement in pointing control systems has been used by J.-E. Blamont to obtain the time variation in a number of broad bands in the ultraviolet between 1200 Å and 2400 Å during the progress of the November 12, 1966, eclipse in Argentina. The French system was launched in a Titus rocket from a portable range established for the purpose in Argentina with primary data by telemetry, although provision was made for photographic recovery for other experiments. The experiments were aligned along the axis of the rocket and a system of nitrogen gas jets and pre-set gyroscopes was used to align the instruments with the Sun. Prior to totality a solar sensor provided information to correct the gyroscopes, which then kept the rocket aligned as the Moon provided the complete eclipse. The system was capable of achieving pitch and yaw accuracies of approximately 15 sec of arc and a roll rate of

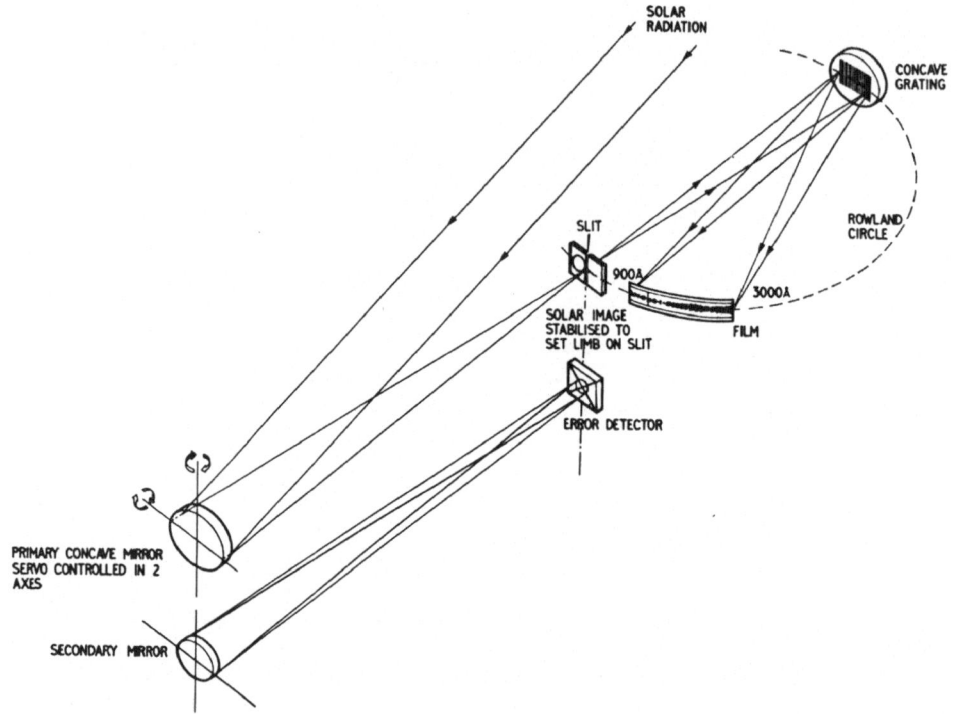

FIG. 3.   *Diagram of the optical system for the Culham-rocket experiment.*

15 sec of arc/sec. During a total eclipse of this type it is possible to obtain spatial resolutions of 1 sec of arc or better, limited by the Moon's irregularities. With control of the precise roll angle and selection of a smooth lunar area it would be possible to achieve spatial resolution in the order of 0·1 sec of arc.

Several 3-axis pointing control systems are nearing completion in the U.S.A.; they will provide pitch and yaw accuracies of better than 15 sec of arc and absolute roll position to several degrees for both solar and stellar observations. Three different systems with somewhat different characteristics have been assembled and tested and are scheduled for test launch in the early autumn. Two of the systems, SPARCS and STRAP, are produced by NASA and the SPCS system is available from Ball Bros. Research Corporation. A summary of the pointing control characteristics is contained in the next review paper (Wilson, 1968).

3. Detectors

The activity in the field of space research has continued to demand the development of improved photomultiplier tubes for the ultraviolet and vacuum ultraviolet regions which are capable of withstanding the launch environments of a variety of rockets.

Photomultipliers for the solar ultraviolet fall into two general classes, the very short-wavelength, windowless devices, and the more conventional discrete stage electron-multiplier tubes with a variety of window and photocathode materials. These latter photomultipliers are now available with a wide range of photocathodes such as KBr, CsBr, RbI and CsI, CsTe, etc., with ultraviolet window materials such as glass, quartz, sapphire, lithium fluoride, and most recently magnesium fluoride. MgF_2 windows represent a significant improvement over LiF in that they can accept a better optical polish, are not susceptible to humidity, are harder, and exhibit a greatly reduced sensitivity to both ultraviolet and energetic particle-radiation damage, while allowing comparable efficiencies above $\lambda 1150$ Å.

The radiation damage indicated in Figure 4 for MgF_2 and LiF corresponds to a predicted maximum flux for January 1966 in a circular polar orbit at 1400 km. The transmissions were measured within hours of radiation and all radiated crystals showed fluorescence which was not detected visually. Of the materials tested, synthetic

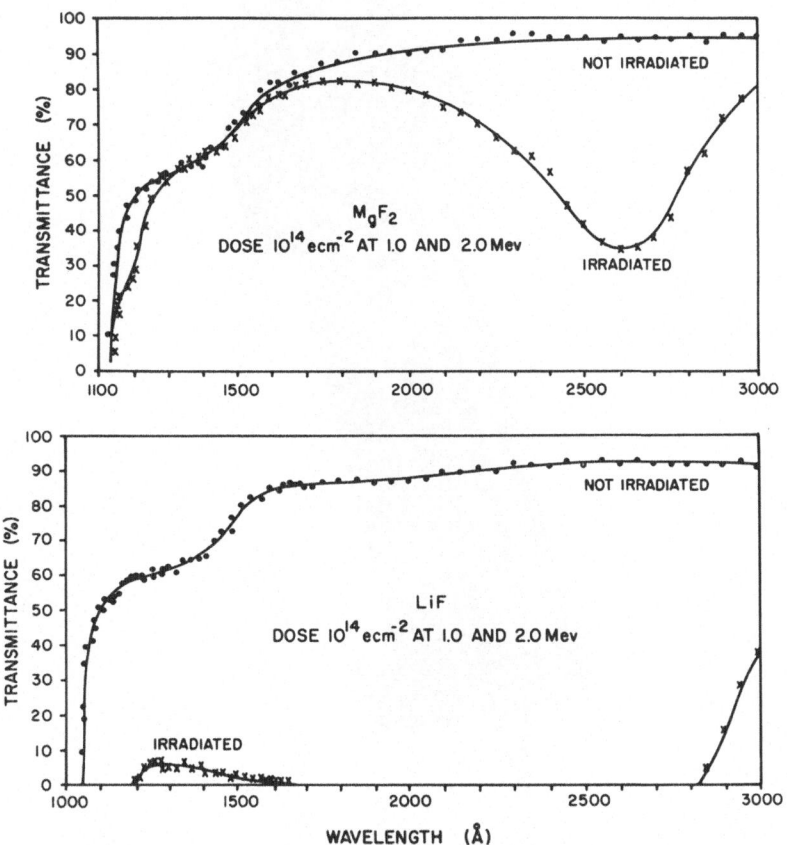

FIG. 4.   *Effect of high-energy electron bombardment on the transmittance of MgF_2 and LiF.*

sapphire and quartz of highest purity showed the least degradation of transmittance with irradiation (Heath and Sacher, 1966). I will not refer in greater detail to the large amount of work which has been done on the radiation effects on glasses for lenses for space applications, since a number of review articles are available (Monk, 1952; Kreidl and Hensler, 1957; Stroud, 1962; Becker, 1967). A wide selection of optical materials is now available in the visible region of the spectrum which do not show appreciable effects from hard radiation in space. A number of optical systems, particularly in the Moon and planetary satellites, have survived long periods in space without appreciable degradation, provided that due care has been exercised in the choice of materials.

One significant improvement in these more conventional photomultipliers for the ultraviolet has been made by Electro-Mechanical Research, Inc. (Figure 5), in which

FIG. 5.   *A photomultiplier with removable side window.*

the window and its housing can be removed in flight in 50 ms by the action of an explosive device, as a once only and irreversible process (Rome, 1964). The conversion of the multiplier from a windowed to a windowless device, once a clean high vacuum has been achieved, has some useful advantages. The inability to remove and replace the window during photometric calibration is a limitation insofar as the detector efficiency cannot be verified during instrument testing. There is no reason why this type of electron multiplier could not be fitted with a high-work function metal cathode, blind above 1300 Å, and still have the very large dynamic range of several megahertz, characteristic of the discrete dynode multiplier.

A photon scintillating device has been developed by the same company and by other laboratories as well, and is illustrated in Figure 6 (Rome, 1964). The incident

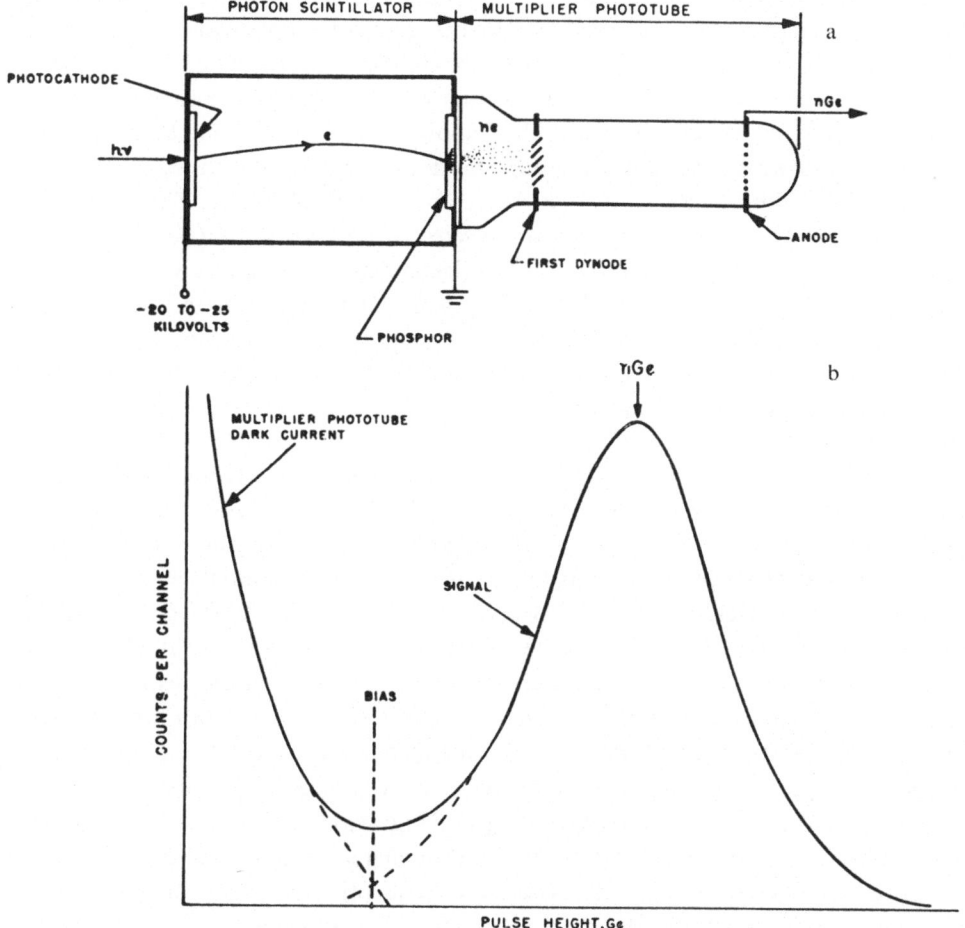

FIG. 6.   (a) *Photon scintillator detector assembly.* (b) *Pulse-height distribution.*

photon releases electrons from a semi-transparent photocathode, which are then accelerated by approximately 20 KV and strike the phosphor with sufficient energy to release a large number of photons within the fluorescence time of the phosphor. The converted photons are then viewed with a conventional photomultiplier. The advantage in this device lies in the output pulse-height distribution, in which the signal exhibits a well-defined peak which permits discrimination against noise pulses in the photomultiplier. The efficiency of detection is increased over that of a single ultra-violet multiplier where the pulse-height distribution requires that some counts be biased off. The device could also be used with a side window or as a windowless device with a metal photocathode, although these latter would require some special development.

The common EUV photomultiplier for space research is the crossed electric and magnetic-field multiplier with continuous dynode strips (Timothy *et al.*, 1967; Hinter-egger, 1961). These devices are produced by the Bendix Company and are illustrated in Figure 7. They are generally referred to as Magnetic Electron Multipliers or MEM's. The tubes are usually provided with a metal photocathode such as tungsten or other inert metal, all of which are blind above 1300 Å (Cairns and Samson, 1966) and sensitive down to the X-ray region. The devices consume little power, and are capable of providing a linear dynamic range up to about 200 KHz. The continuous resistance dynode and field strips are quite sensitive to temperature which produces changes in the gain of the multiplier. Unless the detection system as a whole has been designed to compensate for the effect, which can be accomplished in one of several ways, the detector efficiency will be temperature-dependent. The devices are function-ally quite good, although a considerable amount of work is still required to fully understand their properties and to properly qualify them for extended satellite use with constant efficiency.

The Channel Electron Multiplier (CEM) has received much attention and development for rocket and satellite use (Adams and Manley, 1965; Goodrich and Wiley, 1962; O'Brien *et al.*, 1967). A variety of different designs with and without shaped cathode ends as produced by Bendix are shown in Figure 8. Similar devices are now available from Electro-Mechanical Research, Inc. in the U.S.A. and from Mullard in England. The devices are a hollow tube coated with a resistive material which serves as the dynode and also as the cathode material. The photometric response of the channel multipliers is quite similar to that of the metal photocathodes for the region above 500 Å (Johnson, 1966). Primary photoelectrons are accelerated along the tube by the 3-KV potential, and on striking the wall they release additional secondary electrons, which are in turn accelerated to give a high-gain output pulse. The channel-trons are basically very simple and reliable, although guaranteed linear dynamic range is limited to about 20–60 KHz. The tubes are frequently operated at highest gain so that the output pulses 'saturate' the multiplier, which provides a natural shaping of the anode pulses. However, for the higher intensities this mode leads to the problem

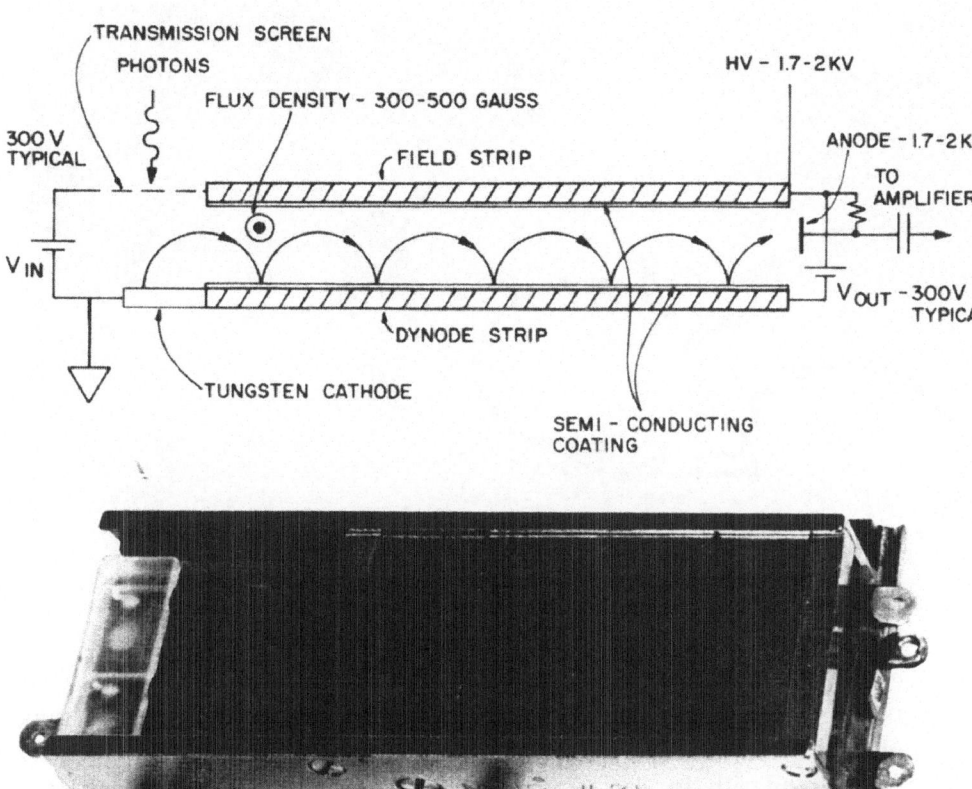

FIG. 7.   *Bendix crossed electric and magnetic-field photomultiplier (MEM).*

of 'fatigue', which shows up as a deterioration in the pulse height distribution which becomes more critical as the intensity × time product increases. This can be overcome to some extent by using the multiplier at lower gains.

A further development in the simple channel multiplier has been the serial use of two channels of different impedances. The lower-impedance second stage allows more power to be delivered, and the linear dynamic range can be increased to about 1–2 MHz for random pulses. These devices are in the developmental stage at the present time.

One of the most interesting advancements in the field of EUV detectors has been the miniaturization and close packing of arrays of channel multipliers. Bundles of multipliers well in excess of 15000 have been produced with capillary size of 250 μm and

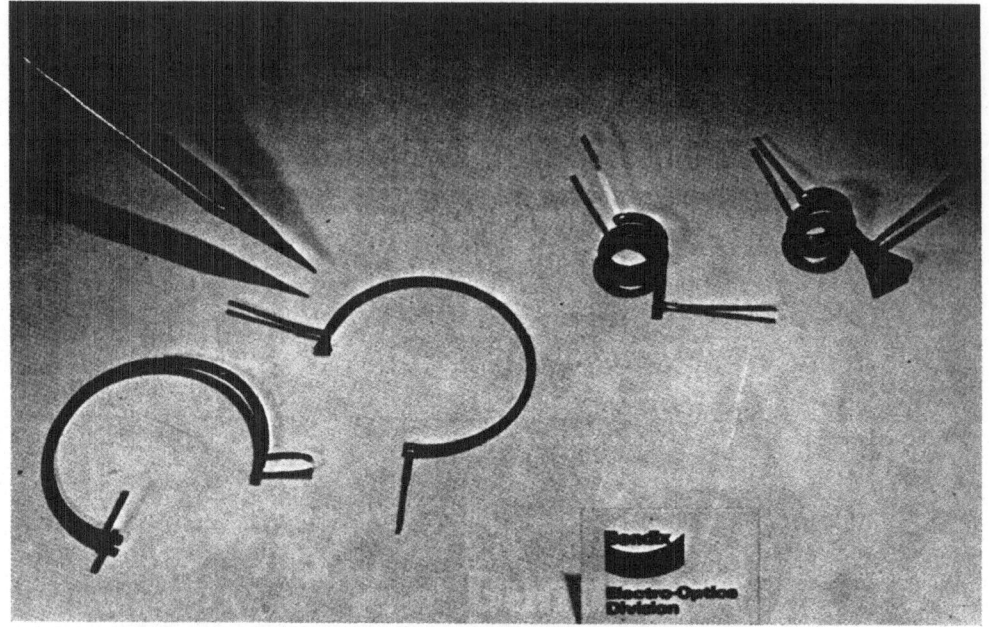

FIG. 8. *Several varieties of Bendix Channeltron electron multipliers.*

even with multiplier tubes as small as 12 μm or less. For each photoelectron resulting from a photon incident on the face of the array, an amplified electron pulse emerges from the anode end with the spatial intensity distribution on the front end preserved. These electron clouds may then be accelerated with a potential of 8–10 KV to strike an aluminized phosphor and film device which will produce a photographic image spot for a single initial photoelectron, or the array can be used with some comparable electronic imaging system. At the present time these arrays are commercially available as a special order.

4. Optical

Above 1200 Å optical efficiencies can be obtained which are quite comparable to visible region reflectivities, and folded optical systems can be used to reduce aberrations and decrease the instrument envelope. The most widely used coating is still aluminum with a thin layer (200–400 Å) of magnesium fluoride deposited in high vacuum immediately following the deposition of aluminum. This prevents the formation of an aluminum oxide layer. Efficiencies between 80 and 90% can be obtained over broad wavelength ranges by varying the thickness of MgF_2 overcoating (Berning *et al.*, 1960; Bates and Bradley, 1967; Hass and Ritter, 1967).

Multilayer coatings for the visible region have certainly progressed in the last few years, particularly for narrow-band transmission filters and non-reflective coatings. At Harvard College Observatory we are currently evaluating Hα interference filters for

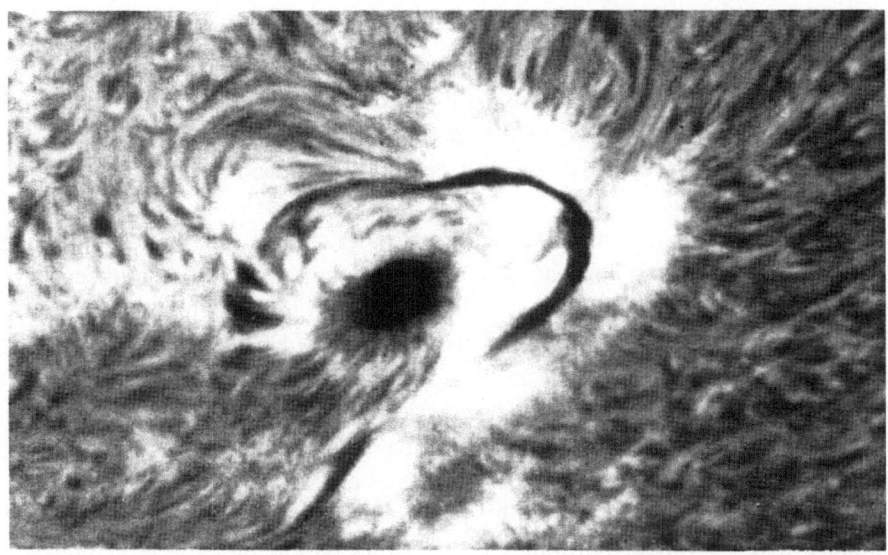

$\Delta\lambda = 0.5\text{Å}$ HALLE FILTER

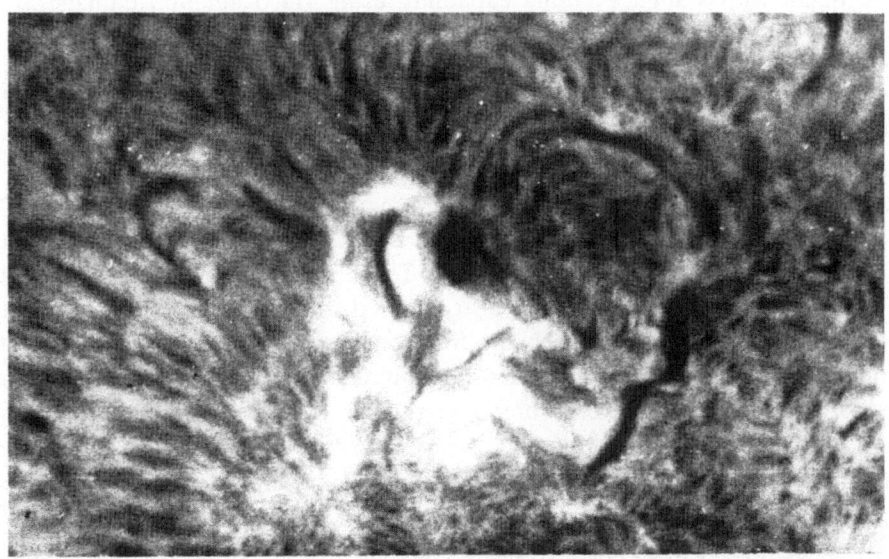

$\Delta\lambda = 0.6\text{Å}$ TYPE "U" FILTER

FIG. 9.   *Photographs of similar solar active regions using a Halle birefringent and a type 'U' interference filter. (Photographs courtesy of Lockheed Solar Observatory.)*

satellite applications. Spectrolab in California are now able to produce narrow-band filters with a halfwidth of 0·5 Å, a uniform field, and transmissions of 20–25%. The filters have been produced down to 0·3 Å halfwidth with a transmission of more than 10%. Several other companies are also engaged in developing similar filters and within the next few months should have been successful. The Spectrolab type 'U' filters are a solid Fabry-Pérot device with interference coatings on each side of a thin substrate. As a result the devices are only a few millimeters thick and are structurally rugged. They are thus much more compact than the birefringent filters, very insensitive to shock and vibration, and have a lower requirement on temperature stabilization.

Figure 9 shows two pictures taken by the Lockheed Solar Observatory of two solar regions, one with an Hα Halle filter and the other with a similar bandpass interference filter. The spatial resolutions are quite comparable and are about 1 sec of arc or a little better. The interference filters have significantly greater transmission than the birefringent filters, although at the present time they are somewhat more limited in the field of view. A report on the progress in this filter evaluation program should be available in a few months. Although no difficulty would be expected, interference filters have not been qualified for extended exposure to high vacuum, but there would be no problem in flying the filters in a sealed, thermostatically controlled oven. Our group at Harvard plans to do this on one of the early manned solar satellites to provide the astronaut with a solar pointing capability using Hα. Radiation damage in the filters does not appear to be a problem.

There have been some advances in making interference filters for the region below $\lambda 3000$. Bates and Bradley (1966) have made filters with 300 Å passbands or less and a transmission of 25% for the region 1800–2400 Å. They are using the filters in high-resolution ultraviolet rocket and satellite interferometers for the Sun.

Miss Fredga (1966) has recently obtained photographs of the Sun in Mg II at 2802·7 Å from a rocket in which a specially developed Solc-type birefringent filter was combined with a telescope and 35-mm camera. The filter had a passband of 4 Å and transmission of 1% or less. The filter also required the development of a servo-controlled temperature system to keep the filter on line. A schematic diagram of the equipment is shown in Figure 10. This does represent the first time that a birefringent filter has been flown in a rocket to obtain pictures in the ultraviolet. Figure 11 shows a reproduction of the results of the flight. The image resolution is approximately 1 min of arc and was limited by jitter on the rocket solar-pointing control system. The development and qualification of these narrowband birefringent filters for the ultraviolet is a significant achievement, and pictures at higher spatial resolution will be possible now that improvements in solar-pointing controls are available.

In the visible region of the spectrum the adoption of the multiple occulting disks in white-light coronagraphs, as suggested by Newkirk, has significantly reduced the problem of diffraction-scattered light in balloon- and rocket-borne coronagraphs (Newkirk and Bohlin, 1964).

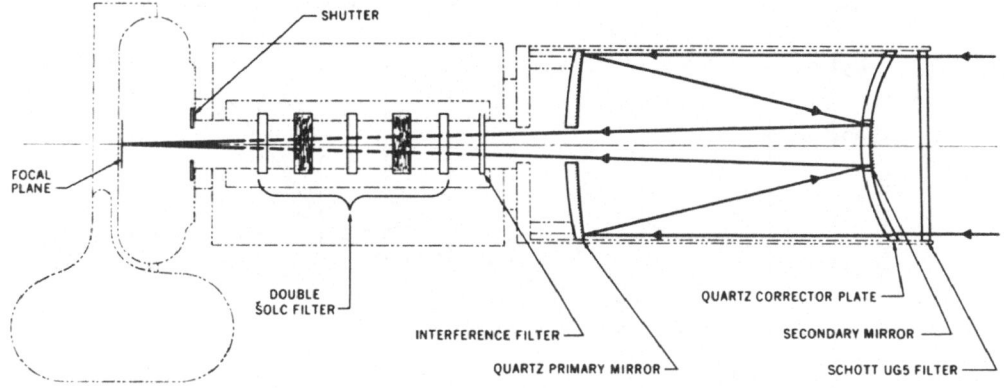

FIG. 10.   *Optical system for a Mg ɪɪ rocket spectroheliograph.*

12 APRIL 1965

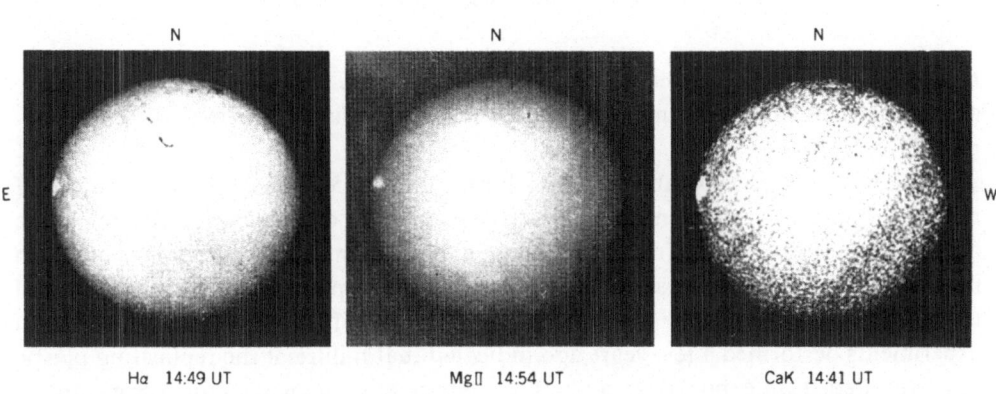

Hα 14:49 UT   Mg ɪɪ 14:54 UT   CaK 14:41 UT

FIG. 11.   *Comparison of the Mg ɪɪ with hydrogen and calcium spectroheliograms.*

For the wavelength region below 1200 Å at near normal incidence, reflectivities are substantially lower than can be obtained above $\lambda 1200$ with Al and MgF_2. The best available coatings are still Au (Canfield *et al.*, 1964), Pt (Reeves, unpublished), and more recently, Ir (Hass *et al.*, 1967).

Figure 12 shows the reflectivities for Au, Pt and Ir from 300 Å to 2200 Å. An additional edge which had been noted in Au has also been found very recently near 400 Å in both Pt and Ir (Hunter, private communication). Hunter's most recent paper on Ir contains data only down to $\lambda 500$, although I have indicated by a dotted line the approximate shape of the curve to 300 Å. The 400 Å edge in Pt does not seem to be as pronounced as it is in Au.

The reflecting films which give the maximum reflectance are partially transparent rather than opaque and are quite sensitive to small amounts of contamination, which greatly reduces the reflectance; however, cleaning will usually restore the initial reflec-

E. M. REEVES

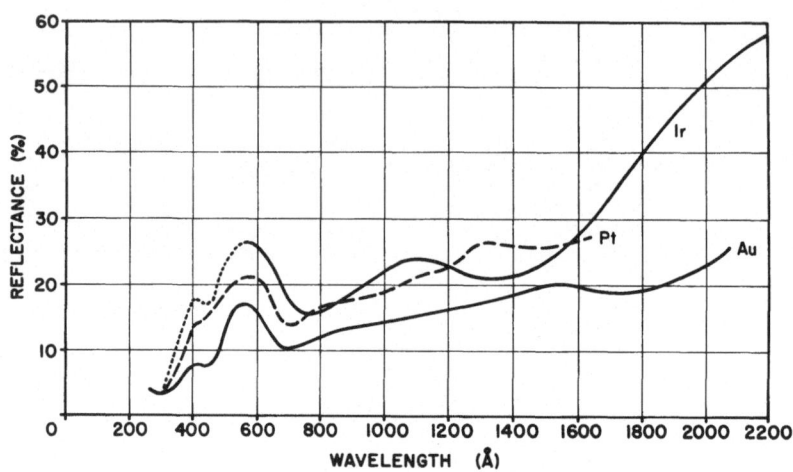

FIG. 12.   *Reflectances of Au, Pt, and Ir in the EUV.*

tance, if ultraviolet radiation has not been too intense. The techniques of preparation are critical, and there is some difficulty in reliably obtaining mirrors with these reported reflectances.

There has been considerable concern during the last 5 years or so over the use of replica gratings for extended-lifetime satellites. The use of replica gratings for space instruments stems not only from the expense and scarcity of original gratings, but also from the fact that, in some cases, the performance of replica gratings exceeds that of the ruled original, both in increased signal and decreased scattered light. Some initial experiments performed a few years ago indicated that failure of the replicating plastic layer resulted from exposure to a simulated space environment. Some six groups in the U.S.A. have been carrying out research programs to obtain better information on the effects of thermal cycling from $-20\,°C$ to $+50\,°C$ at pressures of less than 10^{-8} Torr, irradiation with 1·5 MeV electrons at dose rates up to 10^{15} cm^{-2}, and limited use in a 600 nautical mile polar-orbit satellite. All of the present data, although they are by no means complete, indicate that replica gratings can be qualified for use in satellite instruments (Paul *et al.*, 1967).

The problems associated with optical systems in space are generally quite well known. A recent paper by Becker (1967) reviews some of the general problem areas: outgassing contaminations, meteroid damage, radiation, and temperature effects. The thermal problem is one which is of special importance for solar satellites, particularly in near-equatorial orbits where the Sun's radiation is occulted every revolution, and where the time-variant Earth albedo can cause a transverse bending of the optical system. These problems become particularly important when high spatial and spectral resolutions are required. Up to the present time, however, solar satellite EUV observations have been limited to about 1 min of arc spatial resolution in telescope systems;

here the problems are less severe, and normal optical materials can be used with passive thermal control. In rocket flights, significantly higher spatial and spectral resolutions have been achieved, since thermal distortions can generally be overcome during the short period of a sounding rocket flight by a certain amount of careful design and selection of compensating materials for construction.

For the larger solar satellites now being designed and built, thermal considerations play one of the most important roles of all in instrument design. The heating of the front surface of a mirror at orbital sunrise sets up a time-varying thermal gradient through the mirror blank which causes a change in curvature. Even in the steady state, the change in curvature (γ) is $\gamma = q(\alpha/k)$, where γ is the change in curvature, q is the heat input per unit area, α is the thermal expansion coefficient, and k is the thermal conductivity; the resulting change in focal length (Δf) is given approximately by $\Delta f \approx 2\gamma f^2$, where f is the focal length.

In the ultraviolet particularly, the reflective coating must be optimized for that particular region, with little or no control over the visible and infrared absorption coefficient which determines the heat input. For the Al and MgF_2 films, this effective absorption coefficient is only 7·6%, but for the EUV semi-transparent Pt films it has recently been measured by us to be $0·32 \pm 0·02$ on one particular sample. This absorption coefficient is now being evaluated for various EUV coatings in our own laboratory.

The ratio of the coefficient of thermal expansion (α) to the thermal conductivity (k) determines the susceptibility of the material to distortion effects, and this ratio must be kept small either by having a large thermal conductivity or a small expansion coefficient (see Table 1). For the larger solar telescope mirrors, the use of fused silica or quartz may be completely unsatisfactory, unless somewhat elaborate steps are taken to increase the effective thermal conductivity artificially.

Table 1

Selected parameters for the selection of materials for telescope mirror substrates

Material	$\alpha(\times 10^{-7} °C)$	$k(\text{Watts cm}^{-1}°C^{-1})$	$\alpha/k\ (\times 10^{-7})$
Quartz	5·5	0·0138	400
Be	124	1·58	78·5
Cu	166	3·92	42·3
ULE Silica	$0·2 \pm 0·3 (5–35 °C)$	0·0138	14·5
Cer-Vit	$0 \pm 0·3 (0–50 °C)$	0·0167	0 (30 °C)

The approach used on the Mariner-IV television system (which, however, did not have the high incident solar heating directly through the optical system) was to match the mirror material and the support structure so that the image stayed within the depth of focus as the thermal environment changed. The method requires material of high thermal conductivity in order to reduce temperature gradients. This approach has many advantages; however, the use of metal mirrors such as beryllium requires an overcoating of a hard layer of a material like Kanigen, which is susceptible to 'crazing' or cracking, which gives rise to increased scattered light and distortion of the figure.

Recent experiments by Bennett, Madden, and their co-workers (Bennett, 1963; Bennett and Porteus, 1961; Johnston *et al.*, 1967) have measured the effect of surface roughness on scattered light, particularly at the lower wavelengths, at normal incidence. The effect becomes extremely important as the wavelength decreases below 500 Å and may even limit the lower wavelength observed with a mirror. The effect of surface roughness can also be very important in achieving the highest specular reflectance for the EUV region. At the present time, metal mirrors do not appear to be suitable for resolution in the few seconds of arc region in the solar ultraviolet. This does not necessarily hold true for the wavelength region below about 500 Å where one generally uses grazing incidence optical systems with hyperboloids and paraboloids. At grazing incidence, the efficiency of reflection increases markedly, and the tolerances on the figure are reduced. However, the techniques of grazing incidence have specifically been omitted from this review, since they are very similar to those used for soft X-rays and will be treated by other authors during this Joint Discussion.

An alternate approach to the solution of the thermal problem is to make the system insensitive to thermal effects by choosing a material with near zero coefficient of expansion. Three such materials have recently become available; these are: ULE Fused Silica with an expansion coefficient of $0 \cdot 2 \pm 0 \cdot 3 \times 10^{-7} \, {}^{\circ}\mathrm{C}^{-1}$, ULE Glass Ceramic with $0 \pm 0 \cdot 1 \times 10^{-7} \, {}^{\circ}\mathrm{C}^{-1}$ made by Corning, and Cer-Vit made by Owens-Illinois. Figure 13 shows a plot of the expansion coefficient vs. temperature for Cer-Vit, which indicates

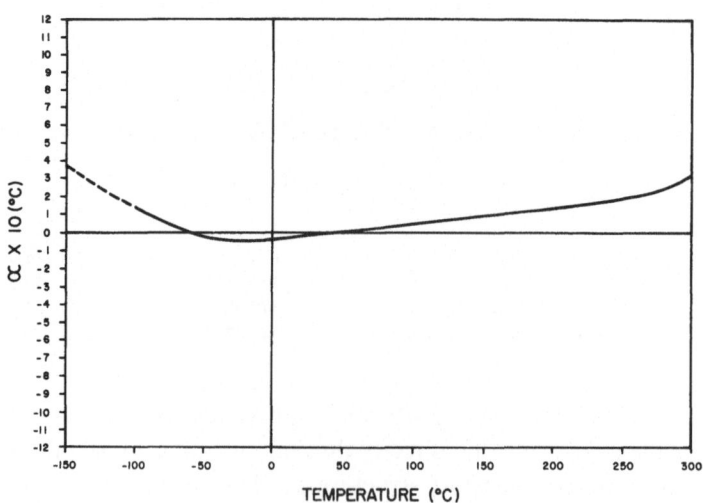

FIG. 13.   *Coefficient of thermal expansion for Cer-Vit as a function of temperature.*

an expansion coefficient close to zero over a wide temperature range. Recent experiments by Dietz and Bennett (1967) have shown that this new material can be polished as smooth as the best fused silica, namely to an r.m.s. surface roughness of better than 8 Å, using a 'Bowl Feed' technique (Dietz and Bennett, 1966). It will also withstand amazing thermal shocks from alternate exposures to ice water and a 350 °C oven, showing negligible deformation of figure within $\lambda/10$. These materials will be used for many of the larger solar satellite optics, both for the telescope mirrors and possibly for structural elements for part or all of the instrument assembly.

I have already referred to the complications attendant on the thermal problem at high spatial resolution. An additional problem which currently is limiting many experiments is the use of recoverable photographic film in long-mission satellite experiments. The frequent passage of near equatorial orbit satellites through the lower part of the South-Atlantic Anomaly causes most ultraviolet films, such as the EUV Schumann-type emulsions (SC5, SWR), as well as 103-0, Pan-X, and Plus-X, to be fogged to a density of 0·2 in a matter of a few days. Red-sensitive films such as SO-375 and Starburst films for Hα photography are much less prone to radiation fogging. Shielding against the energetic protons responsible does not appear to be possible within reasonable weight constraints. On-board development of the film may have to be used to overcome this problem in long-duration manned missions.

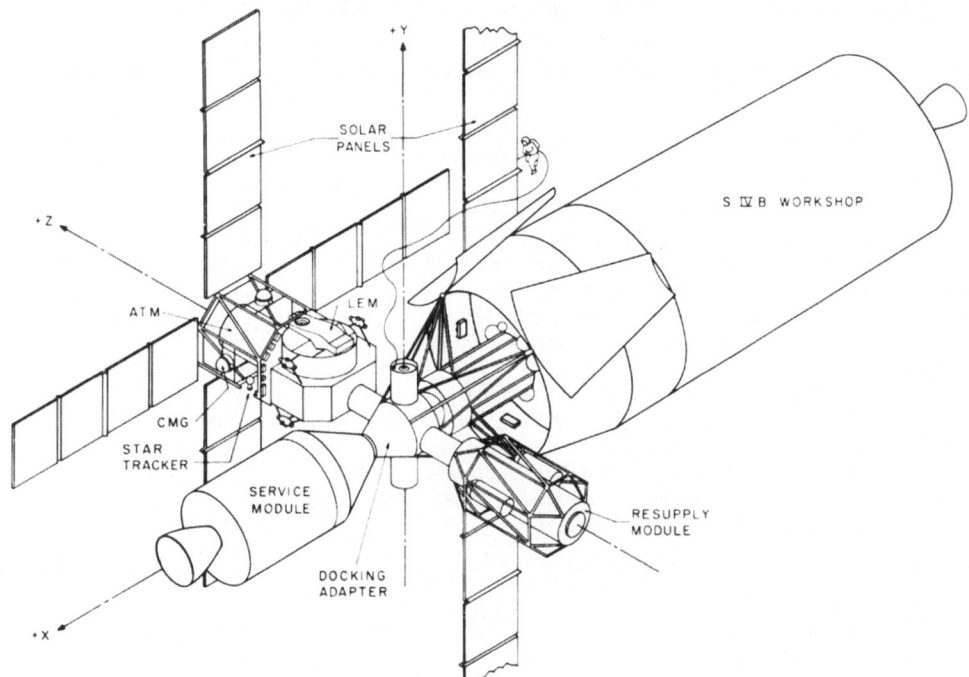

FIG. 14.   *Illustration of the Apollo Telescope Mount orbited configuration.*

The use of the high data-storage capability of normal photographic emulsions presents many highly desirable advantages; however, their use in satellites with film recovery is presently a large outstanding problem, unless very low or purely equatorial orbits are used, or unless emulsions with lower sensitivity to radiation fogging can be developed. Several research groups in the U.S.A. are investigating the quantitative effects of stimulated anomaly radiation fields on commonly used photographic emulsions (Shelton and deLoach, 1967).

One of the 'techniques', if one can refer to it as such, that will shortly exert a great influence on solar space research is that of manned experiments. Figure 14 shows an illustration of the manned Apollo Telescope Mount (ATM), which is currently being built for launch within the next few years. The ATM contains a group of experiments such as a visible light coronagraph (High Altitude Observatory), X-ray telescopes (American Science & Engineering and Goddard Space Flight Center), vacuum ultraviolet spectrographs (Naval Research Laboratory), and spectrometers (Harvard College Observatory), EUV photographic (NRL) and photoelectric (HCO) imaging systems, and Hα-filter telescopes (HCO). The astronaut will guide the experiments, which will have spatial resolutions of the order of 5 sec of arc, to specific locations on the solar disk to an accuracy of several seconds of arc, using an Hα telescope, and instrument slit-jaw displays with the narrowband interference filter described above. Experiments on the quiet Sun, active regions, and flare activity will then be carried out simultaneously over a wide wavelength range. The advent of the manned solar satellite has introduced a significant new capability into the field of solar space research.

References

Adams, J., Manley, B.W (1965)   *Electronic Engineering*, **37**, 180.
Bates, B., Bradley, D.J. (1966)   *Appl. Opt.*, **5**, 971.
Bates, B., Bradley, D.J. (1967)   *J. Opt. Soc. Amer.*, **57**, 481.
Becker, R.A. (1967)   *Appl. Opt.*, **6**, 955.
Bennett, H.E. (1963)   *J. Opt. Soc. Amer.*, **53**, 1389.
Bennett, H.E., Porteus, J.P. (1961)   *J. Opt. Soc. Amer.*, **51**, 123.
Berning, P.H., Hass, G., Madden, R.P. (1960)   *J. Opt. Soc. Amer.*, **50**, 586.
Burton, W.M., Wilson, R. (1965)   *Nature*, **207**, 61.
Burton, W.M., Ridgeley, A., Wilson, R. (1967)   *Mon. Not. R. astr. Soc.*, **135**, 207.
Cairns, R.B., Samson, J.A.R. (1966)   *J. Opt. Soc. Amer.*, **56**, 1568.
Canfield, L.R., Hass, G., Hunter, W.R. (1964)   *J. Phys.*, **25**, 124.
Dietz, R.W., Bennett, J.M. (1966)   *Appl. Opt.*, **5**, 881.
Dietz, R.W., Bennett, J.M. (1967)   *Appl. Opt.*, **6**, 1275.
Fredga, K. (1966)   *Astrophys. J.*, **144**, 854.
Goodrich, G.W., Wiley, W.C. (1962)   *Rev. Sci. Instrum.*, **33**, 761.
Hass, G., Ritter, E. (1967)   *J. Vacuum Sci. and Techn.*, **4**, 71.
Hass, G., Jacobus, G.F., Hunter, W.R. (1967)   *J. Opt. Soc. Amer.*, **57**, 758.
Heath, D.F., Sacher, P.A. (1966)   *Appl. Opt.*, **5**, 937.
Hinteregger, H.E. (1961)   in *Space Astrophysics*. Ed. by W. Liller, McGraw-Hill, New York, p. 34.
Johnson, M.C. (1966)   *Bendix Corporation Technical Report No. 3640.*
Johnston, R.G., Canfield, L.R., Madden, R.P. (1967)   *Appl. Opt.*, **6**, 719.

Kreidl, N.J., Hensler, J.R. (1957)   *J. Opt. Soc. Amer.*, **47**, 73.

Monk, G.S. (1952)   *Nucleonics*, **10**, 52.

Newkirk, Jr.,G.D., Bohlin, D. (1964)   *Appl. Opt.*, **3**, 543.

O'Brien, B.J., Abney, F., Burch, J., Harrison, R., LaQuey, L., Wineicki, T. (1967)   *Rev. Sci. Instrum.*, **38**, 1058.

Paul, F.W., McAndrew, C.E., Diggins, J.C. (1967)   *J. Opt. Soc. Amer.*, **57**, 582.

Rome, M. (1964*a*)   *IEEE Trans. on Nuclear Science*, June 1964.

Rome, M. (1964*b*)   *Appl. Opt.*, **5**, 855.

Shelton, R.D., deLoach, A.C. (1967)   *National Aeronautics and Space Administration Report TMX-53666*. Marshall Space Flight Center.

Stroud, J.S. (1962)   *J. Chem. Phys.*, **37**, 836.

Timothy, A.F., Timothy, J.G., Willmore, A.P. (1967)   *Appl. Opt.*, **6**, 1319.

Wilson, R. (1968)   in the present volume, p. 94.

ULTRAVIOLET ASTRONOMY

R. WILSON

(Culham Laboratory, England)

1. Introduction

Ultraviolet astronomy is in a phase of very rapid development and any review of the new techniques involved is necessarily conditioned by the fact that most of the work is in a preparatory stage. This paper, therefore, will not be restricted to a discussion of those techniques which have undergone the ultimate test of their effectiveness, i.e., an actual space mission, but will include a consideration of those techniques which are still under development. Some omissions are therefore inevitable.

The presentation is divided into three categories: (a) the vehicle and its stabilisation (if any), (b) the optics, and (c) detectors. Category (a) presents the major new technological problems and contributes the greater part of the total cost. Experience in the other categories of optics and detectors had, of course, reached an advanced and sophisticated level at the time of the advent of the rocket, but new and difficult problems have been posed mainly by their operation in a space environment, but also by their use in the ultraviolet region of the spectrum.

2. Vehicle Stabilisation

Many of the technical problems posed by UV astronomy are common to all areas of space research, e.g., power supply, telemetry, etc., and are not dealt with here. However, new problems are posed by the fact that the space vehicle is used as a platform for the observation of remote objects, in contrast to most geophysical experiments where the vehicle is used as a probe. For astronomical studies it is therefore essential to know the attitude of the vehicle with time and for any detailed observations it is also necessary to control the attitude.

A. UNSTABILISED SOUNDING ROCKETS

The pioneering observations in UV astronomy were made from unstabilised sounding rockets in which the sky was scanned by the motion of the vehicle. The limited observing time per star therefore restricted the experiments to measurement of the UV flux in broad bands from the brightest early-type stars.

The first ultraviolet astronomical observations were secured in a rocket flight during

Perek (ed.), Highlights of Astronomy, 94–107. © I.A.U.

1955 by Byram *et al.* (1957) of N.R.L., which was followed by further observations by that group (Friedman, 1962; Chubb and Byram, 1963; Byram *et al.*, 1964). The early observations were made with mechanical collimating devices feeding broad-band selective photomultipliers. These were later replaced by prime focus telescopes, feeding ionisation chambers operating in the range 1200–1500 Å. The photometers were flown in the American Aerobee rocket with a viewing direction perpendicular to the rocket axis, a sky scan being obtained by the roll motion of the vehicle. Using the same rocket, Boggess (1961, 1964) has made observations in the 2000–3000 Å region, using photo-electric photometers fed by Cassegrain telescopes; and Gulledge and Packer (1963) observed bands centred at 2985 Å and 2100 Å with respective widths of 52 Å and 160 Å by employing photomultipliers and interference filters. Observations with the British Skylark rocket have been made in the Southern hemisphere by Alexander *et al.* (1964) using a mechanical collimator to feed a selective photomultiplier giving a band-width of ~ 400 Å centred at 1950 Å. The first dispersing system was flown by Stecher and Milligan (1962) in the Aerobee rocket. The stellar radiation passed through a mechanical collimator onto a large plane grating, the deflected light then being focussed onto an entrance slit by two reflections and measured by a photomultiplier. The spin axis of the rocket was normal to the plane of dispersion and the stellar spectrum was scanned by the motion of the rocket. The operating spectral resolution was set at ~ 100 Å, and observations were limited to stars brighter than about third magnitude.

B. SCANNING SATELLITES

The type of sky scan performed by unstabilised rockets can also be employed from satellites with a consequent increase in total observational time. Further, a reduction in the spin rates of the vehicle will result in a greater observational time per object, and by 'scanning satellites' is meant those vehicles for which some control of angular rates is exercised. This may result in either a slow but unpredictable scan of the sky or a controlled scan over the celestial sphere.

1964-83C: Smith (1967*a*) has flown a photometer in the American satellite 1964-83C and obtained several stellar observations with a pass band ~ 200 Å centred at 1376 Å. The satellite employed a form of magnetic stabilisation which allowed a slow scan of the sky. Because of the complexity of the dynamic effects this scan was not predictable and the attitude was determined from solar sensors and magnetometers followed by star recognition.

Cosmos-51: Dimov and Severny (1965) flew a photometer in the Soviet satellite Cosmos-51 and obtained observations with a pass-band of ~ 300 Å centred at 2700 Å, together with a simultaneous observation in the visible. A form of stabilisation allowed a slow scan of the sky but, as in the case of 1964-83C, the attitude was not predictable and attitude information was determined from ancillary sensors and star recognition.

Venus-3: Ultraviolet observations have been made by Kurt and Syunyaev (1967) from the Soviet deep-space probe Venus-3. The photometers used had wide fields in order to measure the sky background radiation in the spectral intervals 1050–1340 Å and 1225–1340 Å using Geiger counters with appropriate gas and windows.

ESRO-TD1: This European satellite is under development and is planned to operate in a Sun-synchronous orbit. It will be Sun-stabilised with a roll control which maintains a rotational period equal to the orbital period, thus enabling experiments mounted in the appropriate axis to scan a great circle of the sky in each orbit and the precession of $\sim 1°$ per day will allow a systematic coverage of the celestial sphere. This controlled kind of scan will therefore give attitude information, but this will still be supplemented by star recognition.

Two ultraviolet astronomical experiments are to be carried out in this satellite. A combined Edinburgh/Liège (Housiaux, 1965) package consists of a 25-cm telescope feeding a photometer which will measure four pass-bands between 1250 Å and 3000 Å with bandwidths between 100 Å and 200 Å. An infrared photometer is also included for measurements near 3 μ. The second ultraviolet experiment is being prepared by the Utrecht Group (De Jager *et al.*, 1967); this employs a 22–25 cm telescope feeding a grating spectrometer with scanning exit slits and photomultipliers. In this case the experiment includes its own fine stabilisation device which selects a bright star in the field of view and counterrotates the telescope so as to take out the scanning motion of the satellite and thereby allow observation times per object to be substantially increased to 4 min. This allows a much higher spectral resolution to be employed and the exit-slit widths are set at 1 Å, the spectrum being scanned in steps of 0·5 Å. It is expected that stellar spectra will be obtained down to magnitude 4.

SAS: A small astronomical satellite is being developed by NASA (1967) to be launched in a Scout rocket. Present plans call for a gyro-stabilised spacecraft which can be controlled in three axes by magnetic torquing and allow a variation in the type of scan which can be performed. To date no ultraviolet experiment has been assigned to this type of vehicle, the first of which will carry an X-ray experiment.

C. BALLOONS

Although ultraviolet observations down to the Lyman limit of atomic hydrogen at 912 Å are possible only from heights requiring rocket vehicles, observations from high-altitude balloons can extend the accessible spectrum down to 1900 Å. A group at the Geneva Observatory (Golay *et al.*, 1967) has obtained low-resolution objective prism spectra using French balloons and a gondola equipped with spinning rotator stabiliser which allows a sweep of the sky at 0·4° per sec.

D. STABILISED ROCKETS

The full development of ultraviolet astronomy requires space vehicles which can be fully stabilised with respect to the object being observed. For a package consisting of a telescope and grating spectrometer, the degree of stabilisation can be determined by expressing the element of spectral resolution $\delta\lambda$ in terms of the angle $\delta\omega$ it subtends in the telescope field, thus:

$$\delta\omega = \frac{G}{T} D \, \delta\lambda,$$

where G is the diameter of the grating, D its angular dispersion, and T the diameter of the telescope. Typical values are $G/T=0\cdot1$ and $D=25$ arcsec per Å, giving $\delta\omega=2\cdot5$ arcsec for $\delta\lambda=1$ Å. Whereas in the case of scanning systems the observational time per object is limited by the scanning rates, in the case of a fully stabilised system the limitation is imposed only by the trajectory time for a sounding rocket, and by oc-cultational effects for a satellite.

A number of star-pointing stabilisation systems are currently under development for operation in sounding rockets, and these are summarised in Table 1.

The IACS (Inertial Attitude Control System) (NASA, 1967) is based on a 3-axis gyro-reference which is programmed before flight, actuation being achieved by a gas jet system. The STRAP (Stellar Tracking Rocket Attitude Position) and FACS (Fine Attitude Control System) (NASA, 1967) systems are based on the IACS but have additional star trackers which take over the control on acquisition and use an auxiliary low-thrust jet system to give higher pointing accuracies.

The U.K. Starling (Elliott Bros., 1967) system is based on a Moon-pointing control unit which is now in an advanced state of development. This unit will be used for off-setting to the desired star, at which point a star tracker will take over for the final acquisition. Actuation is by means of gas jets.

The German system being developed by Dornier to the requirements of the Max-Planck Institute will use a spinning rocket, the axis being aligned at a required star from an inertial reference. The final acquisition will be made by means of a star sensor and gas jets will be used for actuation.

Of the control units described in Table 1 only the IACS and STRAP systems are currently operational and have undergone a number of successful flights, which have obtained new and important results in UV astronomy. The Princeton group (Morton and Spitzer, 1966; Morton, 1967) have used the IACS system on Aerobee rockets and obtained objective grating spectrograms of a number of bright stars in the range 1200–3000 Å with spectral resolutions between 1–3 Å. This rather good spectral resolution was achieved by attaching the experimental package to a large gyroscopic rotor which gave an additional fine stabilization in the direction of dispersion, the limit cycle being reduced to ±16 arcsec. Objective grating spectra have also been obtained in the 1600–3000 Å range with a resolution of about 10 Å by Boggess (1967) using the IACS/Aerobee. The STRAP system has been flown on an Aerobee rocket by Stecher (1967), who employed a 13-inch telescope with a scanning 3-channel photo-electric spectrometer and obtained a number of stellar spectra over the range 1200–

R. WILSON

Table 1
Rocket star-pointing systems

Star-Pointing System	Rocket	Lateral Pointing Accuracy	Limit Cycle Amplitude	Limit Cycle Frequency	Roll-Pointing Accuracy	Faintest Acquirable Star	Present Status
IACS (Space-General)	Aerobee Veronique	< 3°	± 15 min	$\frac{1}{2}$ cps	< 3°	–	Operational (several flights)
U.S.A. STRAP (Goddard)	Aerobee	30 sec	± 30 sec	$\frac{3}{4}$ cps	2°	3 mag	Developed (3 flights)
FACS (Space-General)	Aerobee	30 sec	± 12 sec	$\frac{1}{2}$ cps	2°	3 mag	Under Development (1st flight 1967)
U.K. STARLING	Skylark	< 2·5 min	± 5–10 sec	3–5 cps	1°	5 mag	Under Development (1st flight 1969)
GERMANY (Max-Planck) (Dornier)	Dragon	1·5 min	–	–	none	3 mag	Under Development (1st flight 1968)

4000 Å with resolutions of 10 Å and 5 Å. Using the same vehicle, Smith (1967b) has flown an objective grating in a Wadsworth mount to obtain a spectrum of Spica to wavelengths below Lyman-α. Carruthers (1967) has obtained objective grating spectra with the IACS/Aerobee. An image intensifier of the electronographic type was used to cover the range 950–1400 Å together with two photometers in the respective ranges 1050–1180 Å and 1230–1350 Å. Courtes and Viton (1967) have used the IACS system in a Veronique rocket to obtain UV sky photographs with an F1 camera in the range 2400–3200 Å.

E. POINTING SATELLITES

Considerable efforts are currently devoted to the development of pointing systems in satellites which will allow the orientation of telescopes to a desired object and subsequent accurate stabilisation for the required period of observation. Such systems are both large and sophisticated, but will present the astronomer with his greatest opportunity for carrying out advanced studies in UV astronomy.

Orbiting Astronomical Observatory (OAO): This series of satellites is currently under development by NASA and involves a number of UV experiments to be launched sequentially in basically the same spacecraft (Rogerson, 1963). The first of these was launched in April 1966 and although it achieved orbit and stabilisation, a fault in the power sub-system prevented any scientific data being obtained (Space Science Board, 1967). The planned operation is as follows: Firstly, the angular momentum is reduced by using rate gyros and gas jets; then the roll axis is aligned on the Earth–Sun line with the aid of Sun sensors. The observatory is then rotated about the roll axis while the star trackers are set at the correct angles with respect to the programmed reference stars. When these are located the spacecraft control is switched to the star trackers and inertial wheels are used for actuation. The observatory is then slewed to the required object whilst the star trackers continue to track until either occultation occurs or they reach the end of their range. The use of six star trackers ensures the continuous acquisition of sufficient reference stars for the operation of the system.

The OAO experiments are planned in order of increasing spectral resolution. A Wisconsin experiment consists of seven telescopes feeding different forms of photometers, allowing broad-band measurements in the region 1110–3000 Å. A Smithsonian experiment (project Celescope) consists of four telescopes feeding Uvicon image tubes and is designed to map the sky in four wavelength bands centred at 1350 Å, 1500 Å, 2250 Å, and 2650 Å. Both these experiments require pointing accuracies of the order of 1 arcmin. The Goddard experiment consists of a 36-inch telescope feeding an Ebert-type spectrometer used in a scanning mode with six photomultipliers as detectors. The instrument covers the range 1050–4000 Å with a best spectral resolution of 2 Å. The required pointing accuracy of 1 arcsec is to be achieved by the use of an error signal derived from a star sensor within the experimental package. The Princeton experiment covers the range 700–3200 Å and is aimed at a spectral resolution of 0·1 Å. The required pointing accuracy of 0·1 arcsec is also to be achieved with the aid of an experimental error signal. A 32-inch Cassegrain telescope is used over the whole available length in the spacecraft, the spectrometer being spaced between the primary and secondary. The spectrometer is used in a Pachen-Runge mounting and the spectrometer is scanned by two carriages, each containing two exit slits and multipliers. Princeton have also been considering more advanced experiments in the field, and a design study has been carried out by Perkin-Elmer (1967a, b) to consider the problems involved in orbiting a 40-inch telescope, operating at its diffraction limit for direct image work and also being used to feed an echelle spectrometer for UV spectral studies with a resolution of 0·1 Å. One of the many problems associated with this is that of fine stabilisation which is required to 0·01 arcsec and it is proposed to achieve this by magnetically supporting the experimental package within the spacecraft.

European Large Astronomical Satellite (LAS): Design studies for a large astronomical satellite to provide a European space observatory have been carried out by ESRO, culminating in a project development plan (ESRO, 1967). This project has not yet been finally approved, but the first experiment has been assigned to spectroscopic studies of objects down to 9th magnitude with variable spectral resolution up to a best value between 0·1–0·3 Å (Culham Lab., 1966). The LAS concept differs from that of the OAO in that the space reference is derived from gyros. Two star trackers (one for redundancy) are used as a means of up-dating the gyro-reference. An on-board computer will be included to carry out many of the spacecraft functions and it is planned to launch the observatory into an equatorial orbit which gives a minimum of radiation background.

F. MANNED SPACECRAFT

The first astronomical observations carried out by man in space were performed by U.S. astronauts during the stand-up extra-vehicular activity periods of Gemini flights 10, 11, and 12. The equipment was designed by Henize and Wackerling (1967) and consisted of objective prism and objective grating instruments. Low-dispersion spectra of a number of objects in the range 2300–4000 Å were obtained. An extension of the use of man for astronomical studies in space is currently being considered by NASA. The Apollo Telescope Mount is to be devoted to solar studies and has already been described by Reeves (1968).

G. MOON TELESCOPE

The possibility of a Moon-based telescope has been considered by Muzjherin *et al.* (1965) with the aim of establishing a proposal within reach of present-day technology. They suggest a 50-cm meniscus Maksutov telescope to be placed in a vertical position on the lunar surface for UV photometric studies. The rotation of the Moon provides a slow scan of a circle of the sky, a television-type tube being used as detector, and the data being telemetered to Earth. A possible movement of the optical axis from the vertical would allow a wider coverage of the sky. It is estimated that such a system could reach magnitudes 8–9.

3. Optics

The optics used in UV astronomy will take several forms but will usually involve a telescope plus ancillary equipment such as spectrometers, photometers, and polarimeters. It is not the intention here to review the whole field of UV optics but to select those developments which are new and linked to a space application either to meet the peculiar environment or the problems associated with operation in the UV.

A. REFLECTING COATINGS

Great advances have been made during the past decade in the development of high-reflecting coatings for use in the UV. A review of these developments and the relevant literature is given by Hunter (1962). Of the several kinds of coatings investigated, five types are particularly valuable for use in UV astronomy; $Al + MgF_2$, $Al + LiF$, Pt, Au, and $Ge + ZnS$. Overcoating of Al with MgF_2 has achieved a reflectivity of about 80% down to about 1200 Å where it drops steeply. Overcoating of Al with LiF has achieved a reflectivity of about 60–70% down to about 1000 Å. Pt and Au give reflectivities of only 20–30%, but this extends over the full range and they are therefore important for observations down to the Lyman limit. $Ge + ZnS$ also gives a reflectivity of about 25% but can give a low reflectivity in the visible for discrimination purposes.

The possible environmental effects on mirrors coated with $Al + MgF_2$ have been investigated by Canfield et al. (1966) with encouraging results, reflectivity being maintained after extended exposure to air, UV radiation, and bombardment with 1 MeV electrons and 5 MeV protons. Other relevant information is supplied by Heath and Sacher (1965), who investigated the transmission properties of a number of optical materials after irradiation to 1 and 2 MeV electrons with a dosage of 10^{14} electrons cm^{-2}. Whereas a number of materials, including MgF_2, were not affected significantly, the transmission of LiF dropped seriously over a wide range in the UV.

B. OPTICAL MATERIALS

Operation in a space environment imposes severe requirements on materials used for optical components. This is particularly so for the large primary telescope mirrors being considered for the major observatory satellites, where the optical performance has to be maintained in the face of severe weight restrictions and a peculiar thermal environment. The resulting requirements are:

(1) The material must be good optically, i.e., it must be optically workable to the desired accuracy, must be stable and capable of carrying a high-reflecting coating.

(2) It must have a low weight, i.e., have a high strength/weight ratio, and

(3) It must be good thermally, i.e., have high conductivity and low expansion.

These requirements are clearly incompatible for existing developed materials. The most essential requirement is (1), since the weight and thermal problems may be solved by other means; e.g., a more powerful rocket booster and a thermostatic thermal design.

For the Princeton OAO experiment, fused silica has been adopted for the telescope mirror. This is a well-tried optical material and the weight problem has been tackled by Corning and Perkin-Elmer by using mirror blanks in an egg crate construction. This gives a mirror whose weight is between $\frac{1}{3}$ and $\frac{1}{2}$ of that of a solid mirror of the same volume. The optical requirement on this particular mirror is for 0·1 arcsec image quality. An easing of the thermal problem may be possible in the future by the use of ultra low-expansion (ULE) silica being developed by Corning (Perkin-Elmer, 1967a, b), which gives an expansion coefficient of $0·2 \times 10^{-7}$ °C near 20 °C compared with 4×10^{-7} for normal silica. Another low-expansion material, 'Cer-vit', is being developed by Owen (1967) of Illinois and a low-expansion glass, Setal, is being developed in the Soviet Union.

Another approach to this problem is afforded by the use of beryllium. This has a very high strength-to-weight ratio and also has good thermal properties. It presents considerable optical problems, however, some of which have been discussed by Barnes (1966). This material has been adopted for the primary telescope mirror in the Goddard OAO experiment, which has a requirement of 1 arcsec for its image quality. The mirror is coated with Kanigen, an amorphous nickel compound which then carries the high-reflecting coatings.

C. SPECTROMETERS

Of the UV payloads discussed above, those which employed dispersing systems have used well-established optical configurations. They are mostly based on concave or plane diffraction gratings and any new major problems posed are mainly of a mechanical or thermal rather than an optical nature. The application of echelle spectrometers will prove valuable in future experiments, since they offer the possibility of high resolution and wide spectral coverage in a compact instrument. Such a system may be used to relieve the pointing accuracy.

Another type of dispersing system which offers advantages in UV astronomy is the Fabry-Pérot etalon which provides high spectral resolution and luminosity, but over a very restricted wavelength range. Since it employs transmission optics its wavelength of operation is restricted by the available optical materials. The shortest wavelength at which such a system has been operated in the laboratory is about 1800 Å where Bradley et al. (1964) used high-quality Spectrosil coated with $Al + MgF_2$ for the interferometer plates, together with an interference filter consisting of layers of Al and MgF_2 deposited on fused quartz substrates.

D. POLARIZERS

Polarization measurements will become important for certain branches of UV astronomy. The present laboratory position is reasonably good for the range 2000–3000 Å where the use of Calcite and quartz allows polarization measurements to be made with a considerable degree of sophistication. Below 2000 Å polarization devices are in a stage of research and development, but considerable progress has been made in recent years. Burton (1964) constructed a polarization analyzer consisting of a LiF plate which reflected radiation on the Brewster angle. This simple system is limited in its efficiency and Walker (1964) used a stack of 6 LiF plates giving polarization efficiencies of 82% at 1200 Å and 65% at 1600 Å with the respective transmissions of 4·3% and 21·5%. More sophisticated devices have been developed by Johnson (1964) and Steimetz *et al.* (1967), who built a Wollaston prism and a double Rochon prism respectively, both in MgF_2. The latter device has a transmission greater than 30% at wavelengths above 1600 Å. As a basis for developing both reflecting and transmitting polarizers, Hunter (1964) has studied the optical properties of a number of materials in the UV. The possibility of using thin films has also been considered by Abeles (1964).

4. Detectors

A. PHOTOGRAPHIC EMULSION

The photographic plate is still an important detector in UV astronomy because of its image storage properties and its great simplicity. In addition to its well-known disadvantages as a detector it requires recovery when used in space vehicles. A number of UV emulsions are available of which the fastest are Kodak Pathé SC5 and SC7.

B. PHOTOMULTIPLIER TUBES

Photomultiplier tubes have now reached an advanced state of development and a wide selection is available commercially which are robust enough for space use and have a wide coverage of spectral response. An extensive literature exists and the reader is referred to the review article by Dunkelman (1962). The types of tube include discrete dynode systems with a variety of photocathodes and windows (Rome, 1964) and magnetic multipliers employing a continuous dynode strip (Heroux and Hinteregger, 1960). The photomultiplier is an excellent detector giving high gain, low noise, and reproducibility. Its only disadvantage for application to UV astronomy lies in its physical size, which limits the number of channels which can be built into any system.

C. CHANNEL MULTIPLIERS

The possibility of very small tubes is offered by the channel multiplier (Goodrich and Wiley, 1961). This consists of a hollow glass cylinder, coated on the inside with a highly resistive material which provides both the continuous dynode and, at the open input end, the photocathode. An electric field applied along the length of the tube causes a cascade of photoelectrons down the tube with subsequent secondary emission. Gains up to 3×10^8 can be achieved (Adams and Manley, 1967) and tube diameters can be as small as fractions of a millimetre. Considerable advances have been made recently by the use of curved tubes which have eliminated the positive ion feed-back produced by electron collisions with residual gas atoms. The open end may be shaped for radiation collection or a slot may be introduced into the side of the tube. The full application of such devices to UV astronomy will require the use of sealed as well as open tubes with a selection of windows and photocathodes. The tubes can also be built in the form of a closely packed channel array giving an imaging system with point by point intensification (Adams and Manley, 1967).

D. IMAGE STORAGE TUBES

The image storage tube combines the major advantages of the photographic plate and photomultiplier tube. Its imaging and storage properties lead to a large information capacity where the electrical characteristic of the signals allows the telemetry of the information and hence its use in satellites. Considerable effort is currently under way for the development of such devices for application over a very wide range of fields and the requirements for their application to UV astronomy are particularly demanding. These requirements can be listed as follows:

(1) A spectral response covering most and preferably all of the wavelength range 900–3000 Å.

(2) A long storage time – from minutes up to several hours depending on the particular use.

(3) A low background to permit long integration times.

(4) A high spatial resolution.

(5) Good reproducibility.

(6) Compatability with spacecraft.

(7) Low weight.

(8) Low power consumption.

(9) Rugged design.

A review of the possible application of image storage tubes to UV astronomy has been made by Boksenberg (1965) and the discussion here will be limited to two devices which have either been flown or specifically assigned to a space experiment. The first is not an electric image tube in the full sense, but is an image intensifier of the electronographic (Lallemand) type flown by Carruthers in a pointing Aerobee to obtain objective grating stellar spectra (Carruthers, 1967). The tube (Carruthers, 1965) uses a KBr photocathode with magnetic focussing and an accelerating potential of 20 kV, the electron image being focussed to a nuclear emulsion. The second device, known as the Uvicon (Skorinko *et al.*, 1961) is being developed by Westinghouse Research Laboratories for the Smithsonian OAO experiment (Project Celescope) (Davis and Rustgi, 1962). The tube is sealed and the imaging and read-out sections are entirely electrostatic. The original tube has been modified to incorporate an SEC (secondary electron conduction) target (Boerio *et al.*, 1966) which is capable of storage times of ~ 24 hrs. (Doughty, 1966). Tubes with various spectral responses through the ultraviolet are obtained by suitable selection of window and photocathode.

References

Abeles, F. (1964)   *Proc. Conf. Photo. Spec. Optics* (Tokyo and Kyoto), p. 517.
Adams, J., Manley, B.W. (1967)   *Philips Tech. Rev.*, **28**, 156.
Alexander, J.D.H., Bowen, P.J., Heddle, D.W.O. (1964)   *Proc. Roy. Soc.*, A**279**, 510.
Barnes, W.P. (1966)   *Appl. Opt.*, **5**, 1883.
Boerio, A.H., Beyer, R.R., Goetze, G.W. (1966)   *Adv. Electronics Electron. Phys.*, **22A**, 229.
Boggess, A. (1961)   *Mem. Soc. Roy. Sci. Liège*, **4**, 459.
Boggess, A. (1964)   *IAU Symposium No. 23*, Liège, p. 173.
Boggess, A. (1967)   *IAU Symposium on Space Spectroscopy*, Dearborn.
Boksenberg, A. (1965)   *ESRO Memorandum SM-5*.
Bradley, D.J., Bates, B., Juulman, C.O.L., Majamdar, S. (1964)   *Nature*, **202**, 579.
Burton, W.M. (1964)   *Internal Culham Memorandum*.
Byram, E.T. *et al.* (1957)   *Astron. J.*, **62**, 9.
Byram, E.T., Chubb, T.A., Werner, W.M. (1964)   *IAU Symposium No. 23*, Liège, p. 177.
Canfield, L.R., Hass, G., Waylonis, J.E. (1966)   *Appl. Opt.*, **5**, 45.
Carruthers, G.R. (1965)   *Report to NRL Progress*, July.
Carruthers, G.R. (1967)   *IAU Symposium on Space Spectroscopy*, Dearborn.
Chubb, T.A., Byram, E.T. (1963)   *Astrophys. J.*, **138**, 617.
Courtes, G., Viton, M. (1967)   Paper 96, COSPAR Meeting, London.
Culham Laboratory (1966)   Design Study for Scientific Package in European Large Astronomical Satellite.
Davis, R.J., Rustgi, O.P. (1962)   *Appl. Opt.*, **1**, 131.
De Jager, C., Underhill, A.B., Emming, J.G., Hammerschag, A. (1967)   *Tech. Phys. Dienst. Report No. 609*, 502.
Dimov, N.A., Severny, A.B. (1965)   *Proc. Int. Astronautical Congress (Athens)*.
Doughty, D.D. (1966)   *Adv. Electronics Electron. Phys.*, **22A**, 261.
Dunkelman, L. (1962)   *J. quantit. Spectrosc. Radiat. Transfer*, **2**, 533.
Elliott Bros. (1967)   *Report T(F)68*.
ESRO (1967)   Project Development Plan for the Large Astronomical Satellite – A European Space Observatory.
Friedman, H. (1962)   *J. quantit. Spectrosc. Radiat. Transfer*, **2**, 547.
Golay, M., Gaide, A., Huguenin, D. (1967)   *IAU Symposium on Space Spectroscopy*, Dearborn.
Goodrich, G.W., Wiley, W.C. (1961)   *Proc. Image Intensifier Symp.* Fort Belvoir, Va., U.S.A.
Gulledge, I.S., Packer, D.M. (1963)   *Astron. J.*, **68**, 537.
Heath, D.F., Sacher, P.A. (1965)   *Goddard Report X-622-65-382*.
Henize, K., Wackerling, L. (1967)   *IAU Symposium on Space Spectroscopy*, Dearborn.
Heroux, L., Hinteregger, H.E. (1960)   *Rev. Sci. Instrum.*, **31**, 280.

Housiaux, L. (1965)   *Second Report Edinburgh/Liège Scanning Experiment, Inst. d'Astrophys.*,
    Liège.
Hunter, W. R. (1962)   *Optica Acta*, **9**, 255.
Hunter, W. R. (1964)   *Proc. Conf. Photo. Spec. Optics* (Tokyo and Kyoto), p. 520.
Johnson, W. C. (1964)   *Rev. Sci. Instrum.*, **35**, 1375.
Kurt, V. G., Syunyaev, R. A. (1967)   *JETP Letters*, **5**, 246.
Morton, D. C. (1967)   *Astrophys. J.*, **147**, 1017.
Morton, D. C., Spitzer, L. (1966)   *Astrophys. J.*, **144**, 1.
Muzjherin, V. M., Nikonov, V. B., Prokofiev, V. K., Chernykh, N. S. (1965)   *Proc. Int. Astro-
    nautical Congress* (Athens).
NASA (1967)   Document NHB8030.1A, Opportunities for Participation in Space Flight Investi-
    gations.
Owen, G. (1967)   *Sky and Telescope*, **34**, 293.
Perkin-Elmer (1967a)   Report No. 8346(I), Princeton Advanced Satellite Study.
Perkin-Elmer (1967b)   Report No. 8688, Princeton Advanced Satellite Study.
Reeves, E. M. (1968)   in the present volume, p. 75.
Rogerson, J. B. (1963)   *Space Sci. Rev.*, **2**, 621.
Rome, M. (1964)   *IEEE Trans. Nucl. Sc.* **NS-11**, 93.
Skorinko, G., Doughty, D. D., Feibelman, W. A. (1961)   *Westinghouse Res. Lab. Sci. Paper
    912-J902-01.*
Smith, A. M. (1967a)   *Astrophys. J.*, **147**, 158.
Smith, A. M. (1967b)   *Report on Commission 44*, IAU Meeting, Prague.
Space Science Board, Nat. Res. Council (1967)   U.S. Space Science Programme, *Report to
    COSPAR*, London.
Stecher, T. P. (1967)   *IAU Symposium on Space Spectroscopy*, Dearborn.
Stecher, T. P., Milligan, J. E. (1962)   *Astrophys. J.*, **136**, 1.
Steimetz, D. L., Phillips, W. G., Winick, M., Forbes, F. F. (1967)   *Appl. Opt.*, **6**, 1001.
Walker, W. C. (1964)   *Appl. Opt.*, **3**, 1457.

DISCUSSION

D. C. Morton: In discussing the possibility of putting optical instruments on the Moon, one fact that we should always take into consideration is the great disadvantage of the Moon's gravitational field with all the old *problems* of flexure.

R. Wilson: What you say is true, of course. It is also true that if you put a man into a large orbiting laboratory you introduce a certain increased cost, and also a certain increase in the complexity of operation. I'm not sure whether the kind of impartial analysis which would be required has really been made in the full sense of evaluating all these different techniques involving man, the Moon, sounding rockets, and unmanned satellites. It is a very difficult assessment to make.

L. Gratton: Could you mention the possibility of an automatic telescope visited infrequently by astronauts?

R. Wilson: Yes. There is the combination of approaches, using the astronaut as the repair agent for an unmanned experiment.

G. J. Odgers: What would be required to do diffraction limited stellar spectroscopy from above the Earth's atmosphere with a line stabilization of the order of hundredths of a second of arc? What are the engineering possibilities, and how does the stabilization compare with the best so far achieved?

R. Wilson: This particular proposal is an advanced one which is not entirely accepted by NASA. A study was carried out by the Perkin-Elmer Corporation for Princeton University in which they reached the conclusion that such stabilization was feasible. They proposed to have magnetic stabilization in which the instrument package is mechanically separated from the spacecraft. There are very attractive advantages to this system, because one of the great problems of an astronomical satellite is the experiment-spacecraft interface. If you can have this separation, then the thermal interface problems improve tremendously.

R. E. Danielson: I would like to add something on the achievement of one arc second pointing.

The biggest problem there is to sense the error since when you have a diffraction-limited telescope you have images which are a tenth of an arc second in diameter. The difficult problem is the thermal problem that you mentioned.

G.A. Newkirk, Jr.: I would like to comment on the problem of maintaining a fairly large aperture optical system in diffraction limit. There is a program now in progress at the Perkin-Elmer Corporation on 'active optics'. The mirror is either composed of several pieces, or one has a large mirror which is rather flexible. One senses the figure of this mirror by an auxiliary system which encircles the mirror and adjustments are made to maintain the appropriate figure. They have just begun and now have a working model for a telescope about 12 inches in diameter.

NEW TECHNIQUES IN X-RAY ASTRONOMY

C. DE JAGER

(University Observatory and Space Research Laboratory, Utrecht, The Netherlands)

ABSTRACT

A review is given of some modern developments in X-ray astronomy. Improved designs of proportional counters and other detectors are described. We further enumerate developments in solar
X-ray spectrographs and heliographs, like the 'gutter' spectrograph, and the paraboloidal-hyper-
boloidal frustrum mirror system, the zone plate technique. We also give a review of instruments and
collimating techniques for the detection of non-solar celestial X-ray sources.

1. Introduction; Optical and X-Ray Astronomy

Astrophysical observational techniques are essentially based on the three following
properties of astronomical information:

(a) The faintness of the photonic flux, which makes necessary the construction of
large light-collecting areas with a low background noise. Only the Sun with its large
flux is an exception.

(b) The smallness of the details, making necessary an angular resolution of the
order of 10^{-5} radians.

(c) The need for detailed spectral information, necessitating a spectral resolution
ranging between $\Delta\lambda/\lambda \approx 10^{-1}$ (broad-band observations, for object recognition and for
statistical information), or to $\approx 10^{-3}$ (detection of strong spectral lines; orienting
astrophysical investigations), or down to $\approx 10^{-5}$ (line profiles, detailed and refined
astrophysical interpretation).

These properties apply to radiation of any wavelength, also to X-rays. An essential
property of X-ray astronomical instruments is that refracting media can not be applied
in this wavelength range, so that all imaging or spectral resolution should be based on
reflection and/or diffraction. Furthermore, roughly speaking, *normal* reflection is
impossible for wavelengths shorter than about 500 Å, while virtually any reflection is
impossible for wavelengths below about 10 Å. This may easily be shown (see e.g.
Mayer, 1964): If i is the angle between the incident rays and the normal to the surface,
we have

$$\sin i = n = 1 - \delta,$$

where the real quantity δ may be approximated as

$$\delta \approx 2.7 \times 10^{10} Z\rho\lambda^2/A. \tag{1}$$

Perek (ed.), Highlights of Astronomy, 108–123. © I.A.U.

By defining the *Bragg angle* or *grazing incidence angle* $\Theta = 90 - i$, one obtains the critical Bragg angle Θ_c above which no reflection will occur, according to:

$$\sin(90 - \Theta_c) = n - 1 - \delta. \tag{2}$$

From Equations (1) and (2) one derives, with $Z/A \approx \frac{1}{2}$:

$$\Theta_c \approx (2\delta)^{1/2} \approx 1 \cdot 3 \times 10^5 \, \rho^{1/2} \lambda. \tag{3}$$

For $\rho = 8$ and $\lambda = 10$ Å one finds with Equation (3): $\Theta_c \approx 1°7$. For 1 Å: $\Theta_c \approx 0°2$.

The wavelength below which normal incidence is impossible may roughly be put at about 500 Å. Thus one may distinguish between two different regions in astronomical X-ray techniques:

EUV optics: $\lambda \gtrsim 500$ Å   normal incidence optics;

X-ray optics: $\begin{cases} 10 \lesssim \lambda \lesssim 500 \text{ Å} & \text{grazing incidence optics;} \\ \lambda \lesssim 10 \text{ Å} & \text{'optics' based on mechanical collimators.} \end{cases}$

In this review we shall successively discuss the following problems:

(a) the construction of large X-ray detectors with a low background noise;

(b) image forming techniques: (1) grazing incidence optics; (2) mechanical collimators;

(c) X-ray spectrographs;

(d) X-ray spectroheliographs.

This review should certainly not be considered as an exhaustive summary of X-ray instrumental techniques; we only wish to discuss *new* techniques and in doing this we prefer to emphasize some modern trends, rather than to give a complete summary of all developments nowadays underway.

2. Detector Developments

The kind of detectors used in X-ray astronomy depends on the energy range of the incident photons. Down to a wavelength of about 10 Å phototubes are still useful. The windowless *magnetic electron multipliers* and the *channeltrons* go even to below 10 Å. At high photon energies, from about 0·2 keV (60 Å) upward, proportional counters may also be used. For photonic energies between a few tens of keV's up to higher energies scintillation counters are the best kind of detectors. At very high energies (hundreds of MeV's) spark chambers and Čerenkov detectors come into the picture, but since this energy range deals with gamma- rather than X-ray astronomy it is outside the scope of the present discussion. Geiger tubes and ionisation chambers find application in a broad energy range. In this section we discuss only the magnetic electron multiplier, proportional and scintillation counters. We further refer to Mayer's (1964) and Boyd's (1965) review papers.

2.1. MAGNETIC ELECTRON MULTIPLIERS AND CHANNELTRONS

The most successfully used version of this type of open multiplier is the Bendix windowless photon and particle detector. It finds many applications in the EUV and the soft X-region. With tungsten cathodes their lack of sensitivity to visible and near ultraviolet radiation offers another advantage.

When a very small light-collecting area is needed the channeltrons are very useful.

2.2. PROPORTIONAL COUNTERS

These are based on photo-ionization of the gas molecules in the tube. If the photonic energy is E, and if a photon loses e electron-volts per ionization, one photon produces on the average E/e ion-pairs. Hence, the relative statistical fluctuation in the photon current is $\sqrt{e/E}$. Normally $e \approx 30$ eV; hence, the spectral resolution for 10 Å photons ($= 1$ keV) is about 0·2. Experimentally Culhane *et al.* (1966) found a fractional standard deviation of $0·167[E(\text{keV})]^{-1/2}$, in the energy range 1–30 keV.

Hence, proportional counters are not suitable for spectral discrimination at wavelengths longer than about 10 Å. However, by a careful choice of gas filling and filters Den Boggende at Utrecht is able to prepare a so-called multi-wavelength, narrowband, X-ray solar spectrometer, consisting of six proportional counters, each sensitive to two different, narrow wavelength bands in the 1–60 Å region. Discrimination between the two bands of each counter and combination of the results from all counters make possible the determination of the solar flux in a dozen of small bands 2 to 3 Å wide below 60 Å.

The use of proportional counters at wavelengths longer than 60 Å poses a number of serious problems, since here the filter windows have to be extremely thin, necessitating both supporting grids and gas flow. At energies above about 50–100 keV these counters lose their efficiency, because the gas filling becomes transparent for photons of such energies.

The essential difficulties in the construction of large proportional counters for the detection of weak fluxes of X-radiation are the wiring of the counters and the choice of the filters, which should prevent the gas from leaking away. If too much gas leaks away, so-called *flow-counters* should be used, in which a constant flow of gas replaces the losses. Furthermore, the counters should be mounted in an anti-coincidence circuit to reduce the effect of counting pulses due to cosmic radiation particles and other sources. A simple version of this kind for the detection of solar radiation below 3 Å was constructed by Culhane *et al.* (1966) and is shown in Figure 1a. This 'pill-box' type of counter consists of a flat cylindrical tube with a window area of the order of 10 cm². Two of these pill-boxes, one of which is windowless, are arranged back to back (Figure 1b) and circuited in anti-coincidence to reduce the influence of cosmic radiation pulses.

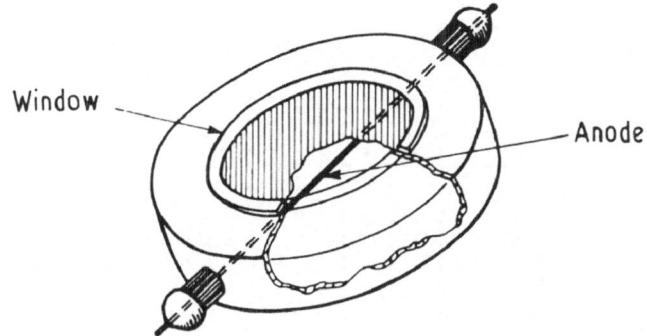

FIG. 1a.   *'Pill-box' counter developed by Culhane et al. (1966), diameter 6 cm; depth 1·2 cm.*

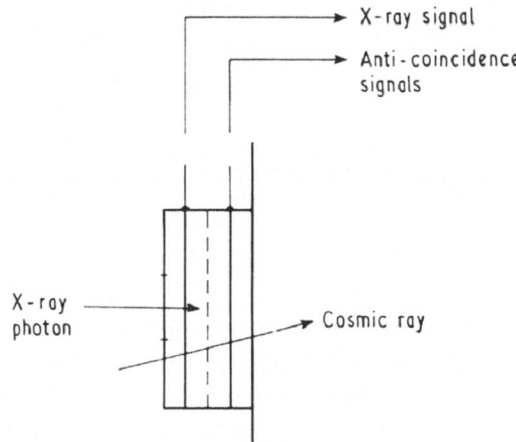

FIG. 1b.   *Operation of the double pill-box detector in anti-coincidence mode.*

Proportional counters with very large areas were constructed or are under investigation by various groups. Labeyrie in Saclay produces counters with Be windows having areas of about 200 cm². These areas can be made as large as this because the window has a considerable thickness. In other cases the areas should be made smaller, and a sufficient collecting area can then be obtained only by combining proportional counters. Thus Austin *et al.* (1967) at the American Science and Engineering Inc. are preparing the payload for an 'X-ray Explorer Satellite' which essentially consists of X-ray counters with dimensions of approximately $56 \times 5 \times 5$ cm³. Several of these are combined to form a sensitive area of 56×51 cm². In order to reduce the background noise, these counters are combined with so-called background counters with a volume of about $56 \times 5 \times 2\frac{1}{2}$ cm³; these are mounted to form a jacket around the X-ray counter, leaving only the X-ray windows exposed to the incoming radiation from the collimators in front of the X-ray detectors.

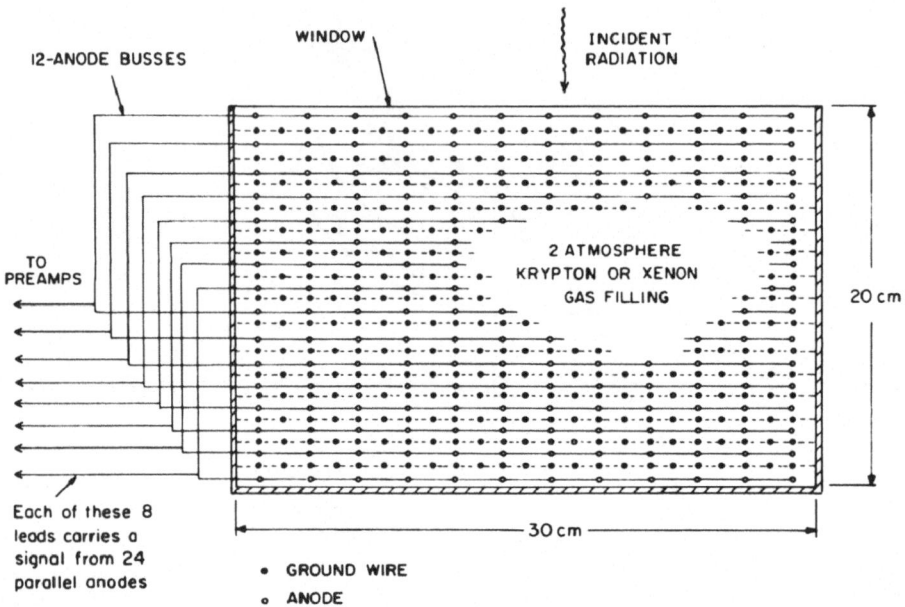

FIG. 2.   *Cross-section of the fluorescence-coincidence proportional counter developed by Stein and Lewin (1967).*

At M.I.T. an interesting detector is in preparation by Stein and Lewin (1967). Their 'multi-electrode fluorescence-coincidence gas proportional counter' is a vessel, 20 cm deep with a surface of 30×60 cm. It will be filled with krypton or xenon at a pressure of 2 atmospheres (Figure 2). A characteristic signature for the detection of a photon with energy above or near to the K-edge of the detector gas (krypton: 14 keV; xenon: 35 keV) can be obtained in the form of the detection of the coincident occurrence of 2 events in 2 different cells. One of these events is the photoelectric ejection of a K-electron by an incident X-ray, producing a pulse at one of the anodes. The other event is the capture of the fluorescent characteristic K X-ray photon of the detector gas itself, which may be emitted subsequent to the K photo process. Since the amplitude of the pulse resulting from this capture corresponds, within the limits of resolution of the counter, to the average K X-ray energy of the detector gas, the coincident detection of the two events constitutes a highly specific signature for a photon which has undergone a K photoelectric interaction. This signature is not easily forged by Compton electrons which result from interactions of high energy photons in the detector: these electrons are distributed nearly uniformly and isotropically in the detector gas, and produce pulses on many of the electrodes. In the energy range 15–80 keV this kind of counter should detect about 25% of the incident flux from a celestial X-ray source, while rejecting over 96% of the background resulting from Compton interactions in the detector.

2.3. SCINTILLATION DETECTORS

In the energy range above 20 to 50 keV, scintillation detectors are more efficient and are certainly a more reliable type of photon detector. Since photons of these energies penetrate rather deeply into the atmosphere, X-ray astronomy in this energy range may be performed by means of balloons. However, in that case sufficient precautions must be taken to eliminate the quite intense background of secondary X-rays resulting from cosmic ray debris in the atmosphere. With balloons, fairly long integration times are possible. An example of a simple but highly successful set-up is given in Figure 3, taken from Bleeker *et al.* (1967). The X-ray counter consists of a NaI (Tl) crystal, 76 mm in diameter by 5 mm thick. The crystal photomultiplier assembly is mounted in a cylindrical tin cup (2 mm thickness) which provides an effective shielding against background X-rays in the energy range concerned. The collimator consists of two coaxial cones of tin (2 mm thick) with a half angle at the top of 18°, which corresponds to the zenith angle of Cyg XR·1 at its meridian passage at the latitude of the observer (52 °N). The inner surface of the cup and the collimator is covered with copper foil (0·3 mm thick) in order to absorb K X-rays which are possibly generated in the tin. The maximum effective area seen by the source is 26 cm². A plastic scintillator guard counter surrounds the detector and the collimator for the rejection of background events associated with charged particles. A rotating tin disk (3 mm thick) in

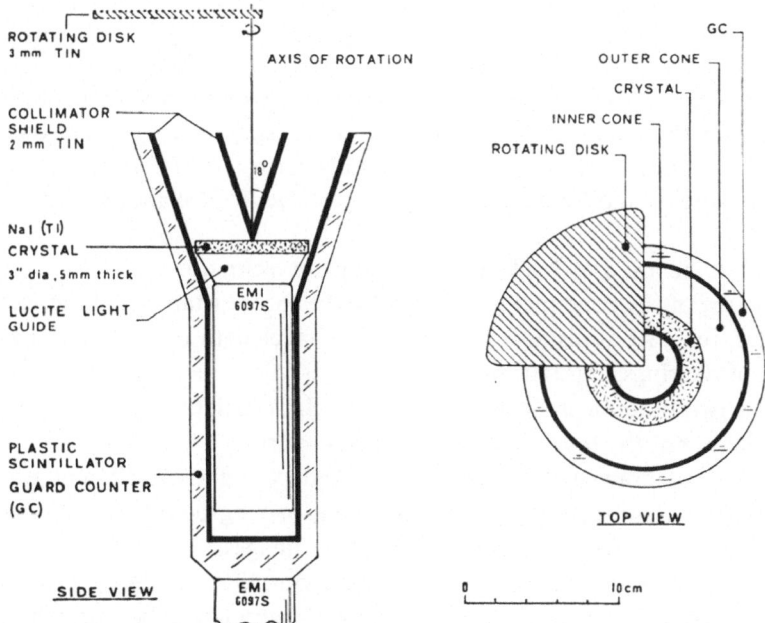

FIG. 3. *Configuration of the X-ray telescope used by the Leiden Group for the measurements of the hard X-ray spectrum of Cygnus XR-1 (Bleeker et al., 1967).*

front of the detector covers a part of the opening angle and revolves around the axis of the collimator. The period of rotating is about 100 sec. Therefore, the disk periodically masks the X-ray source during the time it is in the field of view. Similar detectors with all kinds of minor variations have flown or are going to be flown in rockets and satellites.

3. Image-Forming Techniques

For wavelengths longer than a few Angströms imaging instruments can be based on glancing incidence reflecting or on diffraction. For shorter wavelengths this is impossible, as explained in Section 1, and mechanical collimators must be applied. In Section 3·1 we discuss the Fresnel zone plate technique; in Section 3·2 grazing incidence techniques, and in Section 3·3 mechanical collimators.

3.1. THE FRESNEL-SORET ZONE PLATE

In this device convergence of the incident rays is produced by diffraction at a series of concentric circular diaphragms. If r_1 is the diameter of the inner zone, the diameter of the mth zone is:

$$r_m = r_1^2 m^{1/2}.$$

Furthermore, the focal distance of the first order image is:

$$f = r^{1/2}_1/\lambda.$$

With the Fresnel-Soret zone plate a theoretical resolution should be attainable, comparable to the diameter of the outermost transparent ring. High-quality zone plates are now produced at the Physics Laboratory at Tübingen University by Möllenstedt and at the Technical University of Delft, Holland, by Le Poole; see Figure 4.

However, there are drawbacks to the zone plate technique:

(1) The area of the zone plate is limited by the focal distance and the wavelength; for instance, for a focal distance of 1 m and a wavelength of 50 Å and a zone plate consisting of 100 rings, r_1 is 0·07 mm and the largest diameter r is 0·7 mm. The distance between the two outermost rings is 3·5 micron. So, the light gathering power of this device is small, but the resolution would be 3·5 micron.

(2) The second drawback is the chromatic effect: light of different wavelengths converges at different distances; a careful predispersing or filtering of the incident light is necessary to produce chromatically purer images.

(3) Finally, a considerable part of the light, about 30%, passes as zero-order light, giving an appreciable background haze. This difficulty could be overcome if means could be found to subtract the zero-order image from the first order one by using a suitable pinhole (Einighammer, 1966).

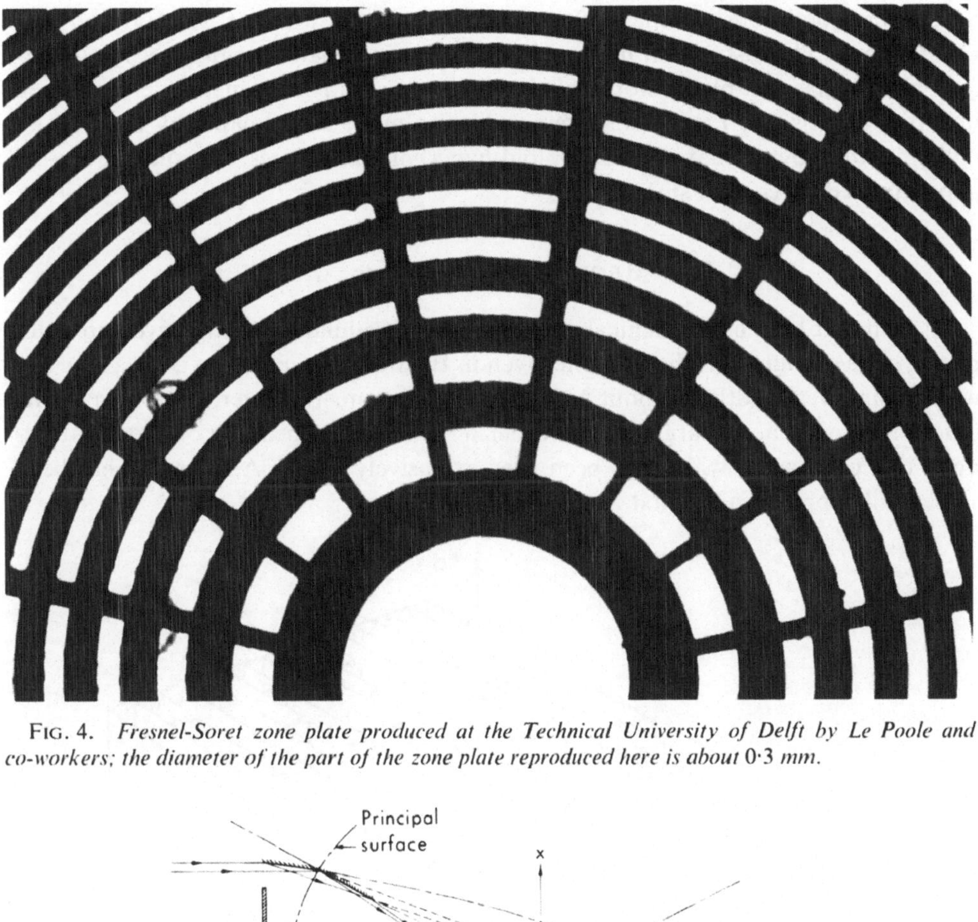

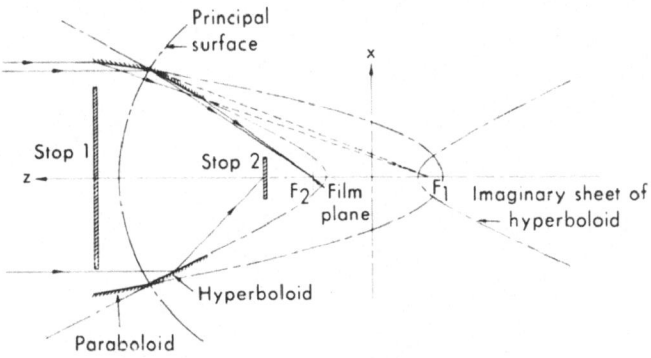

FIG. 4. *Fresnel-Soret zone plate produced at the Technical University of Delft by Le Poole and co-workers; the diameter of the part of the zone plate reproduced here is about* 0·3 *mm.*

FIG. 5. *Paraboloid-hyperboloid glancing-incidence solar X-ray telescope used by Underwood and Muney (1967).*

3.2. GLANCING INCIDENCE OPTICS

Considerable advance in the development of glancing incidence optics has been made by the American Science and Engineering Group (Giacconi *et al.*, 1965*a, b*) and later by the Goddard Space Flight Center Group with Underwood and Muney (1967). With a view to the smallness of the angles of incidence, extreme care must be taken to have a very good polish of the surface, a problem that now seems well solved by the GSFC-group. The instrument makes use of a combined paraboloid-hyperboloid surface while at relevant places stops are introduced to avoid the vignetting encountered in earlier models (see Figure 5). Reference is also made to the paper by Reidy *et al.* (1968), who showed that the paraboloid-hyperboloid surfaces can successfully be approximated by conical surfaces.

3.3. Mechanical collimators

The simplest kind of mechanical collimator is the cellular or the honeycomb collimator, see e.g. Figure 6. The example given in Figure 7 collimates on a slit at the sky with a narrowest width of about $1°5$ at half maximum. However, the width in the other direction is much larger, of the order of 10 to 15°, depending on the construction. This collimator, which has been used extensively by the ASE-group, has been successfully applied in the first X-ray scans of the sky.

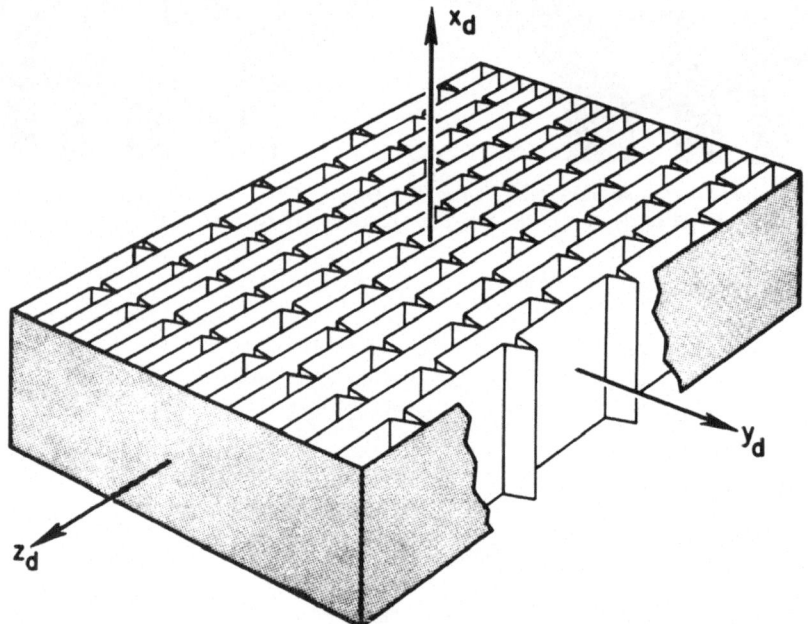

FIG. 6.   *Schematic drawing of a cellular collimator, after Rossi (1966).*

A more sophisticated type of collimator that also scans the sky in a slit mode is the modulation collimator, also called the Oda collimator, after its designer. The principle is illustrated in Figure 7. It consists essentially of two plane grids placed one in front of the other at a suitable distance. The diameter of the wires is nearly equal to the spacing between them. In a parallel beam of rays (left-hand figure) the front grid casts a sharp

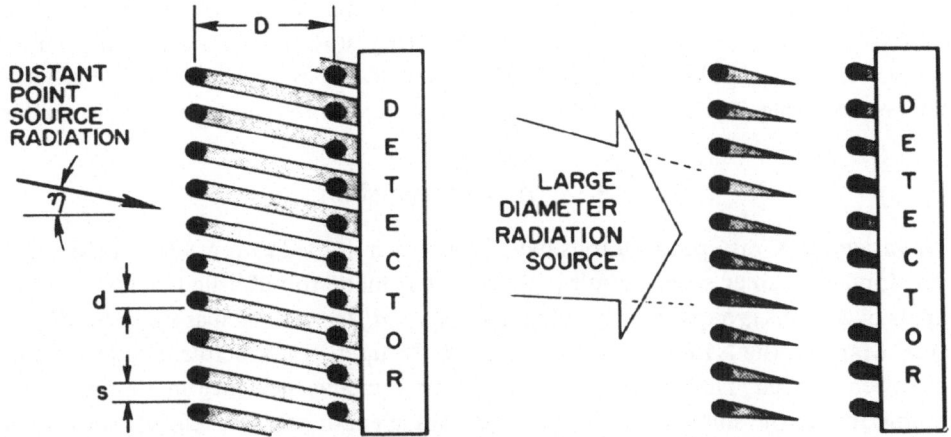

FIG. 7. *Illustrating the principle of operation of the modulation collimator. The drawing on the left indicates the shadowing that obtains with parallel radiation; at the right-hand side the situation is illustrated in the case of a large diameter radiation source. After B. Rossi (1966).*

shadow on the rear grid. The shadow shifts as the orientation of the collimator relative to that of the incident beam changes. The transmission of the collimator changes correspondingly, being a maximum when the shadow of the front wires falls exactly on the backwires and a minimum when it is centered half way between adjacent wires. For a source with a large diameter the illumination of the detector remains always the same (right-hand part of Figure 7). Hence, this collimator is very well suited for determining diameters of cosmic X-sources; collimators with a resolving power finer than 1' are feasible (see e.g. Rossi, 1966).

Of course the system can be extended and brought to a greater degree of refinement: a number of the wire units of the Oda collimator can be placed behind each other in order to have a better resolution in determining the angular size and the location of celestial X-ray sources; this technique has been used successfully by the ASE-MIT-group, who could reduce the angular resolution in one direction to some 10". Further experiments of this kind using stellar pointing rockets have been flown by the ASE-group and are planned for further flights (Gursky et al., 1967).

A further step forward may be introduced by the technique of the sieve-plates in development at the Utrecht Space Research Laboratory, after a suggestion of Le Poole from Delft. The sieve-plates are identical plates containing holes with diameters to be chosen at will from about 20 micron to about 200 micron. The holes are placed in a rectangular pattern at mutual distances depending on their sizes but ranging between 30 and 300 micron. By placing a series of these sieve plates behind each other at well chosen distances a very high collimating accuracy may be attained. The Utrecht Space Research Laboratory is planning to apply this technique in a solar X-ray limb-scanning spectrograph to be flown in the ESRO-TD2 satellite (Heynekamp), in a

118   C. DE JAGER

rocket solar X-ray high-dispersion spectrophotometer (Dijkstra) and in a rocket
celestial X-ray telescope (Heynekamp). Laboratory studies indicate that a pointing
accuracy and an angular resolution of the order of 20″ or perhaps somewhat better
may be attainable. Technically difficult is the problem of the alignment of the plates.

4. X-ray Spectrographs

A variety of X-ray spectrographs have been flown in the last few years; all of these
were applied to solar spectrography. Reference is made to published or forthcoming
papers by various groups, mostly in the *Astrophysical Journal*, such as those by Hinteregger, Manson, the NLR-group, Pounds *et al.*, Neupert. In principle, two techniques
have been applied. Diffraction gratings used in glancing incidence mode now allow
photographing the solar spectrum down to a wavelength of a few tens of Angströms.
At shorter wavelengths spectral resolution may be obtained by using the well-known
property of crystals that the angles of incidence and of emergence are equal for a beam

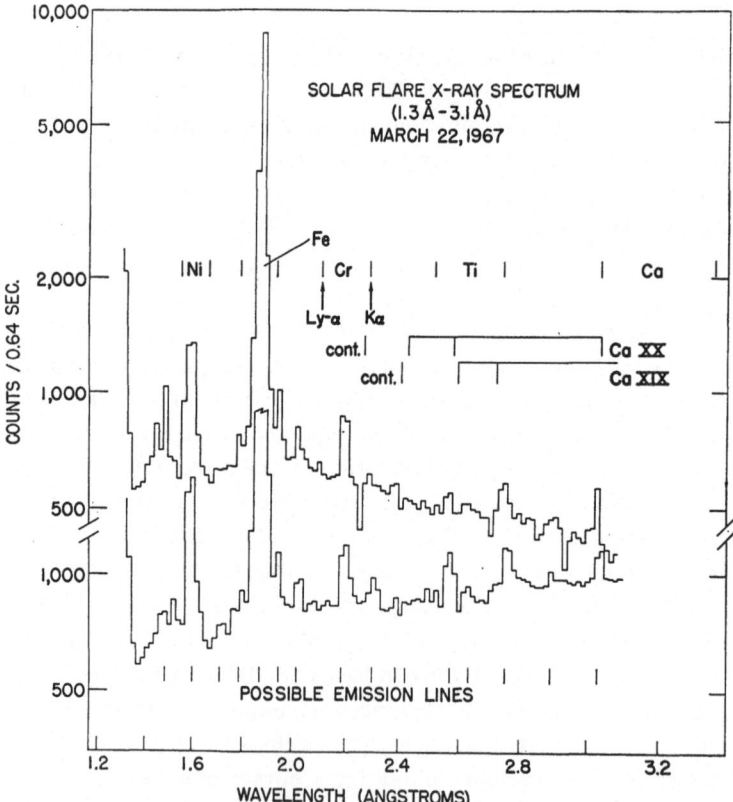

Fig. 8.   *The solar spectrum between 6·3 and 20 Å observed by the OSO-III instrument during a flare
on March 22, 1967, and on previous days when no flare was in progress. After Neupert et al. (1967).*

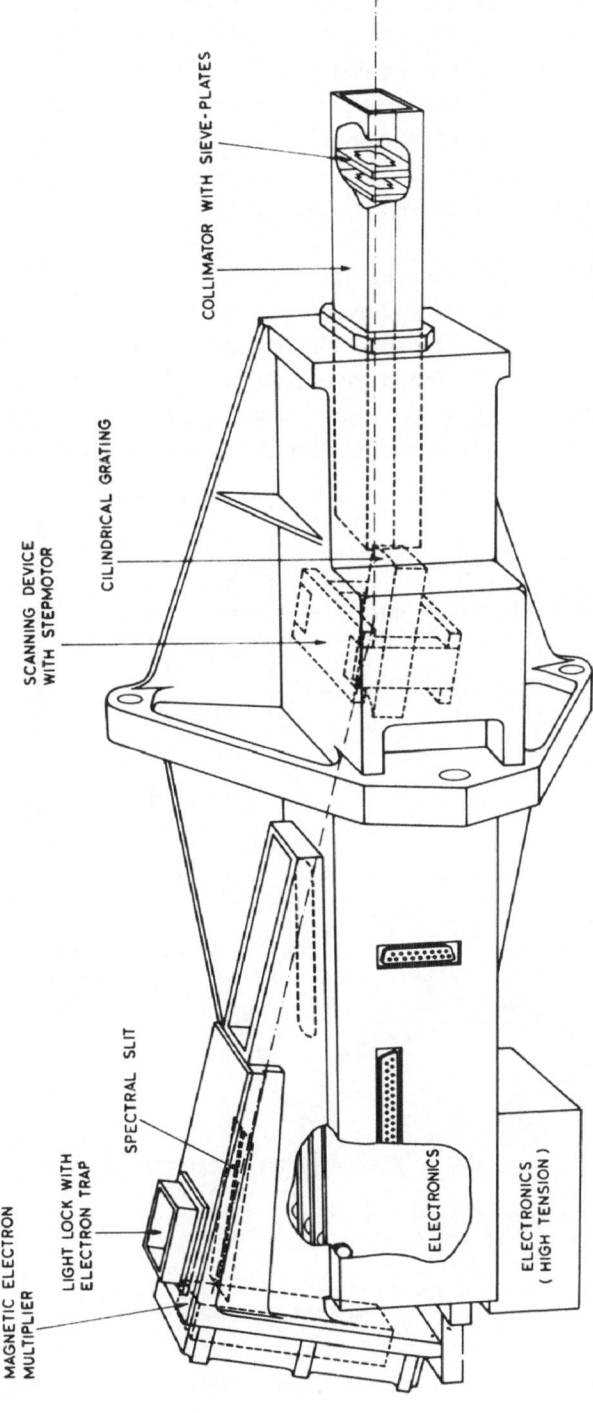

FIG. 9. *Schematical design of the 'gutter' solar spectrophotometer after Werner and Dijkstra.*

of parallel light and that a simple relation exists between the wavelength and this angle. Hence, a crystal acts as a very sharp monochromator in the X-ray region. By turning the crystal with reference to the incident light beam a spectral scanning is obtained. This technique has been applied mostly for wavelengths between about 5 and 30 Å, but has recently been extended by Neupert *et al.* (1967) down to 1·3 Å (Figure 8). In that region, and particularly at shorter wavelengths, spectral discrimination may be further obtained by pulse-height discrimination with proportional counters or scintillation counters.

A new type of solar X-ray spectrograph, under development, is the 'gutter' spectrograph, developed at the Technical University of Delft and the Utrecht Space Research Laboratory by Werner and Dijkstra. The 'gutter' spectrograph makes use of a cylindrical concave grating with the grooves parallel to the cylinder axis. The grating is used in glancing incidence, but in a quasi-normal mode. As shown in Figure 9, solar radiation is collimated by a sieve-plate collimator in such a way that the plane through

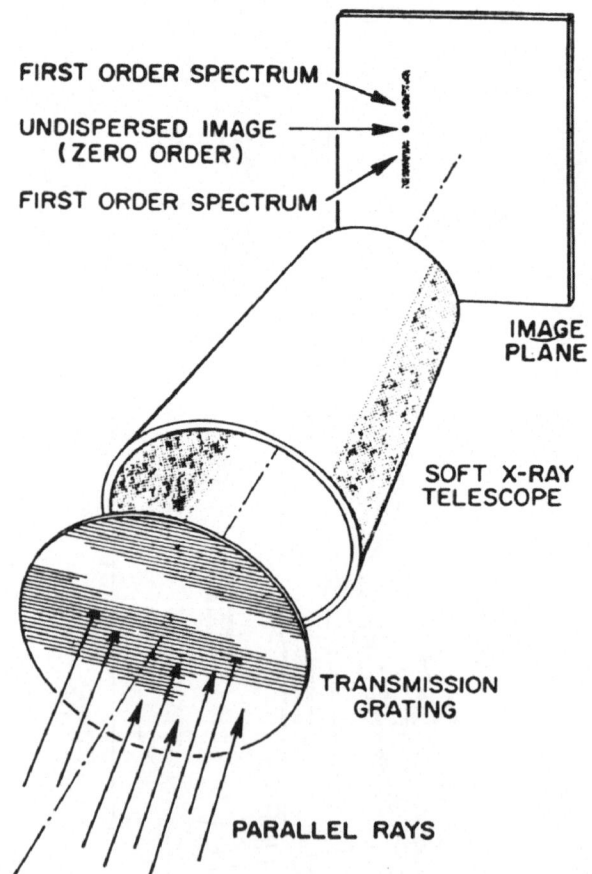

FIRST ORDER SPECTRUM

UNDISPERSED IMAGE
(ZERO ORDER)

FIRST ORDER SPECTRUM

IMAGE
PLANE

SOFT X-RAY
TELESCOPE

TRANSMISSION
GRATING

PARALLEL RAYS

FIG. 10.   *Principle of the soft X-ray slitless spectrometer of Gursky and Zehnpfennig (1966).*

the axis of the grating and the vector representing the direction of the incident radiation are *normal* to the grating. The spectrum then is formed at some distance from the grating above it where the relevant part of the spectrum is selected by the exit slit. Spectral scanning is obtained by slowly turning the grating around an axis coincident wiht the central groove.

A proposal for an image-forming slitless spectrometer for soft X-ray astronomy has been published by Gursky and Zehnpfennig (1966) and worked out by Zehnpfennig (1966). The set-up is very much like the objective grating as used in classical astrophysics, and consists of the combination of a soft X-ray transmission grating and an image-forming soft X-ray telescope, the grating being positioned in front of the entrance aperture of the telescope in order to disperse a portion of the incident radiation before it is collected by the telescope. See Figure 10.

An essential difficulty was the production of the transmission grating. The required structure was a picket-fence type of grating consisting of parallel strips of material of low transmission to the wavelength range of interest, separated by strips of relatively high transmission material. A grating constant of 10^{-4} cm or smaller was required to achieve reasonable dispersion in the first-order spectrum. The applied process (Zehnpfennig, 1966) consists of evaporating a metal such as gold on a thin plastic replica grating at near grazing incidence such that the metal is deposited preferentially in certain regions of each groove. See Figure 11.

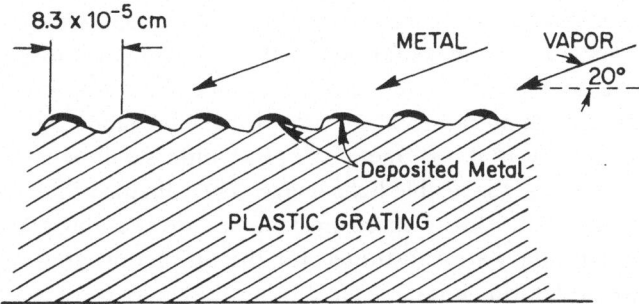

FIG. 11.   *Fabrication of a soft X-ray transmission grating after Zehnpfennig (1966); metal is vapor deposited on a thin replica grating at near grazing incidence.*

The X-ray telescope used in this spectrometer has a focal length of 84 cm, a diameter of 7·6 cm, a collecting area of 2 cm² and a nominal angular resolution of 1 min of arc. Laboratory spectra obtained with this instrument show the zero-order central image of the source and spectral lines with a width at half height of the order of about 0·5 Å. The instrument should be applicable to point sources such as the cosmic X-ray sources. However, a very long integration time and the photographic technique are needed.

Perhaps a large proportional counter aboard a stellar pointing rocket would yield the same result in a shorter interval of time.

5. X-ray Spectroheliograph

We briefly mention the ASE X-ray spectroheliograph, in preparation for the Apollo Telescope Mount (Giacconi, 1967). The instrument should be capable of the following achievements during a 56-day mission:

(a) To obtain images of X-ray flare events with a spatial resolution of the order of a few seconds of arc.

(b) To simultaneously record flare spectra over the range of 2 to 10 Å with a spectral resolution of a fraction of an Å. These observations should be continued during the development of the event and correlated with ground observations. The instrument should be able to obtain solar images with fine spatial resolution during non-flare conditions at selected wavelength intervals. In the Apollo Telescope Mount the instrument is man-operated for coarse pointing while fine pointing is done automatically. The instrument has a length of 280 cm and consists of two concentric, con-focal optical elements. The larger of the two elements has an approximate inside diameter of 31 cm; the smaller element has an inside diameter of about 23 cm. The axial length of these elementsi s 35 cm. There is another, smaller mirror with a focal length of 93 cm and an inside diameter of 8 cm. It is 15 cm long. The images formed by the two large mirrors are recorded by the camera; in addition there is an objective for visible light which also produces an image for the exact location of the features. Also this image is recorded by the camera. The image formed by the small mirror is displayed electronically on a cathode ray tube. The camera will contain 300 m of film of 70 mm format. The camera will operate in either automatic or manual mode; the film magazine should be replaced every 14 days. One of the two shutters will be used for the X-ray image and the second for the visible light image. The camera will record on film the undispersed image and the X-ray spectrum focussed on the film plane by the large X-ray mirror. When filters are substituted for the diffraction grating by controlled action of the astronaut during non-flare conditions, the camera will record broad-band images of the Sun in selected regions of the X-ray spectrum. Simultaneously, on the same film visible light images from the achromatic lens will also be recorded. These images will be used to determine the orientation on the Sun of the X-ray images. In addition, during active periods, there will be visible features such as sunspots which will determine roll orientation.

Acknowledgement

Gratefully I acknowledge the help of many friends and colleagues at the Utrecht Space Research Laboratory and abroad, for commenting on a first draft of the manuscript and for sending me their preprints in advance of publication.

References

Austin, G., Giacconi, R., Gursky, H., Kellogg, E., Kutz, M., Payne, P., Sinnamon, G., Tietsch, R., Waldron, J., Waters, J. (1967)   *Amer. Sci. Engineering* no. **1567**.

Bleeker, J.A.M., Burger, J.J., Deerenberg, A.J.M., Scheepmaker, A., Swanenburg, B.N., Tanaka, Y. (1967)   *Astrophys. J.*, **147**, 391.

Boyd, R.L.F. (1965)   *Space Sci. Rev.*, **4**, 35.

Culhane, J.L., Herring, J., Sanford, P.W., O'Shea, G., Phillips, R.D. (1966)   *J. Sci. Instrum.*, **43**, 908.

Einighammer, H.J. (1966)   *Naturwiss.*, **53**, 272.

Giacconi, R., Harmon, N.F., Lacey, R.F., Szilagnyi, Z. (1965*a*)   *J. Opt. Soc. Amer.*, **55**, 345.

Giacconi, R., Reidy, W.P., Zehnpfennig, T., Lindsay, J.C., Muney, W.S. (1965*b*)   *Astrophys. J.*, **142**, 1274.

Giacconi, R. (1967)   *Experimental Interface Requirements Document for the Apollo Telescope Mount, Experiment S-054 for X-ray Spectrographic Telescope*, ASE 1607.

Gursky, H., Zehnpfennig, T. (1966)   *Appl. Opt.*, **5**, 875.

Gursky, H., Giacconi, R., Gorenstein, R., Manko, H., Waters, J.R. (1967)   *A Program of High Angular Resolution Studies of Celestial X-Ray Sources*, NASA C-R **752**.

Mayer, U. (1964)   *Space Sci. Rev.*, **3**, 781.

Neupert, W.M., Gates, W., Swartz, M., Young, R. (1967)   *Astrophys. J.*, **149**, L 79.

Reidy, W.P., Vaiana, G.S., Zehnpfennig, T., Giacconi, R. (1968)   *Astrophys. J.* (in press).

Rossi, B. (1966)   in *Perspectives in Modern Physics*. Ed. by R.E. Mashale, Interscience Publ., New York, p. 383.

Stein, J.A., Lewin, W.H.G. (1967)   *Rev. of Sc. Instrum.* (in press).

Underwood, J.H., Muney, W.S. (1967)   *Solar Phys.*, **1**, 129.

Zehnpfennig, T. (1966)   *Appl. Opt.*, **5**, 1855.

DISCUSSION

R. Wilson: I want to inquire about the 'gutter' grating spectrograph. What is the spectral range that can be covered, and what is the spectral resolution?

C. de Jager: In our first flight we would like to cover the spectral range between 100 and 200 Å, with a spectral resolution which we hope to be approximately 0·3 Å or perhaps 0·5 Å. This grating will have 1200 lines per millimeter. In a next flight we want to cover the range between 40 and 65 Å.

SOLAR AND INTERPLANETARY MAGNETIC FIELDS AND PLASMAS*

JOHN M. WILCOX

*(Space Sciences Laboratory, University of California,
Berkeley, Calif., U.S.A.)*

ABSTRACT

A few recent observations of interplanetary fields and plasmas are discussed, including the evolution over several years of the longitudinal sector pattern, the corotating filamentary structure that exists within the sectors, guiding of solar cosmic rays along such magnetic filaments, a field-aligned thermal anisotropy of the plasma, and a small component of corotating plasma velocity observed at 1 AU. The solar wind plasma streaming past spacecraft near 1 AU contains a large amount of information about the detailed structure of the Sun.

1. Introduction

The use of satellites and space probes in the last few years to make *in situ* observations of interplanetary magnetic fields and plasmas has produced significant new information about the extended solar corona. Hydrodynamic expansion of the corona with a concomitant extension of the solar magnetic field was discussed theoretically by Parker (1963), and verified in 1962 by the Mariner-2 spacecraft on its flight to Venus. The velocity of the expanding solar plasma (solar wind) is approximately radial with a magnitude near 1 AU of 350–400 km/sec in quiet times, increasing to 600–700 km/sec in disturbed intervals. The flux is about 10^9 ions/cm^2/sec, the principal positive ionic component being protons with an energy of about 1 keV. A varying admixture of alpha particles is present with the average concentration about 5%.

The combination of radial plasma velocity and solar rotation stretches the interplanetary magnetic field into an Archimedes spiral making an angle near the Earth of about 45° with the Earth–Sun direction. The field near the Earth is approximately parallel to the ecliptic, and its average magnitude is about 6γ (6×10^{-5} gauss), with a range from 2γ to about 40γ, the larger values usually occurring during flare-associated disturbances.

2. Evolution of Sector Structure

This review of new observations will begin with a large-scale property of the interplanetary medium, and then proceed to smaller-scale structures. In accord with the

* This paper was prepared while the author was a guest at the Division of Plasma Physics, The Royal Institute of Technology, Stockholm, Sweden.

purpose of this symposium only a few recent observations are discussed; a complete survey is not attempted (for a recent review see Ness, 1967). An unexpected property of the interplanetary magnetic field is shown in Figure 1, which gives the predominant polarity (toward or away from the Sun) of the field observed by IMP-1 during the winter of 1963–64. For several consecutive days the field polarity remains unidirectional. The sector pattern thus defined corotates with the Sun; i.e., the pattern tends to repeat itself every 27 days. Considerable evidence (Wilcox and Ness, 1965) suggests that each sector so defined was a coherent entity, with plasma velocity and field magnitude reaching maxima in the preceding portion and declining in the following portion. Observations from Mariner-2 had previously shown a pattern of recurring maxima in

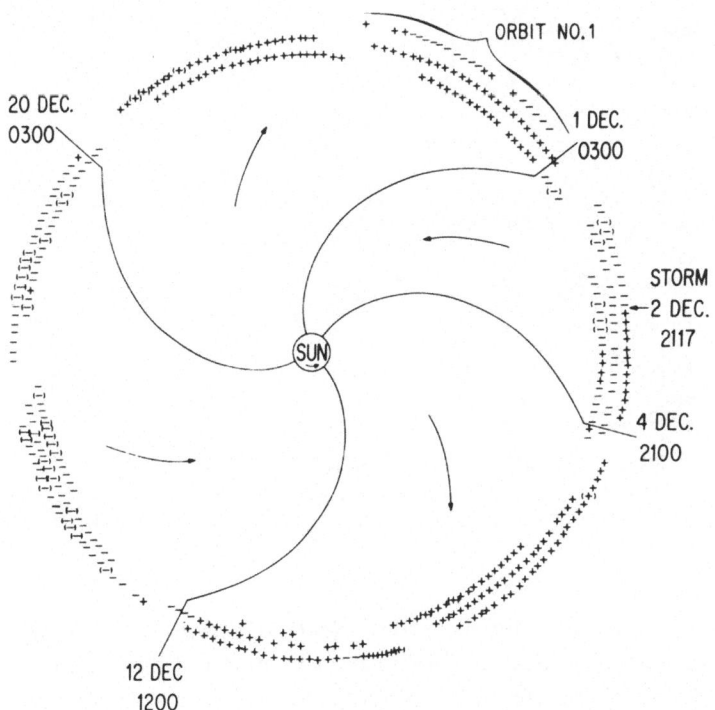

FIG. 1.   *Longitudinal sector structure in the interplanetary medium observed by IMP-1 in 1963–64* *(Wilcox and Ness, 1965).*

the solar wind velocity (see Figure 2); the field at that time was divided into two sectors per solar rotation. It should be emphasized that it is the *pattern* of field and plasma structures that tends to corotate with the Sun; the actual motion of the plasma is approximately in the radial direction.

The evolution of the interplanetary sector pattern during several years is shown in Figure 3. The Mariner-2 and IMP-1 patterns previously mentioned can be seen, as well

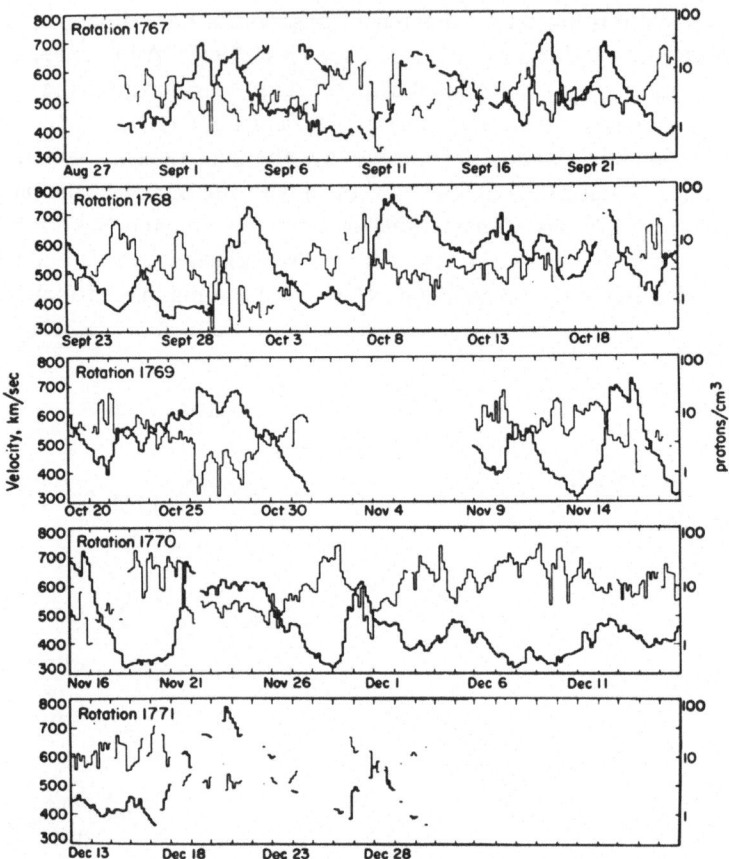

FIG. 2. *Solar wind plasma velocity and proton-number density observed by Mariner-2 in 1962; plotted on a calendar of Bartels 27-day solar rotations to show recurrence properties (Neugebauer and Snyder, 1966).*

as observations by several later spacecraft. Near the end of 1964 IMP-2 observed almost the same pattern as that of IMP-1, and the interpolation indicated in Figure 3 is based on this fact, plus the observed sequences of recurring geomagnetic activity. It is interesting that independent cosmic-ray observations by the Vela satellites (Asbridge *et al.*, 1967) in the middle of 1964 also led to the conclusion that the IMP-1 sector pattern was essentially unchanged at that time.

   The sector pattern shown in Figure 3 appears to be dominated by the old sunspot cycle up to the end of 1964 and the beginning of 1965, and by the new cycle thereafter. Characteristics of the old cycle pattern include a quasi-stationary structure for at least 1 year, and a recurrence interval of 27 days. Comparison of this pattern with the direction of photospheric magnetic fields has suggested that the average heliographic latitude of the interplanetary field observed near the Earth was about 15°N (Wilcox

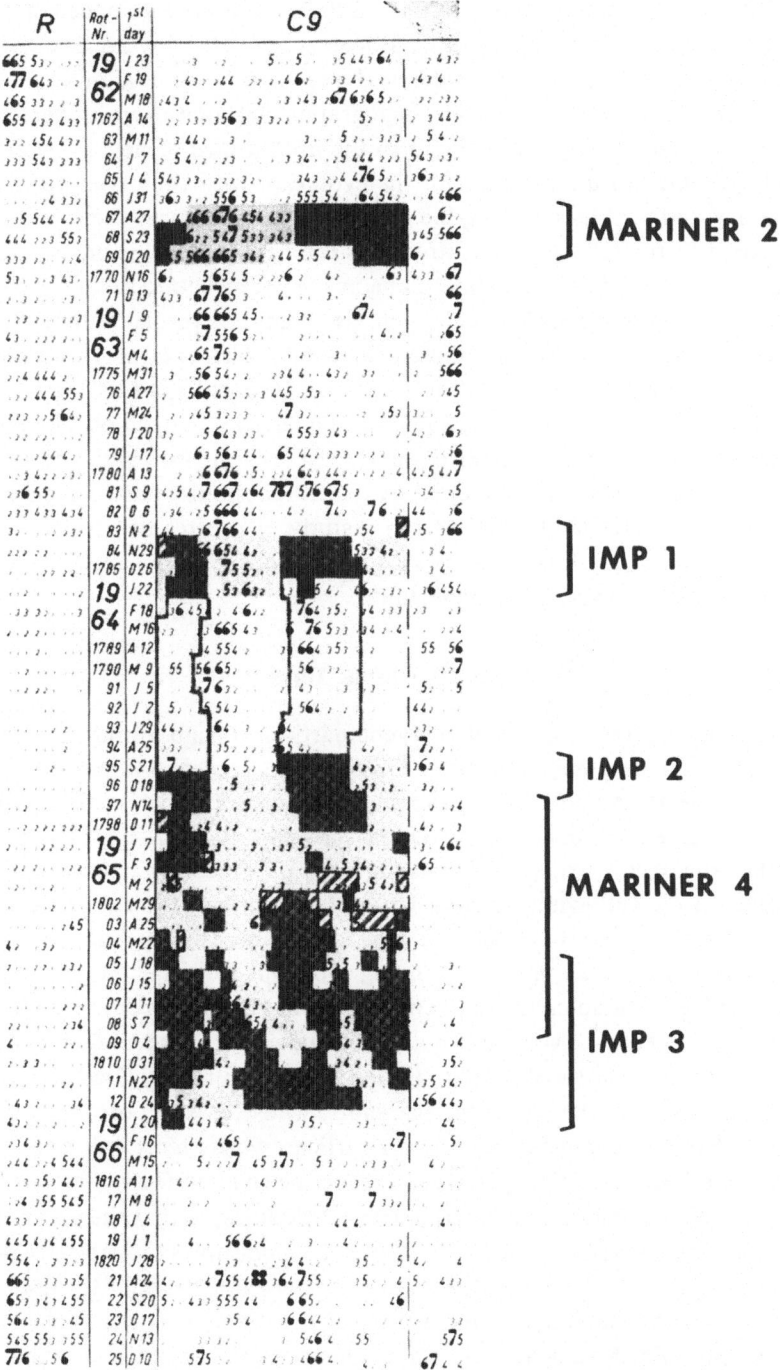

FIG. 3. *Evolution of the interplanetary sector structure from 1962 to 1966, overlayed on a 27-day calendar of geomagnetic activity figure C9. Light shading is field away from Sun, dark shading is toward, and cross-hatching indicates occasional periods of mixed polarity (Ness and Wilcox, 1967).*

and Ness, 1967). This is presumably related to the overall predominance of Northern hemispheric solar activity at this time. Characteristics of the new cycle sector pattern include a more rapid evolution with time, and a recurrence interval of about 28 days. This probably indicates a source at a heliographic latitude even higher that than discussed above, which might be related to the appearance of new cycle activity at higher heliographic latitudes. Although the evolution with time is more rapid, it is interesting that so far at least during the increase of solar activity the sector property has persisted; i.e., for several consecutive days the field polarity is unidirectional.

The significance of the observed interplanetary sector structure as a possible indicator of a fundamental large-scale solar organization cannot yet be evaluated. It seems possible that although individual photospheric field elements at various latitudes respond to the differential rotation, as would be expected (and as they have been observed to do by Wilcox and Ness, 1966), an underlying pattern of the appearance of photospheric magnetic regions exists for at least several years and is related to the interplanetary pattern shown in Figure 3. A related possibility has been suggested by Dodson-Prince and Hedeman (1967) from a study of centres of activity; they also suggest that the tendency for active regions to develop on opposite sides of the Sun approximately 180° away may be significant.

3. Solar Cosmic Rays

Several spacecraft observations of energetic particles have been interpreted in relation to the interplanetary magnetic fields. Observations shown in Figure 4 by Simpson and colleagues on IMP-1 showed that a recurring stream of protons of about 1 MeV energy was contained within one of the sectors shown in Figure 1. The proton intensity followed the usual sector structure of a maximum in the preceding portion of the sector and a decline in the following portion. Understanding the physical processes which have populated this one sector with an appreciable intensity of recurring energetic protons is surely one of the most challenging problems for solar physics.

Our knowledge of the so-called solar cosmic rays has been considerably increased by the elegant experiments of McCracken and colleagues on Pioneers-6 and -7 in 1966. Directional properties of the cosmic radiation have been investigated utilizing a detector that rotates with the spinning spacecraft, whose axis is perpendicular to the ecliptic. The detector responds primarily to particles whose motion is approximately parallel to the ecliptic. The first particles to reach the spacecraft from a solar flare are often very much guided along the interplanetary field lines, so that the observed distribution is quite anisotropic, as shown in Figure 5, which shows the intensity of cosmic rays in the energy interval 7·5–45 MeV observed after three flares. If the flare position is near the middle of the Western hemisphere of the visible disk the particles tend to have a rapid Sun–Earth transit. If the flare is in the Eastern hemisphere the particles tend to arrive after a considerably longer delay. McCracken *et al.* (1967) interpret this

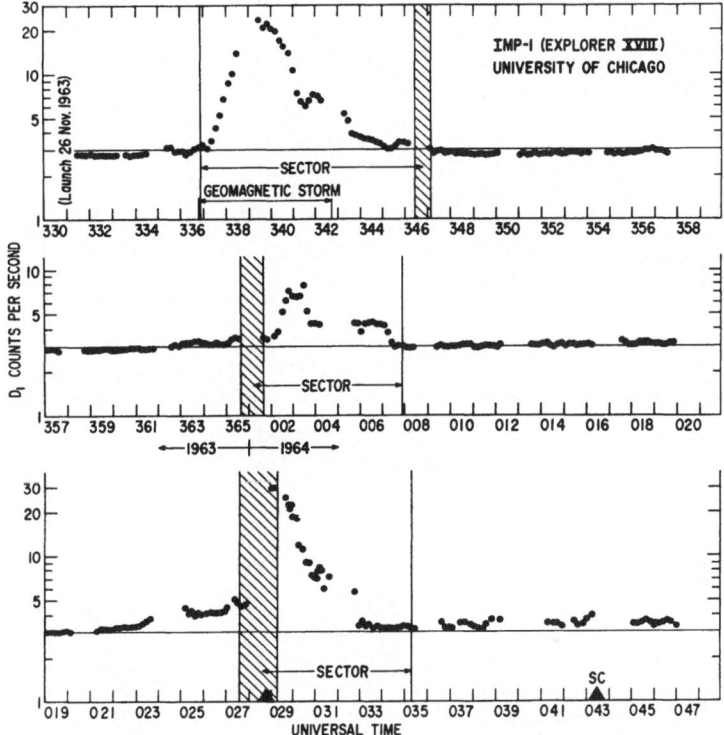

FIG. 4.   *Recurrent streams of 1 MeV protons observed by Simpson on IMP-1 tend to be contained within a single sector of the structure shown in Figure 1. Three 27-day solar rotations are shown (Wilcox and Ness, 1965).*

as being caused by longitudinal diffusion of the flare particles close to the Sun, until they reach the magnetic lines that connect to the spacecraft and propagate rapidly along them. The mean free path for scattering in the extended interplanetary medium is estimated at 1 AU by McCracken *et al.* (1967), and they suggest that earlier discussions of diffusion in this medium may need to be reexamined.

The guiding of solar cosmic rays by the interplanetary magnetic field has been demonstrated by McCracken and Ness (1966), as shown in Figure 6, which compares Pioneer-6 observations of the field direction with the direction of cosmic-ray anisotropy. The field direction observed during 4 hours is shown as the continuous line, while the superposed arrows indicate the direction of cosmic-ray anisotropy (the direction from which the maximum flux is incident on the detector). The guiding of the cosmic rays around several sharp bends in the field is obvious. It is often observed that when the field direction has an abrupt change, the cosmic-ray flux may abruptly increase or decrease. This leads to the concept of discrete filaments of magnetic flux that extend from regions very close to the Sun out to 1 AU (and beyond). Simultaneous observations from two spacecraft at different longitudes by Ness (1966) have

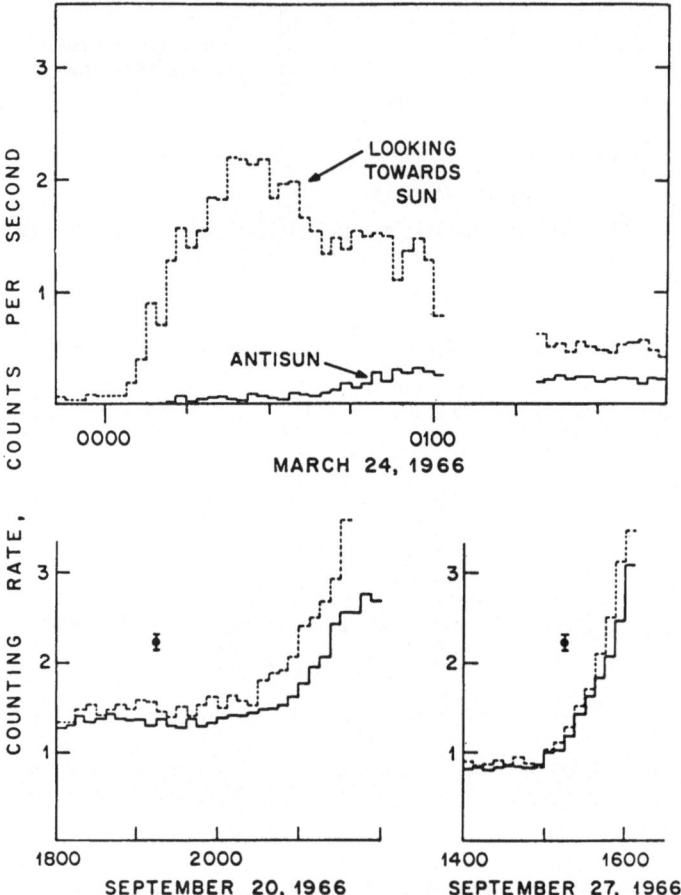

FIG. 5.   *Onset characteristics of three flare effects in quadrants looking toward the Sun and away for energy range 7·5–45 MeV, observed on Pioneers-6 and -7 (McCracken et al., 1967).*

shown that these filaments corotate with the Sun in the same way as do the sectors previously discussed. The cross-section of these filaments is typically in the range of $0·6–5 \times 10^6$ km. Thus at least at times there may be a detailed mapping of small-scale features close to the Sun onto a sphere of radius 1 AU. The study of the Sun through observations of interplanetary fields and plasmas by spacecraft at 1 AU can perhaps be compared with a field-emission microscope, in which the structure of field-induced emission from a microscopic point is magnified onto a large concentric observing sphere.

4. Solar Wind Plasma

Recent observations in the Vela satellite program by the Los Alamos group have yielded important new information about the solar wind plasma. These satellites are

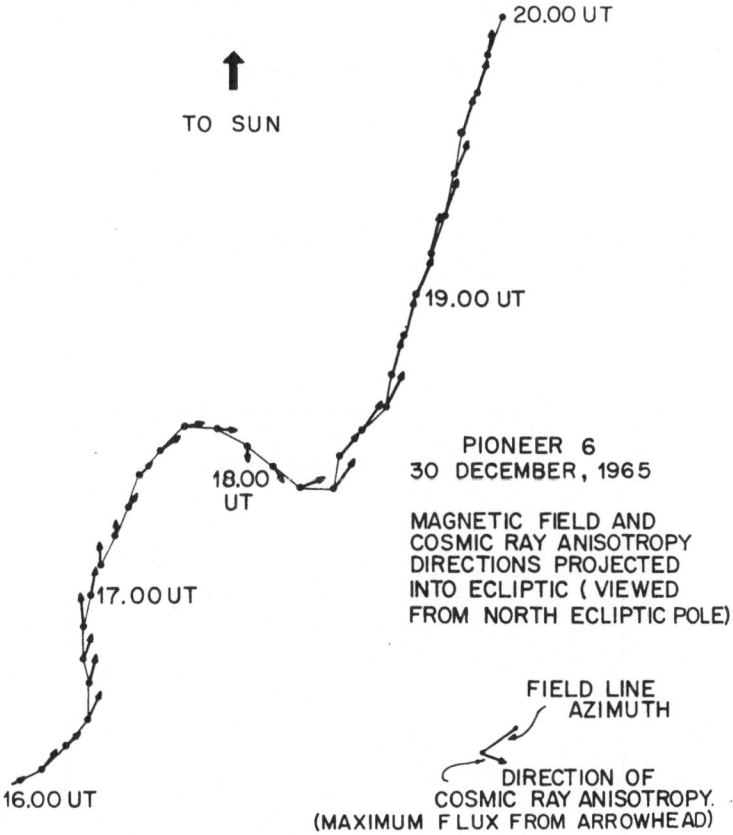

TO SUN

20.00 UT

19.00 UT

PIONEER 6
30 DECEMBER, 1965

MAGNETIC FIELD AND
COSMIC RAY ANISOTROPY
DIRECTIONS PROJECTED
INTO ECLIPTIC (VIEWED
FROM NORTH ECLIPTIC POLE)

18.00
UT

17.00 UT

FIELD LINE
AZIMUTH

DIRECTION OF
16.00 UT   COSMIC RAY ANISOTROPY.
(MAXIMUM FLUX FROM ARROWHEAD)

FIG. 6.   *Comparison of direction of interplanetary magnetic field and direction of arrival of maximum cosmic-ray flux (McCracken and Ness, 1966).*

in circular orbits about the Earth at a distance of 18 Earth radii, and spend about one third of each orbit outside the magnetosphere and magnetosheath. A hemispherical plate electrostatic analyser makes detailed measurements of the incident charged particle flux. Figure 7 shows a typical spectrum and angular distribution. The large peak in the spectrum is the almost monoenergetic stream of protons with energy of slightly less than 1 keV. The small peak at the right is attributed to alpha particles having the same streaming velocity as the protons and therefore twice the energy per unit charge. The angular distribution shows that the particles are coming not quite from the Sun but rather from a few degrees East of it. Since the peaks in Figure 7 have a finite width it is clear that in addition to the streaming energy of the plasma there exists a random energy, which is often described in terms of a 'temperature', whose value has ranged from about 5×10^3 °K to a few times 10^5 °K, with a normal proton temperature of a few times 10^4 °K (Strong *et al.*, 1966). A complete measurement of flux as a function of energy and angle is performed in about 4 min, and a typi-

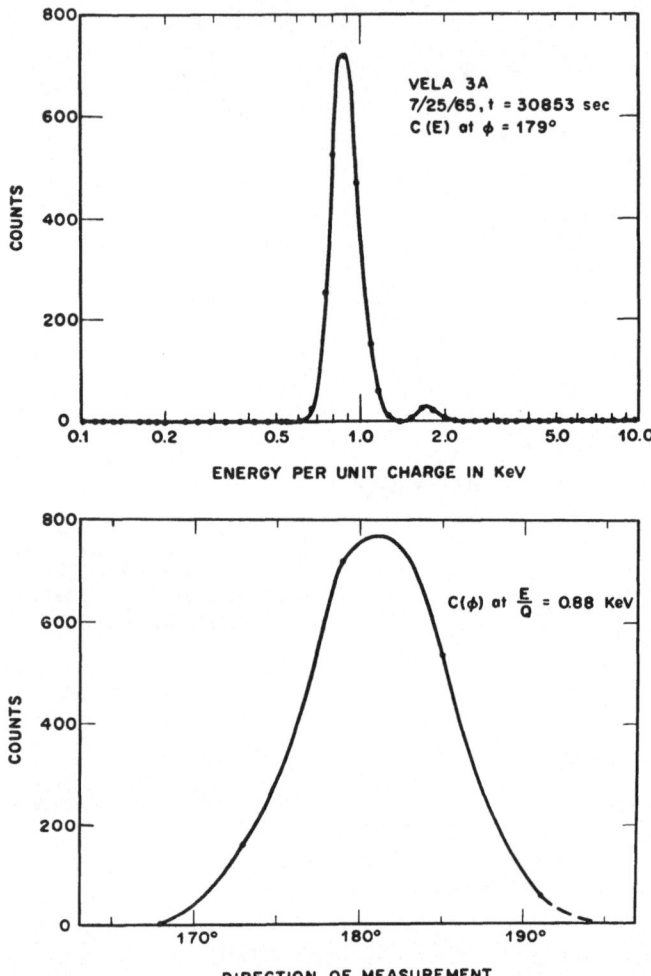

Fig. 7.   *Typical Vela-3 spectrum (counts vs. energy per unit charge in a fixed direction) and the associated angular distribution (counts vs. direction at a fixed energy per unit charge). The direction away from the Sun is labeled 180°, and the peak is a few degrees East of this (Hundhausen et al., 1967a).*

cal result is presented in Figure 8. Contours of observed particle flux are shown as a function of two velocities; the abscissa is the radial velocity V_1 from the Sun and the ordinate is a velocity perpendicular to both V_1 and to the spacecraft spin axis, which is tipped by 55° to the ecliptic. The arrow shows the direction of the interplanetary magnetic field simultaneously measured on the near-by IMP-3 satellite.

The contours in Figure 8 depart from isotropy in two ways. First, the temperature parallel to the field is considerably larger than the transverse temperature, the ratio being about 3 in Figure 8, which is a typical value. Second, the temperature ($9\cdot2 \times 10^4 °$K) along the field in the direction away from the Sun is larger than in the field

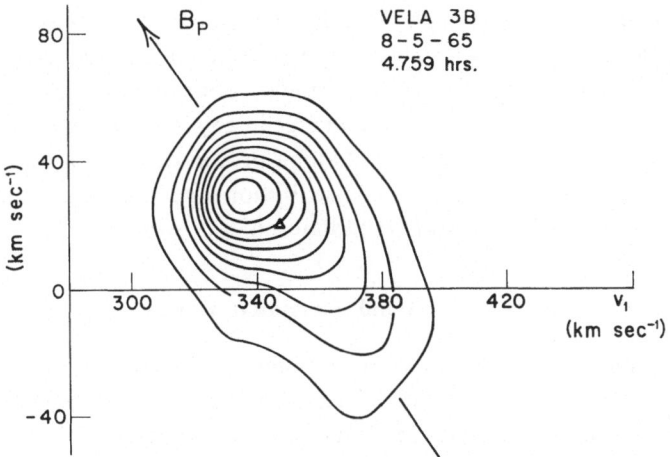

FIG. 8.   *Typical Vela-3 contour map of proton-velocity distribution in the radial direction from the Sun (abscissa) and in a normal direction (ordinate). The arrow is the field direction simultaneously measured on the nearby IMP-3 satellite (Hundhausen et al., 1967b).*

direction toward the Sun (this is observed to be true independently of whether the polarity of the field lines is toward or away from the Sun). Thus the solar wind protons are transporting heat away from the Sun at a rate which is estimated at about 10^{-5} ergs/cm^2/sec. A similar heat transport by electrons would be expected to be several orders of magnitude larger.

The persistence of the alignment between the plasma anisotropy direction and the field direction during a period of 6 hours is shown in Figure 9. The abrupt change in

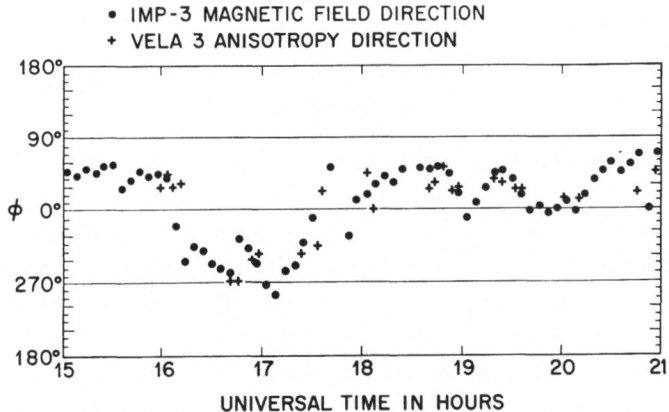

FIG. 9.   *The magnetic-field direction measured on IMP-3 and the direction of maximum plasma-thermal anisotropy measured on Vela-3 (Hundhausen et al., 1967b).*

direction at 1600 UT probably represents the appearance of an interplanetary filament of the sort discussed previously in connection with the cosmic-ray observations of McCracken and Ness (1966). It is worth noting that during the passage of this filament past the satellite there was no significant change in any of the usual plasma parameters (mean velocity, density, temperature and alpha/proton-density ratio). Only the orientation of the anisotropic temperature distribution and the field direction changed during the passage of this filament (Hundhausen *et al.*, 1967*b*).

5. Solar Wind Azimuthal Velocity

The solar wind angular distribution in Figure 7 shows that the solar wind was coming from a direction a few degrees East of the Sun. Individual observations show longitudinal deviations from radial plasma flow over a range of about $\pm 10°$, but the average longitudinal plasma velocity (in the *direction* of corotation with the Sun) is observed to be about 10 km/sec (Strong *et al.*, 1967). This is consistent with the analysis by Brandt (1967*a*) of the orientation of ionized cometary tails. Such values for the solar wind longitudinal plasma velocity at 1 AU suggest that corotation of the extended coronal plasma (this must be clearly distinguished from the corotation of magnetic *patterns* previously discussed in this paper) extends to greater distances than had previously been considered. Dicke (1964) has shown that the loss of angular momentum by the Sun to the solar wind can be computed on the basis of rigid corotation out to a radius at which the Alfvén velocity based on the radial component of the field is equal to the streaming velocity. This is a formal statement; it does not mean that rigid corotation actually exists to this distance. Dicke obtained a typical value of 35 solar radii for this distance, which is consistent with estimates by other authors (Pneuman, 1966; Weber and Davis, 1967; Modisette, 1967; Brandt, 1967*b*). Brandt (1966) has shown that the angular momentum loss in the solar wind could be important in the structure and evolution of the Sun.

6. Summary

In summary, we now have a picture of the streaming solar wind at 1 AU carrying with it much detailed information about the solar structure. The interplanetary magnetic field is of crucial importance even though its energy density is about two orders of magnitude less than the streaming-energy density of the plasma. The interplanetary sector structure is evident in the polarity patterns of the field. Thermal anisotropies in the plasma are field aligned. Cosmic rays from localized regions of the solar atmosphere are guided to 1 AU (and beyond) by magnetic filaments, whose pattern corotates with the Sun. The plasma velocity at 1 AU has a small component in the direction of corotation, leading to important consequences for the extent of corotation near the Sun, the loss of angular momentum, and the structure and evolution of the Sun.

References

Asbridge, J.R., Bame, S.J., Felthauser, H.E., Gosling, J.T. (1967)   *Trans. Amer. Geophys. Union* **48**, 172.
Brandt, J.C. (1966)   *Astrophys. J.*, **144**, 1221.
Brandt, J.C. (1967a)   *Astrophys. J.*, **147**, 201.
Brandt, J.C. (1967b)   *Astrophys. J.*, **148**, 905.
Dicke, R.H. (1964)   *Nature*, **202**, 432.
Dodson-Prince, H., Hedeman, E.R. (1967)   IQSY/COSPAR Symposium, London, July.
Hundhausen, A.J., Asbridge, J.R., Bame, S.J., Gilbert, H.E., Strong, I.B. (1967a)   *J. Geophys. Res.*, **72**, 81.
Hundhausen, A.J., Bame, S.J., Ness, N.F. (1967b)   *J. Geophys. Res.*, **72**, 5265.
McCracken, K.G., Ness, N.F. (1966)   *J. Geophys. Res.*, **71**, 3315.
McCracken, K.G., Rao, U.R., Bukata, R.P. (1967)   *J. Geophys. Res.*, **72**, 4293.
Modisette, J.L. (1967)   *J. Geophys. Res.*, **72**, 1521.
Ness, N.F. (1966)   *J. Geophys. Res.*, **71**, 3319.
Ness, N.F. (1967)   IQSY/COSPAR Symposium, London, July.
Ness, N.F., Wilcox, J.M. (1967)   *Solar Phys.*, **2**, 351.
Neugebauer, M., Snyder, C.W. (1966)   *J. Geophys. Res.*, **71**, 4469.
Parker, E.N. (1963)   *Interplanetary Dynamical Processes*, Interscience Publishers, New York.
Pneuman, G.W. (1966)   *Astrophys. J.*, **145**, 800.
Strong, I.B., Asbridge, J.R., Bame, S.J., Heckman, H.H., Hundhausen, A.J. (1966)   *Phys. Rev. Letters*, **16**, 631.
Strong, I.B., Asbridge, J.R., Bame, S.J., Hundhausen, A.J. (1967)   *Trans. Amer. Geophys. Union* **48**, 182.
Weber, E.J., Davis, L., Jr. (1967)   *Astrophys. J.*, **148**, 217.
Wilcox, J.M., Ness, N.F. (1965)   *J. Geophys. Res.*, **70**, 5793.
Wilcox, J.M., Ness, N.F. (1966)   *Astrophys. J.*, **143**, 23.
Wilcox, J.M., Ness, N.F. (1967)   *Solar Phys.*, **1**, 437.

DISCUSSION

L.W. Acton: What were the cosmic-ray energies on the last slide?

J.M. Wilcox: These particular measurements are around 15 MeV, but in general the range is a few MeV.

R. Michard: What is the relationship between energy and electric charge in the solar wind? Do you think the ratio of the two peaks agrees with the solar abundance of hydrogen and helium?

J.M. Wilcox: It would certainly seem to be related. The average ratio, over many observations is approximately 10%, which is a value one would be willing to accept. However, there are considerable variations, which may be an important clue to the dynamic processes involved in the acceleration processes in the solar wind.

INFRARED ASTRONOMY

FRANK J. LOW

(Rice University, Houston, Texas, and University of Arizona, Tucson, Ariz., U.S.A.)

The Earth's atmosphere transmits infrared radiation through a number of windows. Table 1 lists the seven photometric systems in use at the University of Arizona which are chosen to fit the windows between 1·0 and 25 microns. An absolute calibration (Johnson, 1965; Low, 1966) has been worked out for each wavelength band and, for reference, we include an estimate of our current limiting magnitudes using a 60-inch telescope. At about 1000 microns, observations from the ground are again possible and both our group and workers in Russia (Fedoseev 1963) and England (Baldock *et al.*, 1965) have succeeded in making observations of celestial sources. Between 25 and 1000 microns a few data have now been obtained from stratospheric altitudes by observers using jet aircraft (Low and Gillespie, 1968) and helium-filled balloons (Hoffman *et al.*, 1967). We can anticipate that activity of this sort will increase greatly in the near future. At present, however, most of what we know concerning the nature of celestial objects at infrared wavelengths was obtained with ground-based instruments.

The planets radiate strongly in the infrared and have been studied extensively. The ultra-high resolution near-infrared spectra of Connes and co-workers (Connes *et al.*, 1967) show what can be done from the ground despite telluric contamination. Kuiper (personal communication) has recently succeeded in obtaining lower resolution spectra of Venus and Mars from a jet aircraft, with much less telluric interference. In collaboration with Stein and Gillett at the University of California, San Diego, we

Table 1

Infrared photometric systems in use at the University of Arizona

Photometric Designation	λ_{eff} (μ)	Absolute Flux, Mag = 0		Limiting Magnitude
		$W/cm^2/\mu$	$W/m^2/Hz$	
J	1·25	$3\cdot4 \times 10^{-13}$	$1\cdot77 \times 10^{-23}$	–
H	1·60	$1\cdot28 \times 10^{-13}$	$1\cdot09 \times 10^{-23}$	13·5
K	2·2	$3\cdot9 \times 10^{-14}$	$6\cdot3 \times 10^{-24}$	10·0
L	3·4	$8\cdot1 \times 10^{-15}$	$3\cdot1 \times 10^{-24}$	8·0
M	5·0	$2\cdot2 \times 10^{-15}$	$1\cdot8 \times 10^{-24}$	5·0
N	10·2	$1\cdot23 \times 10^{-16}$	$4\cdot3 \times 10^{-25}$	2·0
Q	22·0	$7\cdot7 \times 10^{-18}*$	$1\cdot02 \times 10^{-25}*$	0·0

* Provisional.

Perek (ed.), Highlights of Astronomy, 136–147. © I.A.U.

have made absolute spectral scans of the planets from 2·8 to 15 microns on the ground, using the 60-inch infrared telescope at Arizona. Figure 1 shows a scan of Jupiter revealing the existence of a window at 5 microns in the Jovian atmosphere. The brightness temperature at the centre of this window is about 200 °K, corresponding to the temperature at the cloud deck. High-angular resolution (~4″) scans of the Jovian disk at 10 and 22 microns were made with the 61-inch planetary telescope at Arizona, showing limb darkening and meteorological features (Figure 2). The Caltech group

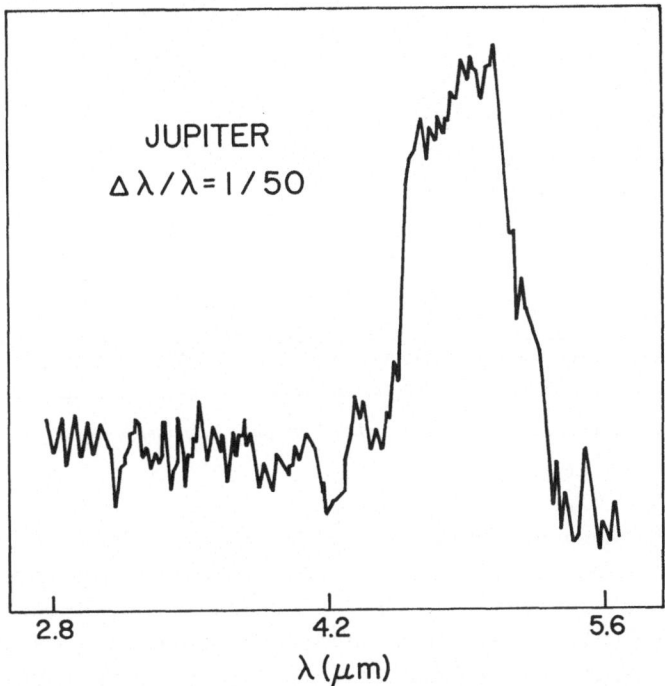

FIG. 1.   *Spectral scan of Jupiter, 2·8–5·6 microns with resolution, $\Delta\lambda/\lambda$, 1/50. Note the relative absence of energy at 3·4 microns.*

(Westphal *et al.*, 1965) made similar studies of Venus at 10 microns, using the 200-inch telescope on Palomar Mountain. They also made interesting 1–10 micron studies of the recent comet Ikeya-Seki (Becklin and Westphal, 1966).

Turning from the planets to the Sun, we can report progress on several efforts to define the Sun's spectral energy distribution. Figure 3 shows the brightness temperature, T_B, plotted against λ from the radio region to 10 microns. Robert Noyes of the Smithsonian Astrophysical Observatory, Jacque Beckers of Sacramento Peak Observatory, and the author observed the solar limb intensity at 22 microns during the total eclipse of November 12, 1966. Center-to-limb studies were carried out by Pierre Léna, using the 60-inch solar telescope at Kitt Peak National Observatory. These

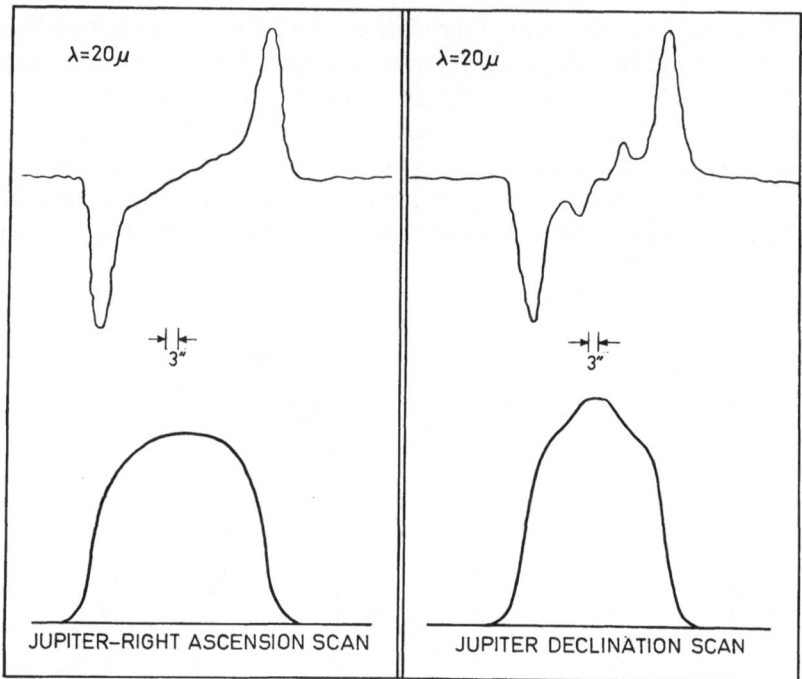

FIG. 2. *Scans of Jupiter at 22 microns in right ascension and declination with an angular resolution of 4 sec of arc.*

results show that the minimum in the T_B versus λ curve has the shape from 10 to 100 microns as shown in Figure 3. The author (Low and Gillespie, 1968) has carried out a series of absolute radiometry experiments on the Sun from an altitude of 44000 feet, in a jet aircraft. The brightness temperature at 1000 microns is in good agreement with the earlier value (Low and Davidson, 1965) obtained from the ground, when indirect corrections were made for atmospheric absorption (Low and Davidson).

Johnson and co-workers (Johnson *et al.*, 1966) have published catalogues giving magnitudes of numerous bright stars in the range 0·3–3·4 microns. Several papers have been published by the Arizona group giving results for 5, 10, and 22 microns (Johnson and Mitchell, 1963; Low and Johnson, 1964; Low, 1966). Bolometric corrections and effective temperatures of cool stars have been obtained by use of the intrinsic colors which are now available (Johnson, 1966). Detailed studies of the spectral energy distribution of stars are of great interest both for understanding the intrinsic properties of stars and for studying interstellar and circumstellar phenomena. Figure 4 shows the absolute spectral energy curve for α Orionis. Note the dip at 5 microns and the excess at 22 microns.

Johnson and co-workers at Arizona (Johnson, 1965, 1968) have carried out extensive studies of interstellar extinction. Figure 5 shows the reddening law they found for

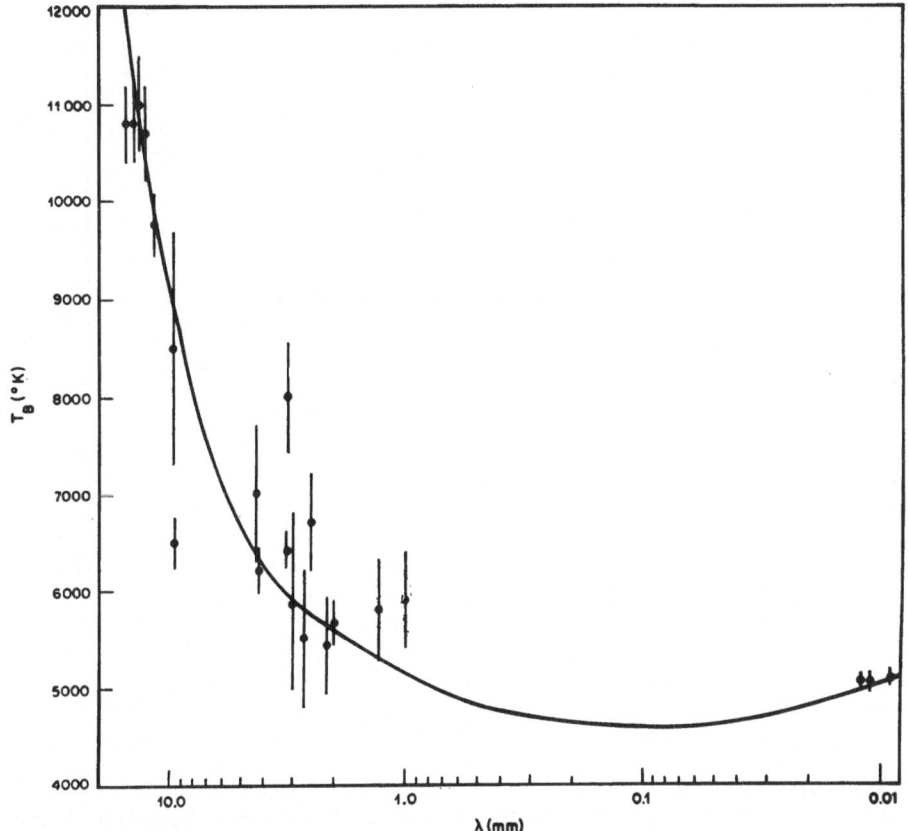

FIG. 3. *Brightness temperature of the Sun plotted against wavelength.*

Cepheus. In most directions the ratio of selective to total extinction, *R*, has a value around 3 to 3·5. Here, however, the value is close to 5, presumably because of interstellar grains which are gray between 3000 and 10000 Å. As infrared techniques are developed it should become possible to deal directly with this problem by studying the thermal emission of interstellar grains over a wide range of wavelengths.

As Figure 5 shows, the interstellar medium becomes transparent in the infrared, revealing the obscured portions of the Galaxy. One of the most dramatic examples of the importance of this fact is the recent discovery of the galactic nucleus in the infrared by Becklin and Neugebauer. Figure 6 is taken from their paper, which is still in press. It shows the striking similarity between the nucleus of our Galaxy and that of M 31. The total visual extinction in the path to the galactic centre is evidently about 25 magnitudes. The position found for the infrared centre agrees with the radio position of the source Sagittarius A.

The existence of infrared stars has been established since Hetzler's (1937) photographic work in the 1930's. The recent sky survey at 2·2 microns carried out by the

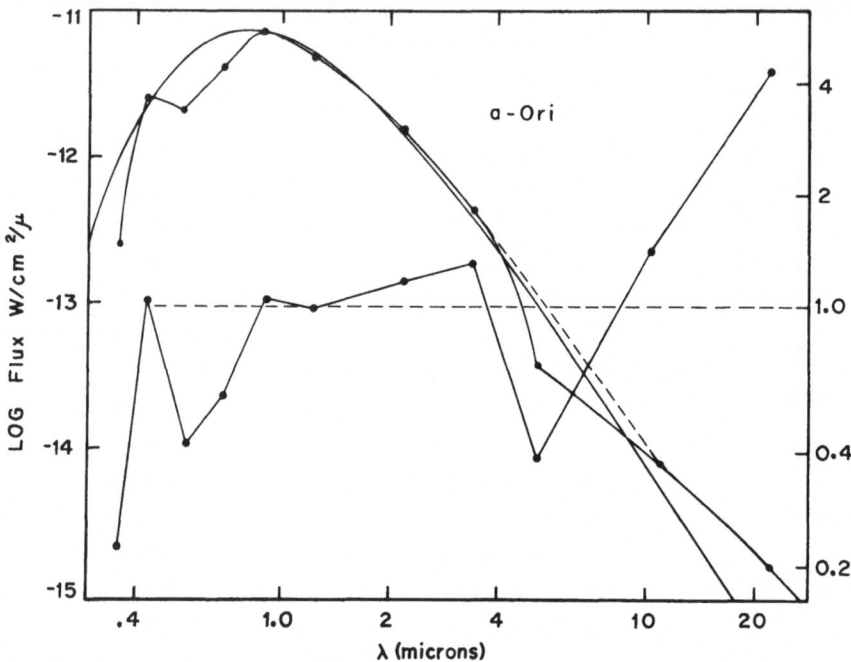

FIG. 4. *Absolute spectral energy distribution of* α *Orionis showing a large excess at the longest wavelengths.*

Caltech group (Neugebauer *et al.*, 1965; Ulrich *et al.*, 1966) is complete down to K magnitude of +3 and contains 5500 objects. Some of the more interesting of these stars have been studied in detail by various workers at Caltech (McCammon *et al.*, 1967), Berkeley (Wing *et al.*, 1967), Arizona (Johnson *et al.*, 1965) and Rice (Low and Smith, 1966). Most of the objects are cool stars ($T_e > 1200\,°K$) or are stars highly reddened by interstellar dust or by a circumstellar envelope (Wisniewski *et al.*, 1967). Other infrared stars have been found by Mendoza (1966) of the University of Mexico while visiting Arizona. Studying various types of cool stars – some found photographically by Haro and Chivara – he discovered large infrared excesses in T Tauri stars. Becklin and Neugebauer (1967) discovered a point source in Orion too faint to be seen by the sky survey, and this in turn led Kleinman and Low (1967) to the discovery at 22 microns of the first purely infrared nebula. Of chief interest is the question. How are these objects related to stellar or planetary formation? Poveda (1965), among others, had already suggested several possibilities. Figure 7 shows the result found by Low and Smith (1966) for R Monocerotis. Here it is not possible to explain the intense infrared continuum by extinction alone. A model was constructed which accounts for the high infrared luminosity and extended continuum as a protostar heating a preplanetary envelope of dust grains which reradiate at very long wavelengths. Figure 8 shows normalized spectral energy curves for five infrared objects;

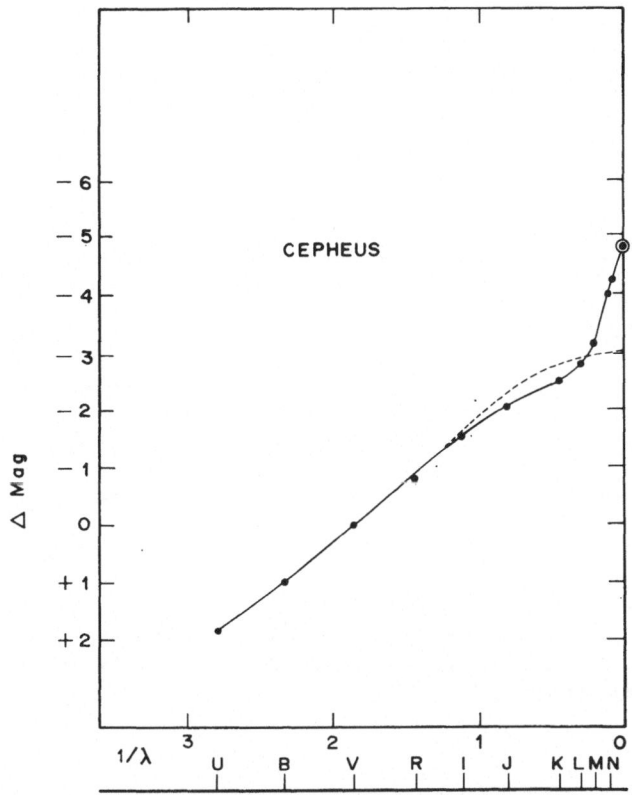

FIG. 5.   *Interstellar reddening law in Cepheus.*

only one, NML Cygnus, appears to be unrelated to the problem of stellar formation. Figure 9 shows the size and position of the newly discovered infrared nebula in Orion, which may be a contracting cloud heated by a cluster of stars forming near its centre (Cameron, 1967; Hartmann, 1967).

The Crab Nebula presents a unique opportunity to study the synchrotron mechanism over an enormous spectral range. Moroz (1964) at the Crimean Observatory was the first to observe the Crab in the near infrared. Figure 10 shows our present knowledge of the spectrum; it includes new data obtained by Becklin at Caltech and Kleinmann at Rice.

The infrared properties of quasars are discussed later at this conference. Here we will only refer to the spectrum of 3C273 (Figure 11), point out that at 1·6 microns we have found it to be highly polarized, and take note of the new work at 1·6 and 2·2 microns on some 36 different quasars (Table 2).

Johnson's (1966) observations of 12 galaxies out to 3·4 microns produced a mean spectral energy distribution which can be explained on the basis of a suitable stellar population. This is not the case for the Seyfert galaxy NGC 1068 observed by Pachol-

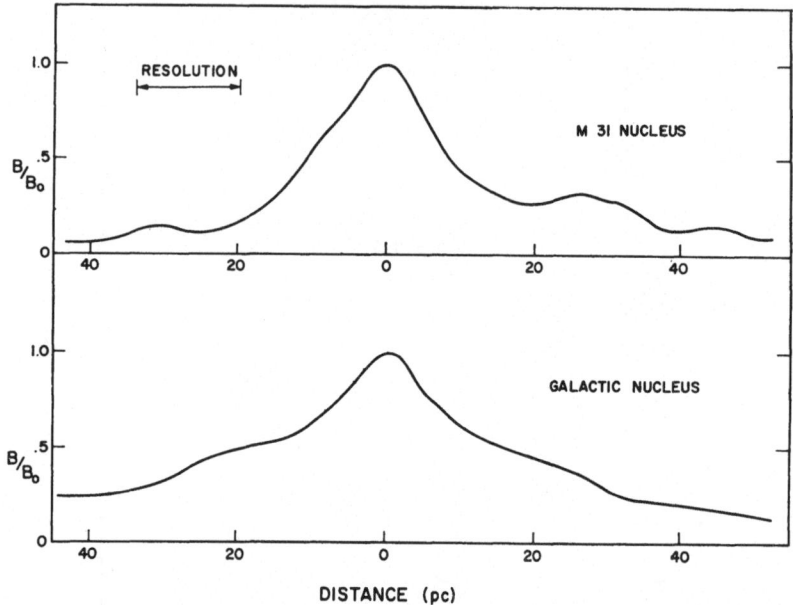

FIG. 6. *A right ascension scan of M31 at 2·2 μ (Neugebauer and Becklin) is compared with a 2·2 μ scan of the centre of the galaxy. The latter was taken along a line 50° to the galactic plane with a resolution degraded to the observation of M31.*

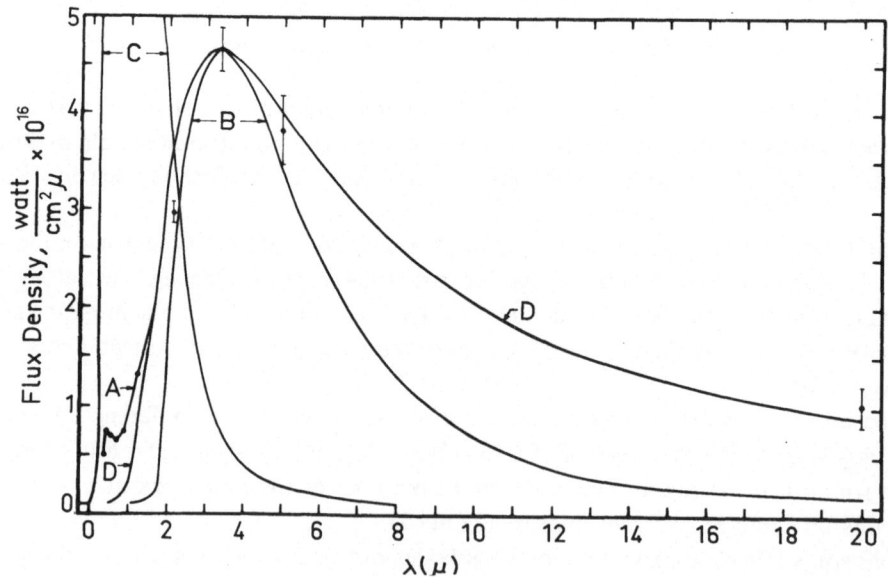

FIG. 7. *Absolute spectral energy distribution of R Monocerotis compared to preplanetary dust model of Low and Smith.*

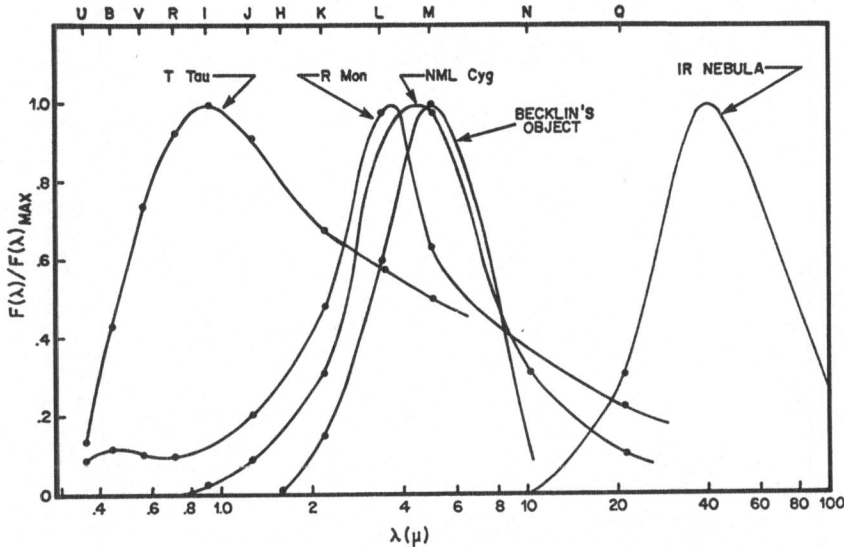

FIG. 8. *Normalized spectral energy curves for 5 infrared objects.*

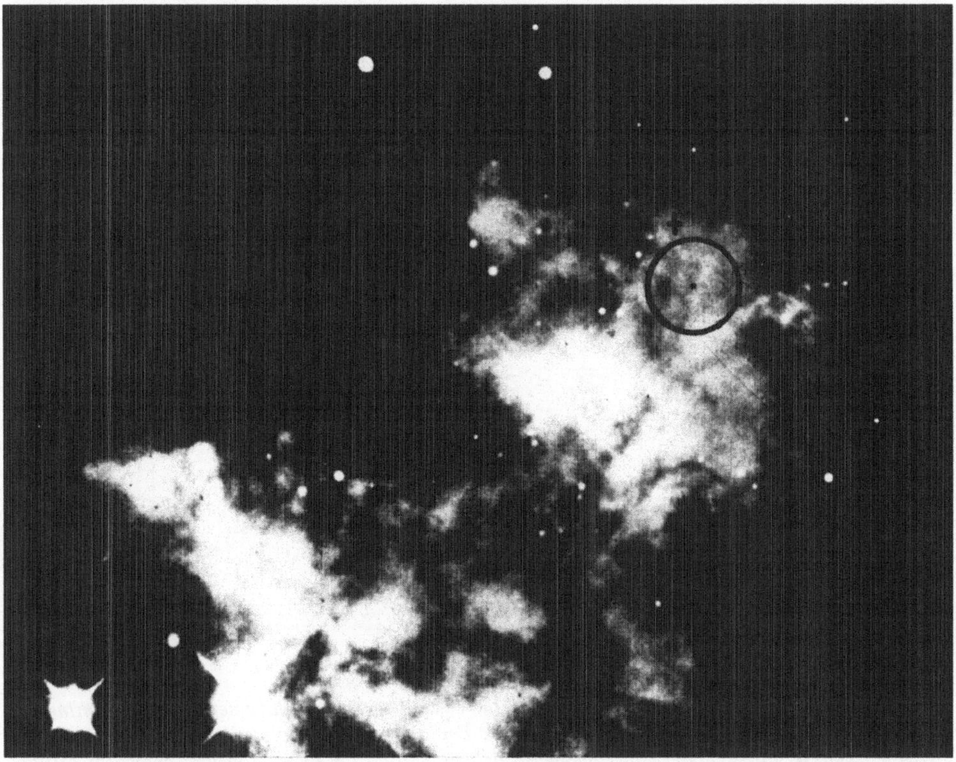

FIG. 9. *Central region of Orion Nebula showing size and location of Infrared Nebula.*

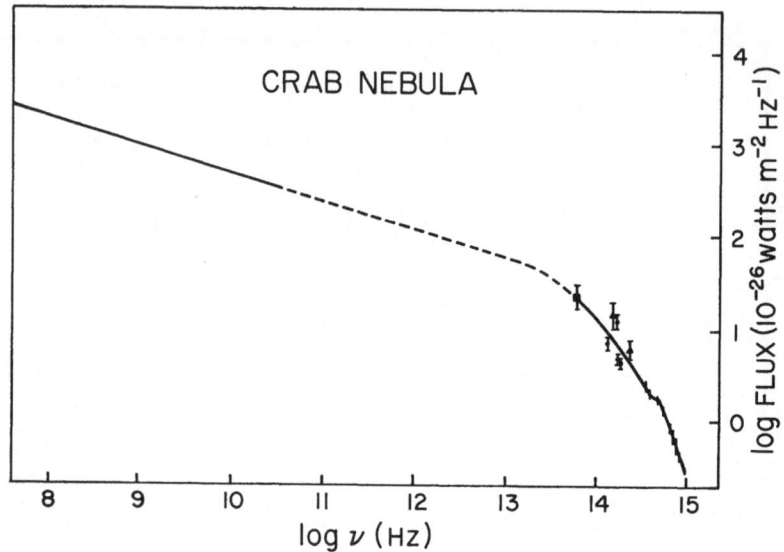

FIG. 10.   *Log S (v) vs. log v for the Crab Nebula. Triangles are data from Moroz, filled circles from Becklin, and filled squares from Kleinmann.*

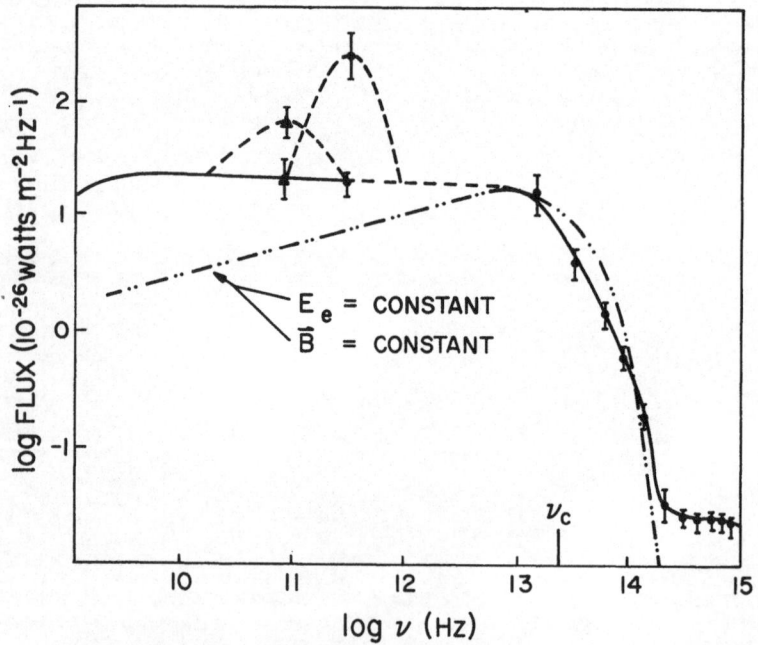

FIG. 11.   *Log S (v) vs. log v for 3C 273, showing flat optical continuum, flat quiescent radio continuum, millimeter-wave outbursts and steep IR component. For comparison, the spectrum of monoenergetic relativistic electrons in a homogeneous magnetic field is plotted for an appropriate choice of critical frequency, 3×10^{13} cps. Data at 10^{11} cps due to Epstein.*

Table 2

Infrared properties of quasars at 1·6 and 2·2 microns

Designation	1·6 μ Flux 10^{-29} $W/m^2/Hz$	Designation	1·6 μ Flux 10^{-29} $W/m^2/Hz$
PHL 658	2·3	*3C 345	8·9
3C 9	0·43	*3C 351	3·6
*3C 48	3·8	3C 371	16
PKS 0405-12	3·3	*3C 380	1·0
PKS 0859-14	⩽ 2·5	3C 403G	2·6
TON 469	4·6	3C 403S	28
*3C 273	30	*PKS 2115-30	1·1
*PKS 1354 + 19	0·80	3C 432	0·45
*3C 298	1·0	PKS 2135-14	3·6
*3C 309·1	0·98	3C 446	2·9
*PKS 1510-08	1·1	*CTA 102	1·1
*3C 323·1	2·6	3C 454	0·31
*TON 256	3·8	*3C 454·3	0·74
*3C 334	1·4		

* 2·2 μ obs. by Neugebauer.

czyk and Wisniewski (1967) with the same apparatus. Figure 12 shows the spectrum of NGC 1068 including 5, 10 and 22 micron observations of Kleinmann and Low (previously unpublished) and the 3·4 mm result of Epstein (personal communication). Neither the optical spectrum nor the infrared spectrum can be produced by stars

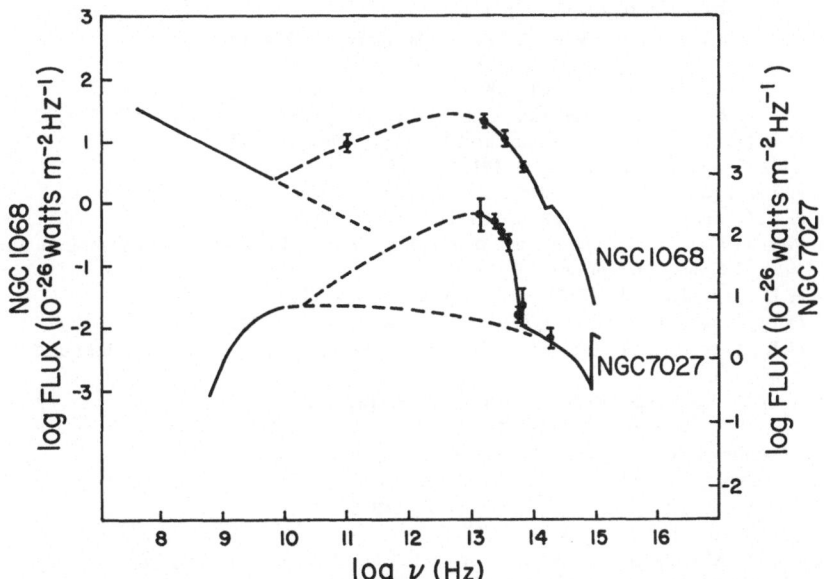

FIG. 12. *Log S (v) vs. log v for the Seyfert galaxy NGC1068 and the planetary nebula NGC7027. Note the similarities between the infrared components in each object. Data at 10^{11} cps due to Epstein.*

alone. There is, however, some question whether to invoke a non-thermal process for the infrared, because the attendant strong radio emission is absent.

The planetary nebula, NGC 7027, has the spectrum shown in Figure 12. Here the 10 micron data are from Gillett *et al.* (1967) and the 1·6, 5, and 22 micron data were recently obtained by the author, using the 200-inch telescope at Palomar Observatory. As is the case for Seyfert galaxies, the radio flux of the nebula is well below the infrared level. It seems plausible that the same mechanism is responsible for the infrared components of both types of object. Although much further work is required to prove this hypothesis, the simplest explanation is that the infrared is thermal reradiation from circumstellar or interstellar grains heated to only a few hundred degrees by a hot nearby source. As was shown in the case of R Monocerotis (Low and Smith, 1966), a large fraction of the available energy can be degraded into the infrared in this way.

In conclusion, it can be stated that infrared astronomy, until recently reserved for those interested in the planets, is becoming a major contributor to the rapidly expanding science of space astronomy.

References

Baldock, R. V., Bastin, J. A., Clegg, P. E., Emery, R., Gaitskell, J. N., Gear, A. E. (1965)   *Astrophys. J.*, **141**, 1289.
Becklin, E. E., Neugebauer, G. (1967)   *Astrophys. J.*, **147**, 799.
Becklin, E. E., Westphal, J. A. (1966)   *Astrophys. J.*, **145**, 445.
Cameron, A. G. W. (1967)   Infrared Conf., London, February.
Connes, P., Connes, J., Benedicts, W. S., Kaplan, L. D. (1967)   *Astrophys. J.*, **147**, 1230.
Fedoseev, L. N. (1963)   *Radiofizika*, **4**, 425.
Gillett, F. C., Low, F. J., Stein, W. A. (1967)   *Astrophys. J.*, **149**, L97.
Hartmann, W. K. (1967)   *Astrophys. J.*, **149**, L87.
Hetzler, C. (1937)   *Astrophys. J.*, **86**, 509.
Hoffman, W. F., Woolf, N. J., Frederick, C. L., Low, F. J. (1967)   *Science*, **157**, No. 3785, 187.
Johnson, H. L. (1965*a*)   Comm. Lunar and Planetary Lab., *No. 53.*
Johnson, H. L. (1965*b*)   *Astrophys. J.*, **141**, 923.
Johnson, H. L. (1966*a*)   *A. Rev. Astr. Astrophys.*, **4**.
Johnson, H. L. (1966*b*)   *Astrophys. J.*, **143**, 187.
Johnson, H. L. (1968)   in *Nebulae and Interstellar Matter*. Ed. by B. M. Middlehurst and L. H. Aller, Univ. of Chicago (in press).
Johnson, H. L., Mitchell, R. I. (1963)   *Astrophys. J.*, **138**, 302.
Johnson, H. L., Low, F. J., Steinmetz, D. (1965)   *Astrophys. J.*, **142**, 808.
Johnson, H. L., Mitchell, R. I., Iriarte, B., Wisniewski, W. Z. (1966)   Comm. Lunar and Planetary Lab., *No. 63.*
Kleinmann, D. E., Low, F. J. (1967)   *Astrophys. J.*, **149**, L1.
Low, F. J. (1966)   *Astrophys. J.*, **146**, 326.
Low, F. J., Davidson, A. W. (1965)   *Astrophys. J.*, **142**, 1278.
Low, F. J., Gillespie, C. M. (1968)   (in preparation).
Low, F. J., Johnson, H. L. (1964)   *Astrophys. J.*, **139**, 1130.
Low, F. J., Smith, B. J. (1966)   *Nature*, **212**, 675.
McCammon, D., Munch, G., Neugebauer, G. (1967)   *Astrophys. J.*, **147**, 575.
Mendoza, E. E. (1966)   *Astrophys. J.*, **143**, 1010.
Moroz, V. I. (1964)   *Astrophys. J.*, **7**, 755.
Neugebauer, G., Martz, D. E., Leighton, R. B. (1965)   *Astrophys. J.*, **142**, 399.

Pacholczyk, A.G., Wisniewski, W.Z. (1967)   *Astrophys. J.*, **147**, 394.
Poveda, A. (1965)   *Bol. Obs. Tonantzintla y Tacubaya*, **4**, 15; *ibid.*, p. 22.
Ulrich, B.T., Neugebauer, G., McCammon, D., Leighton, R.B., Hughes, E.E., Becklin, E. (1966)
     Astrophys. J., **146**, 288.
Westphal, J.A., Wildey, R.L., Murray, B.C. (1965)   *Astrophys. J.*, **142**, 799.
Wing, R.F., Spinrad, H., Kuhi, L.V. (1967)   *Astrophys. J.*, **147**, 117.
Wisniewski, W.Z., Wing, R.F., Spinrad, H., Johnson, H.L. (1967)   *Astrophys. J.*, **148**, L29.

DISCUSSION

Y. Ohman: It would be extremely interesting to hear if one could make polarization measurements of the thermal emission from grains. We found some years ago that the thermal emission of the very narrow metallic filaments are strongly polarized when the filaments are of a size comparable to the wavelength of light. It might be different in the infrared but still this question might still be worth looking into.

F.J. Low: What I referred to was the possibility of observing in the far infrared beyond 25 microns. These observations will give an entirely new handle on the problem of interstellar grains. The condition that you just mentioned would not be met there. The measurements in the infrared are quite possible and I think will become very important in the future.

G.P. Kuiper: You did not say in what wavelength the measurements at the galactic centre were made.

F.J. Low: The measurements were made at a variety of wavelengths, 1·6, 2·2, and 3·4 μm. They have not yet succeeded in making measurements at 5 μm, although there is no doubt it can be measured. Because the galactic centre is an extended source, about 1 min of arc in size, it will be difficult to measure (at longer wavelengths) from the ground. At wavelengths shorter than 1·2 microns no radiation has been detected, and only an upper limit has been determined.

J.A. Roberts: I would like to make an appeal for standardization of the flux units in watts per square meter per micron rather than the standard watts per square meter per unit frequency which is used at radio frequencies. I think it is quite important, since it changes the shape of the curves completely.

F.J. Low: We have unhappily used two systems of units for flux in the infrared, because we lie between the radio and the optical astronomers. The radio astronomers like watts per square meter per Hertz, but it is also useful at times to use watts per square centimeter per micron.

Z. Turto: Do you have any information on the radiation from the H II region in the Orion nebula?

F.J. Low: There are quite a number of results on H II regions which are new. Our measurements on the Orion nebula, that is the H II region and not the infrared nebula which I showed, indicate that the surface brightness is about 10 times higher than the free-free continuum would predict. This is the case in 7027, the planetary nebula. So H II regions in the infrared are going to be very interesting.

S. Plagemann: In your measurements of T Tauri and R Monocerotis, what considerations have you made for the fact that these are variable stars? All your measurements have been spaced very close to each other or done over several lines.

F.J. Low: The observations have been carried out over several months. We have not been able to relate optical variations to infrared variations if indeed, infrared variations do exist. This general subject is one we hope can be explored in the future, and will give a better understanding of the rather complex models.

SOME PROBLEMS AND INSTRUMENTAL FEATURES
OF SUBMILLIMETER ASTRONOMY

A. E. Salomonovič

(P. N. Lebedev Physical Institute, Academy of Sci., U.S.S.R.)

1. Introduction

Modern astronomy includes optical, ultraviolet, infrared and, in recent years, radio, γ-ray, and X-ray astronomy. Such a classification is justified to a certain degree. In fact, the difference in wavelength ranges causes a distinction in the methods and techniques of receiving radiation. Also, the solution of specific problems requires observation in different wavelength regions. In this respect it is possible to describe a new astronomical branch, the submillimeter one.

The submillimeter range is intermediate between the infrared and microwave regions, as shown in Table 1. The boundaries of this region are not very definite. Some authors include in the submillimeter range the wavelengths longer than 50 microns, others, those longer than 100 microns. The long wavelength edge of the range is also diffuse. Formally it is a wavelength of 1 mm, but in some cases the 2-mm or 4-mm wavelengths are also included in the submillimeter range. The measurements of submillimeter receiver performances are sometimes carried out at wavelengths up to 8 mm. The uncertainty of the boundaries is very understandable: their shifting depends on the methods of generation, transmission, and detection of radiation. In this review paper, following Martin's (1962, 1963) terminology, wavelengths between 50 microns and 2 mm will be attributed to the submillimeter range.

Among other branches of astronomy, submillimeter astronomy may perhaps most reasonably claim to go beyond the Earth's atmosphere, since the Earth's atmosphere is practically opaque for space submillimeter radiation. Absorption by atmospheric water vapour prevents any serious astronomical observation from sea level.

In Figure 1 are shown the results of calculations of the absorption by atmospheric water vapour which have been performed by Ževakin and Naumov (1963). Even in the windows of relative transparency at wavelengths shorter than 2 mm, the absorption coefficient turned out to be more than 1 db/km. The results of these calculations are in satisfactory agreement with observational results in the millimeter (Straiton and Tolbert, 1960; Salomonovič, 1964; Wort, 1962; Drjagin *et al.*, 1966) and submillimeter (Drjagin *et al.*, 1966) wavelengths, as well as in the infrared region.

Submillimeter astronomy has appeared possible only in recent years and its development is due to two reasons: (1) the success in semi-conductor physics and quantum

Perek (ed.), Highlights of Astronomy, 148–163. © *I.A.U.*

Table 1

λ mm	λ μm	ν GHz	ν cm^{-1}	Photon energy eV	Spectral range
10	10 000	30	1	0·0001	
8	8 000	37	1·23		Microwave
5	5 000	60	2		(mm)
4	4 000	75	2·5		
2	2 000	150	5		
1	1 000	300	10	0·001	
0·5	500	600	20		Far IR
0·2	200	1500	50		or sub mm
0·1	100	3000	100	0·01	
0·05	50	6000	200		
0·01	10	30000	1000	0·1	IR
0·001	1	300000	10000	1·0	

electronics; (2) the extremely rapid development of space-astronomy techniques which allow us to eliminate completely, or for the most part, the influence of absorbtion in the Earth's atmosphere. However, the progress of submillimeter astronomy could hardly be explained only by new technical possibilities. The main reason is that the solution of certain specific and important problems requires observations in just the range under consideration.

Before proceeding to the techniques involved, I would like to touch briefly on several points.

2. Some Problems of Submillimeter Astronomy

Until the present, the submillimeter window into space has been, and perhaps still is, strictly curtained (if not completely closed). Extension beyond the Earth's atmosphere promises extremely unusual discoveries. The first attempts at submillimeter observations confirm this statement. Even before the wide program of observations began to develop, the astrophysicists tried to formulate some problems which required submillimeter astronomical activity.

A. THE CHARACTERISTICS OF PRESTELLAR MATTER

The expanding universe theory of A. A. Friedman predicted the possible existence of an isotropic electromagnetic thermal radiation with a black-body temperature of several degrees Kelvin. In accordance with the model of a hot universe, developed in the framework of this theory by several authors (Zeldovič, 1966), matter in the prestellar state is specified by high level of entropy. In thermal equilibrium, the density of strong radiation in a compressed hot plasma at an early phase is many times greater than the density of matter. In the process of expansion the number of quanta remains the same but their energy diminishes, causing an increase in wavelength. The density

150   A.E. SALOMONOVIČ

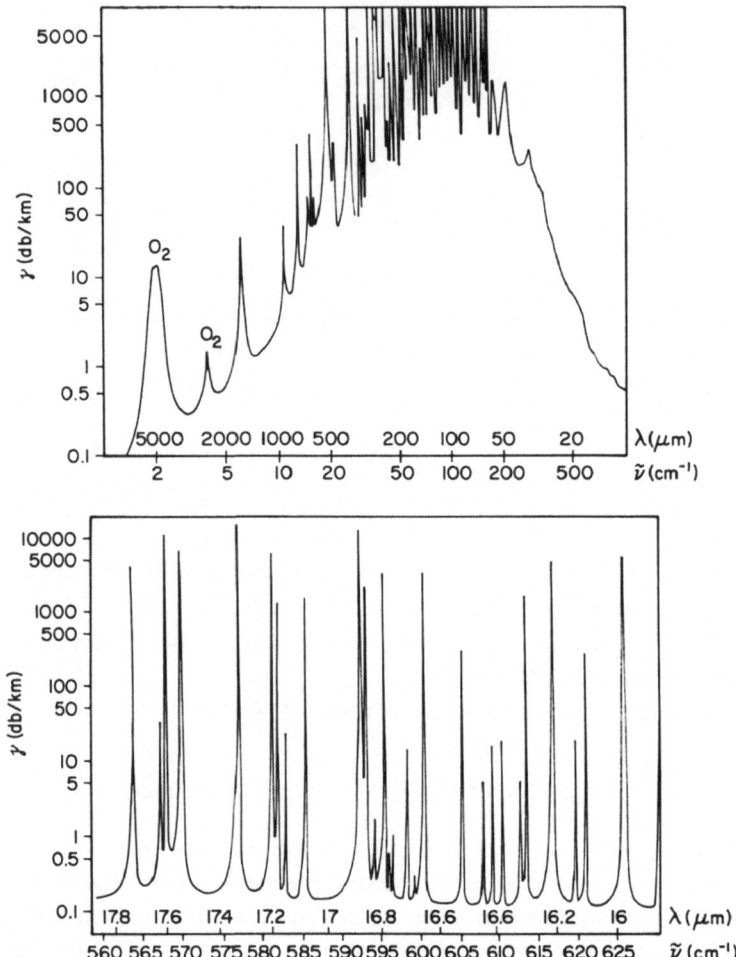

FIG. 1.   *Absorption in atmospheric water vapour (db/km) calculated by Ževakin and Naumov (1963)* *(T = 293°K, P = 760 mm Hg, ρ = 7·5 gm⁻³).*

of this radiation at present was found to be many orders of magnitude greater than that of other sources of radiation (radio galaxies, radio stars) in the wavelength range where the maximum of the radiation curve occurs. Figure 2 shows the predicted spectrum calculated in 1964 by Novikov and Doroškevič for a proposed temperature of 1 °K.

Figure 3 shows some recent observational results. The measurements carried out by Penzias and Wilson (1965) at the wavelength $\lambda = 7\cdot3$ cm, and observations made by Rall and Wilkinson (1966) and Stankevič *et al.* (personal communication) at $\lambda = 3\cdot2$ cm, apparently suggest the existence of such radiation with an approximate blackbody temperature of $\sim 3\,°$K. The same conclusion is supported by the analysis of the rela-

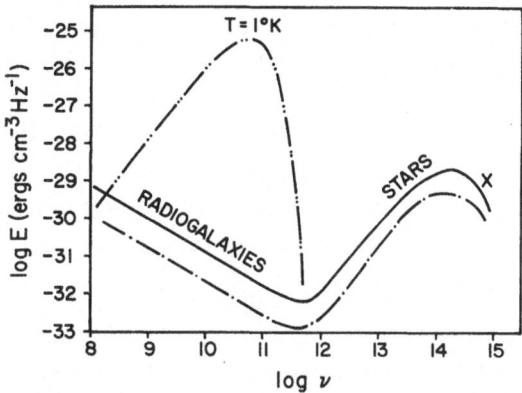

FIG. 2.   *Spectrum of stellar and radio-source radiation together with background spectrum ($T = 1°K$) corresponding to the hot model (calculated by Doroškevič and Novikov, 1964, Doklady AN SSSR, 154, 745).*

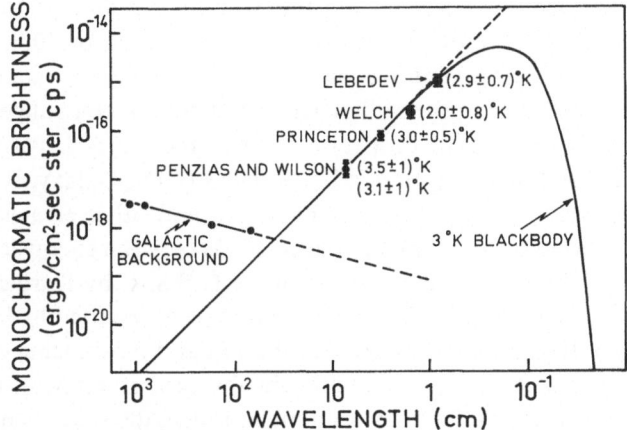

FIG. 3.   *Planck's radiation for $T = 3°K$ and the recent results of background measurements in cm and mm regions (see Field and Hitchcock, 1966; Thaddeus and Clauser, 1966).*

tive intensity measurements of interstellar CN lines ($\lambda = 3874 \cdot 60$ Å and $\lambda = 3874 \cdot 61$Å) giving the temperature of background radiation at wavelength 2·53 mm (Field and Hitchcock, 1966; Thaddeus and Clauser, 1966). In July, 1967, V.S. Stankevič, V.I. Puzanov and the author of this paper (Puzanov *et al.*, 1967) completed an absolute measurement of the cosmic background radiation at a wavelength of 8·2 mm. The measurements were carried out with the help of a sensitive superheterodyne radiometer and artificial black-body screens cooled to liquid-nitrogen temperature. The black-body temperature of the background radiation turned out to be $2 \cdot 9 \pm 0 \cdot 7°K$ (root-mean-square error).

The maximum of the background radiation falls at a wavelength ~ 1 mm. After the discovery of this background radiation, it is very important to make measurements with the high accuracy at wavelengths shorter than 1 mm. These measurements apparently allow us to conclude whether the spectrum of the background radiation is a Planck distribution. Probable features of this spectrum may give some unique information about different stages of the evolution of the universe (Galaxy condensation, etc.).

B. STATE AND CHEMICAL COMPOSITION OF INTERSTELLAR AND INTERGALACTIC MATTER

Submillimeter wavelengths provide the optimum range for investigating the very cold parts of the Galaxy. Measurements of the intensity distribution in this range allow us to detect the regions where, perhaps, gravitational condensation is continuing, which is very important for theories of stellar and planetary cosmogony. In particular, detailed spectral investigation in the millimeter and submillimeter ranges should allow us to detect the existence of molecules and dust in the Galaxy. At wavelengths close to 20 μm one would expect to find the maximum radiation from interstellar dust, under the assumption that the dust temperature is close to 20°K.

Further, it is known that several resonance lines of hydrogen, water, oxygen, and some other molecules occur in the millimeter and submillimeter ranges. The investigation of these lines permits the investigation of the temperature, density, and chemical composition of the coldest and most condensed parts of the Galaxy. It is also known that many lines caused by excited atomic hydrogen and other elements exist in the millimeter range. These lines correspond to transitions between energy levels with large quantum numbers. This effect was predicted in the U.S.S.R. by Kardašev (1959) and discovered also in this country by Soročenko *et al.* (1964). Excited hydrogen lines are a very effective tool for investigating the distribution and movements of regions with strongly ionized interstellar gas. Although the line-brightness temperatures in the submillimeter range are smaller than in the centimeter range, the spectral density must be considerably greater than the thermal radiation density of the galactic continuum. For the brightest nebulae, the expected hydrogen line-flux densities in the 10–0·05 mm range (quantum number $n = 50$–10) are $(3 \times 10^{-18} \text{ wm}^{-2}) - (3 \times 10^{-12} \text{ wm}^{-2})$ (for $\Delta f/f = 10^{-4}$).

C. 'INFRARED STARS' AND QUASARS

In recent years, hitherto unknown sources of electromagnetic radiation have been discovered. These are the 'infrared stars', for which the maximum in the spectrum intensity falls in the wavelength range between 3 and 20 microns, corresponding to a black-body temperature of only 700°K. It is very probable that such sources may be detected in the longer submillimeter range. Radio-astronomical and optical observations show that the maxima of intensity of the strongest sources, the most intense superstars and remnants of supernovas, must be in the submillimeter or infrared ranges. This radiation possesses a diversity of characteristics. In particular, radiation in this range has variable intensities, is strongly polarized, has an unusual spectrum, etc.

Investigations in the submillimeter and infrared ranges may be of decisive importance for revealing the nature of these objects, and also for resolving the related problems (formation and evolution of the galaxies, model universes studied with the help of the most distant sources, mechanisms for the acceleration of cosmic rays, finding of extraterrestrial civilizations). The expected flux densities from the brightest sources in the wavelength range close to 1 mm must be about 10^{-24}–10^{-25} wm^{-2} cps^{-1}, that at $\Delta f/f = 30\%$ gives 10^{-12}–10^{-13} wm^{-2}.

D. PLANETARY ATMOSPHERES

Spectral observations in the millimeter and submillimeter ranges are extremely important for revealing the chemical composition, pressure and temperature distributions in the atmospheres of the planets. The composition and conditions in planetary atmospheres are practically uninvestigated. It is known that in the millimeter and submillimeter ranges there are large resonance lines and bands of molecules such as those to be expected in planetary atmospheres: H_2O (13·5 mm, etc.), O_2 (5 mm and 2·53 mm), CO (2·61 mm, 1·3 mm), NO (1·99 mm and 1·2 mm), etc.

The investigation of the spectral features of planetary radiation will allow us to detect the existence of certain molecules in the atmosphere. Measurements of the shape and intensity of spectral lines will help to determine the height distribution of pressure and temperature. In the submillimeter range, flux densities may be equal to 10^{-20}–10^{-22} wm^{-2} cps^{-1} or 10^{-9}–10^{-11} wm^{-2} in the relative wave band $\Delta f/f \approx 30\%$.

E. SOLAR RADIATION

Submillimeter solar radiation is a source of information about the deepest layers of the chromosphere. In particular, the measurement of spectra, intensity, polarization, and time-dependence of radiation from the limb and active regions on the solar disk, connected with flocculae and spots, gives reliable data about the magnetic fields and electron densities above spots. These measurements help to clarify the nature of bursts connected with chromospheric flares, which play a significant role in geophysical phenomena.

A rather high instrumental resolving power (exceeding 30″) is a necessary requirement for these observations.

3. Instrumental Techniques

The state of submillimeter astronomy techniques is characterized by an intermediate position of this wavelength range. The review by Putley (1963) makes it unnecessary to restate all the methods of submillimeter radiation detecting.

At the present time, superheterodyne very wide band receivers with crystal-mixers at the input are used in the long-wave part of the submillimeter range (Cohn *et al.*, 1963). Optical methods are represented by Goley cells, bolometers of various kinds (including the superconducting type and also germanium cooled to liquid-He temperature). The most promising detectors for the submillimeter range seem to be photoconductive detectors and cooled bolometers with semi-conducting sensitive elements (In-Sb, Ge). Radiometers with such receivers have been developed and used for ground-based astronomical observation in England, U.S.S.R., U.S.A., and France (Low, 1961; Putley, 1965; Rollin, 1961; Popov, 1965; Karlova and Karlov, 1966; Arams *et al.*, 1966; Besson *et al.*, 1965). Figure 4 shows, for example, a submillimeter radiometer with an In-Sb sensitive element developed in the Moscow Institute of Radiotechnics and Electronics by Vystavkin and Popov. A significant success must be pointed out in the field of submillimeter guidance, filtering, and techniques of

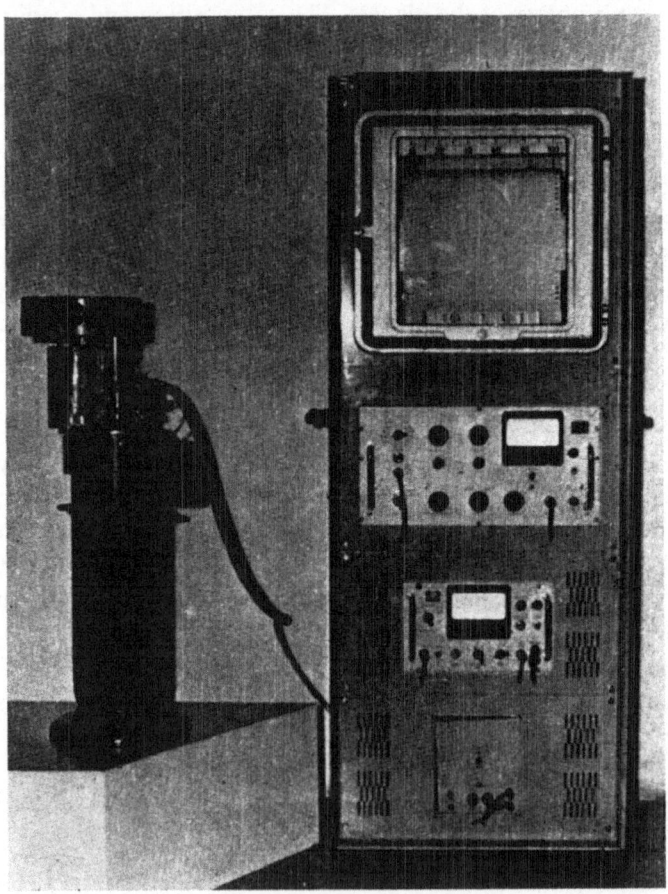

FIG. 4. *A submillimeter radiometer with sensitive element of In-Sb developed in the Institute of Radiotechnics and Electronics (Moscow) by A.N. Vystavkin and E.J. Popov.*

measurements, in particular, in those of polarization and interference (Colleman, 1963; Vinogradov *et al.*, 1967). In Figure 5 is represented a submillimeter Fabry-Pérot interferometer using wire-grid elements, developed in the Moscow Lebedev Physical Institute (in the laboratory led by A. Prohorov) by Natalia A. Irisova and her collaborators E. Dianov and E. Vinogradov.

In submillimeter radiometers, multimode detectors are used instead of the single-mode ones, which were developed for the microwave range. In the submillimeter detectors the linear dimensions are larger than the middle wavelength; furthermore, each element of the detector transforms the radiation incident on it independently of the others. This characteristic causes some modifications in the expressions for antenna directivity and noise-threshold sensitivity of radiometers. These problems have been investigated by a number of workers (Williams and Chang, 1963; Karlov and Prohorov, 1964; Popov, 1965; *et al.*).

For some calculations of sensitivity Planck's representation must be used instead of that of Rayleigh-Jeans, in those cases when the energy of one quantum is of the same order or more than kT. In the opposite case, when we consider a high-temperature submillimeter radiation, it is possible to keep a Rayleigh-Jeans representation. These problems have been investigated in detail by Karlov and Čihačev (1959) from the Lebedev Institute, and by other authors.

The main difficulties in the development of submillimeter-astronomy techniques

Fig. 5.   *Submillimeter Fabry-Pérot interferometer with wire-grid elements developed in Moscow, P.N. Lebedev Physical Institute, by N.A. Irisova, E. Dianov and E. Vinogradov.*

arise from the necessity of taking the receiving equipment out of the Earth's atmosphere, or at least of partially excluding its influence. Naturally, the simplest method consists of elevating the radiometers on mountains, where the humidity is less than 1 gm^{-3} during most of the year. Such attempts were undertaken by several groups, among which is the group (Bastin *et al.*, 1964; Baldock *et al.*, 1965) from the Queen Mary College (England) which conducted observations at an altitude of about 2000 m at wavelengths 1–4 mm. The group from Gorky Radiophysical Institute (U.S.S.R.) has stations on Elbrus and Aragac (Caucasus), and also on East Pamyr (Drjagin *et al.*, 1966; Gorohov *et al.*, 1962; Kisljakov and Plečkov, 1964). During these observations, atmospheric attenuation and submillimeter radiation from the Sun and the Moon have been chiefly investigated. All of these experiments gave very interesting data for the Sun and especially for the Moon in the long-wave part of the submillimeter range. At the same time, the importance of carrying out the observations at the highest altitudes is evident, especially for the wavelengths shorter than 1 mm.

The results of measurements by Fedoseev (1966) indicated the observed attenuation caused by a non-resonant absorption in oxygen may considerably exceed that of water vapour even in the winter. For this reason the calculated attenuations for 1–1·5 mm wavelengths were somewhat underestimated.

From the above it is evident that for submillimeter astronomy the most promising methods are those of space astronomy.

One of the relatively simple techniques of space astronomy consists of elevating the equipment by stratoplanes and balloons. The first published results obtained by such a technique for far-infrared observations (Bater *et al.*, 1967; Woolf *et al.*, 1967) are very interesting. Figure 6 gives the results of the first attempts to measure the water-

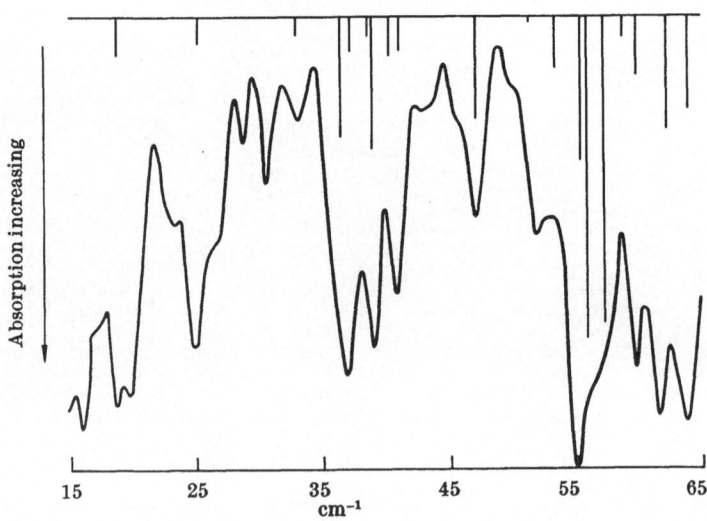

FIG. 6.   *An observed spectrum of the sky at 40000 ft. in the range 15–65 cm^{-1} (from Bater et al., 1967)*

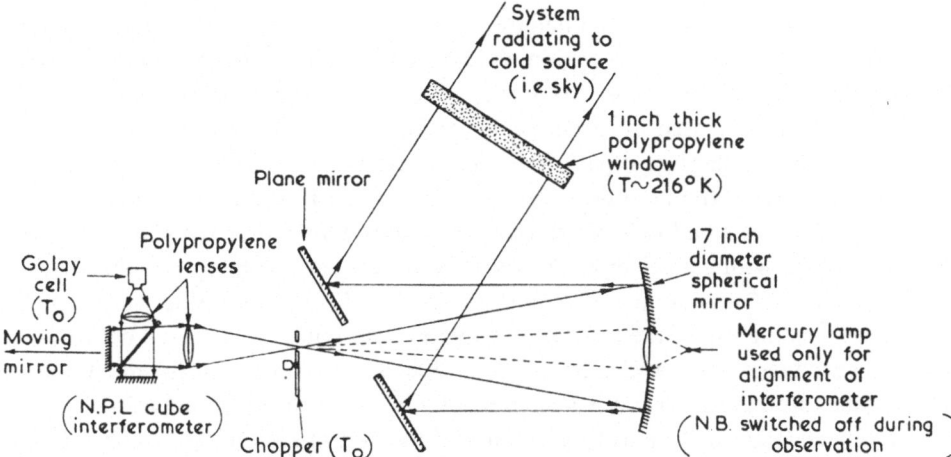

FIG. 7.   *Schematic diagram of the experimental arrangement of the telescope and the modular interferometer (Bater et al., 1967).*

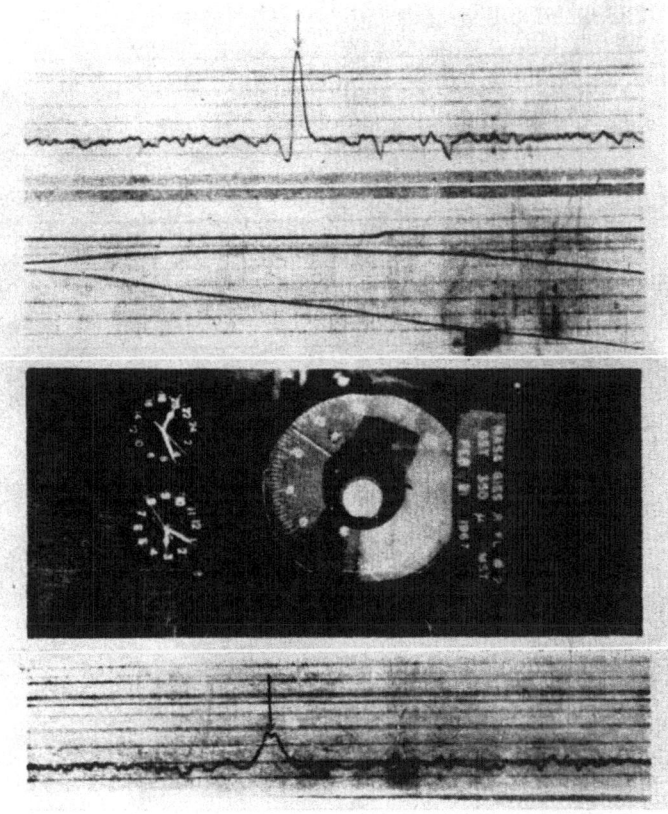

FIG. 8.   *Moon radiation in the submillimeter wave range ($\lambda = 320$ microns) obtained by Woolf et al. (1967).*

vapour spectrum, which were obtained by Gebbie's group on the stratoplane 'Gallileo' at an altitude of 40000 ft. in the region of 15–65 cm^{-1}. The measurements were made with the help of a submillimeter Fourier spectrometer (Figure 7). The balloon technique has been widely used in recent years for infrared astronomy. The first attempt at balloon submillimeter investigations has been made recently by Woolf and Low and their colleagues from the Hoddard Institute for Cosmic Researches (U.S.A.).

In the gondola of a balloon, which was flown up to 98000 ft., was placed a modulated radiometer with a receiver involving a Ge He-cooled bolometer. A sensitivity of 7×10^{-14}w cps$^{-1/2}$ was obtaine dat temperature 1·8°K. An interesting feature of the experiment was the use of the exterior vacuum for pumping the dewar. The measurements were made in the wavelength range of 300–450 microns with the maximum at 320 microns. Modulation was performed by oscillation of the reflecting mirror so that the radiometer recorded intensity gradients of the sky radiation. Although the only source observed was the Moon (Figure 8), an upper radiation-flux limit was estimated which turned out to be equal to $2 \cdot 10^{-23}$ w cm^{-2} cps^{-1}. This first attempt to track the sky in the submillimeter region displays the great possibilities of balloon submillimeter astronomy.

Of much interest also are the explorations in the field of balloon submillimeter

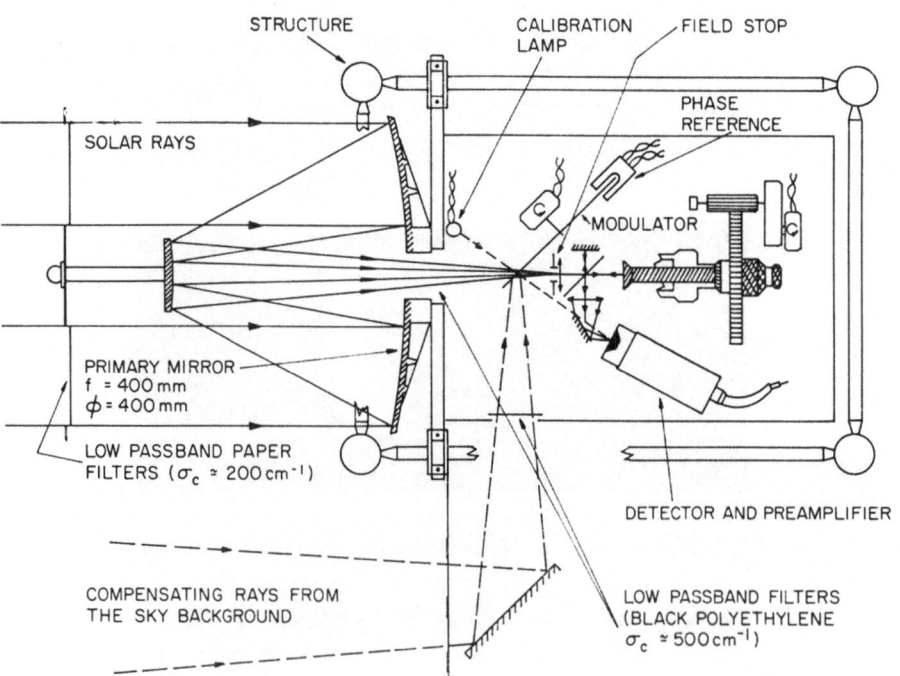

FIG. 9–11. *Submillimeter radiometer and some parts of the installation flown in the gondola of the balloon of the Meudon Observatory (Gay et al., 1967).*

astronomy by J. Gay and others, which started about 2 years ago in the Meudon Observatory (France) under the guidance of Lequeux (Gay *et al.*, 1967). Using hydrogen balloons of the French National Centre for Cosmic Researches, together with a stabilized gondola having a pointing precision 10–20 sec of arc, this group made spectral investigations of the Sun with a resolution of 0.3 cm^{-1} in the range of 50 to 2300 μm. The experiment was planned to obtain a spectrum of residual water vapour of $O^{18} H_2$ and OHD and, possibly, the solar C^+ line at 156 microns. It was intended to make the absolute calibration on the ground, and relative measurements on-board, which were expected to give spectrophotometry of the Sun with an accuracy better than 5%, which would be essential for selecting models of the photosphere.

As a first step this group has used a 40-cm Cassegrain telescope and a Fourier interferometer with a mylar beam-splitter. The detector is of Golay type, built in France by ONERA with a modulation frequency of 10 spc (Figure 9). After synchronous detection the output voltage is converted into variable frequency pulses, sampled in synchronism with the motion of the mirror of the interferometer, and recorded by an on-board tape recorder. The balloon radiometer apparatus of the Meudon Observatory is shown in Figure 10 and 11. The first flight was made on June 24, 1967, with the equipment operating in the wave band 50–300 μm. The measured solar spectrum

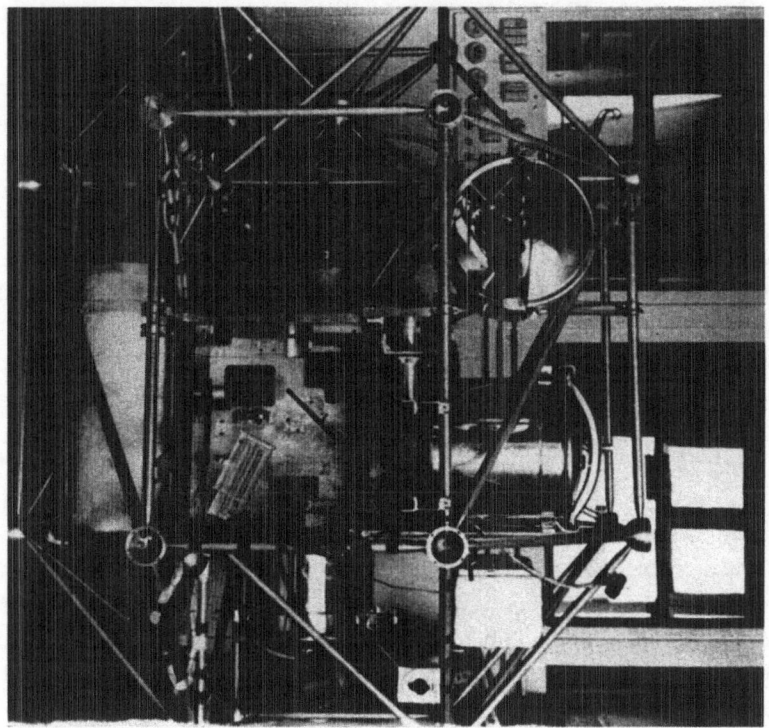

FIG. 10.

Fig. 11.

(after corrections) is shown in Figure 12. For measurements needing higher sensitivity, a helium-cooled bolometer of the Putley type is under study. The measurement of the sky background near 1 mm is under investigation, with a wide field apparatus allowing continuous comparison of the sky temperature with the temperature of the helium bath. Balloon observations of planets and quasars are supposed to be made with a reflector of 1 m to 1·40 m diameter and with a very sensitive detector. Ground observations are also planned at an altitude of 3140 m, above 900 microns and in the 350 micron 'window'. The above-mentioned difficulties increase considerably when satellite technique and rockets are used for submillimeter observation.

Two problems remain to be solved, and will determine the future success of submillimeter space astronomy. These are:

(a) The development of a high-precision mirror antenna suitable for the submillimeter wavelength and supplied with tracking systems;

(b) The use of cooled submillimeter detectors on satellites.

For the first problem, the difficulties may be somewhat smaller than for Earth radio astronomy. The absence of wind and weight loads will make engineering calculations easier. The main difficulties, probably, will be the assembling of highly directive submillimeter antennae as well as tracking after predicted points of the sky. These problems are very similar to those arising in optical space astronomy.

The detector-cooling problem, under rocket and satellite conditions, is a rather complicated one. The requirement for cooling to the temperature of liquid helium (since no semi-conductor or other sensitive elements have been developed operating at a higher temperature) requires a volume filled by liquid helium. The storage of such liquid requires some economical micro-cooler or cryostat capable of storing liquid helium during some long period in the absence of gravity, and in a high-vacuum and

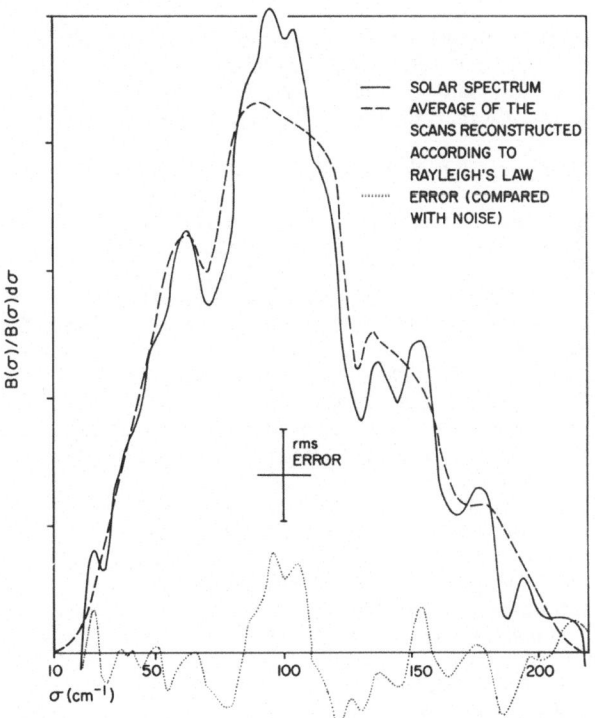

FIG. 12.   *Solar spectrum in submillimeter band obtained by the Meudon group (Gay et al., 1967).*

low-temperature environment, which is also to withstand vibration and overloading during the active part of trajectory. The situation is analogous to that arising in super-conductivity space experiments (for cosmic rays or space relativity investigations). In this connection the results of methodical experiments which have been obtained recently by Lidie V. Kurnosova's group at the Levedev Physical Institute (Anaškin *et al.*, 1967) are of interest. In this experiment the possibility of using superconductive devices on cosmic vehicles was tested. Helium at overcritical pressure (2·4 atm and an initial temperature $\approx 5·2\,°K$) was used as the cryogenic agent. Two superconducting solenoids with magnetometers and six thermometers were installed inside the helium container (volume $\approx 11 \cdot 5$ l). The experiment was carried out on the sputnik Kosmos-140 and resulted in the statement that such a one-phase method of storage at low temperature (up to $\sim 7\,^{\circ}$K) can be used in the conditions of weightlessness.

If these difficult technical problems can be solved, we may hope that a new and promising branch of astronomy, namely submillimeter astronomy, will arise and give us surprising results.

References

Anašk in, O.N., Belickij, B.M., Brodskij, V.B., Kurnosova, L.V., Razorenov, L.A., Sidjakina, T.M. Mihailov, N.M., Fradkin, M.I. (1967)   Lebedev Phys. Inst., preprint, June.
Arams, F., Allen, C., Peyter, B., Sard, E. (1966)   *P.I.E.E.E.*, **54**, 183.
Baldock, R.V., Bastin, J.A., Clegg, P.E., Emery, R., Gaitskell, J.N., Gear, A.E. (1965)   *Astrophys. J.*, **141**, 1289.
Bastin, J.A., Gear, A.E., Jones, G.O., Smith, H.J.T., Wright, P.J. (1964)   *Proc. R. Soc. London* **A278**, 543.
Bater, M., Cameron, R.M., Burroughs, W.J., Gebbie, H.A. (1967)   *Nature*, **214**, 377.
Besson, J., Carro, R., Matteoli, M., Papoular, R., Phillppeau, B. (1965)   *Onde Elect.*, **45**, 107.
Cohn, M., Wentworth, F.L., Wiltse, J.E. (1963)   *P.I.E.E.E.*, **51**, 1227.
Colleman, P.D. (1963)   *IEEE Trans.*, Microw. Theory Techn., **11**, 271.
Drjagin, Yu.A., Kisljakov, A.G., Kukin, L.M., Naumov, A.J., Fedoseev, L.I. (1966)   *Izv. VUZOV, Radiofizika*, **9**, 1078.
Field, G.D., Hitchcock, J.L. (1966)   *Phys. Rev. Lett.*, **16**, 817.
Fedoseev, L.I. (1966)   *Izv. VUZOV, Radiofizika*, **9**.
Gay, J., Lequeux, J., Turon, P. (1967)   Obs. Paris-Meudon, preprint, July.
Gorohov, N.A., Drjagin, Yu.A., Fedoseev, L.I. (1962)   *Izv. VUZOV, Radiofizika*, **5**, 413.
Kardašev, N.S. (1959)   *Astr. Zu.*, **3**, 813.
Karlov, N.V., Čihačev, B.M. (1959)   *Radioteh. Elektron.*, **4**, 1647: 1052.
Karlov, N.V., Prohorov, A.M. (1964)   *Radioteh. Elektron.*, **9**, 2088.
Karlova, E.K., Karlov, N.V. (1966)   *Pribory Tehn. Experimen.*, **2**, 19.
Kisljakov, A.G., Plečkov, V.M. (1964)   *Izv. VUZOV, Radiofizika*, **7**, 46.
Low, F.J. (1961)   *J. Opt. Soc. Am.*, **51**, 1300.
Martin, D.H. (1962)   *Contemp. Phys.*, **4**, 139; (1963) **4**, 187.
Penzias, R.A., Wilson, R.W. (1965)   *Astrophys. J.*, **142**, 419.
Putley, E.H. (1963)   *P.I.E.E.E.*, **51**, 1412.
Putley, E.H. (1965)   *Appl. Opt.*, **4**, 649.
Popov, E.J. (1965)   *Izv. VUZOV, Radiofizika*, **8**, 862.
Puzanov, V.I., Salomonovič, A.E., Stankevič, V.S. (1967)   Rep. 13 Gen. Assemb., IAU, Prague; (1967) *Astr. Zu.,* **44**, 1128.
Rall, P.G., Wilkinson, D.T. (1966)   *Phys. Rev. Lett.*, **16**, 405.
Rollin, B.V. (1961)   *Proc. Phys. Soc.*, **77**, 1102.
Salomonovič, A.E. (1964)   Dissertation, Phys. Inst. Lebedev.
Soročenko, R.L., Borodzič, E.V., Dravskih, Z.V., Dravskih, A.F., Kolbasov, V.A. (1964)   Rep. 12 Gen. Assemb., IAU, Hamburg.
Straiton, A.W., Tolbert, C.W. (1960)   *P.I.R.E.*, **5**, 898.
Thaddeus, P., Clauser, J.F. (1966)   *Phys. Rev. Lett.*, **16**, 819.
Vinogradov, E., Dianov, N.A. (1967)   *J.E.T.P. Lett.* (in press).
Williams, R.A., Chang, W.S.C. (1963)   *IEEE Trans.,* Microw. Theory Techn. **11**, 513.
Woolf, N.J., Hoffmann, W.F., Frederick, C.L., Low, F.J. (1967)   Report on Discussion Meeting on Infrared Astronomy, London, May.
Wort, R.J.H. (1962)   *Nature*, **195**, 4848.
Zeldovič, J.B. (1966)   *Usp. fiz. Nauk.*, **89**, 647. (In this review paper see bibliography of problems.)
Ževakin, S.A., Naumov, A.P. (1963)   *Izv. VUZOV, Radiofizika*, **6**, 674.

DISCUSSION

F.J. Low: The last experiment that you discussed. When was it flown, and was it a complete success?.

A.E. Salomonovič: It was flown this spring, and the results were presented at the Tenth International Conference on Cosmic Rays in Canada.

NEW TECHNIQUES IN SPACE RADIO-ASTRONOMY

J.-L. STEINBERG

(Paris and Meudon Observatories, France)

In spite of the title of this paper, it is not easy to point out many significant developments in the techniques of space radio-astronomy in the last few years. Radio-astronomers are limited in their endeavours by several difficulties which they have not yet
been able to overcome completely. I still remember F.G. Smith, in his 1965 Liège
Symposium talk, saying that "space radioastronomy is in great danger of turning to
geophysics". This statement seems still valid three years later.

However, we might consider that this tendency to deal with geophysical phenomena
more than with observations of real astrophysical significance, is after all, not so bad.
First of all, it is becoming more and more difficult to define the boundary between
geophysics and astrophysics. And second, the Earth environment can be used as a
laboratory where a number of processes which are assumed to take place in celestial
bodies can be studied *in situ*. Some types of radio emissions observed by space telescopes are still to be explained and could very well lead to the understanding of
mechanisms overlooked by the theoreticians. Radiophysical methods make a very
powerful tool to study these phenomena, which occur in a medium that is now being
extensively probed by rockets and satellites which send a continuous flow of data to
the Earth, about 100000 bits per second, according to Parker.

Also, techniques which were introduced many years ago in geophysics, such as the
analysis of radio-propagation phenomena in the ionosphere, are now considered and
used for studying other planetary atmospheres and the corona of our Sun. This can be
taken as a new branch of space radio-astronomy.

We must not, therefore, be pessimistic about the future of space radio-astronomy;
many projects are being planned in many countries and their carrying out will certainly bring important results.

1. Galactic Radio Emission

Smith (1965) has shown that, at frequencies close to 1 MHz, we are likely to observe
emissions of extragalactic origin. This prediction is based on our present knowledge
of the spectra of different components of the radio emission observed from the ground
at higher frequencies. We know what we should measure, in the absence of any
unexpected phenomena.

Perek (ed.), Highlights of Astronomy, 164–175. © I.A.U.

A. MEASURING TECHNIQUES AND PROBLEMS

The difficulties encountered in these measurements of absolute flux densities in space are twofold: (a) Radiation in the Earth's environment or of yet unknown origin, (b) Calibration problems connected with the behaviour of an antenna in a plasma.

(a) *'Abnormal' radiation.* – We shall not discuss here radiation observed in the plasma surrounding the space radiotelescope when $1 > X > 1 - Y^2$ (where X and Y are the familiar plasma parameters), which is not completely understood but is probably of local origin (Harvey, 1965). But 'abnormal' levels have been observed by Huguenin *et al.* (1964) at 700 kHz, by Benediktov and coworkers (1965) at 725 and 1525 kHz, and by Alexander *et al.* (1966) at 2·7 MHz. We have too little information on the characteristics of this radiation to explain its origin. Benediktov *et al.* have shown it to be correlated with the flux of suprathermal particles of energy higher than 100 eV, but the mechanism of production of radio-energy is not completely understood. Huguenin's observations were also taken at fairly high altitudes; they show some bursts occurring sometimes at the same time at 700 and 2200 kHz; but the general level observed at 700 kHz remained stable and very high during the whole flight.

Many observations incorporating some directivity and simultaneous measurements on the surrounding medium are necessary to throw some light on this irritating problem.

(b) *Antenna behaviour in the plasma.* – If we are not making observations at very high altitudes where the operating frequency is very much larger than all resonance frequencies in the surrounding medium, we must investigate the behaviour of our antenna in the plasma.

Extensive theoretical work has been done on the problem in the last few years; the difficulties encountered are formidable, so that a number of simplifying assumptions have to be made, mainly in neglecting the ion sheath which forms on any body immersed in a plasma (Walsh and Haddock, 1964).

The measurement of sky brightness with an antenna in a magnetoplasma has been analyzed by various authors for rather simple cases: uniformly bright sky, plane stratified ionosphere with uniform magnetic field. Two approaches are available: (1) measure the r.m.s. open-circuit noise voltage at the antenna terminal and the antenna impedance: $Z_a = R_a - j/C_a\omega$, then apply Nyquist's theorem (Walsh *et al.*, 1964); or (2) measure the electric field in the medium and the index of refraction of this medium (Budden and Hugill, 1964; Hugill and Smith, 1965). The two methods have been shown to be physically equivalent by various workers (Walsh and Weil, 1967).

However, in the first method, all observers who measured R_a found it much larger than predicted by plasma theory, so that one has to rely on calculations to obtain the real electromagnetic-radiation resistance in the medium from C_a measurements, which lead to the index of refraction. It seems that this method is valid, but will it still be applicable to antennas which can no longer be considered short dipoles and treated by quasi-static theory? The only way out is to understand this 'abnormal resistance'. Experiments by Bramley (1965), Walsh *et al.* (personal communication), Alexander, Stone and Weber (personal communication) and the Meudon Group (1965) have produced data on Z_a. The mechanism which produces this 'abnormal resistance' is most probably electron collection in phase with the h.f. potential on the antenna. This is Bramley's conclusion; however, more experiments will have to be carried out before we completely understand the question.

The 'electric-field' method is probably less sensitive than the first one to the presence of this 'abnormal resistance'. But the design of the 'voltmeter' to obtain the open-circuit noise voltage at the antenna terminals must be based on some knowledge of the upper limit of the antenna impedance in the plasma. The above-mentioned experiments show that the measured impedance can be, in some instances, much higher than predicted.

B. RESULTS

There is little doubt that we shall not be able to 'look' at extragalactic objects on frequencies lower than 1 MHz. From all measurements obtained today, it appears that the apparent brightness of the sky decreases with frequency below about 2 MHz. However, there is still a large scatter between measurements taken between 1 and 3 MHz, and all we can say of the radio emission in this spectral region is that its spectral index is close to zero. Only one set of measurements has been obtained at frequencies lower than 1 MHz, on board Elektron-II, which reached 70000 km altitude (Benediktov *et al.*, 1965). Only this set of measurements shows, indeed, a fall in the integrated brightness with frequency.

Observations have also been obtained by Slyš (1965) at 30 kHz, 210 kHz and 1 (or 2) MHz from several Soviet space probes. They show a very steep *increase* in the brightness with *decreasing* frequency between 500 and 30 kHz. This phenomenon is still unexplained; its existence does not seem doubtful, as it has been measured on several probes carrying different experiments with radio-astronomy receivers; all of these measurements agree to within the experimental uncertainty, ruling out any explanation based on noise from the spacecraft. Similar observations have been taken on board some OGO satellites from 0 to 100 kHz and show also a very high level of noise. But we have no quantitative data on this result.

A tentative explanation based on emission from Jupiter has been put forward by Slyš, but more recent Soviet results do not confirm this hypothesis.

C. PROSPECTS

It is clear that, as in early ground-based radio-astronomy, we need some directivity however small; it would give brightness distributions and would help in rejecting and/ or measuring Earth radio radiation, if any. Brightness distribution would in turn bring some knowledge of the distribution of ionized hydrogen in the Galaxy which is supposed to produce absorption on low frequencies.

Ionospheric focussing has been proposed by many radio-astronomers as a means to provide directivity (Smith, 1961; Jennison, 1961). It appears that this method is difficult to put into practice as no results have yet been published. The University of Michigan Group has proposed a satellite system based on it, and Ariel-III is also partially designed to use the method. Theories of the phenomenon are available. Most of them do not take into account the Earth's magnetic field or the Earth's curvature. *A fortiori*, they treat the ionosphere as a continuous well-behaved medium, neglecting inhomogeneities which are known to be present.

The NASA-GSFC Group is planning to launch a series of RAE satellites with a double travelling-wave V antenna made of four 750-foot booms (Alexander and Stone, 1964). This enormous structure will be gravity-gradient stabilized in a circular orbit at 6000 km altitude. Each boom is made of thin heat-treated metal which, when driven out of a drum in strip form, takes a tubular section. Recent developments include perforations, silver-plating of the outside surface and blackening of the inside, together with some interlocking device. These characteristics insure very low thermal bending and twisting. Ingenious techniques have been devised to measure current distribution on full scale of these antennas and deduce their polar diagrams in vacuum (Cory and Fenwick, 1966). The RAE booms can be connected in different ways to achieve directivity in the upward vertical direction, the vertical downward direction, and also sideways. This system should give a 20 to 30° beam at 4 MHz with a front-to-back ratio of better than 15 db. It is scheduled for flight in early 1968.

One can consider that such systems, of still greater size, could be launched to still higher altitudes if gravity-gradient forces are strong enough for their stabilization. Much larger antennas are being studied, mainly in the U.S.A. to our knowledge, which could produce better directivity and front-to-back ratios. They would rely on small satellites, slaved to a master one, all of them with active stabilization to keep the shape of giant rhombic antennas in space and point them (Walsh, personal communication).

If we want more resolving power, the use of aperture synthesis will be necessary. A proposal for such a system has been produced by the Harvard College Observatory Group and by the NASA-GSFC Group (Huguenin, Alexander, personal communications).

D. CONCLUSION

The data presently available show that below 4 MHz and down to probably 700 kHz the sky brightness falls with frequency. This is interpreted in terms of free-free absorption by ionized hydrogen in our own Galaxy.* Below 0·7 MHz, new phenomena

---

* One is led to an electron measure of 5 pc cm^{-6}.

occur producing high levels of noise with a density increasing as frequency decreases. No reliable theory is yet available to explain the rise of the apparent sky brightness.

Improvements in the understanding of antennas in a plasma (both for short dipoles and more complicated antennas) are necessary to decrease the spread in low-directivity absolute-brightness measurements below some MHz. The need for some directivity (some tens of degrees or better) is widely recognized, and new systems are readied for flight or being designed to achieve this goal. No systems for better than 10° beam-width are being designed to our knowledge, for flight in the next few years.

2. Solar Studies

In the past few years, a number of observations of solar radio radiation below 10 MHz have been made from satellites. Evidence has been obtained that solar bursts of types II and III do, indeed, present spectra extending to frequencies as low as 200 kHz.

A. EXPERIMENTS FLOWN

Most of the published results have been obtained with equipment which was not designed for solar noise observations: Hartz's (1964) pioneering work used the a.g.c. voltage of an ionospheric sounder receiver on board Alouette-I and -II and no on-board calibration or data processing. Slyš (1967) used receivers designed for sky-brightness measurements, and the bit rate available to him was very low. Both Hartz and Slyš observations have been related with ground-based spectral records of solar radio emission.

The only experiment flown specifically to study the Sun has been designed by the University of Michigan Group for flight on an EGO satellite. Results have been obtained in the 2 to 4 MHz band with a frequency-swept receiver; they look promising and correlate well with ground-based observations (Warwick, personal communication).

When all these data are completely reduced, they will bring a mine of information on the solar corona at very high altitudes, on the mechanisms which produce solar bursts and the agents which trigger the radiating processes. There is still a lot to be learned about these processes and new phenomena might very well be discovered; for instance, it appears that some bursts show a low-frequency cut-off whose origin is not yet understood.

However, more refined observations than available at present are required; they are being planned.

B. PROJECTS

We know of at least four projects specifically aimed at solar radio observations at low frequencies. They are:

Project Pilgrim   Harvard College Observatory (G. R. Huguenin)
Project RAE A and B   NASA-GSFC (R. G. Stone and J. K. Alexander)
Project ATS-E   NASA-GSFC (R. G. Stone and J. K. Alexander)
Project Roseau   R.A. experiment from Meudon Observatory, France

The first three projects are designed and built in the U.S.A. The fourth one is a cooperative project between the U.S.S.R. and France; the satellite will be designed and built in France and launched in the U.S.S.R. in late 1971; it is approved and financed.

The Pilgrim and RAE satellites will carry nothing but radio-astronomy experiments.

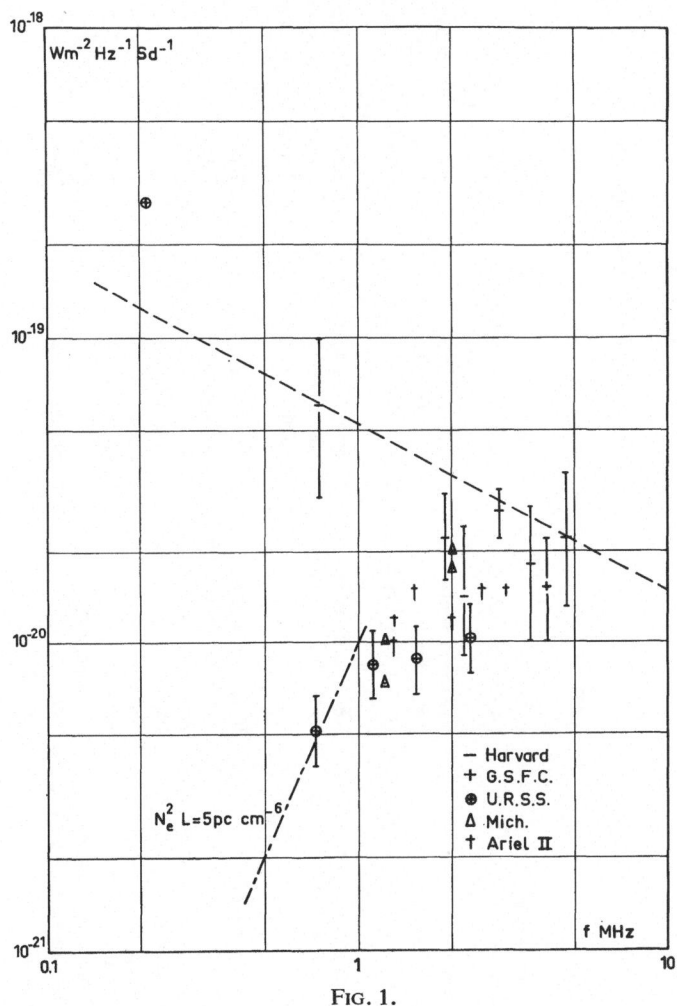

FIG. 1.

Table 1

Solar radio-astronomy satellites

	RAE	ATS-E	Pilgrim	Roseau
Name	RAE	ATS-E	Pilgrim	Roseau
Launch date	1968	1969		1971
Country	U.S.A.-NASA	U.S.A.-NASA	U.S.A.-NASA Harvard Coll. Obs.	CNES-France Intercosmos- U.S.S.R. Meudon Obs.
Orbit	6000 km circular	Synchronous	300–10000 km 37° Inclin.	500–100000 km 52° Inclin.
Antennas	2 V 1 dipole	4 × 43 m booms	16 and 64 m dipoles	2 × 120 m dipoles
Antenna orientation	gravity stabilized	gravity stabilized	Dipoles perpendicular to ecliptic	Dipoles in ecliptic
Frequency range	200 kHz–9 MHz	50 kHz–4 MHz	250 kHz –16 MHz	30 kHz–2 MHz
Frequency distribution	20 fixed f. channels	step sweep 32 freq. or fixed	3 ranges cont. sweep 2 fixed freq. 0·95 and 4·01 MHz	8 switched channels or fixed any freq. 1 fine step swept channel 30–200 or 200–800 kHz
Receiver type	Ryle-Vonberg or Total pwr. Int. calibr.	Total pwr. Int. calibr.	Total pwr. Int. calibr.	Total pwr. Int. calibr.
Ant. measurements	Z^a	Z^a	Z^a at 2 freq.	C^a all freq.
Band width	10	40 kHz	?	1–10–50 kHz
Sweep	None	4-8 s	?	10 s or 5 s on swept channel
Dynamic range	20+20+20	60 dbs	?	80 dbs
Other experiments	Electron trap	None	None	4 main – 1 at a time 3 auxiliary
Data processing on board	None	?	None	Computer on board
Weight	165 kg		?	500 kg

ATS-E will carry other experiments as well (Alexander *et al.*, 1967). Roseau has five experiments on board, only one or two being in use at one time. Table 1 gives a comparison between these four satellite experiments.

A common characteristic is the use of simple dipoles in all projects (we do not consider the RAE V antennas as designed for solar observations). Solar emissions will therefore be modulated by the satellite rotation in both RAE and ATS systems. The Pilgrim project is spin-stabilized around the Sun–Earth axis and will therefore exhibit no such effects. The Roseau satellite, on the contrary, will use spin modulation to get some idea of the direction of solar emissions. Its two crossed dipoles will rotate in the ecliptic plane and can be switched in an attempt to use the dipole directivity to obtain the altitude of radio-emitting regions in the solar corona.

Another common characteristic is the rather poor definition achieved in frequency and time resolution and also in intensity. There is a well-known relationship between bandwidth, time constant (or number of measurements per second), and fluctuations in intensity readings, which is especially annoying in this case where the directivity is poor and the sky background temperature high. On the other hand, near the lower frequency limit of the spectral range covered, the dipole efficiency becomes very small indeed, and therefore the system noise referred to the antenna terminals is no longer negligible.

However, we know so little about solar phenomena at frequencies below 1 MHz that all information, even if rather crude, will be welcome. In the future, better antennas and multichannel receivers will bring more valuable information.

C. RADIO PROPAGATION EXPERIMENTS

Observations of the occultations of radio sources by the solar corona have been made for years to study the atmosphere of the Sun. More recently, scintillations of radio sources through the interplanetary medium have been investigated; this kind of work has led to a better knowledge of the inhomogeneities of the interplanetary gas and to measurements of the angular size of radio objects, particularly quasars (Salpeter, 1967; Cohen *et al.*, 1967).

With the advent of satellites, a group in M.I.T. under Professor Harrington, have now proposed to create an artificial radio source to study the solar corona by analyzing the propagation of coherent radio waves through it. This project, named 'Sunblazer', would use a small satellite carrying a modulated transmitter operating on two frequencies, at present 75 and 225 MHz. The orbit would be close to the Earth's, inside it, so that the satellite would be 'seen' behind the corona for some time. The idea is to build it with high reliability and low cost in order to be able to launch several of them and always to have one in operation at a given time. By measuring the apparent radio position of this satellite, as well as the polarization and intensity of the radio waves received from it, a great deal of information can be obtained on the corona and the interplanetary medium. The effective radiated power from the satellite will, of course, be rather small, so that sophisticated coding on board and receiving techniques on the ground will have to be incorporated in the design.

The experiment will use frequencies which can easily be received on the ground. But it may be necessary to use much smaller frequencies to study, by the same method, the interplanetary medium between the Earth and the limits of the solar system. The receiving end would then have to be on an Earth satellite or on the Moon.

This new technique seems very promising.

3. Planetary Observations

All solar low-frequency satellites are, of course, thought to be able to pick up emissions from the planets, if any. The most likely source is Jupiter. However, no emission from this planet has ever been detected from the Alouette satellites, which have provided, by far, the longest time of observation on low frequencies. All surveys of the data have been unsuccessful.

But Alouette equipment was not designed for absolute measurements, so that only bursts of radiation from Jupiter could be detected with any certainty. On the other hand, the Soviet probes Zond-II, Zond-III and Venus-II carried radiometers operating on 30, 210 and 2000 kHz, with input switching and fed from short dipoles of 3·75 and 2·50 m. Slyš (1966), who was responsible for these experiments, has claimed the observation of Jupiter radiation on 210 kHz on the following grounds:

(1) The 210 kHz radiation varies with the distance between the probe and Jupiter according to an inverse square law. This is based on a comparison between three flights, during which this distance varied widely. This conclusion assumes that Jovian flux density is more or less constant in time.

(2) During the ascent of Zond-II above the Earth's surface, periodic variations of the received signal were noticed, which could be interpreted as due to interference fringes between a direct ray incoming from a discrete source and a ray reflected from the ionosphere. Computation of the zenith distance of the source gives a direction close to Jupiter's at the time of observation.

(3) A drop of 80% of the received 210 kHz signal at Venus-II has been observed to last for some minutes, after which the received level resumed its previous value. This is interpreted as the occultation of a source by the Moon. Again the source is found to be Jupiter, in spite of the fact that the probe, the Moon, and Jupiter were not lined up at the proper time. This is, rather convincingly, explained in terms of ionospheric refraction and leads to a maximum apparent diameter of Jupiter of 10′.

(4) The low-frequency spectrum seems to fit rather well with higher-frequencies spectra of Jupiter as reported by Ellis.

These observations combined three new techniques in space radio-astronomy:

(1) The use of the inverse-squared distance law to prove the origin of some radia-

tion from a discrete planetary source. This, of course, can be used only if there is only one such source in the visible sky.

(2) The observation of interference fringes on the ionosphere.

(3) The observation of the Moon occultation of a source of low-frequency radiation.

However, these results can be criticized in many ways. The observation of fringes by reflection on the ionosphere assumes it to be spherically symmetrical and homogeneous. We cannot be sure of that and we even have evidence, at lower altitudes, that it is not the case. The optical path difference at such low frequencies will be so large that it is difficult to believe that coherence between the two interfering rays could be preserved all the way to the satellite. The same applies to the apparent diameter deduced from the Moon-occultation observations. From what we know of the interplanetary medium, we can compute the probable apparent diameter of a point source of 210 kHz radiation as seen through the scattering medium (Salpeter, 1967; Cohen et al., 1967). It is more likely to be some degrees than some minutes. Of course our knowledge of the interplanetary medium between the Earth and Jupiter is rather scanty; however, all we know of the solar wind theory tends to support apparent diameters of the order of degrees with deep intensity scintillations.

In spite of the impressive and ingenious explanations proposed by Slyš, some doubts remain as to whether Jupiter is or is not the source of most of the 210 kHz radiation measured repeatedly on board three different Soviet probes. Moreover, our Soviet colleagues have carried out new experiments which did not confirm Jupiter's radiation at low frequencies.

The Sternberg Institute Group (Shklovsky, personal communication) has flown 3-channel radiometers and short antennas on two lunar orbiters: Luna-XI and -XII. This kind of experiment has been proposed by several groups in different countries but never carried out before. Slyš used the same frequencies as above. In the first orbiter, the antennas were rotating in or near the ecliptic plane. On the second, the spin axis of the satellite was in the ecliptic plane. These two experiments should have provided direct evidence of the presence of a major localized source of radiation, and its position in space should have been measured both by Moon occultation and by spin modulation. Jupiter was not confirmed to be a bright source on 210 kHz, and no major source has been found on the frequencies 20–30, 210, 1000–2000 MHz.

All these beautiful Soviet experiments do therefore require reinterpretation, and the task does not seem to be an easy one. Our colleagues must, in any case, be credited for a number of 'firsts' in low-frequency radio-astronomy. Other groups in other countries are already planning new experiments along the same lines, especially lunar occultation experiments.

4. Conclusion

A number of new results have been obtained in space radio-astronomy, of which the best established appear to be: the fall in sky brightness below 2 MHz, and the existence of solar bursts with spectra extending down to 200 kHz. None of these results has been obtained with essentially new techniques.

There are indications that interference fringes on the ionosphere and Moon occultation have been observed on 210 kHz, but this is still disputable and the results obtained by these methods have not been confirmed.

Lunar orbiters have been used for the first time in low-frequency radio-astronomy. The orbiting of satellites of the Sun to be used as artificial radio-sources of coherent radiation is being planned and deep-space probes are already in use to study the atmosphere of planets (Fjeldbo and Eshelman, 1965).

What we need next is directivity, mainly to investigate the distribution of ionized hydrogen in our Galaxy and get better solar data. We shall certainly see achievements in that direction in the next few years; but it already appears probable that we shall not be able, on frequencies lower than 1 MHz, to probe very deep in our Galaxy and that only the immediate neighbourhood of the solar system will be visible to us, in this wavelength range. The question then arises whether it will be better to fly by or to land on the objects we want to study or to put up giant radio-telescopes in space or on the Moon.

Acknowledgment

I wish to thank the many colleagues who gave me information on their unpublished work and projects.

References*

Alexander, J. K., Stone, R. G. (1964)   *Ann. Astrophys.*, **27**, 837.
Alexander, J. K. *et al.* (1967)   Specifications for the solar radioastronomy experiment of the ATS-E Satellite, March.
Alexander, J. K., Weber, R. R., Stone, R. G. (1966)   Comm. to 123rd Meeting of the American Astronomical Society.
Benediktov, Ye. A. *et al.* (1965)   *Issledovanija Kosmičeskogo Prostranstva*, Moscow, June 10–15. – NASA Translation TT F-389.
Bramley, E. N. (1965)   *Planet. Space Sci.*, **13**, 979.
Budden, K. G., Hugill, J. (1964)   *Proc. Roy. Soc.*, A277, 365.
Cory, T. S., Fenwick, R. C. (1966)   Comm. to URSI-IEEE. Fall meeting.
Fjeldbo, G., Eshelman, R. V. (1965)   *J. Geophys. Res.*, **70**, 3217.
Fjeldbo, G., Eshelman, R. V. (1965)   *J. Geophys. Res.*, **70**, 3701.
Harrington, J. V. (1965)   Study of a small solar probe (Sunblazer) PR 5255-5. Center for Space Res., M.I.T.
Hartz, T. R. (1964)   *Ann. Astrophys.* IAU Symp. no. 23, Liège 1964, p. 357.
Harvey, C. C. (1965)   *Ann. Astrophys.*, **28**, 248.
Hugill, J., Smith, F. G. (1965)   *Mon. Not. R. astr. Soc.*, **131**, 137.
Huguenin, G. R., Lilley, A. E., McDonough, W. H., Papagiannis, M. P. (1964)   *Planet. Space Sci.*, **12**, 1157.
Jennison, R. C. (1961)   *J. Brit. Inst. Rad. Eng.*, 205.
Meudon Group (1965)   Unpublished results on Sept. 1965 radioastronomy rocket flight.
Salpeter, E. E. (1967)   *Astrophys. J.*, **147**, 433.

* The author does not pretend that this list is complete.

Slyš, V.I. (1965)   *Kosm. Issled.*, **3**, 760.
Slyš, V.I. (1966)   *Kosm. Issled.*, **4**, 924.
Slyš, V.I. (1967)   *Astron. Zu.*, **44**, 94–97.
Smith, F.G. (1961)   *Mon. Not. R. astr. Soc.*, **122**, 527.
Smith, F.G. (1965)   *Mon. Not. R. astr. Soc.*, **131**, 145–153.
Walsh, D., Haddock, F.T. (1964)   IAU Symp. no. 23 Liège, p. 335. *Astronomical Observations from Space Vehicles,* C.N.R.S., Publ. Paris.
Walsh, D., Haddock, F.T., Schulte, H.F. (1964)   *Space Research*, IV. Ed. by P. Muller, North-Holland Publ. Co., Amsterdam.
Walsh, D., Weil, H. (1967)   *Radio Science*, **2**, 225.

B. JOINT DISCUSSION OF COMMISSIONS 28, 33, 34, AND 44
X-RAY ASTRONOMY

(Monday, August 28, 1967)

Organizing Committee: S. B. Pikelner (Chairman), B. J. Bok, H. Friedman,

L. Gratton, H. M. Johnson

Contents:

INTRODUCTION

S. B. PIKELNER
(Sternberg Astronomical Observatory, Moscow, U.S.S.R.)

Observations in the invisible regions of the spectrum have become more common in astronomy during the last 20 years. After radio-astronomy, astronomy of ultra-violet, X- and γ-rays appeared. X- and γ-astronomy are still only beginning to develop, even now they give important information which cannot be obtained with former methods. High-energy quanta are produced with high-energy electrons. Therefore X-rays are the emission of high-speed particles which reveal themselves weakly in the other part of the spectrum.

Our programme is heavy today, and I will only remind you of the list of objects which are or may be X-ray sources. Black-body thermal emission is expected from neutron stars, which having a temperature higher than $10\,000\,000\,°K$ and radius about 10 km can be observed only in X-region. Besides the emission from the stars themselves we can expect the emission from their transparent high-temperature envelopes formed, as was suggested by Zel'dovič, in the process of accretion of the gas. The temperature of a stationary shock wave in falling gas can be very high. Similar waves of lower temperatures may be found around the white dwarfs. The possibility of formation of shock waves in close binary systems was discussed, in application to the Sco XR-1. Thermal emission of transparent high-temperature envelopes may be expected in supernovae, especially of type II. Here the peak temperature behind the shock is higher than $10^8\,°K$, the radius R is about 1 ps, the spectrum of free-free emission in the transparent envelope should be flat. Besides, the linear emission of hydrogenlike ions N, O, Mg and others, excited with electron impacts and recombination emission of the same ions are expected. Observations of X-rays from the envelopes allow us to improve knowledge of physical conditions and particularly the kinetics of cooling. Similar phenomena of larger scale may be expected after explosions of more massive bodies as in galaxy M 82 and also in quasars, radiogalaxies, and galactic centers.

The exploration of thermal X-ray emission of intergalactic gas would be of high importance. Even now its upper limit shows that the temperature is no more than a few million degrees if the density of the gas is close to the critical one. But the absence of absorption Lyman-α line in spectra of very distant quasars, is an argument in favor of very high temperature and apparently of low density of the gas.

Non-thermal emission of high-energy electrons-bremsstrahlung and synchrotron one is connected with cosmic rays and their sources. In sources with flat spectra as

Perek (ed.), Highlights of Astronomy, 178–179. © *I.A.U.*

in the Crab nebula there are high-energy electrons and the spectrum of synchrotron emission is continued up to the X-ray region. In other sources and radiogalaxies usually there are no even optical spectra, therefore electrons producing X-rays are not the tail of general energy spectrum but appear due to some secondary processes, for instance due to nuclear collisions which give chains $\pi^{\pm} \to \mu^{\pm} \to e^{\pm} \to X$ or $\pi^{0} \to \gamma$. These processes are taking place in interstellar gas of the Galaxy too, but the intensity of producing X-ray background is much lower than the observed one.

There are low-energy cosmic rays (10–100 MeV) in the interstellar gas, with the number density about $10^{-7} \, \mathrm{cm}^{-3}$. They can strip inner electrons of C, N, O and others and produce characteristic X-ray linear emission. The direct collisions of such cosmic rays with electrons and bremsstrahlung of the faster stripped electrons may give the continuum background. Its intensity may be comparable with the observed background but special investigations are necessary here. The similar process in the metagalactic plasma may give us an upper limit of number density of low-energy cosmic rays in intergalactic medium. The last process of formation of X-rays is an inverse Compton effect for optical and relict radio quanta. It needs rather high-energy electrons about 10^7 and 10^9 eV correspondingly. This process may be effective in a powerful radiative field near quasars. Besides the uniformity of black-body radio-emission at the enormous distances allows us to evaluate an upper limit of density of relativistic electrons in metagalactic space. Even now it is shown that the density of such electrons is several orders of magnitude lower than that in the Galaxy. It is an argument against the metagalactic theory of the origin of cosmic rays. Similar evaluation was inferred by Ginzburg and Syrovatskii from the upper limit of γ-rays as the electrons which transform radioquanta into X-rays, transform optical quanta into γ-rays. The accuracy of such evaluations is dependent also on the accepted model of the Universe. The improvement of experimental data on X-rays which we shall hear today and in the future, allows us to obtain a number of interesting results.

SURVEYS OF X-RAY SOURCES

HERBERT FRIEDMAN

(E.O. Hulburt Center for Space Research, U.S. Naval Research Laboratory,
Washington, D.C., U.S.A.)

My assignment this afternoon is to review briefly the status of the observational work that has been accomplished since the last IAU discussion of X-ray astronomy which took place at the Liège Symposium 3 years ago. Essentially, the methods of observation have not changed in any significant way. All of the information has come from small rockets, on the order of the Aerobee and the Skylark in size, and from packages carried on balloons. The quality of the surveys varies according to the objectives of each experiment. For example, my colleagues and I, at the Naval Research Laboratory, have attempted to perform broad-sky surveys to fill the picture of the general distribution of sources in space and, as a result, have had to sacrifice precision in position and high resolution. At the other extreme, Dr. Giacconi and his colleagues, at American Science and Engineering, Inc. and the Massachusetts Institute of Technology, have concentrated the available flight time on individual sources and have achieved remarkably high positional accuracy for some of the stronger sources – sufficient to associate them with optical objects. Dr. Fisher and his colleagues at Lockheed have taken an intermediate approach and used fan beam scans of limited regions of the sky, which give more precise position and better resolution in one dimension than the broad NRL surveys.

The broad-sky surveys that have been conducted by the NRL group use a large-area detector (~ 1000 cm^2) baffled by a honeycomb which offers an 8-degree field of view at half maximum and a triangular transmission pattern. Such surveys along the galactic equator in the general direction of the galactic centre indicate so many sources that it becomes very difficult to resolve the envelopes of unresolved sources into their discrete components, and much subjective judgment enters into the analysis. Recently, Dr. Giacconi and his colleagues have scanned the galactic centre region with much higher resolution. Essentially, I believe that they find all of the sources listed in the earlier NRL surveys and several additional sources.

The obvious thing about a map of sources is that most are positioned close to the galactic equator. There seem to be two major concentrations, one in the general direction of the galactic centre and the other in the general region of Cygnus and Cassiopeia. In a 1965 survey, we detected two high-latitude signals: one was in the vicinity of M 87 with an uncertainty of about 1·5°, and the other source was in Leo and is designated Leo XR-1. With a distribution of this type, one can attempt some

Perek (ed.), Highlights of Astronomy, 180–183. © I.A.U.

crude statistical analysis. The average displacement of 'galactic-centre' sources from the galactic equator is about 3·5°. The average separation of Cyg-Cas sources from the galactic equator is about 7°. If we assume that these are distributed like the general distribution of stars, with a mean separation from the galactic plane of about 150 parsec, we may estimate distances for these two distributions. The galactic-centre cluster appears to be about 2500 parsec distant and the Cyg-Cas group about 1300 parsec. These results suggest that the galactic-centre sources lie predominantly in the Sagittarius arm of the Galaxy and the Cyg-Cas sources in the Cygnus/Orion arm. However, when one looks at the data in more detail, there are discrepancies with such an analysis. There are evidences of either differences of spectral composition of the sources or in interstellar absorption. If the differences are attributed to interstellar absorption, then they either do not lie in the same arm or the interstellar gas is highly non-uniform. Dr. Giacconi's group has reported evidence for spectral differences between the galactic-centre group and the Cyg-Cas groups which may indicate that the latter group is much closer.

From comparisons of surveys made by different groups at different times, one finds evidences of differences in flux, from which we may conclude many of the sources are highly variable. In some cases the evidence is clear-cut as, e.g., when the observations have been made by the same group using essentially the same equipment at different periods of time. Where the observations have been made by different observers, there is always the problem of intercomparing the absolute fluxes observed when there are great uncertainties in the spectrum and these are reflected in the conversions to true fluxes with different techniques of observation.

During the past year, the AS&E group have conducted a detailed survey of the collection of sources in the Sagittarius region and also in the Cygnus region with higher sensitivity and higher positional definition than the earlier surveys. By offsetting pairs of detectors and comparing the signals, they gained evidence for the position of the source along the direction of the slit pattern. In a typical experiment they have also scanned sources with a fan beam in two directions to obtain precise position. In Cygnus, they found that, where the NRL group had initially observed the sources Cyg XR-1 and Cyg XR-2, there certainly are strong sources, and their positions are only slightly displaced from the centres of the NRL uncertainty diagram. In addition, two other sources were found which were not picked up in the NRL surveys, but what is most important, where the NRL survey of 1965 showed a source at the position of Cyg A, they found no evidence within a factor of 10 of the flux claimed by the NRL group. From the fact that the NRL and AS&E positions do agree so well on sources like Cyg XR-1 and Cyg XR-2, there can be little doubt about the solution of the aspect problem in the earlier NRL survey and the existence of a source in the direction of Cyg A. The fact that the Cyg A source did not appear a year later in the AS&E survey must mean that there was a source there originally in 1965, which declined in flux to the point where it was not observable a year later. Whether the source was Cyg A or some galactic object is, of course, not indicated by these observations.

In April 1965 the NRL group found a signal which roughly fit the position of M 87. In order to substantiate or improve the evidence for the existence of such a source, in May 1967 the NRL group used a stabilized rocket which was programmed to do a very slow scan of a region about 14° long, across Virgo, so that it would pass through the positions of 3C 273 and M 87. The field of view was a fan beam, 1° in the scan direction by 8° in the direction normal to the scan. Position was determined by photography of the star field during the course of the flight. We devoted the entire time of the Aerobee-rocket flight to cover this small region and thus gain the maximum statistical accuracy and the greatest sensitivity of detection for this resolution. Unfortunately, the flight was a little low and the scan was terminated just slightly past the peak of the signal from M 87. The rocket was already so low in the atmosphere that an absorption correction due to atmospheric attenuation had to be applied. What the record shows are signals from the intersections of the fan beam with the directions of 3C 273 and of M 87 within a tenth of a degree, which was the accuracy of the position measurement of the survey. In addition, there are three other sources which appear at about the 3 Σ level and they do not coincide with any well-known radio sources or quasars. To compare this result with the observation of 1965, which was made with a coarse collimator, one has to recognize that the coarse collimator covered so large a field of view that it would have lumped together the M 87 source and all the others. When we take account of the broad field of view, the signals observed agree within the statistical error of the measurements.

There are many important questions that concern the 'diffuse' X-ray background. Is there a truly diffuse background? Is the flux consistent with the interaction of cosmic rays in meta-galactic or intergalactic space with the 3 °K background, producing X-rays by inverse Compton scattering, or is it simply the integral of radiation from all galaxies? I think that from the observations made thus far one cannot decide how much of the background is diffuse and how much is unresolved discrete sources. Having improved the resolution substantially in the May 1967 survey of Virgo, we find more discrete sources, but the resolution must be much further improved before one can interpret the nature of the background.

With regard to spectra, the source which has been most frequently studied and for which the best data exist is Sco XR-1. I believe that the speakers who follow will go into detailed discussion of the significance of these data. A fairly good fit can be made with a bremsstrahlung distribution of 5×10^7 °K except for a flux measurement near 50 Å by the NRL group. Šklovsky has presented an interesting accretion model which takes into account that measurement as well as the measurement at higher energy. It is urgently necessary that the experimenters redetermine, with high quantitative accuracy, the flux of long wavelengths.

With the techniques that have been used in the observations that have been discussed so far, one can hopefully go quite a bit further. For instance, Šklovsky has argued the importance of studying the Centaurus A source, which is a powerful radio galaxy, so close to us and therefore so widely distributed in angle that even with 1 ° collimation one could separate fluxes from the radio plasmons, and from the optical region at the centre of this radio galaxy. The flux which was observed from the direction of 3C 273 is about 2×10^{-10} ergs cm^{-2} sec^{-1} in the 1–10 Å range and, theoretically, one might expect a flux very close to that from the radio plasmons of Centaurus A, as a result of the scattering of the relativistic electrons on the cosmological background radiation.

I would like to just briefly mention what might be done by the time of the next IAU Assembly if sufficient priority were given to the conduct of X-ray astronomy. In the U.S.A. there are opportunities to take advantage of Apollo-system hardware, and designs have been made to carry very large area detectors in such a system. Plans are being developed to carry up to 100 square feet of X-ray detector. With such a large aperture it would be possible to produce an X-ray catalog of sources, that would be analogous to the 3C catalog of radio sources, in the time of one flight. Some experimenters are concentrating on the development of large mirror systems. Dr. Giacconi's group in particular have been trying to promote opportunities for experiments with a very large reflecting telescope which would offer both high sensitivity and high resolution for studies of discrete sources. All of these things are technically possible and may come to pass sooner than we think.

DISCUSSION

S.I. Syrovatsky: What is the angular resolution of the large array of detectors which you mentioned at the end of your paper?

H. Friedman: 1° by 4°.

R.J. Weymann: What is the flux from M 87?

H. Friedman: The flux from M 87 is $8 \cdot 7 \times 10^{-10}$ ergs cm^{-2} sec^{-1} in the 1–10 Å band if we assume a power law spectrum of index 1, we would get roughly the same result with a bremsstrahlung spectrum of 5×10^{7} °K. That would make the luminosity $1 \cdot 1 \times 10^{43}$ ergs/sec. The signal from the direction of 3C 273 would be $1 \cdot 8 \times 10^{-10}$ ergs cm^{-2} sec^{-1}. If we place it at 500 mpc, the luminosity would be $5 \cdot 5 \times 10^{45}$ ergs/sec.

F.J. Low: What is the certainty of your identification of 3C 273? In other words, what is the chance that one of these other sources you showed might be somewhere in a fan beam?

H. Friedman: I think the answer is evident in the data I showed. In the scan direction the source appears in the scan field within a tenth degree of the known position of 3C 273. If it were displaced at right angles to the center of the scan pattern then the flux would decrease according to the triangular pattern. If it were 4° off, the observed flux would be down a factor of 2. If it were off much more than 4°, its true flux would be so great that it would have been observed at the displaced position in earlier surveys of lesser sensitivity.

RECENT RESULTS ON EXTRAGALACTIC COSMIC X-RAY SOURCES FROM ROCKET AND BALLOON FLIGHTS*

G. W. CLARK

(Massachusetts Institute of Technology, Cambridge, Mass., U.S.A.)

Two experiments carried out recently at M.I.T. gave results which bear on the problems of extragalactic X-ray sources. One of these is the work of a group under the direction of Hale Bradt (Bradt *et al.*, 1967), who used an attitude-controlled Aerobee rocket to scan a portion of the sky which included the radio galaxy M 87 (Virgo A). Among their detectors were two banks of argon-filled, 2 mil beryllium window proportional counters, each with an effective area of 350 cm^2 and a mechanical collimator giving a $2° \times 20°$ FWHM field of view. The fields of view were crossed

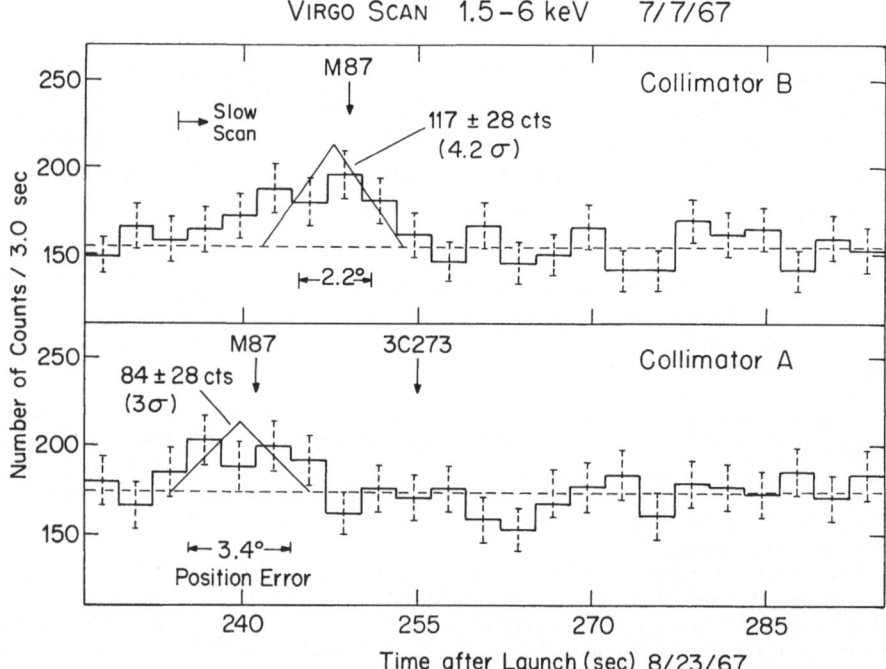

FIG. 1. *Counting rates in counters A and B during a scan of the region near M 87.*

* This work was supported in part by the National Aeronautics and Space Administration under Contracts NSR22-009-129 and NsG-386, and in part by the U.S. Atomic Energy Commission under contract AT(30-1)-2098.

Perek (ed.), Highlights of Astronomy, 184–187. © I.A.U.

so that their long directions made an angle of 60° with one another. Aspect was determined to within 5 min of arc by star photography.

Figure 1 shows the numbers of counts from the two detectors in successive 3-sec intervals during the part of the flight when the detectors scanned the Virgo Cluster. The data show a 4·2 σ and a 3·0 σ peak in the two channels centered at the positions indicated by the triangles which represent the angular responses of the collimators.

As shown in Figure 2 the radio galaxy M 87 lay close to the intersection of the centre lines of the two fields of view at the times corresponding to the centres of these peaks, and well within the 6·5 square degree region of uncertainty within which the source responsible for the observed peaks most probably lies. Considering the low density of X-ray sources in this region of the sky, and the unique character of M 87 among the possible X-ray sources in this uncertainty region, Bradt and his co-workers conclude that the most likely source of the X-ray peaks they observed is, indeed, M 87, with a counting rate of $0·5 \pm ·01$ cts cm^{-2} sec^{-1}, corresponding to an energy flux of about 5×10^{-10} ergs cm^{-2} sec^{-1} in the energy range from 1·5 to 6 keV (2–8 Å). It should be noted that the ratios of the radio, optical and X-ray luminosities of the extragalactic source M 87 implied by this observation are not qualitatively different from those of the Crab Nebula.

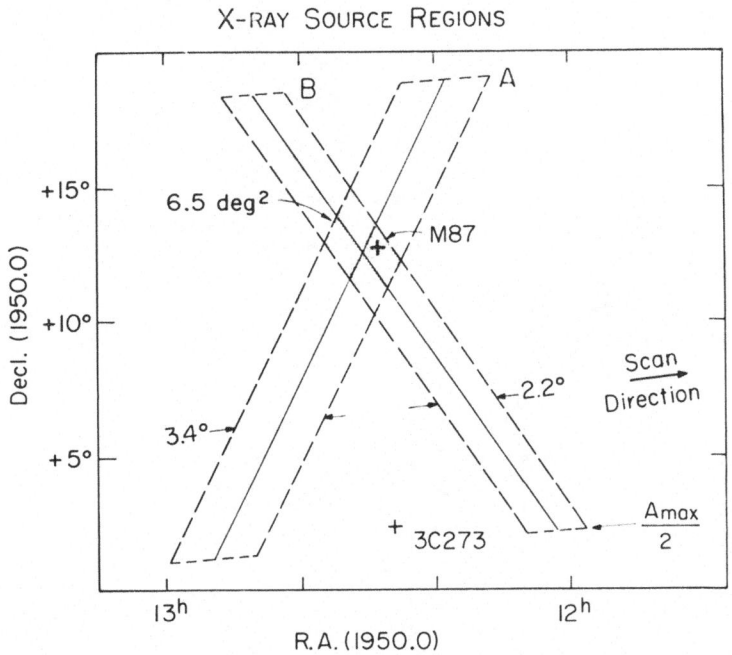

FIG. 2. *Probable position (intersection of solid lines) and uncertainty region (shaded area) of the source responsible for the observed peaks in the counting rates in counters A and B.*

In order to compare the observed X-ray counting rate with that reported by the NRL group (Byram *et al.*, 1967) for the source which they earlier identified as M 87 on the basis of a single observation with an 8° FWHM circular field of view, Bradt *et al.* calculated the relative response of the detectors used in the two experiments and found that the NRL sensitivity was between 1·5 and 2 times greater for a wide range of plausible spectral assumptions. Applying the maximum correction they find a value of $0·08 \pm ·02$ cts cm^{-2} sec^{-1} which is more than two standard deviations below the NRL result of $0·2 \pm ·05$ cts cm^{-2} sec^{-1}.

The second result is a negative one which contradicts the recent report of X-rays from an extended source identified as the Coma Cluster by Boldt *et al.* (1966). The total intensity of this source was reported to be 0·6 times that of the hard X-ray source in Cygnus in the approximate energy range 20–50 keV. This report stirred great interest because of the possibility that the missing mass of the Coma Cluster might be in the form of an intergalactic gas heated by the galactic motions to a high temperature and emitting thermal bremsstrahlung in the form of X-rays (Garmire, 1967; Felten *et al.*, 1967).

Lewin, Smith and I have completed an X-ray survey of the Northern sky above

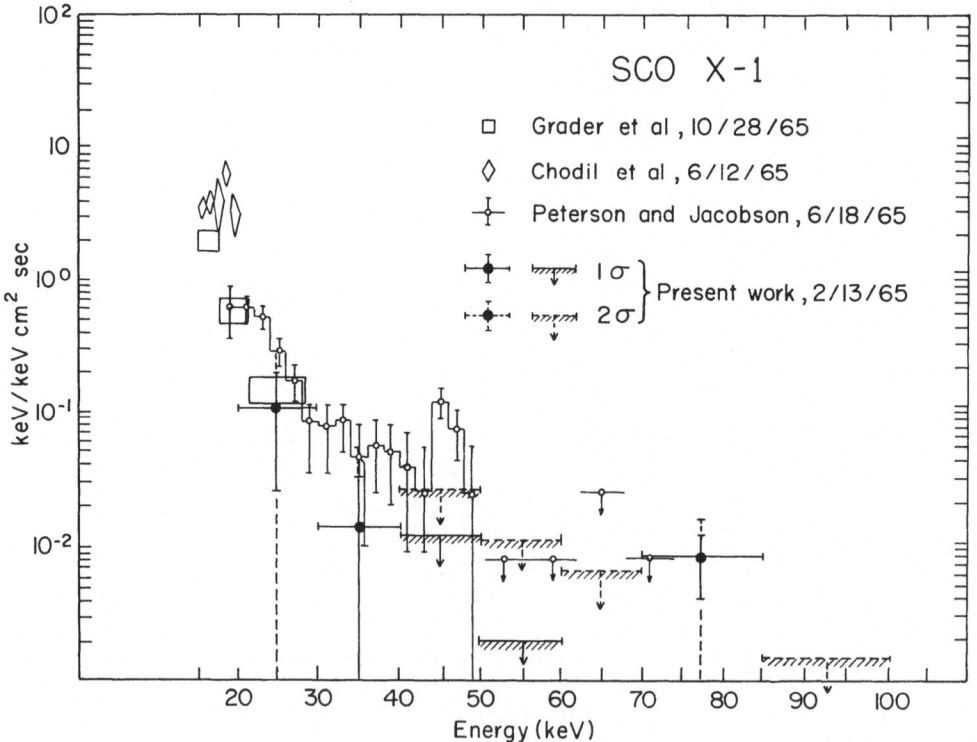

FIG. 3. *Data on the high-energy X-ray spectrum of Sco X-1.*

20 keV (Clark *et al.*, 1968; Lewin *et al.*, 1967) by balloon observations using a 400 cm^2 NaI(Tl) detector in a rotating gondola. No evidence of X-ray emission from the region of the Coma Cluster was found. In the same flight we observed the hard X-ray source in Cygnus, which we identified as Cyg X-1. We were therefore able to set an upper limit of 0·15 on the intensity ratio (Coma/Cyg X-1) for X-rays above 20 keV at the top of the atmosphere. We therefore consider the value of 0·6 found by Boldt *et al.* for this ratio to be in error. In spite of this, the great interest of the missing mass problem will continue to stimulate the search for X-rays from the Coma Cluster at lower and lower limits of detectability.

In the same survey we scanned many times over Sco X-1 and established upper limits on the X-ray intensity from 40 to 50 keV which lie substantially below the values reported earlier by Peterson and Jacobsen (1966), as shown in Figure 3. There is an obvious need for extended observations of Sco X-1 to determine whether the cause of these different observations is actually a time variation in the source.

References

Boldt, E., McDonald, F. B., Riegler, G., Serlemitsos, P. (1966)   *Phys. Rev. Letters*, **17**, 447.
Bradt, H., Mayer, W., Naranan, S., Rappaport, S., Spada, G. (1967)   *Astrophys. J. (Letters)*, **150**, 199.
Byram, E. T., Chubb, T. A., Friedman, H. (1967)   *Science*, **152**, 66.
Clark, G., Lewin, W., Smith, W. (1968)   *Astrophys. J.*, **151**, 21.
Felten, J. E., Gould, R. J., Stein, W. A., Woolf, N. J. (1967)   *Astrophys. J.*, **146**, 955.
Garmire, G. (1967)   private communication.
Lewin, W., Clark, G., Smith, W. (1967)   *Astrophys. J. (Letters)*, **150**, 153.
Peterson, L., Jacobsen, A. (1966)   *Astrophys. J.*, **145**, 962.

A COSMIC X-RAY SURVEY IN THE SOUTHERN HEMISPHERE

B. A. COOKE, K. A. POUNDS and E. A. STEWARDSON

(Dept. of Physics, University of Leicester, England)

1. Introduction

An unstabilised Skylark rocket (SL 118) was launched from Woomera, South Australia at 20:20 local time on April 10, 1967. The rocket carried a large-area proportional counter to perform a high-sensitivity survey for cosmic X-ray sources in the Southern sky. The flight was successful and during the period of observation with the rocket above the absorbing atmosphere, from 80 to 350 sec after launch, a total of 50 scans across the sky were obtained. In addition to the predominant roll motion of the rocket, a slow precession about a flat (70° half angle) cone provided at least two separate looks at every part of the celestial hemisphere. At the time of the flight, the Milky Way was almost perpendicular to the horizon and a region around the Galactic equator from Scorpius to Taurus via Centaurus, Carina, Puppis and Orion was visible.

The proportional counter had an effective photon collection area of 295 cm^2 and the energy band from 0·9 to 5 keV was analysed in flight into five discrete energy bands. The counter window of 6 μm melinex and gas-absorption path of 5 cm-atmos argon-methane ensured a high photon-detection efficiency over this waveband. Geometric collimation of 30° by 30°, being purposefully rather broad to ensure observation of all the available sky, was provided by an aluminum egg-box array mounted before the counter. An interesting aspect of the electronics was the first flight of a new method of discrimination against particle background (Mathieson and Sanford, 1963), in which pulses due to particles were rejected by sensing their relatively long rise-time in comparison with those resulting from the X-ray photon absorptions.

2. Flight Data

The flight data showed a remarkably clear discrimination between the Earth and sky observations on each pass, the count rate changing by a factor of about 20 on crossing the horizon. The initial examination of the data has provided information on the sky background and on several discrete sources and these results are summarised below.

Perek (ed.), Highlights of Astronomy, 188–191. © I.A.U.

A. DIFFUSE BACKGROUND

On each of the 50 separate rocket scans, some period or periods without significant discrete X-ray sources are seen. Normally, with such a broad collimation, the most reliable diffuse background fluxes will be obtained in the high galactic latitude regions which are relatively free of strong, discrete sources. From an average of many passes at high latitude, the following background figure is thus obtained:

$$\text{Diffuse X-ray intensity (2–5 keV)} = 2 \cdot 1 \times 10^{-8} \text{ erg/cm}^2\text{-sec-ster.}$$

This figure is in good agreement with other values (Gould, 1967), some of which have been derived after making appreciable corrections for the particle contribution to be subtracted from the measured count rates. It is clearly not possible with the present data to distinguish between a truly diffuse, isotropic background, and an isotropic distribution of small, discrete X-ray sources.

B. SCORPIUS X-1

This well-known source was observed during a number of passes near rocket apogee (165 km). At this altitude the source, lying $5°$ below the horizontal, was some $7°$ above the X-ray horizon. The corresponding line-of-sight air path, of 10^{-4} gm/cm^2, requires a small correction at the low-energy end of the spectrum to account for terrestrial absorption. Taking this into account the observation gives:

$$\text{Sco X-1 intensity (2–5 keV)} = 1 \cdot 3 \times 10^{-7} \text{ erg/cm}^2\text{-sec.}$$

This figure may be compared with values of $2 \cdot 6 \times 10^{-7}$ erg/cm^2-sec obtained by the LRL group in October, 1965 (Grader *et al.*, 1966) and $1 \cdot 3 \times 10^{-7}$ erg/cm^2-sec by the same group in May 1967 (Chodil *et al.*, 1967). The agreement between the Leicester and most recent LRL measurements is interesting and, since instruments from the same group are involved, the differences observed from the earlier LRL flight may well be real.

C. TAURUS X-1

The Taurus X-ray source was observed on three separate occasions during the flight, with the source well above the Earth's horizon. The observed flux was essentially unchanged on each occasion, with a mean value of:

$$\text{Tau-X-1 intensity (2–5 keV)} = 1 \cdot 9 \times 10^{-8} \text{ erg/cm}^2\text{-sec.}$$

This value is in good agreement with other measurements (Gould, 1967).

D. CENTAURUS X-2

The most outstanding observation of the flight was of a very strong source in

Centaurus. This has been called Centaurus X-2 because of a previous identification of a much weaker source in the same constellation by NRL. Almost certainly Centaurus X-2 is the same source as that seen in the flight from Woomera one week earlier and named Crux X-1 (Harries *et al.*, 1967). The source was observed on several occasions throughout the flight of Skylark 118 and the best source position obtained is:

$$\text{Right Ascension} = \ \ 200°$$
$$\text{Declination} \quad = -60°.$$

The maximum error in position does not exceed $2°$. However, it may be considered as doubtful whether this precision will be sufficient to warrant an optical search, since the source is located in a dense region of the Milky Way. The observed intensity of Centaurus X-2 on April 10, was:

$$\text{Cen X-2 intensity (2--5 keV)} = 1·6 \times 10^{-7} \text{ erg/cm}^2\text{-sec}.$$

This may now be compared with a more recent observation by the LRL group (Chodil *et al.*, 1967), which, for the same energy band, gives an intensity of only $2·6 \times 10^{-8}$ erg/cm^2-sec. Though the LRL measurement was made with the source very close to the horizon, the possibility of significant atmospheric absorption seems to be ruled out by the appearance of the published LRL spectrum. Thus, it may be concluded that the Centaurus X-2 source decreased in intensity by a factor of 6 in a 6-week period. This probably represents the clearest indication yet of the variability of a cosmic X-ray source. A study of the spectrum of Centaurus X-2 shows that this is a significantly steeper function of energy than Scorpius X-1 and, moreover, on April 10 Centaurus X-2 was the strongest source in the sky at an energy below 3 keV.

E. CARINA AND ORION REGIONS

The galactic plane between Centaurus X-2 and Taurus X-1 was scanned many times during the experiment. No definite source observations were made in this region. The highest count rate occurred in the direction of Carina-Vela, with a value some 10% above the background measured at high galactic latitudes. With 95% confidence it is possible, therefore, to place an upper limit of 0·5 photons/cm^2-sec to the total flux of X-rays (2–5 keV) received from any sources in the Carina-Vela direction, and slightly lower limits to fluxes from the galactic plane between Puppis and Orion. The interest in this measurement derives from the fact that the extension of the local spiral arm is believed to lie in the direction of Carina or Orion. Clearly the population of X-ray sources is here considerably lower than in the Cygnus extension of the same spiral arm. The total (2–5 keV) flux previously observed from the Cygnus direction is roughly 1·5 photons/cm^2-sec (Gursky *et al.*, 1967) and this would have been readily detected by the present instrument.

F. LARGE MAGELLANIC CLOUD

Several scans across the LMC revealed no significant X-ray flux from this direction and, with 95% confidence, an upper limit of 0·5 ph/cm^2-sec in the 2–5 keV band can be placed on this possible source.

References

Chodil, G., Mark, Hans. Rodrigues, R., Seward, F., Swift, C.D., Hiltner, W.A., Wallerstein, G., Mannery, E.J. (1967)   *Phys. Rev. Letters*, **19**, 681.

Gould, R.J. (1967)   *Amer. J. Phys.*, **35**, 376.

Grader, R.J., Hill, R.W., Seward, F.D., Toor, A. (1966)   *Science*, **152**, 1499.

Gursky, H., Gorenstein, P., Giacconi, R. (1967)   *Astrophys. J.,* **150**, L75.

Harries, J., McCracken, K.G., Francey, R.J., Fenton, A.G. (1967)   *Nature*, **215**, 38.

Mathieson, E., Sanford, P. (1963)   in *Proc. International Symposium on Nuclear Electronics*, *Paris*, p. 65, published by I.A.E.A., Vienna.

RESULTS FROM DETAILED STUDIES OF X-RAY SOURCES

R. GIACCONI

(American Science and Engineering, Cambridge, Mass., U.S.A.)

Since the results of early surveys revealed the existence of cosmic X-ray sources (Giacconi and Gursky, 1965), the major effort of X-ray astronomy has been devoted to the investigation of their nature. With the exception of the X-ray source in the Crab Nebula (Bowyer *et al.*, 1964), an object well known prior to the discovery of its X-ray emission, the galactic X-ray sources which have been observed do not appear to coincide with conspicuous visible or radio objects.

The attempt to study their nature has proceeded along two main directions:

(a) Accurate determinations of celestial coordinates for the X-ray emitters permit the identification of their visible and radio counterparts. When a likely candidate is discovered, the techniques of optical astronomy yield information regarding the spectrum, size, distance, temporal variations and polarization of the radiation emitted by the object.

Table 1

Position and intensity of observed X-ray sources

	l^{II}	Δl	b^{II}	Intensity 2–5 keV	Other Designation
GX-14·1	−14·1°	±0·25°	0±10°	0·11 cts/cm²−sec	
GX-12·9	−12·9	±0·3	>10	0·33	
GX-10·7	−10·1	±0·1	0±5	0·90	Lockheed-3
GX-5·6	−5·6	±0·2	0±10	0·11	
GX-2·5	−2·5	±0·15	0±10	0·40	
Sco X-1[a]	−0·9	−	23·8	19±5	Sco XR-1
GX+2·6	2·6	±0·1	0±5	0·68	
GX+5·2	5·2	±0·1	−1±4	1·24	Lockheed-4
GX+9·1	9·1	±0·1	2·5±4	0·84	Lockheed-5
GX+13·5	13·5	±0·1	−4±8	0·37	Lockheed-6
GX+16·7	16·7	±0·1	4·0±3	1·25	Lockheed-7
GX+36·3	36·3	±0·2	9°±8	0·15	Ser XR-1
GX+48·7	48·7	±0·2	0±15	0·19	
Cyg X-1	71·4	[b]	3·1	0·40	Cyg XR-1
Cyg X-2[a]	87·4	−	−11·3	0·36	Cyg XR-2
Cyg X-3	80·0	[b]	0·7	0·13	
Cyg X-4	82·9	[b]	−6·4	0·05	

[a] Coordinates of Optical Counterpart.
[b] Errors are given in Giacconi *et al.* (1967a).

Perek (ed.), Highlights of Astronomy, 192–201. © *I.A.U.*

(b) The properties of the X-ray emitter are investigated by careful measurement of the angular size of the X-ray emitting region, its detailed spectrum and its polarization and variability.

The most fruitful approach to date has been the one under (a). By means of accurate positional determinations we have been able to identify the optical counterparts of Sco X-1 and Cyg X-2 in addition, of course, to the Crab Nebula. It must, however, be realized that for the majority of the X-ray sources presently known, this approach

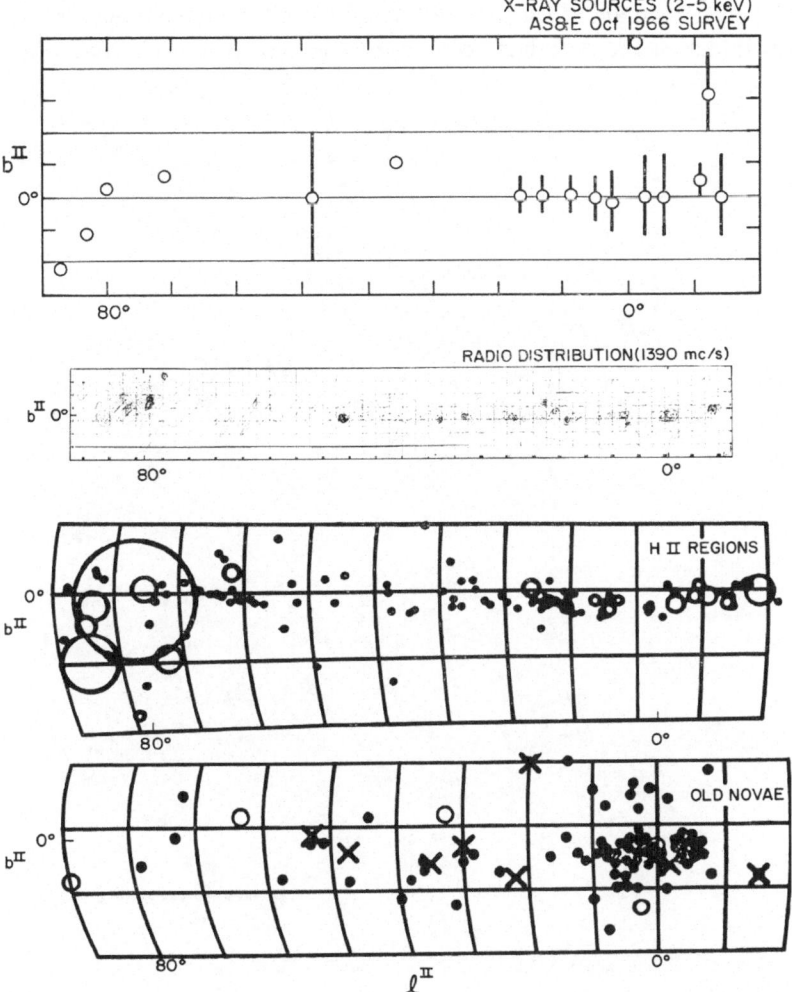

FIG. 1. *Comparison of the distribution of X-ray sources with known galactic objects. The radio distribution is taken from Westerhout, the H II distribution is that prepared by Plaut, and old-novae distribution is that of Sharpless. The angular scale is identical in all four plots. The X-ray positions shown are based on a survey of the galactic equator by the ASE group from a sounding rocket flown on October 11, 1966.*

may not be possible due to the relative low brightness of their optical counterparts and the fact that the sources appear to be located in regions of high obscuration.

For a number of the presently known sources positions have been accurately determined in at least one direction. Table 1 summarizes the results for sources whose position is known with accuracies of a few minutes of arc in at least one direction. Inspection of Table 1 shows that for only five of the sources listed, i.e. Sco X-1, Cyg X-1, Cyg X-2, Cyg X-3, and Cyg X-4, in addition to Crab, do we know positions in two dimensions with minutes of arc accuracy (Gursky *et al.*, 1967). Only for these sources have we been able to search with confidence for optical counterparts.

The remainder of the data has been used to search for correlations with singular

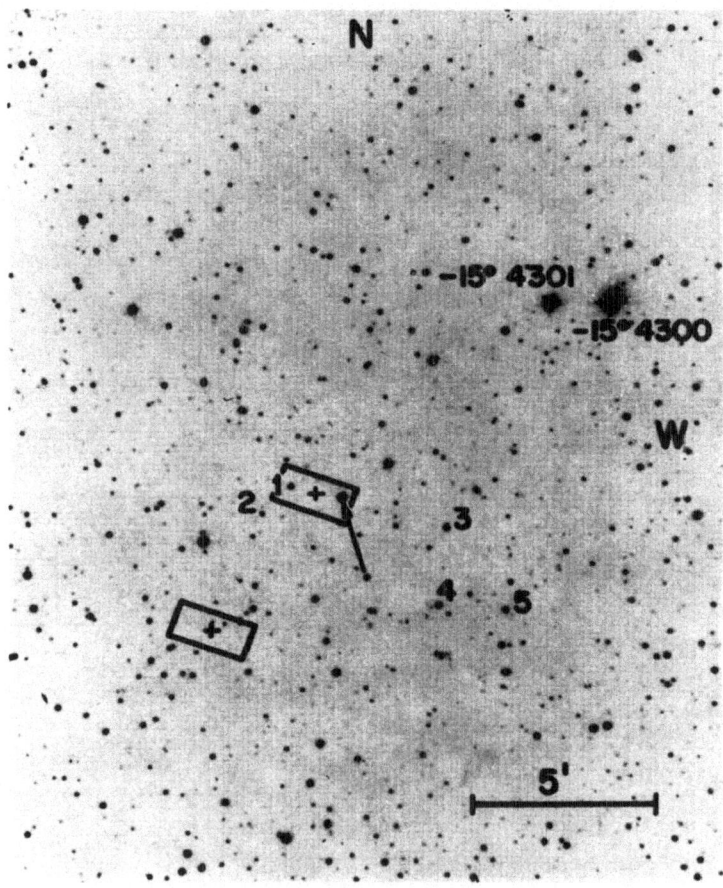

FIG. 2.   *Photograph of the region containing the X-ray position of Sco X-1 reproduced from a Palomar Sky survey print. The two equally probable X-ray positions as determined by Gursky et al. (1966) are marked by crosses surrounded by a rectangle of 1 by 2 min of arc. The object identified as the optical counterpart of the X-ray source is shown by the arrow. Other marked stars were used for comparison photometry.*

features of the galaxy. We find that the distribution of the sources in galactic longitude coincides with the distribution of HII regions and of OB associations. We conclude that most of the observed X-ray sources occur in the spiral arms of our galaxy. The remarkable correlation is shown in Figure 1.

For those sources for which positions are accurately known, a search for optical counterparts has been undertaken. This search has been successful in the case of

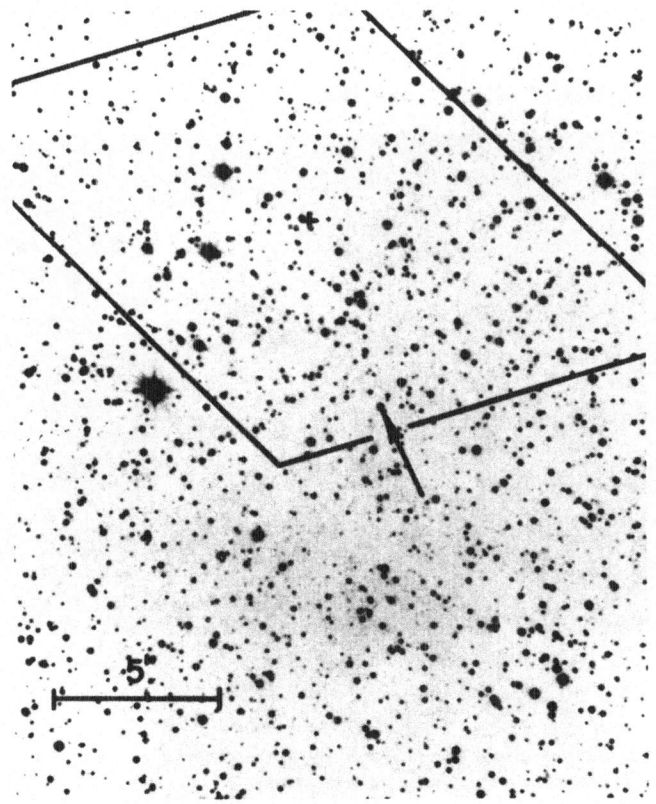

FIG. 3. *The region containing the X-ray source Cyg X-2 taken from a Palomar Sky survey print. X-ray position is marked by the cross and the area of uncertainty is the trapezoid. The optical candidate is marked by an arrow. The elongation of the optical image of the candidate object is caused by a neighboring 19th mag star and is not associated with the candidate object itself.*

Sco X-1 (Gursky *et al.*, 1966; Sandage *et al.*, 1966) and Cyg X-2 (Giacconi *et al.*, 1967*a, b*). In Figures 2 and 3 the portion of sky containing these sources is shown together with the X-ray location and the optical candidate. In Figure 4 the location of the X-ray source in Crab (Oda *et al.*, 1967) is shown superimposed on a visible light photograph. For Cyg X-1, Cyg X-3 and Cyg X-4 no optical candidate has been identified. For Cyg X-1 and Cyg X-3 spectral measurements of their X-ray spectra have been carried out.

In brief, a summary of the results of the optical and X-ray measurements of the more intensively studied objects are as follows:

Sco X-1, Cyg X-2: Both objects exhibit a relatively soft X-ray spectrum. The spectral shape can be described by an exponential with characteristic temperatures of 5×10^7 for Sco X-1 and $36 \pm 6 \times 10^7$ for Cyg X-2. No absorption features have been detected in the X-ray spectra at the lowest energy detected (~ 1 keV). Both

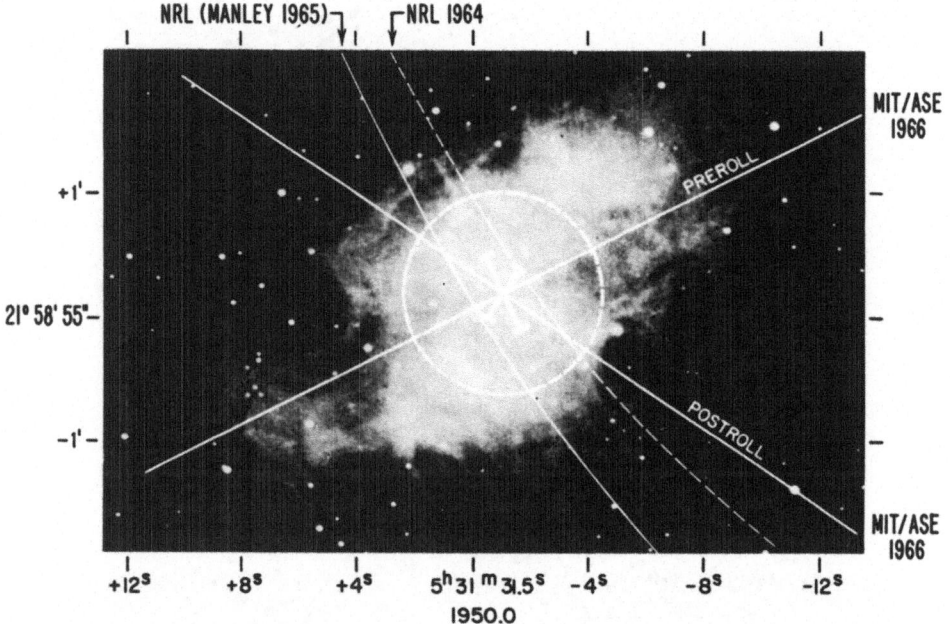

FIG. 4. *Summary of results pertaining to the position of the X-ray source in the Crab Nebula. The dashed curved line is the line of position obtained from a lunar occultation experiment by Bowyer et al. (1964), and the solid curved line is the same result corrected by Manley and Ouellette for parallax due to the motion of the rocket. The intersection of the two MIT/AS&E lines of position is at (1950·0) $\alpha = 5^h\ 31^m\ 30^s$, $\delta = 21°\ 59.1'$. The 100" diameter circle is an idealization of the source region. The origin of the coordinate system is the Southwest component of the central double star.*

sources are at relatively high galactic latitude. The optical candidates appear as blue starlike objects whose magnitude agrees with the one predicted by extrapolation from X-ray to visible according to an exponential spectral law. It follows that the energy output in X-rays exceeds by a factor of 1000 the combined optical and radio outputs. The visible light spectrum (Sandage *et al.*, 1966) from Sco X-1 exhibits characteristic emission lines of hydrogen and higher Z elements in an excited state. Both the line emission and continua appear to vary rapidly with time. The spectrum of Cyg X-2 shows HeII in emission (Lynds, 1967) and also appears to be variable. The distances of Sco X-1, estimated on the basis of the Ca K absorption feature, as well as on the absence of proper motion (Luyten, 1966), yield distances greater than about $\frac{1}{2}$ kpc. No generally accepted model of either of the two objects yet exists. It is, however, generally accepted that the X-ray emission occurs in a medium optically thin to its own radiation. In Figure 5 the measured X-ray spectrum of Sco X-1 is shown.

The two objects (Sco X-1 and Cyg X-2) have several common characteristics and may be representative of a class of celestial X-ray emitters which includes a large fraction of all known X-ray objects.

Crab Nebula: An X-ray spectrum of the Crab Nebula is shown in Figure 6. The

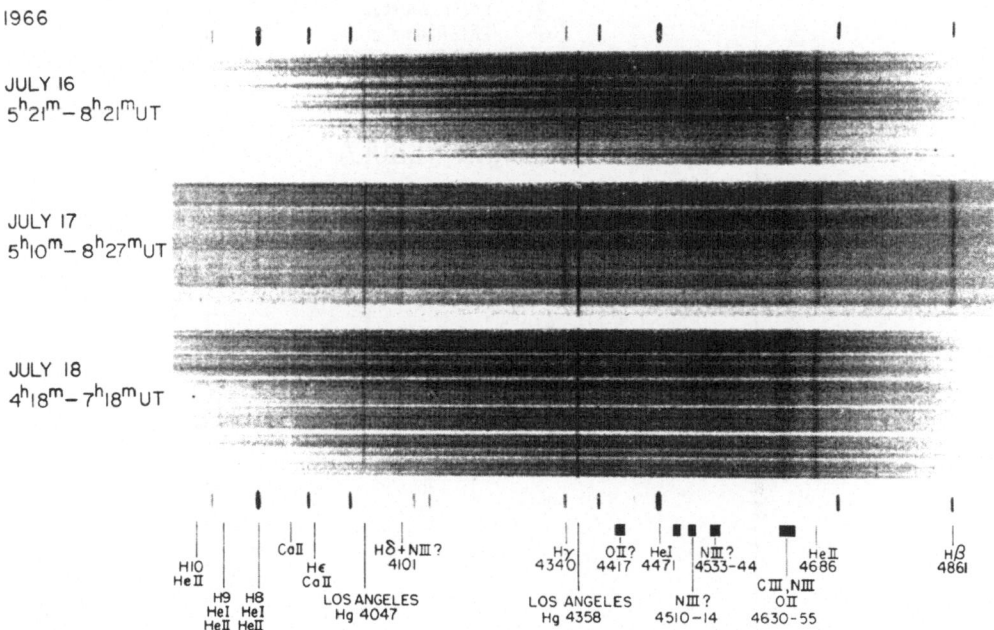

FIG. 5.   *Spectra of the Sco X-1 optical counterpart, obtained by A. Sandage at Mount Wilson and Palomar Observatories. From top to bottom, the exposures were on July 16, 1966, from 5:21 to 8:21 Universal time; July 17; 5:10 to 8:27; and July 18, 4:18 to 7:18.*

fact that a smooth extrapolation from the radio and visible to the X-ray spectrum can be made (Woltjer, 1964) is believed to strongly favor a common synchrotron emission process by relativistic electrons. The spectrum can be described by a power law spectrum of the form $E^{-\alpha}$ where α has the value of 1·1. The X-ray source in Crab is extended (~ 2 min of arc). It has been accepted that Crab may be representative of X-ray emission from supernova remnants. It should be noted that the Kepler supernova has not been observed to be an X-ray emitter. Tentative evidence exists for X-ray emission from Cas A and Tycho's supernova (Friedman *et al.*, 1967).

Cyg X-1: No visible object has yet been identified as the optical counterpart of Cyg X-1. This is believed to be due to the great degree of obscuration in that direction in the sky; however, Cyg X-1 could be an extended object and have low surface brightness. No measurement has been made of its angular size. Its X-ray spectrum can best be fitted by a power law spectral shape ($E^{-\alpha}$) with an exponent of $\alpha = \cdot 7$. No absorption feature has been observed (Gorenstein *et al.*, 1967). The X-ray spectrum of Cyg X-1 is shown in Figure 7. Cyg X-1 appears to differ significantly from

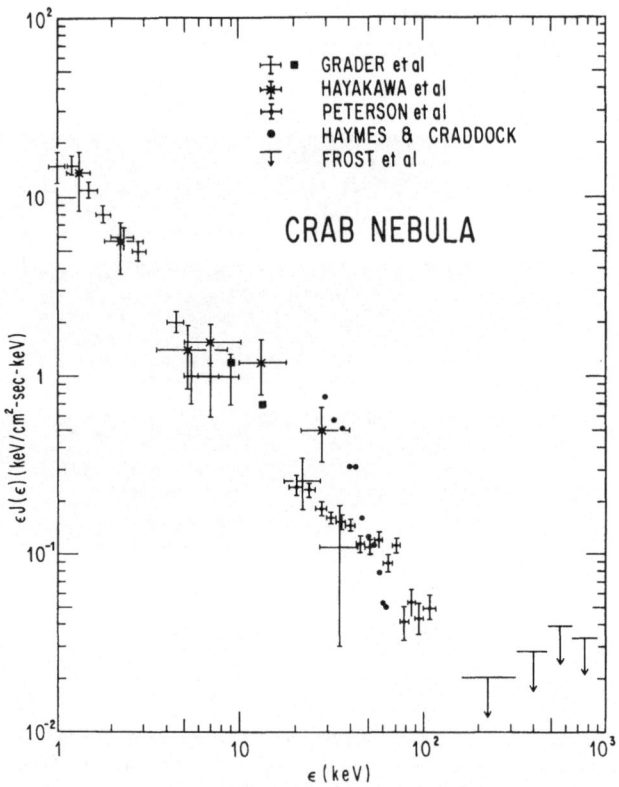

FIG. 6.   *A compilation by Gould of measurements of the spectrum of the Crab Nebula. The measurements of Grader and Hayakawa were obtained during sounding rocket flights. The others (Haymes, Peterson and Frost) were obtained during balloon flights.*

Crab mainly in the absence of any observed radio emission. While the two sources have comparable X-ray intensities, their radio emission differs by a factor of at least 500. This may indicate yet another type of X-ray emitter different from either supernova remnants or Sco X-1 like objects.

Cyg X-3: No visible counterpart of Cyg X-3 has yet been observed. This is also presumably due to obscuration since the object lies along one of the densest regions of the galaxy. Cyg X-3 is situated in the middle of the Cygnus-X radio region. However, the position is known accurately enough to exclude a coincidence with any of the known radio sources. The most interesting feature of the X-ray emission from Cyg X-3 is that we have found evidence of absorption at energies of the order of 2·5 keV. This can be interpreted as absorption at the source or in the interstellar medium. The data (Clark *et al.*, 1968) are insufficient to decide between a black-body spectrum and an exponential spectrum with absorption features. The X-ray spectrum of Cyg X-3 is shown in Figure 8.

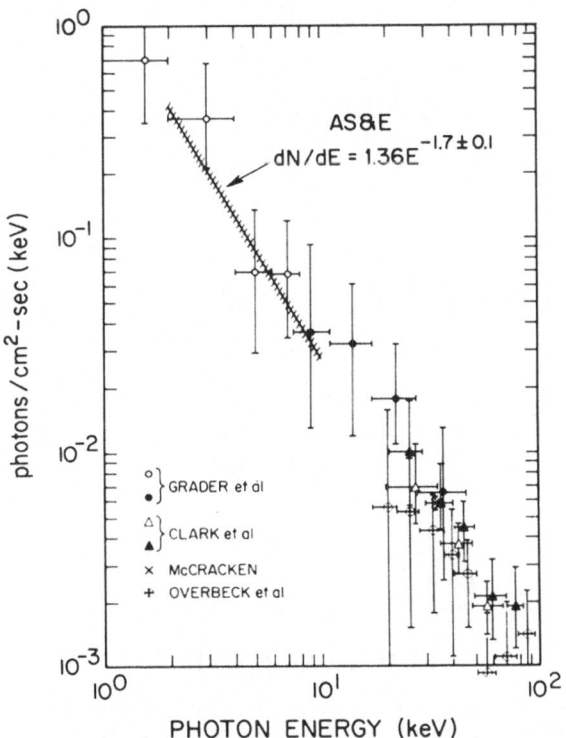

FIG. 7.   *A compilation of several rocket and balloon measurements on the spectrum of Cyg X-1 for a broad range of energies. The ASE result was obtained from the best fit of a power law spectrum to the data obtained in a rocket flight on October 11, 1966.*

We find evidence of absorption also in the spectra of sources in the Sagittarius-Scorpius region. The high degree of correlation between the observed X-ray absorption and the degree of optical obscuration for several sources strongly suggests that this absorption occurs in the interstellar space. Using accepted values of interstellar matter, density and cosmic abundance of elements we find that the degree of observed attenuation is consistent with the qualitative assignment of X-ray sources to the galactic spiral arms, although the data seem to require a greater X-ray absorptivity than conventionally computed (Gursky *et al.*, 1967).

The several X-ray sources observed would then be situated at distances of several kiloparsec from us and would mainly occur in regions of high obscuration. Several different types of celestial X-ray emitters appear to exist and we are beginning to distinguish between Sco X-1, Crab and Cyg X-1-like objects.

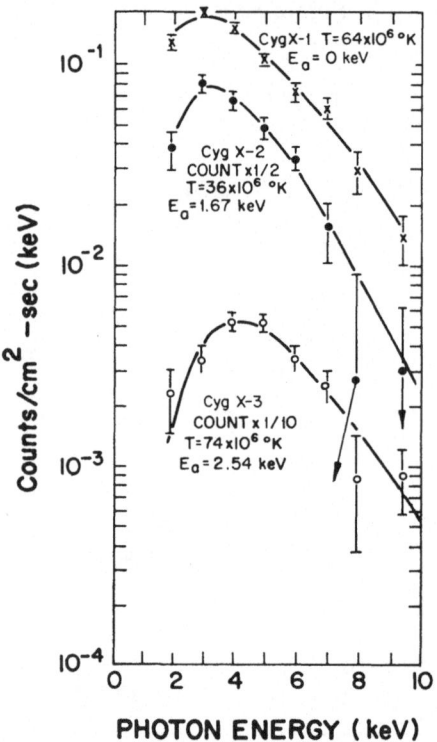

FIG. 8.   *The histograms of counting rate vs. energy are shown for three X-ray sources in Cygnus. The curves are continuous histograms which result from the calculated response of the detectors to an assumed photon distribution,* $dN/dE \sim (Ea/E)^{8/3} \exp(-E/KT)/E$. *Various factors affecting the performance of proportional counters and the finite width of each energy bin are included. The quantities Ea and T were left as free parameters. The values quoted are associated with the curve representing the best fit.*

More refined measurements of the spectral characteristics, positions, angular size, structure, polarization and time variations of the X-ray emission from galactic objects are being carried out. The observation of characteristic X-ray lines in emission or of X-ray absorption edges would contribute significantly to our understanding of the nature of X-ray objects. It is likely that real progress in X-ray astronomy in the future will largely depend on this and other refinements of the X-ray measurements themselves.

References

Bowyer, S., Byram, E.T., Chubb, T.A., Friedman, H. (1964)   *Science*, **146**, 912.

Clark, G.W., Lewin, W.H.G., Smith, W.B. (1968)   *Astrophys. J.,* **151**, 21.

Friedman, H., Byram, E.T., Chubb, T.A. (1967)   *Science*, **156**, 374.

Giacconi, R., Gursky, H. (1965)   *Space Sci. Rev.*, **4**, 151.

Giacconi, R., Gorenstein, P., Gursky, H., Waters, J.R. (1967*a*)   *Astrophys. J.*, **148**, L119.

Giacconi, R., Gorenstein, P., Gursky, H., Usher, P.D., Waters, J.R., Sandage, A., Osmer, P., Peach, J.V. (1967*b*)   *Astrophys. J.*, **148**, L129.

Gorenstein, P., Giacconi, R., Gursky, H. (1967)   *Astrophys. J.*, **150**, L85.

Gursky, H., Giacconi, R., Gorenstein, P., Waters, J.R., Oda, M., Bradt, H., Garmire, G., Sreekantan, B.V. (1966)   *Astrophys. J.*, **146**, 310.

Gursky, H., Gorenstein, P., Giacconi, R. (1967)   *Astrophys., J.* **150**, L75.

Luyten, W.J. (1966)   I.A.U. Astronomical Telegram, Circular # 1980.

Lynds, C.R. (1967)   'Spectroscopic Observations of Cyg X-2'. Preprint.

Oda, M., Bradt, H., Garmire, G., Spada, G., Sreekantan, B.V., Gursky, H., Giacconi, R., Gorenstein, P., Waters, J.R. (1967)   *Astrophys. J.*, **147**, 855.

Sandage, A., Osmer, P., Giacconi, R., Gorenstein, P., Gursky, H., Waters, J.R., Bradt, H., Garmire, G., Sreekantan, B.V., Oda, M., Osawa, K., Jugaku, J. (1966)   *Astrophys. J.*, **146**, 316.

Woltjer, L. (1964)   *Astrophys. J.*, **140**, 1309.

THE X-RAY SPECTRA OF CYGNUS XR-1 AND
THE CRAB NEBULA

Laurence E. Peterson

(University of California, San Diego, La Jolla, Calif., U.S.A.)

In this paper we wish to present briefly the latest results which have been obtained on the hard X-ray spectra of two strong sources in the Northern skies. These observations, which have been discussed in detail previously (Peterson *et al.*, 1967), were made from balloons launched at Palestine, Texas, to 3 gm/cm^2 atmospheric depth during September 1966. The Crab Nebula and the Cygnus XR-1 were observed to have a differential number power law spectra with an index of about -2 over the 20–200 keV range. Both sources have the same intensity within about 10%. The Crab Nebula has been observed on two occasions, one year apart, and showed no change in intensity over this range at about a 5% significance level.

These observations were made using a thin scintillation detector shielded with an active anti-coincidence collimator. The detectors are mounted on an alt-azimuth gimbal in the balloon gondola and referenced within about a degree to the local vertical and magnetic field. During the observation the detectors are fixed in azimuth North or South at the elevation of the object. In this manner a source makes a meridian transit through the aperture of the detector. Data for the background subtraction is obtained before and after the transit. Good events from the central detector are telemetered from a 128-channel pulse-height analyzer in a digital format; additional rate monitoring and housekeeping data are also telemetered.

The observation of the source Cygnus XR-1 was made on September 13, 1966. The detector used in this observation had an effective area of 50 cm^2 and angular response, full width at one-half maximum (FWHM), of 8·4°. The resolution of this detector was about 60% at 123 keV. Our results are shown in Figure 1 after correction for atmospheric absorption, with a power law fitted to the data over the range 20–200 keV. Data points obtained from rocket observations by Grader *et al.* (1966) and balloon observations by the group at M.I.T. (Clark *et al.*, 1967), and the Dutch group (Bleeker *et al.*, 1967) are also shown. Our points tend to fall slightly below the aggregate of the other measurements, though not beyond the statistical significance quoted by these workers. The two points at the lowest energy in our data have large systematic effects because of the poor detector resolution and large atmospheric correction which are not accounted for by the statistical errors shown. These results indicate clearly that the X-ray spectrum of Cygnus XR-1 has a power law characteristic like the Crab Nebula.

Perek (ed.), Highlights of Astronomy, 202–205. © I.A.U.

Our observations on the Crab Nebula are shown in Figure 2. The new observation reported here was obtained on September 20, 1966 with the same detector used in our initial observation of the Crab Nebula one year previously. The detector area was 10 cm², its aperture 24° FWHM and resolution was about 40% at 30 keV. The

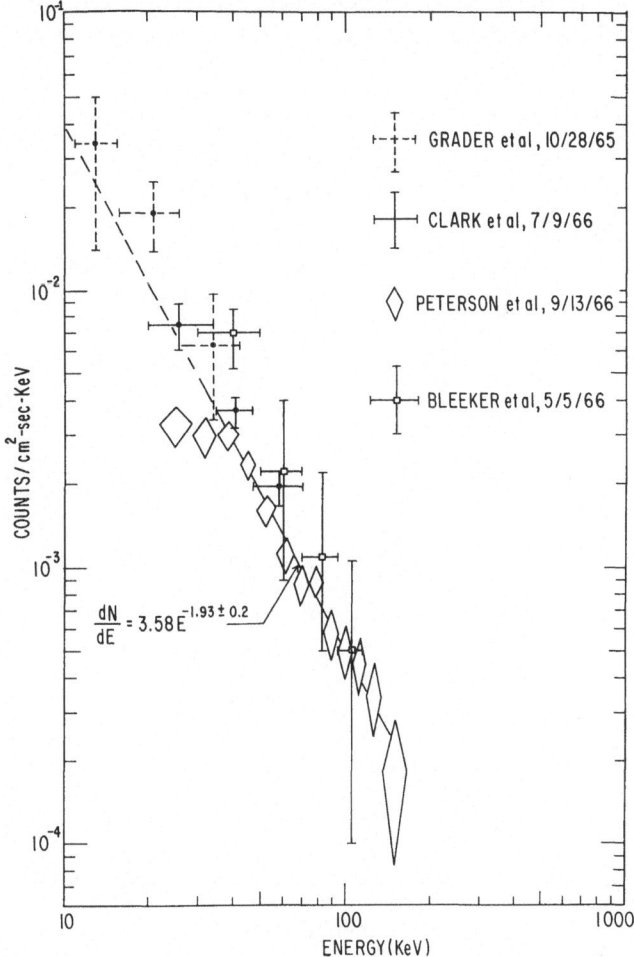

FIG. 1. *This and previous measurements of the spectrum of the source Cygnus XR-1. Our data are best described as a power law whose differential energy index is* $\alpha = 0.9$ *over the 30 to 140 keV range.*

results obtained during the most recent observation is shown by the diamonds in Figure 2. Also shown is a solid line which is a least squares fit to the data obtained on September 23, 1965. Clearly, there has been no obvious change over the 1-year period; the integrated flux of the Crab Nebula over the 20–100 keV range, 0·205 counts/cm²-sec, being constant to ±5% at 3 σ confidence level. Since our observations were made with the same apparatus, systematic errors are minimized. Also shown are the data points obtained by George Clark in June 1964.

It is remarkable that the Crab Nebula and Cygnus XR-1 have a hard X-ray spectrum identical in magnitude and slope to within about 10%. This suggests a comparable distance and a similar X-ray producing mechanism. This is different from Scorpius XR-1 whose exponential spectrum suggested emission from a hot, thin gas at $50 \times 10^6 \, ^\circ$K

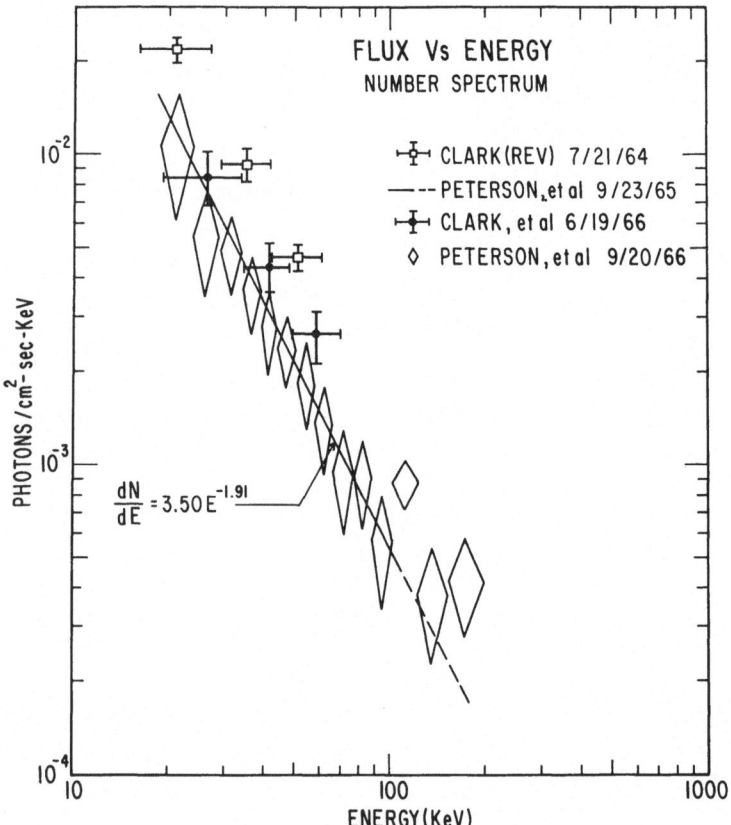

FIG. 2. *Measurements of the Crab Nebula X-ray spectrum. Our results indicate the flux was constant to within ±5% over a 1-year interval.*

(Peterson and Jacobson, 1966). Radio emission from the Crab Nebula is very different from that of Cygnus XR-1, which is presently unidentified with any obvious radio source. Comparisons at optical wavelengths cannot be made because Cygnus XR-1 is located in an obscured region of the galaxy (Giacconi *et al.*, 1967).

Acknowledgments

The author acknowledges the contribution of his associates, A.S. Jacobson, R.M. Pelling, and D.A. Schwartz in this work. The research was supported by the National Aeronautics and Space Administration under Grant NsG-318 and Contract NAS 5-3177.

References

Bleeker, J.A. M., Burger, J.J., Deerenberg, A.J. M., Scheepmaker, A., Swanenburg, B.N., Tanaka, Y. (1967)   *Astrophys. J.*, **147**, 391.
Clark, G.W., Lewin, W.H.G., Smith, W.B. (1967)   M.I.T. Lab. for Nuc. Sci. Preprint.
Giacconi, R., Gorenstein, P., Gursky, H., Waters, J.R. (1967)   *Astrophys. J.*, **148**, L119.
Grader, R.J., Hill, R.W., Seward, F.D., Toor, A. (1966)   *Science*, **152**, 1499.
Peterson, L.E., Jacobson, A.S. (1966)   *Astrophys. J.*, **145**, 962–965.
Peterson, L.E., Jacobson, A.S., Pelling, R.M., Schwartz, D.A. (1967)   *Canadian J. Phys.* (in press).

OPTICAL AND RADIO INFORMATION

HUGH M. JOHNSON

(Lockheed Missiles and Space Company)

The six or eight optically identified X-ray sources comprise starlike objects and extended supernova remnants in the Galaxy, well as as a radio galaxy and a quasar. Both X-ray and radiofrequency radiation penetrate the entire galactic plane, but only two or three galactic radio sources have been identified with X-ray sources. This has led Hayakawa *et al.* (1) to postulate that detectable X-ray sources are not farther than 1 kpc. However, other studies suggest that there is a cluster of a few intrinsically bright sources actually near the galactic nucleus and a scattering of weaker sources near the sun (2, 3).

The distances of X-ray sources can be estimated from extinction by interstellar gas or intergalactic gas on spectra above 10 Å, but the method ultimately depends on the radio and optical data of the gas. Conversely, interstellar densities of certain elements with large photo-ionization cross-sections may be determined from the absorption of X-rays, after calibration of source distances by the methods of optical astronomy. As soon as Tau XR-1, Cas XR-1, and Vir XR-1 were identified with optical objects, their distances were established. The optical distances can lead to the estimation of power outputs, linear sizes, volume emissivities, and ultimately to physical mechanisms. The most recent synthesis of the Crab Nebula (4) incorporates an X-ray source of thermal bremsstrahlung. Scargle (5) has observed an excess brightness of 0·12 mag at 3500 Å, which is only 2 times larger than the model's prediction. Further accounts of the optical and radio propreties of the Crab, Cas A, Vir A, and 3C 273 are given elsewhere.

The optical identifications of Sco X-1 (6) and Cyg X-2 (7) do not give concomitant distances because these 13th and 16th mag 'stars' do not clearly fall into classes for which the method of parallax measurement is obvious. The distance of Sco X-1 has been estimated by hypothesizing a relationship with the Galactic Spur (8) or with surrounding dust clouds (9). If they are relevant, these phenomena must contain clues to the history as well as the distance of Sco X-1. Wallerstein (10) has measured the equivalent widths of interstellar CaII H and K in the spectrum of Sco X-1 and nearby comparison stars. On the assumption that the radiation field of Sco X-1 has not altered the surrounding medium so as to increase the number of Ca^+ ions, he finds that Sco X-1 is more distant than the comparison stars, or $\geqslant 270$ pc. The nil proper motion (11, 12) also stands against very small distance. Unless the distance is small, the neutral-hydrogen data in the direction of Sco X-1 (13) combined with

Perek (ed.), Highlights of Astronomy, 206–209. © I.A.U.

photo-ionization cross-sections (14) suggest that the optical depth is as large as $\tau = 12$ at 50 Å.

The bright, blue irregular variable very near the position of Sco X-1 (6, 11) has remained secularly stable back to 1895, the date of the first known plate of the field by Barnard. In 1966–67 $+0 \cdot 07 \leqslant B \text{-} V \leqslant +0 \cdot 26$ and $-1 \cdot 15 \leqslant U \text{-} B \leqslant -0 \cdot 64$ for $12 \cdot 14 \leqslant \leqslant V \leqslant 13 \cdot 09$, with little correlation between the variables in these ranges (6, 15, 16, 17). D. E. Mook (18) observed on 41 consecutive nights at Cerro Tololo, and W. A. Hiltner made further observations there. Their work suggests that the light curve varies in character with no established periodicity, that the amplitude increases with decreasing wavelength of passband, and that $B \text{-} V$ and $U \text{-} B$ decrease as V magnitude brightens. Bursts of $0 \cdot 2$ mag in as little as 90 sec are noted in continuous monitoring. Other observations and autocorrelation tests also fail to demonstrate periodicity in the light curve of Sco X-1 in the range 2–300 sec (19). A discussion of the possible disparity between the observed V magnitudes and the magnitude to be extrapolated from the observed X-ray data would involve the interstellar extinction and much faith in the absolute photometric accuracy of the X-ray data. The optical polarization is nil (11) within $0 \cdot 015$ mag (20).

The spectrum of Sco X-1 in the blue resembles the spectrum of an old nova (6, 11, 21). Weak emissions of H, He I, He II, O II, C III, and N III are broadened several angstroms with no simple dependence on wavelength or ion. Radial velocity is not easy to observe in Sco X-1; there is a measurement of 0 ± 100 km/sec on June 27, 1966 (11). The reporter's spectrograms of April 18, 1967, reduce to -300 ± 100 km/sec, but the question of the value and the variability of the radial velocity calls for many observations.

Spectrophotometry of Sco X-1 (22, 23) shows that the continuum in the range 3200–7500 Å can be either the black-body radiation of a thick hot gas or the bremsstrahlung of a thin hot gas. The observed gradient is $\langle \text{d } \mathbf{AB}/\text{d}(1/\lambda) \rangle = 0 \cdot 25$ mag microns in terms of magnitudes $\mathbf{AB} = -2 \cdot 5 \log S - 48 \cdot 55$, where S is absolute flux density in ergs cm^{-2} sec^{-1} (c/s)$^{-1}$. This corresponds to $T = 6.2 \times 10^4 \, ^\circ$K in the spectrum of bremsstrahlung, but, if the X-ray bremsstrahlung $T = 5 \times 10^7 \, ^\circ$K extrapolates into the optical range, we can account for the gradient by the hypothesis of interstellar reddening equal to $E_{B-V} = 0 \cdot 11$ mag. A Rayleigh-Jeans distribution would require greater reddening to fit the gradient, namely about $E_{B-V} = 0 \cdot 55$ mag. Changes in the gradient or color can be explained by changes of temperature, especially in a thick gas at frequencies near the peak of radiation. However, magnitude is not correlated with d $\mathbf{AB}/\text{d}(1/\lambda)$ precisely, and there is no systematic departure from linearity in the graph of $\mathbf{AB}$ vs. $1/\lambda$ (23). Variable monochromatic emissions should produce little effect on colors since their total equivalent width is about 1% of the continuum in the B passband, and this is a larger percentage than in U or V. Also, brighter monochromatic emissions should decrease $B \text{-} V$ but increase $U \text{-} B$, whereas these color indexes were observed to vary in the same sense (18). None of the data, including one at $1·6 \mu$ (24), supports the presence of any cool photospheric component in the spectrum of Sco X-1. A dwarf star cooler than K5 V or a hotter star sufficiently below the main sequence could go undetected (23). Until binary nature is indicated by observations, theorists might tell us how to have a single 'star' of $M_v = +8$ but $L = 500 L_\odot$ living for at least three-score and ten years so quietly that astronomers never noticed it.

The equivalent widths of the monochromatic emissions of Sco X-1 (6, 21, 23) suggest that the line-emitting gas is much cooler than the X-ray emitting gas at $5 \times 10^{7} °K$, that the cool gas occupies a much smaller volume than the very hot gas, that the Balmer series is radiatively excited, that the He/H ratio is high, and that some characteristic of Of spectra and novae spectra are present (23). Variability of the line emissions with respect to each other and with respect to the continuum has been studied only enough to say that it is complex in Sco X-1. H. Spinrad has observed variations of Hα relative to the adjacent continuum in short time intervals, and J.C. Golson has observed some variation of Hβ, HeII 4686, and adjacent continua.

Sco X-1 and Cyg X-2 are the only two identified X-ray sources which are not observed to be radio sources. For Sco X-1 flux densities of $<0·09 \times 10^{-26}$ W m^{-2} (c/s)$^{-1}$ at 4170 Mc/s (25) and $(0·01 \pm 0·016) \times 10^{-26}$ W m^{-2}(c/s)$^{-1}$ at 15·3 Gc/s (26) have been set as limits, significantly below the extrapolated value of 0·07, or 0·7 if the Gaunt factor increases by 10 between 2 Å and 2 cm. Sco X-1 must become optically thick at radiofrequencies.

Several other very tentative identifications of X-ray sources have been suggested (e.g. (27)). But in order to bind X-ray and optical astronomy together, a larger share of X-ray observations must be devoted to accurate positional work. Simultaneous optical and X-ray observations of identified sources should be done regularly, especially with improved X-ray photometric accuracy.

References

1. Hayakawa, S., Matsuoka, M., Yamashita, K. (1966) in *Space Research*, Spartan Books, Washington, 6, 68.
2. Johnson, H.M. (1966) *Astrophys. J.*, 143, 261; (1967) *Astron. J.*, 72, 806.
3. Morrison, P., Sartori, L. (1965) *Phys. Rev. Letters*, 14, 771.
4. Sartori, L., Morrison, P. (1967) *Astrophys. J.*, 150, 385.
5. Scargle, J.D. (1967) *Astron. J.*, 72, 826.
6. Sandage, A.R., Osmer, P., Giacconi, R., Gorenstein, P., Gursky, H., Waters, J., Bradt, H., Garmire, G., Sreekantan, B.V., Oda, M., Osawa, K., Jugaku, J. (1966) *Astrophys. J.*, 146, 316.
7. Giacconi, R., Gorenstein, P., Gursky, H., Usher, P.D., Waters, J.R., Sandage, A., Osmer, P., Peach, J.V. (1967) *Astrophys. J.*, 148, L129.
8. Shklovsky, I.S. (1964) *Astr. Cirk.* (Kazan), No. 298; (1965) *Astr. Zu.*, 42, 287.
9. Johnson, H.M. (1966) *Astrophys. J.*, 144, 635; *Proc. 4th 'Texas' Symposium on Relativistic Astrophysics*, Gordon and Breach, New York (in press).
10. Wallerstein, G. (1967) *Astrophys. Letters*, 1, 31.
11. Johnson, H.M., Stephenson, C.B. (1966) *Astrophys. J.*, 146, 602.

12. Luyten, W.J. (1966)   *I.A.U. Circ.*, No. 1980.
13. McGee, R.X., Murray, J.D. (1961)   *Australian J. Phys.*, **14**, 260.
14. Bell, K.L., Kingston, A.E. (1967)   *M.N.*, **136**, 241.
15. Hidajat, B., The, P.-S. (1967)   *Pub. A.S.P.*, **79**, 260.
16. Mumford, G.S. (1966)   *Astrophys. J.*, **146**, 962.
17. Hardie, R.H. (1967)   *Pub. A.S.P.*, **79**, 173.
18. Mook, D.E. (1967)   *Astrophys. J.*, **150**, L25.
19. Lawrence, G.M., Ostriker, J.P., Hesser, J.E. (1967)   *Astrophys. J.*, **148**, L161.
20. Hiltner, W.A., Mook, D.E., Ludden, D.J., Graham, D. (1967)   *Astrophys. J.*, **148**, L47.
21. Ichimura, K., Ishida, G., Jugaku, J., Oda, M., Osawa, K., Shimuzu, M. (1966)   *Pub. Astr. Soc. Japan*, **18**, 469.
22. Code, A.D. Unpublished.
23. Johnson, H.M., Spinrad, H., Taylor, B.J., Peimbert, M. (1967)   *Astrophys. J.*, **149**, L45.
24. Low, F.J. Unpublished.
25. Yokoi, H., Sato, T., Morimoto, M. (1966)   *Pub. Astr. Soc. Japan*, **18**, 472.
26. Kellermann, K.J., Pauliny-Toth, I. Unpublished.
27. Minkowski, R., Johnson, H.M. (1967)   *Astrophys. J.*, **148**, 659.

THEORETICAL REVIEW

WALLACE TUCKER*

(Cornell University)

First, I will discuss the interaction of the radiation from an X-ray source with the surrounding matter. Then the X-ray emission mechanisms will be considered, and finally, a few of the models which have been proposed will be briefly reviewed.

1. The Interaction of the Radiation from an X-Ray Source with Its Environment

One aspect of X-ray sources which can be studied more or less independently of the origin and detailed physical state of X-ray sources is the extent to which their radiation interacts with any surrounding matter. The X-radiation will ionize and heat this matter, and some of the X-rays will be absorbed.

An investigation of these effects is of practical interest for several reasons. First, emission lines which are characteristic of a cool gas with a temperature of the order of 10^4–10^5 °K have been observed in both the Crab Nebula and Sco X-1. Obviously, an understanding of the excitation conditions produced in the cool gas by the X-ray will improve our understanding of the objects. Secondly, observations of the modifications of the X-ray spectrum which are produced by the absorbing effects of matter between the source and the observer can provide information concerning the interstellar and intergalactic medium. Observational evidence of such absorption has been discussed by Giaconni (1967). Finally, the interaction of the X-rays with the surrounding matter must be taken into account when constructing theories of the origin of the X-ray sources. For example, accretion models encounter severe difficulties, both with respect to the absorption of X-rays and the heating of the accreting gas.

Johnson (1967a) has considered the ionizing effects of the X-rays on the cool gas in Sco X-1, Cas A, and Vir A. He concluded that the observed conditions in some cases, but not all, were compatible with those produced by radiative excitation. However, the temperature in this case would be rather low, of the order of 10 000°. Then, in the case of Sco X-1, it is very difficult to explain the relative strengths of the N III (4641) and Hβ emissions, which seem to require a temperature of 100 000°. The ionization equilibrium would then have to be established by electron collisional ionization and radiative and dielectronic recombination (Tucker, 1967a).

* Present address: Rice University, Houston, Texas.

Perek (ed.), Highlights of Astronomy, 210–215. © I.A.U.

Detailed work on the ionization and heating of a gas excited by X-rays has been done by Williams (1967) and Tarter (1967). Williams considered a power law spectrum for the ionizing radiation, Tarter an exponential one. Their general results are the same; (i) each element occupies a number of different stages of ionization at any one time, and (ii) the temperature in the gas heated by X-rays will generally be between 10000 and 20000°, provided that the K shells of ions are still intact. Calculations of this type are very useful because the observation of just one line sets limits on the X-ray source parameters; the radiation cannot be so intense as to ionize the species which is observed. Thus, in the case of Sco X-1, the observation of a N III line implies

$$\frac{L}{NR^2} = \frac{16\pi F}{N\delta^2} \ll 10^{-2}.$$

Here L is the luminosity of Sco X-1, N the electron density in the cool region, R is the radius of the X-ray source, F the flux received at Earth, and δ the angular diameter. Setting $F = 5 \times 10^{-7}$ erg cm^{-2} sec^{-1} yields $\delta \gg 3 \times 10^{-2}/n^{1/2}$. On the other hand, the *upper* limit on the angular diameter appears to be ≈ 1 sec of arc (Johnson, 1967b), so $N \gg 4 \times 10^7$ cm^{-3}. In the case of 3C 273, the observed emission lines and the observed X-ray flux (Friedman, 1967) imply that $\delta \gtrsim 1$ sec of arc, considerably greater than the angular diameter of the *radio* source (Cohen *et al.*, 1966).

The X-ray absorption coefficient of a cold gas has been calculated by Felten and Gould (1966). They also discussed the information about the interstellar and intergalactic medium which could be obtained from observations of the absorbing effects. They pointed out that by measuring the spectral discontinuity due to an absorption edge, one could obtain a measurement of the amount of matter in the line of sight which would be independent of the source spectrum. The increase in intensity across an edge should be $e^{\Delta\tau}$ where $\Delta\tau \approx 2 \times 10^{-22} \int N \, dl$, according to the most recent calculations of the absorption coefficient by Bell and Kingston (1967).

2. X-Ray Emission Mechanisms

Observations of the X-ray spectra and theoretical arguments have led to the generally accepted conclusion that only synchrotron radiation and bremsstrahlung from a high temperature, low-density gas are feasible mechanisms for producing most of the keV X-rays in the discrete sources. Neutron stars are no longer considered to be a direct source of keV X-rays, in view of their rapid cooling rates (a few years or less for a temperature of 10^7°) and the absence of any observational evidence of a black-body spectrum. Compton scattering and non-thermal bremsstrahlung have been ruled out because of their low efficiency.

In the case of bremsstrahlung, however, an important contribution may arise in the 100 keV range. It can be shown that, if the gas is heated by non-thermal particles with energies of about 100 keV, an observable non-thermal bremsstrahlung flux (three or four orders of magnitude above the 100 keV flux from a 50 million degree gas) is produced in the region 50–100 keV. The 'non-thermal tail' starting around 35 keV, which was observed by Peterson and Jacobson (1966) in Sco X-1, may well be a manifestation of this effect.

A. BREMSSTRAHLUNG FROM A HOT GAS

In attempting to construct a 'hot-gas' model for X-ray sources, a number of formidable problems are encountered. Perhaps the greatest of these is the origin of the hot gas. Although there are evidently a number of ways to heat a plasma under astrophysical conditions, it is by no means obvious how to produce a hot plasma having the high temperature and large energy content necessary to explain X-ray sources. Sartori and Morrison (1967) have argued that a hot gas is produced any time electrons are accelerated, the energy of the hot gas being approximately 100 times that of the relativistic gas. Sturrock (1966) has also mentioned such a possibility in connection with his 'flare' theory of quasars. Ginzburg (1967) has advanced similar ideas but he sets $W_{th} \approx W_{rel}$, a much more modest requirement which would not lead to observable fluxes in general.

Another problem is the short cooling time for a compact source ($t_c \approx 6 \times 10^{14}/N_e$ sec for a temperature of 50 million degrees). If $N \sim 10^8$ cm^{-3}, as seems to be required for Sco X-1, then the cooling time is less than 1 year, and a continuous, powerful source of heating is needed. The expansion times are also small for compact sources so the hot gas must be confined by some means.

Because of these difficulties, Ginzburg (1967) has suggested that, whereas the extended sources with long cooling times ($\sim 10^5$ yr) are likely to be hot gases, the compact sources are probably synchrotron emitters. On the other hand, the energy requirements are more severe for extended sources and particles would be thermalized more quickly in the compact sources. For these reasons, I take the opposite viewpoint, *viz.* that sources in which the thermalization mean free-path is less than a certain critical value are hot gases, but in the other extreme the electrons become relativistic and emit X-ray synchrotron radiation.

B. SYNCHROTRON RADIATION

The existence of radio and optical synchrotron sources leads us to believe that this mechanism may also be responsible for the X-rays in some discrete sources. This belief is reinforced by the measurements of the X-ray spectra of the Crab Nebula and Cyg X-1 (Peterson, 1967), which show a power-law dependence of the flux on the frequency. The synchrotron hypothesis has its difficulties, too. Principal among these are (i) the extremely high energy electrons ($\sim 10^{12}$ eV) required to produce the X-rays in magnetic fields of the order of 10^{-3} or 10^{-4} gauss, and (ii) the short radiative lifetimes of the order of a few years ($\tau_s(\propto v^{-1/2}\ H^{-3/2})\approx 30$ yr for $H=10^{-4}$ gauss, $v=10^{18}$ c/s). However, the amount of energy required in the form of relativistic electrons is in general about two orders of magnitude less than the energy required by the hot-gas models.

C. DISTINGUISHING BETWEEN HOT GAS AND SYNCHROTRON MODELS

There are several means by which it may be possible to decide between the two possible mechanisms. The detection of the deviations from the bremsstrahlung spectrum which are produced by recombination edges and line emission would establish the hot-gas nature of the source. The jump due to the Ne^{+10} recombination edge should be about 30% at $10^{7\,\circ}$, about 5% at $5\times 10^{7\,\circ}$, for a gas having the cosmic abundances (Tucker, 1967b). The line emission from Mg, Si, S and Fe may also be detectable.

The observation of polarized X-rays would establish the synchrotron nature of the source, but a negative result would not conclusively rule out the synchrotron hypothesis.

A study of the time variations of the X-ray emission could also provide information concerning the emission mechanism, The spectrum of a composite hot-gas model would become flatter, a synchron spectrum steeper, as a result of radiation losses. Also, for extended sources the cooling time for a hot gas is so long that no time variations would be expected, as is possible with the synchrotron mechanism.

3. Theoretical Models for X-Ray Sources

A. SCO X-1

The strongest X-ray source in the sky, Sco X-1 has an exponential spectrum characteristic of a hot gas with a temperature of 50 million degrees. It has been identified optically with a flickering, blue, star-like object. Optical observations show that its optical continuum radiation is consistent with the extrapolation of a bremsstrahlung spectrum from X-ray to optical frequencies (Johnson, 1967b).

Šklovskij (1967) has proposed a model of Sco X-1 according to which it is a binary system composed of a neutron star accreting matter from an unstable companion. The difficulty with this and the other binary star models which have been put forth is that very high densities ($\sim 10^{16}$ particles/cm^3) are needed to get the required energy-conversion rate. Then the cool gas in front of the shock wave will quickly absorb the X-rays, so that it is very difficult to construct a realistic binary star model.

Another possibility is that a hot gas is produced by the dissipation of magnetic energy. Assume that the X-ray emission from Sco X-1 is due to an optically thin hot plasma having the cosmic abundances, and that the plasma is confined by a magnetic field which is anchored to a central object. Then it can be shown (Tucker, 1967) that the radius of the X-ray source is between about 10^{12} and 10^{15} cm and the electron density is between about 10^6 and 10^{10} particles per cm^3. If the distance is greater than 500 pc as suggested by Wallerstein (1967), then $R_x \sim 5 \times 10^{14}$, $N_x \sim 10^7$ cm^{-3}, and the radius of the central object must be $\sim 10^{14}$ cm, in order to have both confinement of the hot gas and stability of the central object. All this suggests that Sco X-1 may be a protostar in the process of shedding its magnetic field, as originally suggested by Manley (1966). In this model the energy of the magnetic field is degraded into thermal energy by means of flare-like events. Since the source is much more extensive and massive than the solar corona, most of the non-thermal energy normally associated with flares would be transformed into thermal energy by means of ionization, generation of plasma waves, etc. An observable amount of optical synchrotron radiation and high energy non-thermal bremsstrahlung ($\gtrsim 100$ keV) could quite possibly be produced.

Synchrotron models of Sco X-1 have been proposed by Ginzburg (1967), and Manley (1966). The difficulty with such models is that a flat spectrum ($N(E)dE = = KdE$) is needed to explain the absence of strong optical synchrotron emission. In addition, no optical polarization has been observed.

B. THE CRAB NEBULA

The first radio source and the first X-ray source to be optically identified, the Crab Nebula is known to be a synchrotron emitter at radio and optical frequencies. The shape of the optical spectrum is disputed and the X-ray spectrum is still uncertain. Nevertheless, one is tempted to join the optical and X-ray spectrum and say that the spectral index between about 10^{14} c/s and 10^{18} c/s is constant and equal to about $-1\cdot1$. The radio spectral index is $-0\cdot27$, so the change in spectral index is about $-0\cdot83$, which can be understood in terms of continuous injection varying with time as $(t - t_0)^{2/3}$, where t_0 is the present age of the nebula. Šklovskij (1966) has contended that the spectral index in the optical is $-0\cdot8$, and results from continuous injection at a constant rate. The UV flux is then very large, and the spectral index must change again before X-ray frequencies. Two difficulties with this model are the large UV flux which would produce too much ionization in the filaments (Williams, 1967), and the lack of an explanation for the shape of the X-ray spectrum.

Another possibility is that there are two distributions of electrons in the nebula. Then a number of models can be constructed which fit the observational data (Tucker, 1967). However, special conditions must prevail, such as a flat electron spectrum and Fermi acceleration, or a low energy cutoff around 10^{12} eV. The latter case is interesting since a continuous input of energy is not needed to explain the X-ray emission.

Hot-gas for the Crab Nebula has been proposed by Sartori and Morrison (1967), and by Hayakawa *et al.* (1966). However, in light of the observations, which indicate that the power-law spectrum holds at energies up to 200 keV (Peterson, 1967), such an explanation is very unlikely.

References

Bell, K., Kingston, A. (1967)   *Mon. Not. Roy. astr. Soc.*, **136**, 241.
Cohen, M., Gundermann, E., Hardebeck, H., Harris, D., Salpeter, E., Sharp, L. (1966)   *Science*, **153**, 745.
Felten, J., Gould, R. (1966)   *Phys. Rev. Letters*, **17**, 401.
Friedman, H. (1967)   Joint Discussion on X-Ray Astronomy.
Giacconi, R. (1967)   ibid.
Ginzburg, V. (1967)   *Sov. Phys. Uspehi*, **9**, 543.
Hayakawa, S., Matsuoka, M. Sugimoto, D. (1966)   *Space Sci. Rev.*, **5**, 109.
Johnson, H. (1967*a*)   *Canadian J. Phys.* (in press).
Johnson, H. (1967*b*)   Joint Discussion on X-Ray Astronomy.
Manley, O. (1966)   *Astrophys. J.*, **144**, 1253.
Peterson, L. (1967)   Joint Discussion on X-Ray Astronomy.
Peterson, L., Jacobson, A. (1966)   *Astrophys. J.*, **145**, 962.
Sartori, L., Morrison, P. (1967)   *Astrophys. J.*, **150**, 385.
Šklovskij, I. (1966)   *Sov. Astron. – A.J.*, **10**, 6.
Šklovskij, I. (1967)   *Astrophys. J.*, **148**, L1.
Sturrock, P. (1966)   *Nature*, **211**, 697.
Tarter, C. (1967)   Ph.D. Thesis, Cornell University.
Tucker, W. (1967*a*)   *Astrophys. J.*, **149**, L105.
Tucker, W. (1967*b*)   *Astrophys. J.*, **148**, 745.
Wallerstein, G. (1967)   Preprint.
Williams, R. (1967)   *Astrophys. J.*, **147**, 556.

X-RAYS RADIATION OF A NEUTRON STAR
AS A RESULT OF GAS ACCRETION

Ja. B. Zel'dovič

The accretion of gas on the surface of a neutron star is a source of energy, which can be converted into X-rays radiation. Accretion can ascertain a far greater lifetime of an X-ray star than the radiation of internal energy.

The source of gas could be a cloud of intergalactic gas. If the neutron star is a component of a binary system, the gas could come from the other (non-neutronic) star.

The idea of accretion and evaluation of energetic balance were given by Zel'dovič (1964) and Zel'dovič and Novikov (1964, 1965) and Salpeter (1964). Shortly the idea of accretion was mentioned by Šklovskij (1967).

The motion of gas and radiation spectra idealised case of spheric symmetry are studied now by N. Shakuro and Zel'dovič.

As known the rate of accretion dM/dt on the surface of a star with mass M from a gas cloud, resting at infinity with the density ρ_0 and sound velocity a_0 is given approximately by

$$\frac{dM}{dt} = 30\ G^2 M^2 \rho_0 a_0^{-3}.$$

The energy given per gram of the gas is $\varphi(R) = GM/R$ (R=radius of the star). For a neutron star with $M = 1 \cdot 5 M_\odot$, $R = 10^6$ cm, $\varphi(R) \sim 0 \cdot 2 c^2 \sim 2 \times 10^{20}$ is far greater than the nuclear energy per gram. The luminosity L is

$$L = \varphi\ \frac{dM}{dt} = 30\ G^3 M^3 \rho_0 a_0^{-3} R^{-1}.$$

As a first approximation one could calculate the surface temperature T_c from $L = 4\pi R^2 \sigma T_c^4$; but we will prove that the radiation spectra corresponds to a higher temperature.

The flow of energy L acts upon the falling gas, repulsing it from the star. The X-rays undergo Compton scattering. The repulsion equals the gravitational attraction independent of the distance if $L = L_c = 6 \times 10^4\ M$ (Eddington's limit).

So the foregoing formula are true only in the case when they give $L < L_c$, for this it is necessary that $\rho < \rho_{0c}$. Given $T_0 = 100°$, $a_0 = 10^5$ cm/sec, M; R see above, we obtain $\rho_{0c} = 10^{-21}$ g/cm^3. At $\rho_0 > \rho_{0c}$ there is an automatic regulation of flow so that $L = L_c$.

Perek (ed.), Highlights of Astronomy, 216–219. © *I.A.U.*

Below is given a detailed analysis of the case $\rho_0 < \rho_{0c}$. We take $\rho_0 = \frac{1}{3}\rho_{0c} = 3 \times 10^{-22}$ g/cm^3. Then $dM/dt = 3 \times 10^{17}$ g/sec $= 5 \times 10^{-9} M_\odot$/year, the radiation could last some 10^8 year before the relativistic collapse of the neutron star.

In this case $L = 6 \times 10^{37}$, $T_c = 1\cdot7 \times 10^7 = 1\cdot5$ keV. At the surface of the star the velocity of infalling gas is $U = 2 \times 10^{10}$ cm/sec, its density $\rho = 1\cdot2 \times 10^{-6}$ g/cm^3; $n \sim r^{-1/2}$, $\rho \sim r^{-3/2}$, $\int_r^\infty \rho \, dr = 2\cdot4$ g/cm^2. The optical depth for Compton scattering $r = 1$, so the interaction of radiation with falling gas does not affect strongly the spectra and the visible diameter of the source.

The infalling gas is a stream of protons gravitationally accelerated up to energy 150–200 MeV. There is no usual shock wave. The protons strike the atmosphere of the neutron star and beat the atmosphere upon all the depth corresponding to the protons' range. Their range is of the order of $y_0 = 20$–40 g/cm^2 and depends chiefly upon nuclear scattering. Let us denote $y = \int_r^\infty \rho \, dx$, y is the total mass of the atmosphere above a given point. The energy given by protons per gram of atmosphere substance, ω erg/g $\times$ sec is $\omega = Q/y_0$, $y < y_0$; $\omega = 0$, $y > y_0$, where $Q = L/4\pi R^2 = 5 \times 10^{24}$ erg/cm^2. The flow of electromagnetic radiation is given through its volume density E by

$$e \, \frac{\text{erg}}{\text{cm}^2 \times \text{sec}} = -\frac{c}{3x} \frac{dE}{dy}, \quad x = 0\cdot38 \text{ cm}^2/\text{g}.$$

Due to the independence of Compton cross-section from the energy of quanta, the formula for e is independent from the spectra of the radiation but only from the overall density E. Taking for the energy transfer from protons the first approximation, one obtains

$$e = Q \frac{y_0 - y}{y_0}, \quad y < y_0; \quad e = 0, \quad y > y_0.$$

Accounting for the condition on the top of the atmosphere, we obtain

$$y = 0, \quad E = \frac{\sqrt{3}}{2} \frac{Q}{c}$$

$$0 < y < y_0, \quad E = \frac{Q}{c} \left\{ \sqrt{3} + xy_0 \left[3\frac{y}{y_0} - \frac{3}{2}\left(\frac{y}{y_0}\right)^2 \right] \right\}$$

$$y > y_0, \quad E = \frac{Q}{c}\left(\sqrt{3} + \frac{3}{2}xy_0\right) = 20\frac{Q}{c} = \text{const} = 3 \times 10^{15} \text{ erg/cm}^3.$$

At the depth under the layer where the protons are brought to rest, $(y > y_0)$ there is established full thermodynamic equilibrium with the temperature which can be found from

$$E = aT^4, \quad a = 7\cdot8 \times 10^{-15}, \quad T = 2\cdot5 \times 10^7, \quad y > y_0.$$

Inside the layer $(y < y_0)$ one has to calculate the balance of energy of electrons, becoming ω erg/g $\times$ sec and giving up this energy by bremsstrahlung on protons and by 'comptonisation, – the energy transfer to quanta by Maxwellian electrons (Kompaneec, 1955; Weyman, 1965).

$$\omega = \frac{Q}{y_0} = 5 \times 10^{20} \sqrt{T}\rho\left(1 - \frac{T'}{T}\right) + 6 \cdot 5 T E\left(1 - \frac{T''}{T}\right).$$

Here T is the electron temperature, T' is a measure of (brems-)absorption of quanta by electrons in the field of protons, T'' characterises the energy given by quanta to the electrons by Compton scattering; T' and T'' depend not only from T but also from the spectra of the radiation, but as a first approximation we shall assume $T' = T'' = T_r$, where $T_r = \sqrt[4]{E/a}$. The radiation density is given as a function of y above. The density is found trivially through $P = RT\rho/\mu = yg = yGM/R^2$. (The radiation pressure and the impulse of infalling protons are much smaller than the weight of the layer.)

We obtain e.g. for $y = y_0 = 30$ g/cm^2, $P = 6 \times 10^{15}$, and at $T = 2 \cdot 5 \times 10^7$, $\rho = 1 \cdot 5$ g/cm^2. The equation of energy balance is used to find the temperature of electrons T. In the case considered it varies from

$$T = 2 \cdot 6 \times 10^7 \quad \text{at} \quad y = y_0 - 0 \quad (\text{compare } T = 2 \cdot 5 \times 10^7 \text{ at } y = y_0 + 0)$$

to

$$T = 10 \times 10^7 \text{ at the top}, \quad y = 0.$$

Note that the temperature on the top does not depend in this approximation from the infalling flow of the gas and the overall luminosity L; the temperature in the depth is proportional to $\sqrt[4]{L}$.

In the layer taken as a whole, approximately one half of the energy is expended for comptonisation, the other half for bremsstrahlung.

The comptonisation markedly alters the spectrum of radiation as compared with blackbody at $T(y_0) = 2 \cdot 5 \times 10^7$. The mean energy of quanta is raised some $1 \cdot 5$–2 times. The difference is even more remarkable if we compare the real spectrum with the blackbody at $T_c = 1 \cdot 7 \times 10^7$.

The energy of protons in our example is slightly less than necessary for creation of pi-meson. Due to the helium content of the gas, one can anticipate some creation of pions which shall give 60 MeV quanta through $\pi^0 \to 2\gamma$ and 120 MeV through $\pi^- + p = n + \gamma$.

Now we are working on the calculation of spectrum at $\rho_0 < \rho_{0c}$; on the spherically symmetric solution for $\rho > \rho_{0c}$ where the radiation drag on the flow is important; on the stability of spheric-symmetric solutions. Further we expect to analyse the non-symmetric motion when the gas has sidewise motion or in the case of a binary.

In the case of a relativistically collapsing star it was shown previously that in the spheric-symmetric case there will be no external radiation. All the energy of the falling gas will be buried in the star. But the sidewise motion of the gas will give a shock (of the type considered by Salpeter (1964)) outside the Schwarzschild sphere, and this shock gives gamma and X-rays. We hope to investigate this problem also in details.

Note added in proof. The plasma instabilities can shorten the proton range, and increase top temperature. The amount of π^0 and gamma is diminished.

The accretion on white dwarfs was considered by Cameron (I.A.U. Trans., vol. XIII), and Cameron and Mock (1967) *Nature*, **215**, 464.

References

Kompaneec, A.S. (1955)   *JETP* (U.S.S.R.), **31**, 876.
Kompaneec, A.S. (1957)   *Sov. Phys. Jetp.*, **4**, 730.
Salpeter, E.E. (1964)   *Astrophys. J.*, **140**, 796.
Škovskij S.S. (1967)   *Astrophys. J.*, **148**, L1; *Astron. J. (russ.)*, **44**, 930.
Weymann, R. (1965)   *Phys. Fluids*, **8**, 2, 12.
Zel'dovič, Ja.B. (1964)   *Dokl. Akad. Nauk* (U.S.S.R.), **155**, 67.
Zel'dovič, Ja.B., Novikov, I.D. (1964)   *Dokl. Akad. Nauk* (U.S.S.R.), **155**, 1033.
Zel'dovič, Ja.B., Novikov, I.D. (1965)   *Usp. fis. Nauk*, **86**, 447; English translation: *Sov. phys. usp.*, **8**, 522.

DISCUSSION

G.G. Fazio: Have you got to a calculation on the expected intensity of gamma rays in the π_0 mechanism?

Ja.B. Zel'dovič: It is thought that it is of the order of 10^{-3} if the density is as small as 10^{-21} so that protons have energy of free fall.

P.J.A. Gaposchkin: What percent of π_0 mesons decay and what percent will interact?

Ja.B. Zel'dovič: 100% will decay and none will interact due to the very short decay time.

THE COSMIC X-RAY BACKGROUND

R. J. WEYMANN

(Steward Observatory, University of Arizona, Tucson, Ariz., U.S.A.)

Evidence for a diffuse, isotropic non-terrestrial component of X-ray radiation has been accumulating since the very beginnings of X-ray astronomy. This evidence now seems secure beyond reasonable doubt, especially in the light of recent experiments, e.g. one in which a very clean separation from the cosmic ray and terrestrial background has been achieved (Pounds, 1967), and another in which a counting rate proportional to the solid angle of the sky subtended to three different detectors was found (Matsuoka *et al.*, 1967).

In this very brief review I wish simply to discuss the current ideas on the mechanisms for the production of the background and the extent to which the present observations can discriminate between them and to consider the information obtainable from the very soft X-ray region, which so far has been relatively little exploited.

The status of the observations has been summarized in Fig. 1 of a recent excellent review by Gould (1967). More recent observations do not change the over-all picture except that the tendency for the lower energy range (1–20 keV) to have a slightly flatter number-spectral index ($\sim 1 \cdot 7$) than the higher energy range (20 keV – 1 meV: $\sim 2 \cdot 2$) seems strengthened. All groups agree that except for a group of partially unresolved sources in the direction of the galactic centre, any anisotropy present must amount to less than about 10%. In particular, in the plane of the galaxy no firm evidence for a general brightening has been found.

As is almost self-evident from Gould's Fig. 1, thermal bremsstrahlung cannot explain the whole range of the background, and accordingly attention has centered on non-thermal processes – in particular Compton scattering of photons by high-energy electrons. The high degree of isotropy of the background, together with the demonstration by Felten and Morrison (1966) that galactic electrons cannot account for the background, suggest that an extragalactic origin be sought, and these have divided into two classes:

(a) A smooth continuous background arising from universal extragalactic high-energy electrons interacting with the $3\,^{\circ}K$ photons.

(b) Many discrete extragalactic sources.

One could try to differentiate between these two possibilities along any of the following three lines:

(1) the shape of the observed spectrum,

(2) the total-power requirements,

Perek (ed.), Highlights of Astronomy, 220–227. © I.A.U.

(3) the degree of isotropy of the radiation.

The continuum models proposed can be subdivided according to the mode of providing the high-energy electrons: Felton and Morrison suggest direct replenishment by active radio sources, whereas Gould has suggested secondary production via interaction of a postulated intergalactic cosmic-ray component with intergalactic matter, the origin of the high-energy protons being unspecified. In either case, one requires a production spectrum for the electrons which, following the steepening due to Compton losses, gives rise to the observed X-ray spectrum. On the basis of the clustering of radio spectral indices about the value 0·7, one is led to an index of 2·2 in the (number) spectral index for the X-rays; estimates for the secondary production mechanism based upon the observed cosmic-ray spectrum lead to about the same value for the spectral index, which lies nicely in the observed range.

This is taken to be a strong point in favor of the continuum models; on the other hand, some sources in our own galaxy, notably the Crab, have indices of about 2·2 in the 1–100 keV range – also just in the range of the observed spectral index. No data on the spectral index of M 87 in the X-ray region itself are available as yet, but the combined radio, optical, and X-ray data are consistent with an index of 1·7 – again *not incompatible* with the data over the low-energy X-ray range.

At the present time, therefore, we feel that there is no basis for discriminating between the two models on the basis of spectra.

Regarding the total power requirements, and neglecting for a moment evolutionary considerations, both the discrete and continuum models run into some difficulties. The input of relativistic particles into intergalactic space was estimated by Felten and Morrison by supposing that the entire store of these particles (whose total energy store is estimated by assuming equipartition of the field and particle energies) is turned over to intergalactic space every 10^6 years. Taking a typical store of high-energy electrons to be 10^{57} ergs and using a density of strong sources not very much different from that recently given by Schmidt (1966), one finds that, over 10^{10} years, the store of particles fails by about 2 orders of magnitude to explain the X-ray flux; as remarked by these authors, the assumption of a field lower than the equipartition value increases the energy of relativistic particles and hence the input. But the other side of the coin is that the total energy requirements, already very large – 10^{59} ergs in relativistic particles if 100 times the electron energy in heavy particles is assumed, turned over 10^3 times during the lifetime of 10^9 years estimated by Schmidt (1966), amounts to 10^{62} ergs – go up even further.

The situation is equally uncomfortable for the discrete source models. We consider three types of objects radiating in the 1–10 Å band at three different levels:

(1) Objects with a flux 10^{46} ergs/sec – the flux inferred for 3C 273 if at its cosmological distance (Friedman, 1967),

(2) Objects with a flux of 2×10^{43} ergs/sec – based upon M87 (Friedman, 1967),

(3) Objects with a flux of 2×10^{39} ergs/sec which might be appropriate for objects

like our own galaxy – equivalently we might suppose that each gram of ordinary condensed matter radiates at a rate of 10^{-5} erg/g.

The specific intensity in the 1–10 Å band seems to lie between about 3×10^{-8} and $1 \cdot 0 \times 10^{-7}$ ergs/cm^2/sec/ster, and this is related to the source density and luminosity per source by the relation

$$I \sim k \frac{c}{H_0} n_0 \frac{L_\nu}{4\pi},$$

where H_0 is the Hubble constant, n_0 the density of sources, and L_ν the luminosity of the source.

The constant k depends upon the spectral index and the cosmological model; for $q_0 = \frac{1}{2}$ and a photon spectral index of 2·5 it is $\frac{1}{3}$; for an index of 2·0 it is $\frac{2}{5}$. Table 1 summarizes the density of various classes of objects required to produce the observed flux, together with the estimated actual densities. For purposes of a discussion below we also give the required number of objects per square degree out to a redshift of 0·6.

Table 1

Class of objects	Required density	Estimated density	Required no./ deg.^{2} to $Z = 0.6$
Ordinary condensed matter	$1 \cdot 3 \times 10^{-29}$ to 4×10^{-29} gm/cm^3	7×10^{-31} gm/cm^3	$1 \cdot 2 \times 10^6$ [a]
Strong radio sources like M 87	$1 \cdot 9 \times 10^{-4}$ to $5 \cdot 8 \times 10^{-4}$ obj./(mpc)^3	7×10^{-6} obj./(mpc)^3	$1 \cdot 1 \times 10^2$
Quasars	4×10^{-7} to $1 \cdot 2 \times 10^{-6}$ obj./(mpc)^3	2×10^{-9} obj./(mpc)^3	$2 \cdot 3 \times 10^{-1}$

[a] Based upon each object having $M = 10^{11} M_\odot$.

Evidently all three major classes fail by about 1·5–2·0 orders of magnitude to account for the flux – about the same margin as the Felten-Morrison hypothesis. Other sources could of course be added to the list – e.g. Sandage (1967) estimates that the number of 'radio-quiet quasars' are 500 times more numerous than the quasars themselves – but the very fact that they are radio-quiet suggests a deficiency of high-energy electrons and hence, perhaps, of X-rays.

Before considering possible modifications to the foregoing due to evolutionary effects, we consider a third possible discriminant: suppose that the background does in fact consist of discrete sources – under what circumstances might we expect to 'resolve' the sources? We consider some numerical examples based upon the following model:

(a) The sources are all identical and are randomly distributed throughout the universe, assumed to be characterized by $q_0 = \frac{1}{2}$,

(b) The (energy) spectral index is 1·5,

(c) Each source radiates at a constant rate, but there are no sources beyond $z = 15$.

How might one distinguish between two models in which nearly all the flux comes from a continuum (but with a few nearby resolved sources) from one in which it all comes from discrete point sources? Figure 1 shows the cumulative probability distributions for flux levels from random directions in the sky, normalized to the *median* flux level. Due to instrumental errors and photon statistics, even a smooth continuum would be 'broadened' from a step function – this is represented by the integral of a gaussian distribution with a probable error of 10% and labeled 'GS'. The other curves are the distributions in flux (uncorrected for instrumental effects) for cases in which the density of sources is such that 2, 16, and 128 sources *out to a redshift of 0·6* are contained in the beam. This example suggests that for N much larger than 128 ($N \geqslant 500$) it would be very difficult to distinguish between the 'discrete' and 'nearly

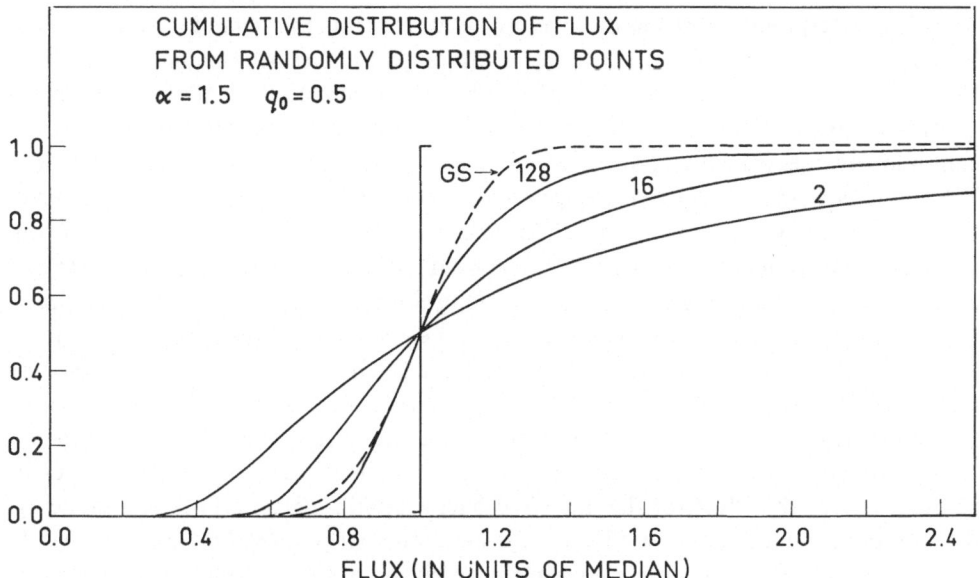

FIG. 1. *The probability that a random sample of sky will yield a flux less than a certain value, in units of the median value of the flux. Point sources radiating with a spectral index of 1·5 in a flat cosmological model were considered. The curves labeled 128, 16, and 2 correspond to the mean number of sources out to $z = 0·6$ in the beam. The curve 'GS' is a Gaussian distribution with a probable error of 10%.*

continuous' models on the basis of resolvability. Realizing that background experiments have been carried out with solid angles ranging from 10^1 to 10^3 square degrees, one sees from Table 1 that even a class of M 87-type objects could not be resolved if they existed in numbers sufficient to explain the background, although the quasar-like objects could be.

Let us now consider what sort of evolutionary considerations are required to avoid the difficulties with both the discrete and continuum models.

This question has recently been discussed by, among others, Bergamini *et al.* (1967). Their work may be summarized as follows:

(a) Rather than make semi-empirical estimates of the X-ray flux from various classes of objects as we have done above, they proceed directly from the ratio of synchrotron to Compton radiation arising from the 3° radiation, which is known if the magnetic field is specified.

(b) Specific account is taken of the increase in temperature and energy density of the 3°K radiation field – this has the consequence that the ratio of Compton radiation to synchrotron radiation increases sharply with increasing red-shift. On the other hand, other things being equal, the severe Compton losses will cause the break in the electron-energy spectrum, and hence in the Compton and synchrotron spectrum, to occur at rather low energies for these distant sources.

(c) The interpretation of the radio-source counts seems to demand a rather strong increase in the activity of a class of strong sources with increasing red-shift (Longair, 1966; Schmidt, 1967), and this will further enhance the contribution of distant sources.

(d) The specific model used for the radio sources is one in which electrons are continuously injected with a power-law spectrum but then escape after a fixed time T_0. The magnetic field in the sources is supposed constant in time and must be a few microgauss. The confinement time is chosen in such a way that the break in the synchrotron spectrum occurs outside the observed region, where no break is observed. If it were to occur for very low frequencies, uncomfortably long confinement times would be required.

The size of these evolutionary effects can be seen by considering sources with $H = 10^{-6}$ gauss, $T_0 = 3 \times 10^6$ years, injection spectral index for the electrons of 2.4, and a flat ($q_0 = \frac{1}{2}$) cosmological model with no sources beyond $Z_{max} = 4$. Consider three situations: (a) No account is taken of either the increase in the 3° radiation field, or increased activity of the sources, (b) no increase in the activity of the sources is considered, but the effects of the increased background radiation, both in enhancing the Compton radiation and in affecting the electron-energy spectrum, are considered, (c) the probability of sources occurring in a given sample of objects increases as $(1+Z)^{5\cdot7}$; 5·7 and the cut off $Z_{max} = 4\cdot0$ are chosen simply because these parameters were found by Longair to be consistent with the radio-source counts.

At about 1 keV, model (b) yields an intensity about 15 times that of model (a), while model (c) in turn yields about 1500 times the intensity of model (b). Finally, if we arbitrarily divide all sources into two classes – those which evolve as $(1+Z)^{5\cdot7}$ and those which do not – then this model accounts for the observed background flux if only 2% of the radio emission per unit volume at the present epoch comes from the strongly evolving class. Longair tentatively identified his strongly evolving sources with the QSS, and both the steep index of 5·7 and the figure of 2% are not inconsistent with recent results of Schmidt (1967). This model predicts a spectral index of 1·7 in the soft X-ray region and 2·2 in the hard region with a broad transition region centered about 3 keV.

If the QSS are indeed cosmological, it seems possible therefore that the X-ray background may be produced by these objects or their immediate successors, although the model used above is certainly much too simple. It will be interesting to repeat the resolvability calculation described above for this model, for the larger contribution by the numerous distant objects will make resolution more difficult. Moreover,

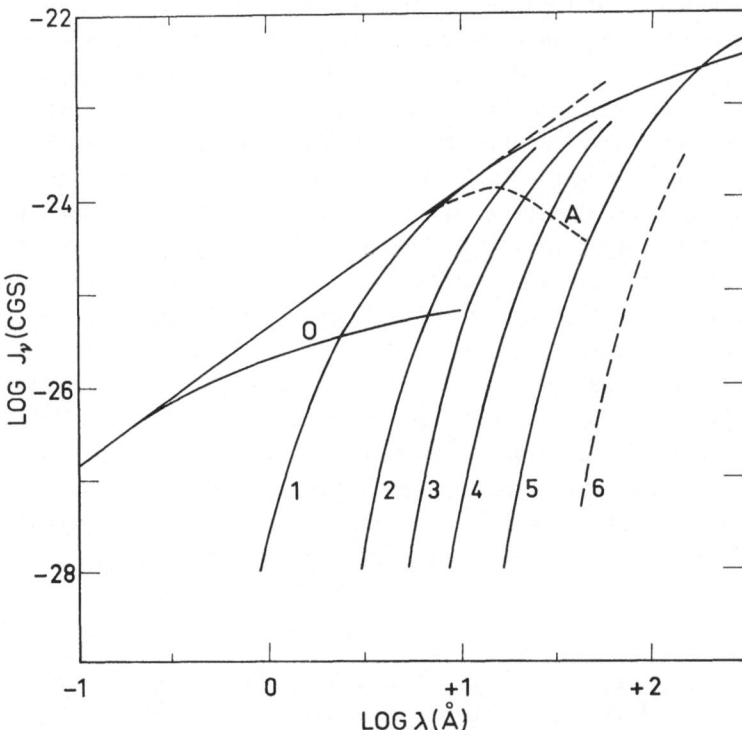

FIG. 2.   *Emission and absorption by intergalactic gas in the soft X-ray region. Curves 1–5 show the emission expected from a hot gas with present density 10^{-5} particles/cm³ with five different thermal histories all satisfying the Gunn-Peterson criterion on the absence of Ly-α absorption at $z \sim 2$. Curve 6 does not satisfy this criterion. Curve 'A' is the background resulting from a uniform distribution of neutral hydrogen in which is embedded a uniform distribution of X-ray emission with a spectral index of 1·5. Curve 'O' is an approximate representation of the observations.*

the electrons will continue to Compton scatter after being disgorged from the sources as in the Felten-Morrison model, and a quasi-continuum background will arise which may make a significant contribution to the total background.

Let us now turn to information which might be obtainable from observations of the X-ray background in the soft X-ray region – from, say, 1·0 to 0·25 keV.

In the first place, it will certainly be of interest to see whether the gradual leveling off of the flux, suggested by the most recent experiments, holds up. But there are also interesting possibilities for obtaining information about intergalactic matter in either neutral or ionized form. Aside from its intrinsic interest, the question of intergalactic matter touches directly upon the mechanism for the background proposed by Gould which requires on the order of 10^{-5} particles/cm^3.

Detection of thermal bremsstrahlung from a hot, ionized gas with densities around 10^{-5} particles/cm^3 might be achieved. Curves 1–5 of Figure 2 show the emission from an intergalactic gas with several plausible thermal histories, all satisfying the Gunn-Peterson restriction on Ly-α at $Z \sim 2 \cdot 0$. Curve 6 is the emission from a case which does *not* satisfy this restriction. Evidently the critical observational requirement is to push the observations to very low energies, well into the region where galactic absorption is expected to become important. The possibility of detecting neutral hydrogen by X-ray absorption has been frequently discussed. Although not nearly as sensitive as the Ly-α absorption in the QSO, the Ly-α test is still plagued by the uncertainty in the local vs. cosmological issue. Curve 'A' in Figure 2 shows the resulting background for a uniform distribution of sources radiating with an energy spectral index of $1 \cdot 5$ (straight line with dashed extension), intermixed with a uniform distribution of neutral hydrogen. Note the decrease is not exponential, but would tend to become so if the emission comes mostly from distant sources as described above. For pure hydrogen, optical-depth unity occurs for a photon whose final wavelength is 10 Å; for a normal admixture of helium this is shifted to 6 Å. This model evidently cannot explain the apparent flattening of the background, indicated by 'O', but obviously much more work with moderate spectral resolution needs to be done in the soft X-ray region before any conclusions about intergalactic matter can be drawn.

References

Bergamini, R., Londrillo, P., Setti, G. (1967)   preprint.
Felten, J.E., Morrison, P. (1966)   *Astrophys. J.*, **146**, 686.
Friedman, N. (1967)   Paper read at Joint Discussion on X-Ray Astronomy, I.A.U., Prague.
Gould, R.J. (1967)   *Amer. J. Phys.*, **35**, 376.
Longair, M.S. (1966)   *Mon. Not. Roy. astr. Soc.*, **133**, 421.
Matsuoka, M., Oda, M., Ogawara, Y., Hayakawa, S., Kato, T. (1967)   preprint.
Pounds, K. (1967)   Paper read at Joint Discussion on X-Ray Astronomy, I.A.U., Prague.
Sandage, A.R. (1967)   Invited discourse, I.A.U., Prague.
Schmidt, M. (1966)   *Astrophys. J.*, **146**, 7.
Schmidt, M. (1967)   Private communication.

DISCUSSION

L.W. Acton: What do you mean by 'evolution model'?

R.J. Weymann: As we observe objects farther away we observe some systems at an earlier stage of evolution and presumably at earlier stages they are stronger X-ray emitters.

Anonymous: What was the value you used for the intensity of the diffuse cosmic X-ray flux?

R.J. Weymann: I used a value of 10^{-7} ergs/cm^2 sec^{-1} between 1 and 10 angstroms. I believe that is the value that Dr. Friedman has quoted you.

Ja. B. Zel'dovič: What areal resolution is required to tell the difference between the continuous and discrete models?

R. J. Weymann: The point of the statistical analysis was really to find out how small does the detector scan beam have to be in order to be able to see if these really are discrete sources. It is my guess that even 1 square degree will not be sufficiently small.

C. JOINT DISCUSSION OF COMMISSIONS 27, 29, 35, AND 36
THE LITHIUM PROBLEM

(Monday, August 28, 1967)

Organizing Committee: K.H. Böhm (Chairman), W.K. Bonsack,
J.L. Greenstein, G.H. Herbig, H. Reeves, A.B. Severny
Discussion Chairman: W.A. Fowler
Secretaries: W. Deinzer, H. Reeves

Contents:

THE OCCURRENCE OF LITHIUM IN STARS

G. H. HERBIG

My task is to outline our present observational knowledge of the occurrence of lithium in stellar atmospheres. On account of the limited time, I shall not attempt to include a description of the situation in post-main sequence stars. Also for shortness of time, forgive me if I do not stop at each point to give due credit to the astronomers who have contributed to that topic.*

The youngest stars we know which are cool enough to exhibit neutral Li are the T Tauri stars, which are in the early stages of contraction toward the main sequence. All these objects that have been adequately observed are very abundant in Li: the range is between 50 and 400 times the so-called 'solar abundance', a convenient unit which corresponds to a H/Li ratio by number of atoms of about 10^{11}. It is significant that the average Li content of chondritic meteorites, which are often considered to be samples of the non-volatile, unprocessed material of the original solar nebula, is about 150 on this system, within the range observed in the T Tauri stars.

As the T Tauri stars contract toward the main sequence, the surface convection zone may be deep enough to circulate Li to a depth where it can be depleted by proton impact. But there is no completely convincing evidence that this does indeed take place. The most direct check would obviously be to observe stars between the T Tauri phase, which is believed to occupy only the first 10–20% of the total contraction time, and the main sequence. Although there must be many such stars, no means has as yet been devised to find them, and so this important test has not been made.

There is now, however, a large amount of information on Li in main sequence stars, and it is clear that such objects contain much less Li than do the T Tauri stars. The pattern of Li abundance vs. spectral type along the main sequence is shown in Fig. 1 of the *Ann. Astrophys.* reference. Between the early F and late G-type dwarfs, there is a major fall-off in the amount of Li; in K and M dwarfs, there is no Li detectable. Since the depth of the surface convection zone increases toward later types, the obvious explanation of this pattern is that it is simply the result of varying amounts of Li destruction by convection. It is the task of the theoretician to decide whether convective circulation is equal to this task, and if so, whether all the effect took place during contraction, or whether destruction must have gone on, on a much longer time scale, following arrival on the main sequence.

* Most of the major papers in the field are listed in *Ann. Astrophys.* 1966, **29**, 593.

Perek (ed.), Highlights of Astronomy, 230–232. © *I.A.U.*

It is clear from the data that at each spectral type there is a major spread in Li abundance: in the early F's, it is at least a factor of 10. In such early-type stars the surface convection zone is shallow, no major convective destruction is expected, and so one is inclined to regard this spread as a dispersion in the original, starting values of the Li content. But this factor of 10 contains also the effect of the evolutionary mass dispersion in a sample of field stars having the same spectral type. This particular source of confusion can be reduced by examination of stars all belonging to the same cluster, where the mass spread should be small. Such results suggest that the real spread in the starting Li abundances is only over a factor of 5.

For the sake of completeness, I should mention that pure convection may not be the only mechanism which affects the surface Li abundance. It is necessary also to consider the effect of mass loss, which in effect peels off the Li-rich surface layers and replaces them with Li-poor material from below. This phenomenon seems to be of special promise in understanding the variation in the abundance ratio of Li^6/Li^7 from star to star.

Where has this Li come from in the first place? Can it be that there is this much Li in the interstellar material from which stars are made? We are coming closer to a decision on this point, but the best observations to date are not yet quite adequate. The best place to search for interstellar Li I is surely in a cool interstellar cloud which is known to contain T Tauri stars. An excellent example is the Sco-Oph dark clouds. There is now available a new analysis of the interstellar line spectrum of ζ Ophiuchi, which is seen through a substantial thickness of these clouds. The analysis for Li is best done differentially with respect to K I and Na I which have similar ionization potentials to Li I. The ionization correction can be made with some degree of confidence since good laboratory measurements of the photo-ionization cross-sections are now available. No interstellar Li I resonance line can be seen on good coudé spectrograms of ζ Oph, but the upper limit is still greater than the chondritic Li content by a factor of 9. Until this upper limit can be pressed down by another order of magnitude, we are unable to say whether there is enough free Li in the interstellar clouds to account for the abundances found in the T Tauri stars.

Whether or not Li is present in the interstellar material, it can be argued that this amount of Li was not formed by nuclear processes in the T Tauri stars after they became luminous. This point rests entirely on FU Orionis, which has been interpreted as a pre-T Tauri star very near the beginning of its Hayashi track. Despite the fact that this star, which must be completely convective, flared up less than 30 years ago, it contains 80 times the solar Li abundance. The production of such an enormous amount of Li in such a short time by proton spallation of light elements would require a surface proton flux of 10^{11}–10^{12} times the present solar value, which seems to be inadmissable. Other possible sites where the Li may have originated will be discussed by later speakers in this program.

DISCUSSION

Feast: This seems a suitable opportunity to mention some work on the small but very interesting group of C-S stars of which UY Cen is probably the brightest example. Eight of these stars have now been examined at coudé dispersion and lithium found in all of them (in UY Cen itself measurements by Mr. R. Catchpole suggest that the lithium is mainly Li[7]). In seven of the eight stars Li I 6708 is roughly the same strength as Ca I 6572 (an equivalent width of about 0·5 Å) indicating a lithium abundance of perhaps 10 times the solar value. However, in the remaining star Li I 6708 is exceptionally strong (equivalent width ∼ 3 Å) although the rest of the spectrum is quite similar to that of the other stars. The abnormal lithium strength places this star in the same class as the three previously known lithium stars (WZ Cas, WX Cyg and T Ara). These three stars are all otherwise normal carbon stars, and it is very interesting therefore to find that extreme lithium overabundance can also occur in stars of lower C/O ratio.

Underhill: Is the relationship between line strength and abundance implicit in the use of 'lithium abundance' based on the simplified LTE theories of line formation?

Herbig: Yes.

L.H. Aller: Several recent investigations (Hollis Johnson, B.J. O'Mara) show that deviations from LTE are not very large for the sodium D-lines and are in fact quantitatively explained. The lithium and sodium lines have been handled theoretically in the same manner by Dr. Herbig. The effects of non-LTE give errors of maybe 20–30% and we are concerned with deviations of a factor of 50 or so! Non-LTE effects cannot be important in this context.

Biermann: For the question of mixing by convection or circulation the state of rotation of the stars would be of interest. Has the correlation of Be-, Li-abundance with rotation been investigated?

Herbig: Not as yet. The Li I $\lambda6707$ line becomes too wide to be seen at all if the rotational velocity becomes greater than about 50 km/sec, so such a test will be possible only at the lower rotations.

Hack: R. Lynds has found a non-identified line at λ 6708 in the solar spectrum. How much can this line, if present in stellar spectra, affect the Li abundance?

Herbig: In two dwarfs of spectral types F 8 (β Vir) and G 8 (τ Cet), there is no Li I line at $\lambda6707$ detectable with a dispersion of 4 Å/mm. Therefore in these stars we can examine the background on which the Li I line is superimposed for possible blending lines. At that dispersion, no other lines can be seen near 6707–8 Å, which corresponds to an upper limit on the equivalent width of perhaps 5 mÅ. Since in F-G dwarfs the equivalent width of the lithium line goes up to 100 mÅ, I hardly think that the contribution of a line weaker than 5 mÅ need concern us in the present rather rough analyses.

EVIDENCE FOR LITHIUM DESTRUCTION AND SYNTHESIS IN MAIN-SEQUENCE AND SUBGIANT STARS OF ABOUT SOLAR MASS

M. W. FEAST

(Radcliffe Observatory, Pretoria, South-Africa)

Herbig's extensive work on lithium abundances in solar-type stars on and near the main sequence, suggested to him that for these stars the lithium abundance might be some simple, decreasing function of age. If this were the full story we should expect a low lithium abundance in subgiants of about solar mass. However, already in Herbig's work there were one or two stars with lithium abundance about 10 times the Greenstein-Richardson solar value whose trigonometrical parallaxes showed them to be possible subgiants. Other subgiants with high lithium abundances have been found by Wallerstein and at Pretoria. These results suggest that solar-mass stars are able to synthesise fresh lithium in or just prior to their subgiant phase.

A study (Feast, 1966) of the Li^6/Li^7 isotope ratios also provided evidence for lithium synthesis during evolution off the main sequence. Freshly synthesised lithium is expected to have a high value of Li^6/Li^7 (close to 0·5, the laboratory spallation ratio). Older lithium which has undergone depletion by convection to regions of high temperature should have a lower isotope ratio. In fact the few stars with a high ratio (all of them close to 0·5) are all subgiants well above ($\sim 1\overset{m}{\cdot}4$) the main sequence. Of course, lithium preserved in old stars in which convection has been inhibited by magnetic fields (as Wallerstein has suggested) would also be expected to have a high Li^6/Li^7 ratio, but we should also expect to find similar high Li^6/Li^7 stars on the main sequence in that case, and so far none have been found. Main sequence stars seem to contain old depleted lithium which is consistent with Herbig's idea that this is the remnant of an earlier, extremely lithium-rich stage.

A detailed study of the total lithium abundances yields some interesting results. Stars with known lithium abundance which also have absolute magnitudes and colours (C) corrected for blanketing (Eggen, 1964) have been plotted in an HR diagram and evolution studied using Iben's (1967) theoretical tracks. For the present purpose the region with C between 0·50 and 0·65, which is roughly centred on Iben's one solar-mass track, is of particular interest. This track, suitably shifted, has been used to define the region shown in Figure 1. Only stars between the two tracks shown will be considered and those parts of the tracks that lie below the Hyades' main sequence are considered to be on the main sequence at the same colour. For any star in this region the distance above the main sequence (ΔM) is a simple measure of its state of evolution. Some of the main-sequence stars in this region may of course be going to evolve to places outside the region in which we are interested, but this will not affect any of the conclusions drawn.

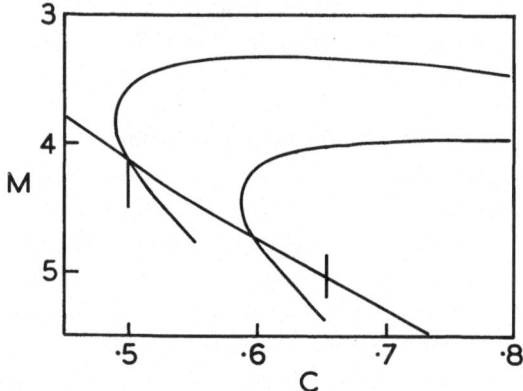

FIG. 1. *HR diagram. The Hyades' main sequence is shown. The tracks are Iben's one solar-mass track displaced to define the main sequence limits in C of 0·50 and 0·65. In the discussion of stars above the main sequence, only those stars between the two tracks are considered.*

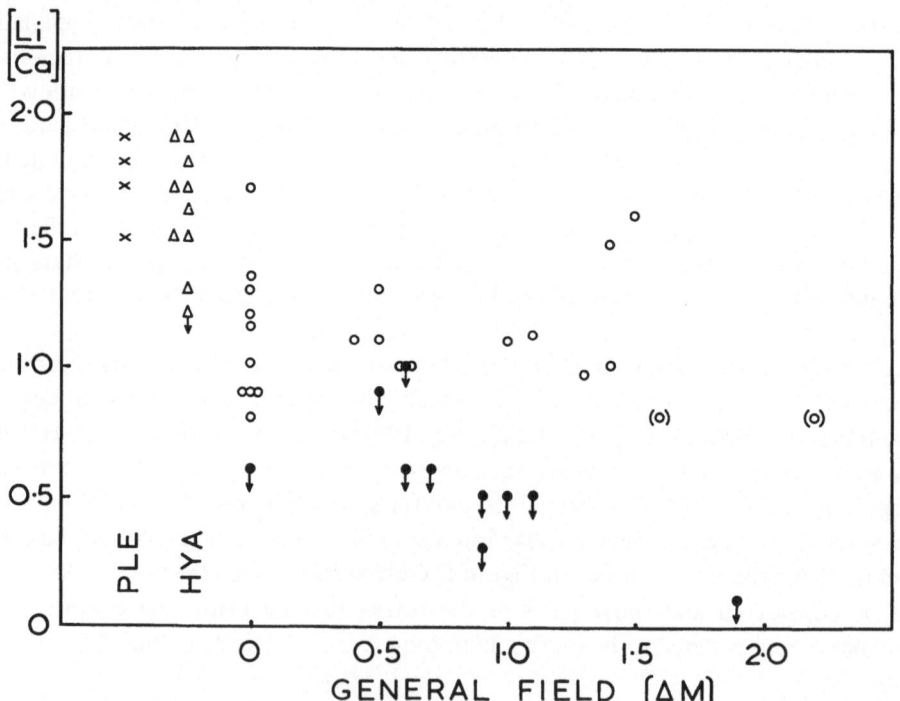

FIG. 2. *Lithium abundance as a function of distance above the main sequence for stars in the region defined by Figure 1. Arrowed points denote upper limits only. See text for further discussion.*

For field stars in this region, Figure 2 shows the log of the lithium abundance plotted against distance above the main sequence. The main-sequence Pleiades and Hyades stars in this region are also shown and arrowed points denote upper limits. Perhaps the most important feature of this diagram is the indication that low lithium (say less than 0·6 in the log) is preferentially concentrated between the main sequence $\Delta M = 0$ and the subgiants at $\Delta M \sim 1\overset{m}{.}4$. It is interesting that the one possible case of low lithium on the main sequence is 20 Leo Min which is a main sequence star in Eggen's paper but which others have considered to be above the main sequence. The greater average abundance of lithium in the Pleiades and Hyades stars is consistent with the field stars being on the average older and more depleted in lithium. The absence, or low frequency of low lithium on the main sequence suggests that depletion is not anything like complete during main-sequence life. The stars of intermediate ΔM appear to demand considerable post-main-sequence depletion. The subgiants, $\Delta M \sim 1\overset{m}{.}4$ on the other hand contain only lithium-rich stars consistent with the idea of synthesis here. The more evolved subgiants such as μ Her A with $\Delta M \sim 2\cdot0$ (the two bracketed points are just outside the region studied) can possibly be interpreted in terms of Iben's ideas on convective dilution as Herbig and Wolff have suggested (though if e.g. δ Eri has passed through a post-main-sequence lithium production phase, it is difficult to see how convective dilution can have produced the observed more or less pure Li^7).

These conclusions are supported by the histogram of Figure 3. Here the stars have been divided into high lithium (solid line) and low lithium (dotted line) and divided

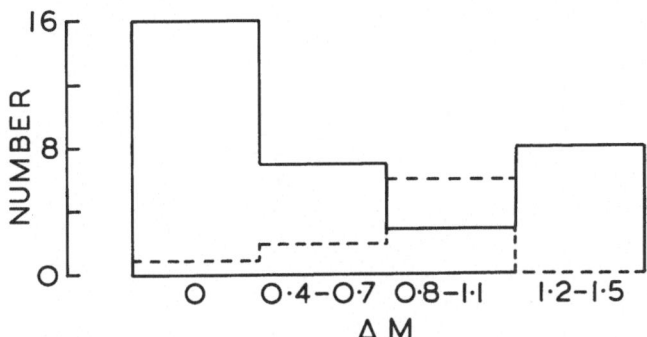

FIG. 3. *Lithium abundance as a function of distance above the main sequence for field stars. Solid line, lithium-rich stars ([Li/Ca] > 0·6). Dotted line, lithium-poor stars.*

into groups according to distance above the main sequence. The data differ from the last slide in that I have included several additional stars (including unpublished work) for which there are, as yet, no detailed abundances available but for which the line strengths give an adequate measure of the class to which the star belongs. 20 Leo Min is still the only low lithium star on the main sequence. In the intermediate region there is a clear fall in high lithium and a rise in low lithium and there are still, as yet, no subgiants with $\Delta M \sim 1\overset{m}{.}4$ with low lithium.

Thus, for stars in this range the lithium abundances as well as the isotope ratios suggest a scheme in which there is both pre-main-sequence and post-main-sequence depletion with fresh synthesis in the subgiant region.

References

Eggen, O.J. (1964)   *Astron. J.*, **69**, 570.
Feast, M.W. (1966)   *Mon. Not. Roy. Astr. Soc.*, **134**, 321.
Iben, I. (1967)   *Astrophys. J.*, **147**, 624.

OBSERVATIONS OF BERYLLIUM IN STARS

ANN MERCHANT BOESGAARD

Since a considerably higher temperature is necessary to consume Be atoms by (p, α) reactions than Li atoms, it has been thought that many clues to the understanding of the synthesis and destruction of the light elements could be found by studying Be abundances and the Li/Be-abundance ratios. Calculations have been made by Bodenheimer (1966) to determine the amount of convective depletion of Be during pre-main-sequence evolution. He found that a star reaching the main sequence at K5 would have destroyed 15% of its original Be; a K2 dwarf would have depleted its Be by 2%. For early K stars main-sequence depletion subsequent to this would be very slight. If there are no complicating effects such as magnetic fields and unusual patterns of circulation of stellar material, the observed Be abundances in stars of type G and earlier should be the initial pre-stellar or proto-star abundances. The Be II resonance lines lie far in the ultraviolet at $\lambda 3131$. But despite the very difficult nature of the

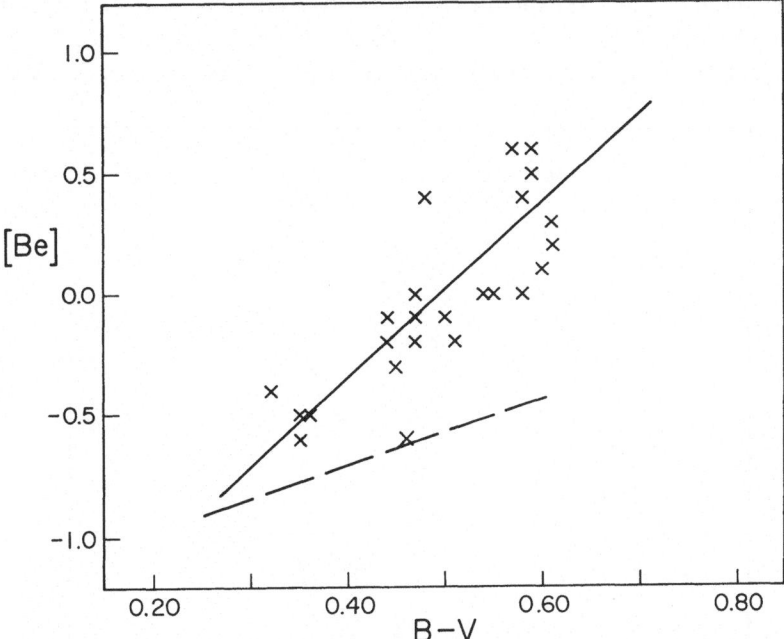

FIG. 1. *The logarithmic abundance ratio of stellar beryllium to solar beryllium as a function of B-V for dwarf stars only.*

Perek (ed.), Highlights of Astronomy, 237–242. © I.A.U.

observations, there are now Be abundances determined in about 60 dwarfs and sub-giants of types A, F, and G and in 2 K giants.

I shall now review these Be observations. Bonsack (1961) has observed 2 normal A stars to have solar Be, while Sirius had 40 times less Be than the Sun – perhaps due to its white-dwarf companion in some way –, and α^2CVn, a magnetic variable had 25 times the solar Be. In those peculiar A stars in which blending is not a serious problem, Sargent *et al.* (1962) have found 4 out of 15 stars to have strong Be lines and estimate the Be abundance in these four stars to be 100 times the solar value. The Be-rich stars are all Mn peculiar A stars, and all have abnormally strong P_{II} lines.

In the stars of spectral type F-G both Li and Be can be observed. Now that there are Be observations in close to 30 of these dwarfs, a very interesting trend shows up, which can be seen in Figure 1. Here the log of the beryllium abundances relative to the Sun are plotted as a function of B-V for dwarf stars only. The dashed line represents the observational limit. Values on or below this line can only be upper limits. The three stars which gave only upper limits are not shown here. The points are a combination of observations by Conti and Danziger (1966), by Conti (1968), and by myself (Merchant, 1966). This reverse correlation is not what would have been anticipated and it seems that it can not be explained away as observational error or inappropriate ionization correction. The hotter stars should show stronger Be_{II} lines for a given abundance than the cool ones – but in many cases the reverse is observed. The posi-

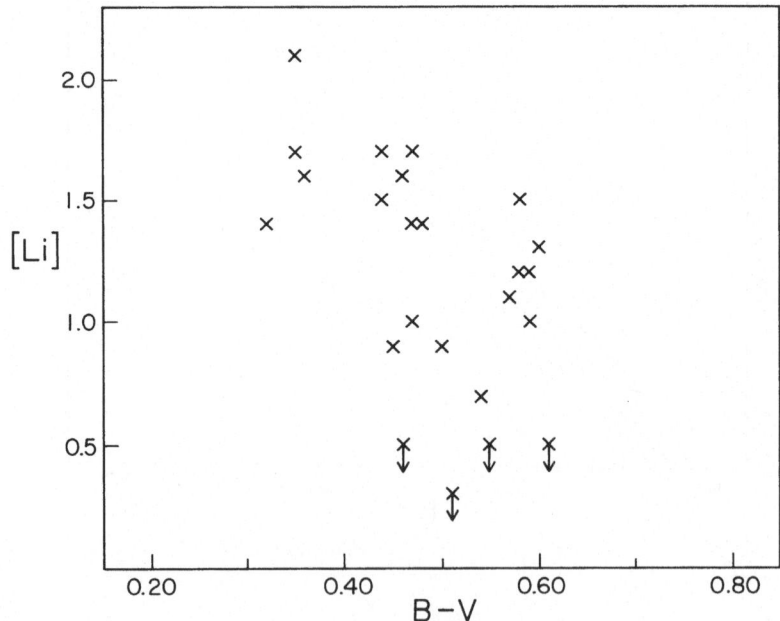

FIG. 2.   *The logarithmic abundance ratio of stellar lithium to solar lithium as a function of B-V for the same stars as in Figure 1.*

tioning of the continuum was done with consistency, and the UV blending and blanketing does not change much over this B-V range in this wavelength region.

Since it is thought that ordinary convective depletion will affect cooler stars *more* than hotter stars, we must either look for a new depletion mechanism or postulate variations in initial composition which might be a function of mass. In the latter category, perhaps the cooler stars had more violent activity leading to more spallation reactions or a longer time during which the Li, Be, B producing activity took place. The energy spectrum of the bombarding protons could be different.

Figure 2 shows the Li observations in these same stars. Here there is more scatter but the trend is clearly in the opposite direction. If all these stars started out with the same Li/Be ratio and nothing has happened to the Be, then the Li depletion has been more severe than previously thought. If the energy spectrum of the bombarding protons is variable as a function of stellar mass, then the Li/Be ratio would not necessarily be constant. Perhaps the theoreticians or experimentalists can shed some light on these alternatives.

Figure 3 shows an H-R diagram adapted from recent work by Conti (1968). The different symbols represent various ranges of values of the Li/Be ratio. These are non-logs. The Sun is at B-V $= 0.62$, $M_v = 4.8$ with Li/Be $= 0.08$. To the left of the vertical line there is no Li *dilution* according to Iben's calculation. By the time the star reaches the vertical dashed line, eighty-five per cent of the Li is diluted. So with the exception of δ Eri (at B-V $= 0.93$, $M_v = 3.7$) there is no dilution effect in these stars. Conti interprets the distribution of Li/Be points on this H-R diagram as due to an age effect. The time-scale for main-sequence Li burning for the G stars is $5 \times 10^7 - 5 \times 10^8$ years determined from observations in the Pleiades and the Hyades. Since so few stars lying to the right of the Hyades track show appreciable Li destruction, Conti concludes that the time-scale for Li burning in the main-sequence F stars is 5×10^8 to 5×10^9 years. That is, the destruction *rate* is inversely dependent on temperature or mass.

In summarizing the main features of the Be observations we can say:

(1) The peculiar A stars are peculiar in their Be content too, but the Mn stars have a greater tendency for high Be abundance than other A stars.

(2) The Be abundances in F and G dwarfs are inversely correlated with temperature. This may be due to:

(a) a new depletion mechanism,

(b) difference as a function of stellar mass in the length of time or intensity of the spallation activity, and/or

(c) difference as a function of stellar mass in the energy spectrum of the bombarding protons.

I have been asked to make some statement on the observability of boron. Boron so far has had an uncertain history. The infrared B I feature at $\lambda 8668$ observed in the Sun is thought to be CN; the B III suggested identification on O stars is now thought unlikely; the BH identification in the solar disk is probably in error although BH may

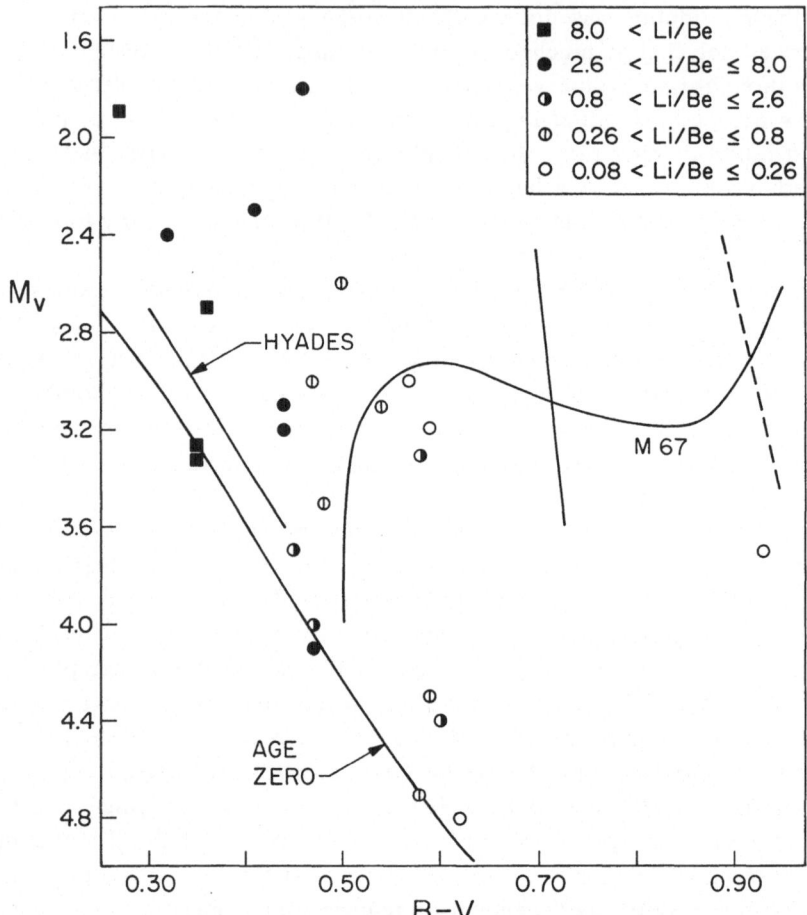

FIG. 3.   *An H-R diagram in which the stars with positive observations of lithium and beryllium are shown with symbols corresponding to their Li/Be abundance ratio. The zero-age main sequence, the Hyades, and M 67 tracks are shown.*

be present in sunspots. The 0,0 band of BO at $\lambda 4227$ has been suggested in β Peg by Davis (1947). The small Frank-Condon factor for this band, 0·04, and the expected low abundance of B perhaps make this identification questionable. But BO and BO_2 should certainly be looked for in stars.

There are two possible observations of atomic B: a high-excitation line of B II occurs at λ 3451·4 and might appear in late B and A stars. However, it is quite badly blended with 2 lines of Fe II. Perhaps the best chance for B observations is in the rocket ultraviolet where lines of B I occur at λ 2497·7 and λ 2496·8 Å. Observations made so far give upper limits for the sun. This would be a profitable region to observe since nearby at λ 2494·6 are 3 lines of Be I. The Be lines are present, but weak and unresolved by observations so far of the sun.

References

Bodenheimer, Peter (1966)   *Astrophys. J.*, **144**, 103.
Bonsack, Walter K. (1961)   *Astrophys. J.*, **133**, 551.
Conti, Peter (1968)   *Astrophys. J.*, **151**, 567.
Conti, Peter, Danziger, I.J. (1966)   *Astrophys. J.*, **146**, 383.
Davis, D.N. (1947)   *Astrophys. J.*, **106**, 28.
Merchant, Ann (1966)   *Astrophys. J.*, **143**, 336.
Sargent, W.L.W., Searle, L., Jugaku, J. (1962)   *Publ. Astron. Soc. Pacific*, **74**, 408.

DISCUSSION

Underhill: The fact that standard spectrum analyses of Li and Be lines give a variety of abundances, whereas those for Na or Fe give normal abundances in many stars, may possibly be affected by the particular intensity distribution in the far UV and X-ray regions of these stars. Li and Be are elements 3 and 4, Na is element 11; in the latter case many electrons must be removed and perhaps the distribution of ionic species follows the predictions from Saha's Law better than in the case of Li or Be with only few electrons to remove and thus possibly more sensitive to a particular line emission spectrum in the far UV.

Conti (in answer to Dr. Underhill): It seems to me more than a coincidence that it is just those light elements, Li and Be, which are easily destroyed in F and G stars, that are anomalous. In fact, they are the only greatly anomalous elements in these stars. Rather than interpret these anomalies in terms of non-LTE, it seems more realistic to consider them as nuclear processes.

Müller: Concerning the B abundance in the solar atmosphere I should like to mention that D.L. Lambert, E.A. Mallia, and B. Warner recently gave an upper limit of 4×10^{-9} derived from the infrared B I line at λ 11662·47 that might be present. They stress, however, that the solar B abundance may be an order of magnitude smaller. The best way to determine the solar boron abundance would be to study the BH bands in sunspots.

Delbouille: As a comment to Dr. E. Müller's remark, we can say that, from observations made by Dr. Roland and myself at the Jungfraujoch of the possible B line at λ 11 662 Å, Grevesse and Boury conclude that $B < 2·5$ on log $H = 12·00$ scale.

Müller: Concerning the solar Be abundance, I should like to point out that the Be I and Be II lines used in the abundance determination fall in the wavelength region which (1) is very crowded and (2) is way down in the UV (around λ 3100–3300 Å) where the continuum is highly uncertain. Consequently, the solar beryllium is, for the moment, still quite doubtful.

Boury: We have redetermined solar Be abundances through Be I and Be II lines. We find, on the log $N_H = 12·00$ scale, 2·33 from Be I lines and only 1·24 from Be II lines. Be I lines are very weak and much perturbed.

Reeves: What is the definition here of Be solar units?

Boesgaard: We used Mutschlecner's determination of 6·9 Be atoms per 10^6 Si atoms. This number can be readily converted *back* to the log $H = 12·00$ scale through the GMA value for Si.

Aller: It is not probable that a reliable solar Li abundance can be obtained unless one takes into account the line profiles – and more particularly the detailed energy distribution in the spectrum. One must 'synthesize' the spectrum by computing the superposed absorption by the continuum and many lines of different elemental origins and excitation potentials. Successful application of this procedure requires very accurate line-profile data and also centre-limb variations. It is necessary also to know the absolute flux in the solar spectrum at some point in your wavelength interval. The method has been applied by John Ross with encouraging results to lines of Ag and Pb in the solar spectrum.

Bidelman: I believe that it is likely that B II is present in the strange A-type star HR 6870 = HD 168733, but Dr. Aller has been studying this star and may know more about this possibility.

Aller: The spectrum of this star is terribly crowded with unidentifiable lines. When Steve Little completes his measurements of the spectrum, we may be able to say something of more confidence about the B spectrum.

Boesgaard: Do you know anything about the Be II lines in that star?

Aller: The declination of HR 6870 is — 36°, so the far UV is very difficult to observe at the latitude of Lick Observatory.

Reeves: Please look at upper limits of B/Be ratios in stars.

LITHIUM OBSERVATIONS IN THE SUN

EDITH A. MÜLLER

The determination of the lithium abundance in the solar atmosphere is essentially based on the Li I resonance doublet at λ 6707·761 and 6707·912 Å. These two lines form a very faint absorption feature, the central depth of the stronger component being of the order of 1% of the continuum. The violet component, which is also the stronger of the two, occurs near the red wing of a faint solar line of unknown origin, and the lines appear to be blended with other faint lines including possibly the doublet of the Li^6 isotope (the isotopic shift being 0·160 Å). No other line of Li I has been detected in the Fraunhofer spectum of the undisturbed solar disk. This is nothing surprising, because practically all lithium is expected to be ionized in the photosphere on account of its low ionization potential (Xion = 5·37 e.v.). In sunspot spectra the lower temperature reduces the degree of ionization of lithium and causes a strengthening of the Li I lines. In fact, the Li I resonance lines which appear as a very faint absorption feature on disk spectra are about 50 times stronger in spot spectra. Furthermore, the very weak feature at λ 6103·6 Å was identified by Dubov (1964) and by Schmahl and Schröter (1965) as due to the 2s ^{2}S–3d ^{2}D transition of Li I. Both the resonance doublet and the faint feature at 6103·6 Å have been used by the above-mentioned authors to derive the lithium abundance in spots.

Since in the following paper we shall hear more about the lithium abundance in sunspots, I shall restrict myself to present the results obtained from disk spectra. In

Table 1

Determinations of the solar lithium abundance

Authors	Total W_λ (mÅ)	Solar Model	$\log \varepsilon_{Li}$[a]
Greenstein and Richardson (1951)	3·5	Milne-Eddington	1·26
Claas (1951)	2·9	Claas	1·08
Dubov (1955)	1·74	···	0·93
Goldberg et al. (1960)	2·6	Aller-Pierce-Elste	0·96
Utrecht (1960)	2·0	–	–
Mutschlecner (1963)	7·4	Mutschlecner	1·54
Lynds (1965)	< 1·6	–	–
Peach (1967)	< 0·61	$T = 5300°$, $\xi = 1·4$ km/sec	< 0·38

[a] On the basis of $\log \varepsilon_H = 12·00$ for the hydrogen abundance.

Perek (ed.), Highlights of Astronomy, 243–246. © *I.A.U.*

particular, I wish to stress the following two main difficulties that are encountered when determining the lithium abundance in the undisturbed photosphere:

(1) The identification of the Li I resonance lines which is hampered by the fact that the absorption feature is faint and composite, and consequently introduces great uncertainties in the equivalent width measurements of the lithium feature.

(2) The choice of the photospheric model which by its temperature distribution in the line-forming region influences the abundance results, inasmuch as the Doppler-broadened lithium lines are quite sensitive to the temperature.

In order to illustrate these two points the Table 1 collects the results of the equivalent width measurements, the photospheric models used, and the lithium abundance derived by different authors. The equivalent widths are given for the whole lithium feature which includes both the Li^7 doublet and the corresponding isotopic Li^6 doublet. The large disagreement in the equivalent widths is mainly due to identification differences, i.e. to how much of the observed absorption feature each author attributes to lithium. Apparently, as time goes on, that is with increasing spectral resolution, one finds less and less lithium in the Sun.

A high resolution spectrum of the Li I absorption feature observed at the centre of the solar disk with the Kitt Peak solar spectrograph was published by Lynds (1965). On this spectrum, two distinct small absorptions occur in the wavelength region of the Li^7 and Li^6 resonance doublets. However, Lynds points out that the wavelengths of the lithium lines do not coincide with the observed features and questions the presence of the lithium-absorption lines. This poor wavelength agreement was already noted by Greenstein and Richardson (1951), who interpreted it as due to pressure effects. Lynds suggests that 1·6 mÅ is the maximum value of the total equivalent width which may be attributed to lithium on the solar disk. Recently, Peach (1967) re-investigated the lithium feature on high-resolution low-noise spectra. He found no direct evidence of the lithium lines in the observed absorption feature. Consequently, he gives an even smaller upper limit for the equivalent width which may be due to lithium.

Since Lynds' observation of the lithium feature is the only high-resolution spectro-photometric recording which has been published so far, we used it as a basis to derive the lithium abundance employing different photospheric models. The computations were carried out for the entire blend composed of the two Li^7 and the two Li^6 lines, the composite line having an equivalent width of 1·6 mÅ. The calculations were performed with the fine analysis program of Baschek et al. (1966) and its FORTRAN translation by Peytremann. For the Li^7/Li^6 isotopic abundance ratio the following three values were used: 2, 11 (corresponding approximately to the terrestrial ratio), and 20. The four photospheric models employed and the resulting lithium abundances are given in Table 2. It is evident that different abundance results are obtained from different models. The abundance values are upper limits, inasmuch as 1·6 mÅ is an upper limit of the total lithium equivalent width according to Lynds. The theoretical

Table 2

The solar lithium abundance derived for $W_\lambda = 1.6$ mÅ with different atmospheric models

Model	$\log \varepsilon_{Li}$
Holweger (1967)	0·88
Mutschlecner (1963)	0·80
U.R.M. (average column) (1964)	0·62
Heintze (1965)	0·49

profiles best reproduced the expected lithium contribution to the observed absorption feature when setting the Li^7/Li^6 isotopic abundance ratio equal to 20. Any large value of this ratio fits just as well. On the other hand, the small ratio of 2 produces two close lying but distinct absorption lines of almost equal central depths which is not observed. We may conclude that, if the light Li^6 isotope is present, it is at least 20 times less abundant than Li^7. It is interesting to note that a similar conclusion was reached by Schmahl and Schröter from sunspot spectra.

That the question of the solar abundance is by no means solved was made clear a few days ago during a session of Commission 12. L. Delbouille presented a spectrographic recording made at the Jungfraujoch of the region between λ 6705 and 6711 Å, in which he showed that numerous unidentified faint lines may be due to the red bands of CN. Some of these CN lines fall right on the so-called lithium-absorption feature. Consequently, the lithium abundance and the Li^7/Li^6 abundance ratio must be revised, taking into account the CN line blending. It may be that we end up having barely any or no lithium at all present in the undisturbed photosphere.

Acknowledgements

The assistance of D. Messerschmidt (Hamburg) and E. Peytremann (Geneva) in part of the computations and the temporary use of the CERN (Geneva) and the Hamburg University (thanks to Dr. G. Traving) high-speed computers are gratefully acknowledged.

References

Baschek, B., Holweger, H., Traving, G. (1966)   *Abhandlg. Hamburger Sternwarte,* **8**, No. 1.
Claas, W.J. (1951)   *Rech. astr. Obs. Utrecht,* **12**, part 1.
Dubov, E.E. (1955)   *Astr. Cirk.,* **159**, 11.
Dubov, E.E. (1964)   *Izv. Krym. astrofiz. Obs.,* **32**, 26.
Goldberg, L., Müller, E.A., Aller, L.H. (1960)   *Astrophys. J., Suppl. Ser.* **5**, 1.
Greenstein, J.L., Richardson, R.S. (1951)   *Astrophys. J.,* **113**, 536.
Heintze, J.R.W. (1965)   *Rech. astr. Obs. Utrecht,* **17**, (2), 1.
Holweger, H. (1967)   *Z. Astrophys.,* **65**, 365.
Lynds, C.R. (1965)   *Astrophys. J.,* **142**, 396.
Mutschlecner, J.P. (1963)   Thesis, University of Michigan.

Peach, J. V. (1967)   Abstract of papers, 124th meeting of the Amer. Astr. Soc., p. 101.
Schmahl, G., Schröter, E. H. (1965)   *Z. Astrophys.*, **62**, 143.
Utrecht (1960)   'Photometric Catalogue of Fraunhofer Lines ($\lambda\lambda$ 6600–8770)', *Rech. astr. Obs. Utrecht*, **12**, part 2.
U.R.M. (1964)   'Utrecht Reference Model' see: Heintze, J. R. W., Hubenet, H., and De Jager, C. (1964)   *Bull. astr. Inst. Netherlands*, **17**, 442.

DISCUSSION

L. Delbouille: Recently, high-resolution and low-noise spectra have been secured by Miss G. Roland and myself. The identifications by W. S. Benedict of many lines due to CN in the λ 6708 Å region of the solar spectrum seriously complicate the situation. It seems impossible to explain this region by fitting in the Li I lines only or even the Li I and the CN lines.

J.P. Mutschlecner: It should also be noted that an additional observational difficulty is the location of the local continuum. In my work, e.g., it was possible to obtain a considerable greater equivalent width due to my different interpretation of the continuum. The unidentified line at about 6707·5 also contributes to the uncertainty in the lithium lines and it would be good if it could be identified. The work of Mark Daehler (so far unpublished) might also be mentioned. In a careful re-observation and analysis he concluded that no lithium exists in the disk spectrum.

W. A. Fowler: Is there any lithium in the Sun?

L. Delbouille: As experimentalists, we try to obtain the best possible spectral tracings, but we are quite reluctant in attempting to derive an abundance of lithium for the Sun.

Mrs. Ch. Moore–Sitterly: Although there is justifiable criticism of the Li I identifications in the disk spectrum, there is no question that the resonance lines are conspicuous features in the spot spectrum. The low ionization potential of Li I makes it reasonable that Li I should be present only in the spot spectrum. Its absence from the disk spectrum may not be puzzling.

G.H. Herbig: I hardly think there can be any question that lithium exists in the solar surface. It may be marginal in the disk, but λ 6707 becomes very strong in spots ($W_\lambda \approx$ 50–100 mÅ) and I would not question the identification there.

H.E. Mitler: Since there is so much question as to the correct lithium abundance in the Sun, may there not be some question as to the stellar observations, especially in the low-lithium cases?

M. W. Feast: It may be as well to point out that the strength of the lithium line in the lithium lide stars is so much greater than in the Sun that we do not have in this case the kind of problems that the solar observers have.

H. Reeves: The lithium in sunspots may be formed right there by increased activity and may not be just strengthened over the disk spectrum by temperature effects. In this case we should expect $Li^7/Li^6 \simeq 2$ and possibly some Be^7.

Miss E. A. Müller: Due to temperature effects the Li I and the CN lines will be enhanced over sunspots compared to the disk spectrum, but they are expected to be strengthened differently and, hence, one might be able to separate the components. Another way of disentangling the Li I and the CN lines is to study the centre-to-limb variation of the Li I + CN absorption feature and of other CN lines in neighbouring wavelengths pertaining to the same band. The CN lines are expected to have a different centre-to-limb behaviour than the Li I lines. So far, all observations discussed here refer to the centre of the solar disk. It seems to me that it would be extremely important to secure high-resolution low-noise observations of the Li I + CN absorption feature at various positions on the solar disk and as far out to the limb as possible. If lithium is formed in sunspots, as suggested by Dr. Reeves, then one might find abundance differences in different spots depending on their activity. Be^7 is probably extremely difficult to detect because the Be I and Be II lines observed in the solar spectrum occur in a big jungle of strong lines due to various atoms and molecules.

ON THE LITHIUM ABUNDANCE IN SUNSPOTS AND THE UNDISTURBED SOLAR ATMOSPHERE

E. Dubov, V. Prokof'ev, and A. Severny

The difference between abundances of such elements as Li, Be, B in active regions on the Sun and in the undisturbed atmosphere, if it exists at all, could be considered a very important indication on the possible nuclear reaction in active regions and on the rate of diffusion of elements inside magnetic fields of sunspots. One of the authors from the considerations of $\lambda 6707$ Å has an estimated (Dubov, 1955) Li abundance in the undisturbed solar atmosphere as $\log N(\mathrm{Li}) = 0.93$ (taking $\log N(H) = 12.0$), which is close to the later results (Goldberg et al., 1960). In sunspots we can expect that the first doublet $\lambda 6103$ of Li diffuse series appears. The preliminary consideration (Dubov, 1964) showed that Li abundance in sunspots could be several times larger than the above estimate for the undisturbed atmosphere. This result forced us to consider as carefully as possible the region of $\lambda 6103$ in eight spectra

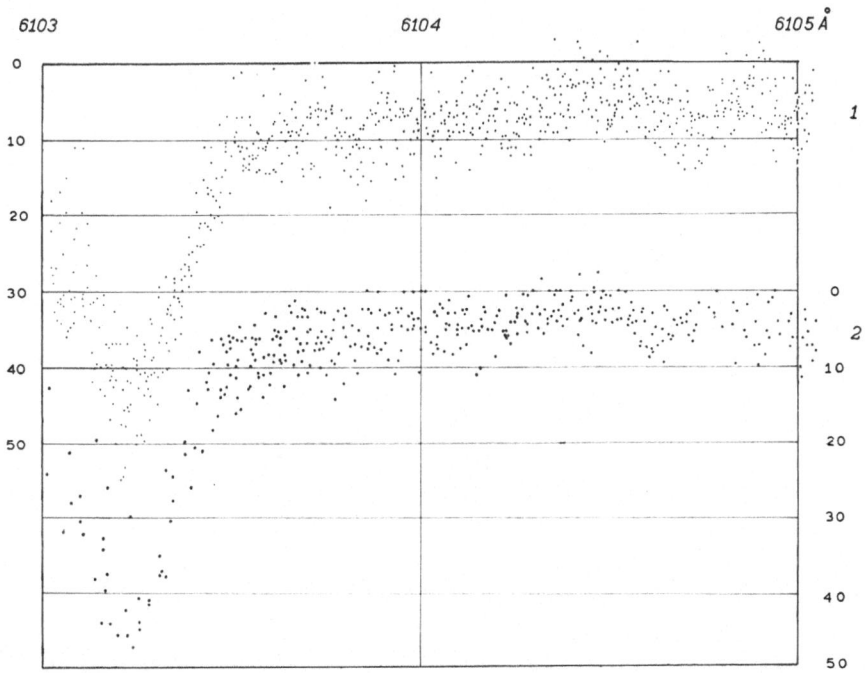

Fig. 1.   Summary of all points of used spectra of different authors: 1 = Prokofiev, 2 = Severny.

Perek (ed.), Highlights of Astronomy, 247–250. © I.A.U.

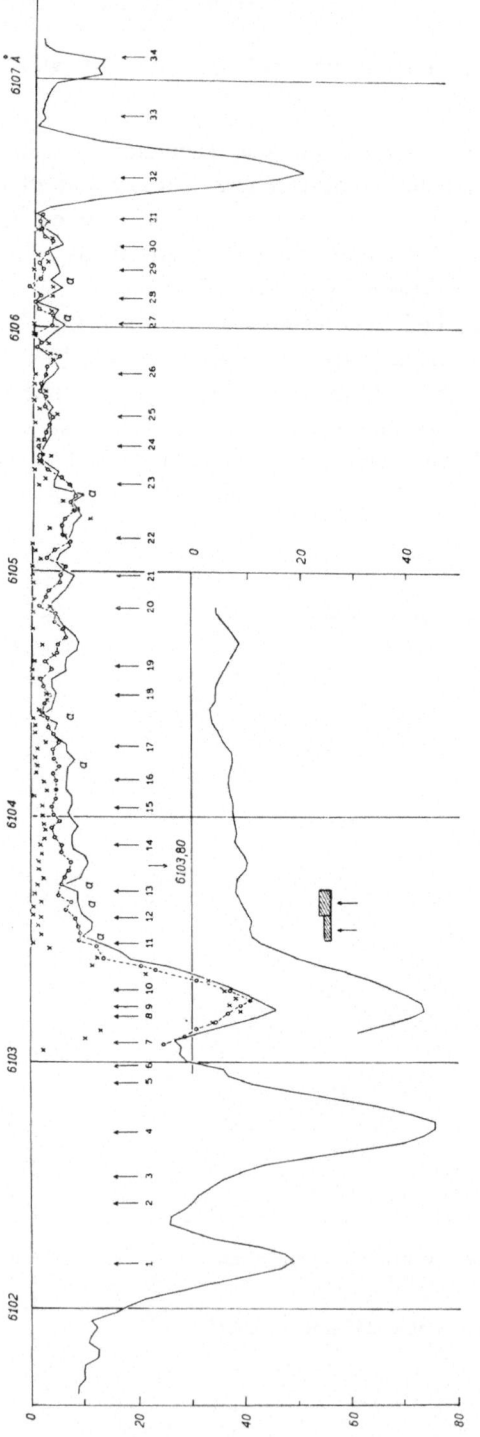

FIG. 2. *The final curves; 1 = Smoothed curve.*

of four sunspots taken with the aid of the echelle-grating spectrograph of the Crimean observatory with a dispersion of 1 Å/mm and measured resolving power 0·03 Å. The consideration of the Zeeman pattern of this line shows that at the field strengths we had in sunspots (2000–3000 gs) the line should be diffuse band with width ∼0·23 Å, and taking into account the broadening of the line the halfwidth of the line has not to be smaller than 0·25 Å.

The careful spectrophotometry made by Prokofiev and Severny (1967) by two independent methods is shown in Figure 1 for all sunspot spectra we had at our disposal. In the same way the spectra of the undisturbed atmosphere were measured, and the mean profile of the region in question is shown in Figure 2 (crosses=undisturbed atmosphere; open circles=sunspot 1st method; solid line=sunspot 2nd method; arrows denote the positions of all lines known up to now in the spectrum of solar atmosphere; 12 corresponds to the line Fe II $\lambda 6103·59$). We see good agreement between different methods of reducing of spectrograms, and minor differences are connected with the different setting of the level of continuous spectrum. Figure 3 shows the same results obtained for different spots separately.

Careful inspection and analysis of these results shows that there is no indication, neither in the mean profile nor in the profile for separate sunspots, of the presence of some appreciable line in the place where we should expect the Li line $\lambda 6103·69$ (Kiasatpoor, 1965) (13 in Figures 2 and 3). The difficulty with the search of the Li line $\lambda 6103·65$ is increased by the influence of blends in this region (e.g. Fe II 6103·59) and due to some uncertainty with the position of Li line. If we adopt the laboratory system of wavelengths (the dashed rectangular areas in Figures 2 and 3), which is hardly adequate, the observed small depression in the region $\lambda 6103·65$ (taking into

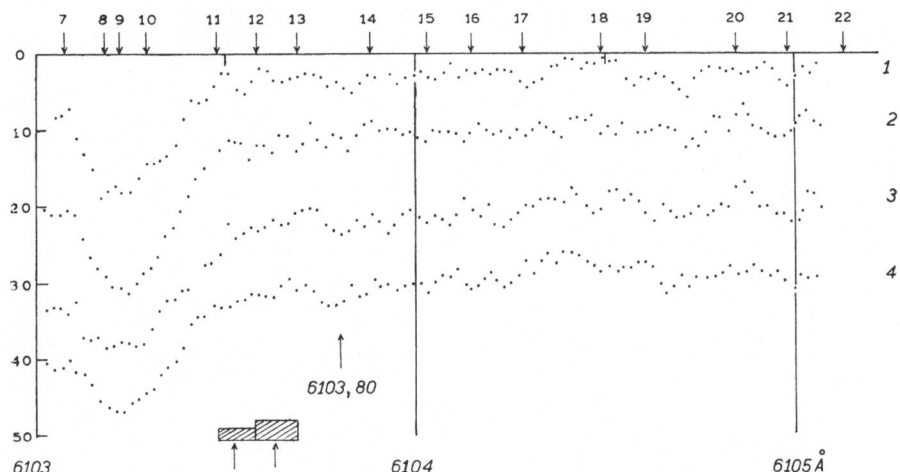

FIG. 3. *The mean curves for different spots: 1 = 18.09.1960; 2 = 2.09.1961; 3 = 5.09.1961; 4 = 6.09.1961.*

account the influence of blend (Fe II-line) would be less than 1·0 mÅ; this is the mean for all eight spectra of four spots instead of 3·3±2·5 mÅ by Dubov and Hromova (1964) and found for one sunspot, which value includes possibly the influence of blends. This agrees with the estimate of the Li abundance by the same line (1·1 mÅ) made by Schmahl and Schröter (1939) and the estimate made by Dubov (1967) from the consideration of the Li-resonance line λ6707.

The abundances from equivalent widths have been derived by the method developed by Dubov and Hromova (1964).

Therefore at the present time we can hardly suspect some appreciable difference in Li abundance between undisturbed atmosphere and active regions on the Sun.

References

Dubov, E. (1955)   Astron. Cirkular No. 159, p. 11.
Dubov, E. (1964)   Izv. Krym. astrofiz. Obs., 32, 26.
Dubov, E. (1967)   Izv. Krym. astrofiz. Obs., 36, 87.
Dubov, E., Hromova, T., (1964)   Izv. Krym. astrofiz. Obs., 31, 247.
Goldberg, L., Müller, E.A., Aller, A.H. (1960)   Astrophys. J., Suppl. 5, No. 45.
Kiasatpoor, A. (1965)   Georgetown Obs., Monograph No. 19.
Prokofiev, V.K., Severny, A.B. (1967)   Izv. Krym. astrofiz. Obs., 36, 90.
Schmahl, G., Schröter, E.H. (1939)   Z. Astrophys., 18, 284.

THE FORMATION OF Li, Be, B ISOTOPES BY THE SPALLATION OF CNO

E. Gradsztajn

(Institut de Physique Nucléaire, Orsay, France)

In 1955 Fowler *et al.* (1955) suggested that the elements Li, Be and B could be produced at the surface of some stars by the spallation of CNO induced by electromagnetically accelerated particles.

Since that time many observational data were gathered, not only on the lithium elemental abundance (Hunger, 1957; Bonsack and Greenstein, 1960), but also concerning the isotopic ratio $^7Li/^6Li$ (Herbig, 1964; Wallerstein, 1965). Then, the abundance of beryllium in many stars was measured (Bonsack, 1960; Conti and Danziger, 1966; Merchant, 1966) and the problem was, and still is to explain both $^7Li/^6Li$ and Li/Be ratios. In addition to spallation, different possibilities may be invoked, such as the inheritance from galactic gas enriched with matter remaining from a previous generation of stars, or from the big bang; but which can be checked most easily at the present time is the spallation assumption. Indeed, simultaneously with the development of the observations on stars, a program for measuring and calculating the production rates of the light elements has been in progress at Orsay. In 1960 the first experimental result concerning the production of ^6Li in ^{16}O by 156 MeV protons was obtained. But it was only in 1965 that the number of experimental and theoretical data on spallation was sufficient to make a valid comparison with abundances observed on the stars as well as in the solar system or in cosmic rays.

Below I shall present recent measurements on the production of beryllium and boron.

Table 1 shows the experimental results obtained by Bernas *et al.* (1965) for lithium and by Yiou *et al.* (1968) for beryllium and boron at various energies.

The cross-sections of Table 1 have been measured by mass spectrometry. This technique, combined with the isotope dilution method, allows one to determine very small amounts (down to 10^{-14} g or 10^{-12} g) of the Li, Be, B isotopes, although difficulties arising from the need of a high sensitivity and from contamination have to be overcome. The details of the procedure are described in the original papers.

While experiments were slowly progressing, we have performed theoretical calculations which could describe the spallation reactions on light targets and thus obtained the needed cross-sections. These calculations are based on the idea that when the incident proton is striking the nucleus, the reaction takes place in two steps (Serber, 1947). The first step is the so-called intranuclear cascade, where a few nucleons are knocked out of the target nucleus. The residual nucleus is left excited and will disintegrate. This disintegration is the second step of the reaction. Details concerning the calcula-

Table 1

Experimental cross-section in millibarns

Target			^{12}C			^{16}O	
Ep (MeV)	44	50	156	550	156	600	19 GeV
^{6}Li[a]	10·6	9·2	9·8	7·4	10 ± 2	$13·6 \pm 3·5$	
^{7}Li[a]	10·8	7·2	7·8	5·6	$8·5 \pm 2·5$	$12·4 \pm 2·5$	
^{7}Be[b]	22	25·5	12·1	11	$5·4 \pm 1$	$6·5 \pm 2$	$6·5 \pm 2$
^{9}Be			0·8[c]		$1·7 \pm 0·5$	$2·4 \pm 1·2$	$2·2 \pm 1·1$
^{10}Be			0·3[c]		$0·4 \pm 0·2$	$0·6 \pm 0·4$	$0·6 \pm 0·5$
^{10}B[d]					11 ± 3	12 ± 5	
^{11}B[d]					25 ± 8	25 ± 12	
^{11}B/^{10}B[d]					$2·3 \pm 0·4$	$2·1 \pm 0·6$	

[a] Accuracy: 20% for Li in ^{12}C.
[b] ^{7}Be in ^{12}C is reproduced from Cumming (1963).
[c] Preliminary results.
[d] ^{10}B and ^{11}B include contributions from ^{10}C and ^{11}C respectively.

tions of the cascade may be found in Rudstam (1956) or Metropolis et al. (1958).

The disintegration of the excited residual nucleus can be calculated by the 'evaporation model' as in Rudstam (1956) or Dostrovsky et al. (1958), or by the 'break-up model'. For light targets such as carbon and oxygen only the break-up model has been used. Table 2 shows results obtained by Epherre and Gradsztajn (1967).

Now, if we compare the calculated results with the available experimental data we see that the agreement is fairly good and that the calculated data may be used with some confidence when experimental cross-sections are missing.

From both experimental and theoretical values, we see that the production cross-section of ^{10}Be is much smaller that for ^{9}Be. This is very unfortunate, because, due

Table 2

Calculated cross-sections in millibarns

Target			^{12}C				^{16}O	
Ep (MeV)	70	100	156	200	70	100	156	200
^{6}Li	18	15·8	13·7	13·4	16·7	16	13·7	13
^{7}Li	10·3	10·1	9·9	11·1	5·5	7·5	8·4	8·8
^{7}Be	17·6	14·6	12·2	11·8	15·6	15	13·2	12·5
^{9}Be	2·7	2·9	2·9	3.1	3·5	4	3·6	3·9
^{10}Be	0·5	0·7	0·8	0·9	0·33	0·46	0·66	0·85
^{10}B[a]	28·6	24·4	24	23	22·8	22	16·4	17·2
^{11}B[a]	87·5	71	55	49	31·2	29·7	24·1	23·5

[a] ^{11}B and ^{10}B include contributions from ^{10}C and ^{11}C respectively; direct ^{11}B is calculated; ^{11}C is experimental (Cumming, 1963).

to its half life of a few million years, this isotope could have been used e.g. to determine the time of diffusion from the surface where spallation is supposed to take place, into the convective zone.

^{7}Be, on the contrary, has a short half life (54 days) but a higher cross-section than ^{9}Be. Using this fact Reeves (1967) has suggested that it may be possible to observe ^{7}Be if it is retained at the surface of the star for a certain time.

Now, since there are as yet no observations concerning boron, the interesting data are the ratios ^{7}Li/^{6}Li and Li/^{9}Be.

Combining the experimental and calculated values we obtain for the formation ratios at energies > 50 MeV:

$$^7\text{Li}/^6\text{Li} \approx 2$$
$$\text{Li}/^9\text{Be} \approx 15 \text{ for oxygen}$$
$$\simeq 20\text{--}30 \text{ for carbon.}$$

From 50 MeV to much higher energies (GeV), the cross-sections of interest change only very slightly, if at all, but below this range, the situation may be quite different. Indeed, the thresholds of the reactions producing ^{7}Li, ^{6}Li and ^{9}Be are different (25, 32 and 34 MeV, respectively in ^{12}C), so that if the energy spectrum of the protons at the surface of a star is concentrated in the low-energy region. This may lead to observations where ^{7}Li/^{6}Li can be higher than 2 and Li/Be higher than 30.

However, as suggested by many observational data, we may assume that Li and Be are produced at the surface of stars at high energies, with the well-defined values of these ratios, as given earlier, and then try to explain the observational data by looking for processes destroying the lithium.

Nevertheless one should remember that the possibility exists to produce any value of ^{7}Li/^{6}Li higher than 2 and Li/Be higher than 20 or 30. This doubt will probably be eliminated when observations on boron isotopes become available: indeed, as ^{11}B and ^{10}B are produced by reactions with thresholds different from Li and Be, this will give additional parameters needed to determine the proton spectrum of the observed star.

References

Bernas, R., Epherre, M., Gradsztajn, E., Klapisch, R., Yiou, F. (1965)   *Phys. Rev. Letters*, **15**, 147.
Bonsack, W. K., Greenstein, J. L. (1960)   *Astrophys. J.*, **131**, 83.
Bonsack, W. K. (1960)   *Astrophys. J.*, **133**, 551.
Conti, P. S., Danziger, I. I. (1966)   *Astrophys. J.*, **146**, 383.
Cumming, J. B. (1963)   *Am. Rev. Nucl. Sci.*, **13**, 261.
Dostrovsky, I., Fraenkel, Z., Rabinowitz, P. (1958)   *Phys. Rev.*, **118**, 791.
Epherre, M., Gradsztajn, E. (1967)   *J. Phys.*, **28**, 745.
Fowler, W. A., Burbidge, G. R., Burbidge, E. M. (1955)   *Astrophys. J.*, Suppl., **2**, 167.
Herbig, G. (1964)   *Astrophys. J.*, **140**, 702.
Hunger, K. (1957)   *Astron. J.*, **62**, 294.
Merchant, A. (1966)   *Astrophys. J.*, **143**, 336.

Metropolis, N., Bivins, R., Storm, M., Miller, J. M., Friedlander, G., Turkevitch, A. (1958)   *Phys.*
    Rev., **110**, 185.
Reeves, H. (1967)   Unesco Conf. on the Origin and Abundance of the Elements, Paris, May 1967.
Rudstam, G. (1956)   Thesis, Uppsala.
Serber, R. (1947)   *Phys. Rev.*, **72**, 1114.
Wallerstein, G. (1965)   *Astrophys. J.*, **141**, 311.
Yiou, F., Baril, M., Dufaure de Citres, J., Fontes, P., Gradsztajn, E., Bernas, R. (1968)   *Phys.*
    Rev., **166**, 968.

DISCUSSION

Fowler: Those of us engaged in experimental nuclear research are filled with admiration for the beautiful work underway by the workers at Orsay. Their ability to detect submicroscopic amounts of non-radioactive materials is just fantastic and they deserve every accolade we can give them. *(Applause.)*

Conti: The highest Li/Be ratio observed in stars is of the order of 20 by number. This is in good agreement with the predicted spallation results. All values smaller than this could be interpreted as selective Li depletion in individual stars.

WHAT CAN WE LEARN FROM Li AND Be STELLAR OBSERVATIONS?

H. Reeves

I want to state a few questions which are raised by the observations of stellar Li and Be, and to see how far we can go in giving answers.

I do not believe that we can, at the present time, give very definite answers to most of these questions and build up a unique model in which all observations are properly understood. We are still faced with different possibilities which can not be excluded, but can only be classified according to their degree of 'plausibility'. Although we are led to choose one of these possibilities in the light of present data, it seems wise to keep the other possibilities in the back of our mind, and to assess their merit. Considering this, a few alternate models will be described and discussed.

A. Were the L elements Li, Be, B generated in the star in which they are observed?

The following observations suggest that at best, a small fraction of the L elements observed in a star was already present in the galactic gas at the birth of the star.

(1) The upper limit of interstellar Be, $n(\text{Be})/n(\text{H}) < 7 \times 10^{-11}$, is more than ten times smaller than the largest values observed in stars (10^{-9}). It would seem improbable that 90% of the Be would be locked in grains. (Available upper limits on interstellar Li and B are too high to be of any interest. More sensitive determinations are needed.)

(2) In the star T Tauri, the gas component with positive velocity has less than ten times the Li content of the star itself. This was considered so far as one of the strongest arguments in favor of the 'individual star' origin of Li. Herbig tells us that the matter may not be that simple (see the Discussion) since the Li-devoid gas shell appears to come from the star itself.

(3) Observations reported by Feast at this Conference suggest a regeneration of Li after a star has left the main-sequence.

(4) The largest $^7\text{Li}/^6\text{Li}$ ratios observed so far are approximately equal to the proton-induced spallation ratios $(\simeq 2 \cdot 5)$ reported by Dr. Gradsztajn. (The alteration mechanisms, to be discussed later, always destroy ^6Li faster than ^7Li (Figure 1).)

In the same figure, we see that the ratio of the destruction rates of Li and Be varies with proton energies, hence a similar argument cannot be made with observed Li/Be ratios.

Other theories of Li, Be, B formation involve either a primordial origin (say from the Big Bang), or a gradual enrichment from Supernovae.

Recent work on the Big Bang (Wagoner *et al.*, 1967) suggests a primordial yield

Perek (ed.), Highlights of Astronomy, 255–260. © *I.A.U.*

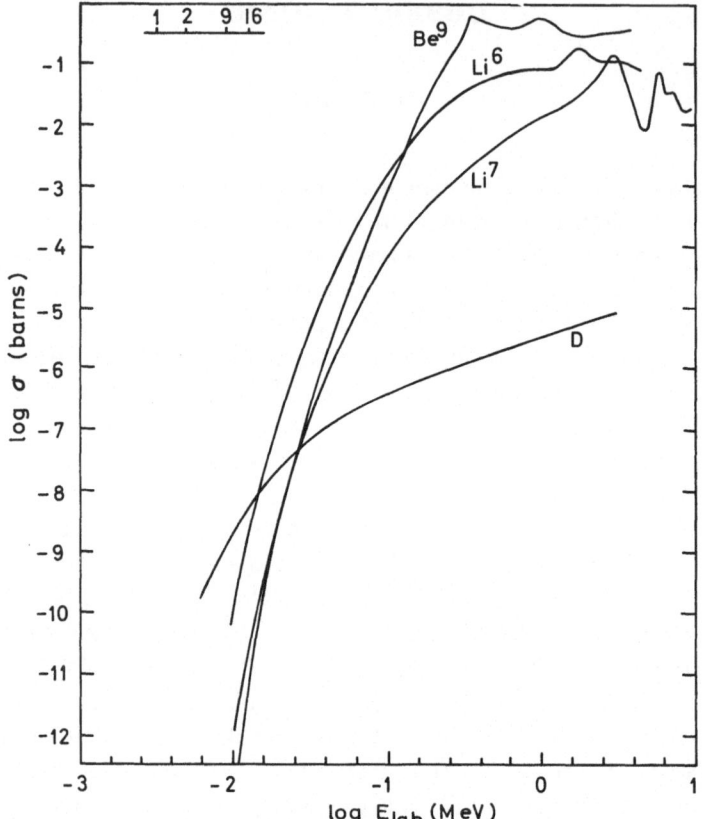

FIG. 1. *Experimental cross-sections for the destruction of D, ⁶Li, ⁷Li, ⁹Be, as a function of the proton incident energies. Note the very small scale of the ordinates. The insert in the upper left shows the temperature in million degrees, for which the Gamow energies of Be are those of the abscissa. For instance, at $T_6 \simeq 10$, the Gamow energy of Be(or Li) is approximately 10 keV.*

of enough ⁷Li, but of not enough ⁶Li and ⁹Be. The existence of many stars with ⁷Li/⁶Li $\simeq 2$, hence shows, that Big Bang contribution is at best small.

Recent observations of Be, reported by Conti at this Conference have important implications on this problem. When plotted against spectral class, the Be abundances are found to increase with decreasing surface temperature (hence deepening surface convective zone). As we shall argue later, Be atoms are most likely not destroyed in these stars. These observations may show a relationship between the intensity of the generation mechanism and the spectral class (thereby implying a mass effect, hence an 'individual star' origin). In view of our previous discussion, we are led to prefer this possibility.

Another possibility has been suggested by Conti (preprint): since late type stars are on the average older than early-type stars (the stars observed here are field stars), we may have here an age effect which could be explained in terms of a gradual destruction of the beryllium of the galactic gas (e.g. through stellar burning). This interpretation is made somewhat unlikely by the fact that several stars with no detectable Be (and also the solar system) do not seem to fit very well in this pattern. The question could be settled by a study of Be in various clusters.

At any rate, the implication of this observation is that the pattern of gradual enrichment of element abundances by nucleosynthesis in stars or supernovae does not apply to beryllium.

We now discuss briefly the importance of alphas and secondary neutron-induced reactions.

The contribution of alpha-induced spallation is probably small. The ratio of accelerated alphas to protons is generally expected to be small (and is observed to be small in the Sun and in the galactic cosmic rays). Furthermore, a systematic analysis of spallation rections (Audouze *et al.*, 1967*a*) has shown that the relative yields of different isotopes is influenced mostly by the nuclear properties (binding energy, isotopic spin, density of levels) of the final product, so that we expect the cross-section ratio to be quite similar. Preliminary experimental results (Cuer preprint) corroborate this view.

Secondary neutrons will play no role if the mean proton energy in the flare is less than one hundred MeV. At higher energies, and if the flare medium is dense enough ($> 10^{12}$ particles cm^{-3}), the formation ratio of Li/Be decreases slowly from about 20 to about 3 at one GeV. (The ^{7}Li/^{6}Li ratio is insensitive to neutrons.) Hence, part, but not all of the variation of Li/Be with spectral class could be explained by making the assumption that (1) neutrons do not decay in flight, and (2) the mean proton energy increases with decreasing surface temperature. We consider this hypothesis as unlikely, but we do not reject it at once.

B. How is the intensity of spallogeneration related to stellar evolution, and what is the influence of stellar mass, rotational velocities, and age?

The large abundances of Li observed in T Tauri stars, and in particular in FU Orionis, suggest that the very early stages of stellar evolution witness a large part (if not most) of the electromagnetic activity of a star. If we are correct in interpreting the Be spectral class correlation as a mass effect, then we are prevented from pushing back the moment of irradiation too far in the past. In terms of proto-cluster evolution, the irradiation must clearly take place at a moment when the fractionation in stars is already well established, and each star must already know of what mass (hence of what spectral class) it is going to be.

From the data presented today, there seems to be a regenerescence of spallation reactions as the star leaves the M.S.: both the total Li and the ^{7}Li/^{6}Li ratio appear to increase before they decrease again as the star becomes a red giant. A coherent pattern emerges; spallation reactions are associated with major structural changes in the star (initial gravitational contraction; post M.S. contraction). These changes must bring important disturbances to the state of the atmosphere and, in turn, these disturbances must accelerate large particle fluxes, when coupled with the surface magnetic field and stellar rotation. In this sense, the problem of Li and Be is not a purely 'superficial' one anymore (in both senses of the word), but must be deeply related ro the problem of stellar evolution.

C. How and where were the spallogenic element abundances further altered? How is the alteration mechanism related to mass-rotational velocity, age, surface properties, etc?

Herbig has pointed out that the total stellar Li abundance appears to decrease with stellar age, with an average lifetime of 10^9 years. A similar conclusion can be drawn from a comparison between the Hyades and the Pleiades. The fact that varying Li/Be (as mentioned before, only part of the variation could come from secondary neutron-induced reactions) and ^{7}Li/^{6}Li ratios are observed shows that the depletion must be, for an important part, due to nuclear destruction (and not only to dilution). In stellar surfaces (and interiors), spallogenic element abundances can only be altered by proton-capture reactions (alpha-capture probabilities are reduced by Coulomb repulsion, neutron-capture occurs predominantly on hydrogen and has no effect on Li or Be). The fact that all ^{7}Li/^{6}Li ratios observed are larger than the proton formation ratio is coherent with this affirmation; as seen from Figure 1, the capture cross-section of ^{6}Li is at all energies considerably larger than the ^{7}Li capture cross-section.

It has been assumed so far that Li depletion occurs mostly in stellar surface convective zone (where the temperatures do not exceed a few million degrees, hence Gamov energies do not exceed 5 or 10 keV). We must, however, consider the possibility that Li is destroyed by the spallogenerating proton flux itself. The abundance equation of an element L must indeed be written as:

$$\frac{dn_L}{dt} = {}^n{}_{CNO} \int_Q^\infty \Phi(Ep)\,\sigma_{sp}(CNO \to L)\,dE$$

$$- n_L \int_0^\infty \Phi(Ep)\,\sigma_{des}(L)\,dE - n_L \int_0^\infty n(Ep)\,\sigma_{des}(L)\,v\,dE.$$

We recognize first the formation term, proportional to the high-energy part of the proton flux $\Phi(Ep)$; the surface destruction term and the convective zone destruction term, proportional to the proton density $n(Ep)$, (with a M.B. energy distribution). The main difference between these last two terms is that $\Phi(Ep)$ has a higher mean energy, but a much lower density than $n(Ep)$.

The relative importance of the two destruction terms is closely related to the location of the most probable energy at which stellar lithium is destroyed. The very low

$n(\text{Li})/n(\text{CNO})$ observed in stars ($\leqslant 10^{-5}$) rules out the energy range over a few tens of MeV; important destruction in this range would imply important destruction of CNO and larger $n(\text{Li})/n(\text{CNO})$ ratios.

A more probable region is the MeV and sub-MeV region; if the proton spectrum keeps on increasing with decreasing energy (as suggested by the solar cosmic-ray spectrum), we may expect important effects from the first term, at a 'Gamow' energy defined by the maximum of the integrand. Even for extremely steep spectrum, the Gamow energy cannot be below 25 keV.

In Figure 1, we note that at $Ep < 25$ keV, $\sigma(\text{Be}) < (^7\text{Li})$

and at $Ep > 25$ keV, $\sigma(\text{Be}) > (^7\text{Li})$.

This crossing of the cross-sections around 25 keV is most useful, since it permits a clear separation between surface destruction ($E_G \geqslant 25$ keV), and convective-zone destruction ($E_G \leqslant 25$ keV). The presence of ^6Li will not perturb this separation very much.

Conti and Danziger have measured Li/Be ratios in a number of stars, and have shown that this ratio is always smaller than the formation ratio ($\simeq 20$). This observation shows that the most probable destruction energy is below 25 keV, and that stellar Li destruction occurs mostly in convective zones. This conclusion is of course strengthened by the correlation between Li/Be ratio and spectral classes (such a correlation would not *a priori* be expected in the case of surface destruction).

D. What can we say about energies involved in the process, and in particular about the efficiency of the accelerating mechanism?

If a star contains n_L atoms of a spallogenic element, the energy lost by the star in spallation is simply $E_L^I = n_L Q_L \simeq 10^{43}$ ergs in typical stars. (The notation E^I refers to the fact that this mechanism belongs to the realm of the first law of thermodynamics.) This loss is small and unimportant.

On the other hand, the amount of energy which must have been used in acceleration is given by:

$$E_L^{II} = n_L \left\langle \frac{\varepsilon n_H}{n_{\text{CNO}} \, \sigma(\text{CNO} \rightarrow \text{L})} \right\rangle \simeq 10^{48} \text{ ergs},$$

where ε is the stopping power of the gas, and the mean value is taken over the proton spectrum. The notation E^{II} (second law of thermodynamics) refers to the fact that this energy is not lost by the star, but largely recuperated through the electronic collisions. This energy is significant in that it characterizes the efficiency of the accelerating mechanism. Its bearing on the theory of stellar evolution has been briefly discussed by Audouze *et al.* (1967*b*).

References

Audouze, J., Epherre, M., Reeves, H. (1967*a*)   *Nucl. Phys.*, **A97**, 144.
Audouze, J., Gradsztajn, E., Reeves, H. (1967*b*)   *Les Congrès et Colloques de l'Université de Liège*, **41**, p. 299.
Wagoner, R.V., Fowler, W.A., Hoyle, F. (1967)   *Astrophys. J.*, **148**, 3.

DISCUSSION

Fowler: Did you notice that several stars with depleted Li have $^6\text{Li}/^7\text{Li} \simeq \frac{1}{2}$, which is the *unmodified* spallation value? I have suggested that we are seeing Li, which has not been involved in the convection. Do you have an alternative explanation?

Reeves: I fully agree with you. If we try to imagine what actually did happen, and what information we can extract from the observed $^6\text{Li}/^7\text{Li}/\text{Be}$ value, then we are led to several possibilities. These possibilities are based on the idea that the spallogenic elements observed are really the sum of two contributions: an 'old' one, in which all Li has been destroyed, but some Be has survived, and a 'new' one, in which all three isotopes are present in their formation ratios. With the formation ratios quoted by Dr. Gradsztajn, it is of course easy to evaluate the fractional importance of each contribution. Let me mention three possibilities.

(I) Some Be would have been present in the original stellar gas. According to our previous discussion, this is unlikely, but can certainly not be ruled out at this time.

(II) The first contribution may have been produced during the very early stages, followed by a thorough destruction of Li during the Hayashi convective period, and followed by a second irradiation somewhat later, e.g. at the time when the star settles on the M.S. The variation in the $^7\text{Li}/^6\text{Li}$ ratio would then show the extent of ^6Li burning since the star is on the M.S.

(III) Let us suppose that some spallogeneration is still going on at the surface of these stars. If the average diffusion time 't_d' for the spallogenic elements to sink from the optical region into the convective zone is not too short or too long, then we could observe in the spectrum both a 'recent' contribution of elements which have not yet sunk in the convective zone, and an 'old' contribution from all elements previously generated and diluted through the convective zone (assumed to be hot enough to burn Li, but not Be). The critical time-scale for the diffusion time to be such that both contributions are comparable is $t_d \simeq xT$, where x is the dilution factor between the irradiated mass and the convective mass, and T is related to the age of the star. Typical values of x range from 10^{-7} to 10^{-9}, so that the required t_d would vary from a small fraction of a year to several years, according to the star under consideration. I have questioned several experts on atmospheres. None of them dismissed such periods as unlikely. As mentioned before, in such cases, ^7Be and ^{10}Be could well be observed. Observers should be on the look.

Masevič: If all the Li is destroyed at the bottom of the outer convective zone, the theoretically expected abundances will include all uncertainties of the theory of the convective zone (mixing lengths, etc.). That will make the relation between the observed Li abundances and the evolution still more difficult to investigate, at least for M.S. stars.

Reeves: According to Miss Merchant, *all* observed Li/Be ratios are smaller than the proton-induced formation ratio, hence I am afraid that we have to face the difficult situation that you are describing.

Herbig (after an unreported remark of Dr. Danziger): I believe that the Bonsack-Greenstein observation (i.e. that in T Tauri there is no Li line in the displaced shell spectrum) does not tell us anything about pre-stellar material. The shell lines are violet-displaced, so the material is rising from the star, and so this is material that has already been processed. Of course, why the Li is not seen in this rising gas is another problem.

THE MIXING OF LITHIUM*

E. A. Spiegel

(Physics Department, New York University)

1. Introduction

The observational status of the problem of lithium abundances has been thoroughly detailed in the preceding papers in this symposium, and it is clear why we must consider how matter is mixed from outer convection zones to inner, hotter regions. The need for appropriate mixing mechanisms has also been nicely brought out by Herbig and Wolff (1), and Böhm (2) has summarized the role of convective mixing. Conventional mixing-length models for the outer convection zones seem to give qualitatively reasonable results for the depletion during pre-main-sequence contraction (3, 4, 5) but do not completely account for the observations, and it seems inescapable that main-sequence depletion of lithium must be considered (1). I shall therefore simplify the discussion of mixing by concentrating on main-sequence models in the following outline of some possible mixing processes, though most of the remarks to be made should apply generally to other phases. I shall also pretend that there is one principle mechanism (or combination of them) that must be found, though stars in different evolutionary phases, or with different masses, may deplete lithium quite differently. Further, I shall use the Sun as an illustration in general since we know some important details about it that are not always known for other stars.

As to the mixing mechanisms themselves, I shall attempt to organize the discussion by considering four problems. These are:

(1) the depth of outer convection zones,
(2) the penetration of convective motions into radiative cores (overshooting),
(3) the effects of rotation and rotational braking, and
(4) the effect of mass loss.

The first two of these are crucial no matter what the ultimate mixing mechanism is, since they determine the depth to which material is mixed by ordinary convection. In this way they fix that the additional distance material must be carried before lithium is effectively destroyed. The other two are rather uncertain since they depend on incomplete theories or uncertain observational results, but it is certainly worth noting their possible relevance.

* Prepared mainly during the tenure of a National Science Foundation Senior Postdoctoral Fellowship (1966–67) at the Department of Applied Mathematics and Theoretical Physics, Cambridge.

2. The Depths of Convection Zones

The structures of convection zones have been carefully calculated using Mrs. Böhm-Vitense's (**6**) version of mixing-length theory for the relevant spectral types (**7**). In this approach the bottom of the convection zone is the depth at which the star becomes locally stable against convection according to the Schwarzschild criterion (**8**). The familiar uncertainty about the choice of mixing length is normally resolved by adjusting the mixing length so that the stellar radius comes our right when complete stellar models are calculated. This does not necessarily guarantee a correct value for the depth of the convection zone, but it at least makes the models seem reasonable.

Even granting that the mixing length has been chosen correctly, we must allow for some smaller uncertainties in the calculations. The depth of a convection zone is quite sensitive to various parameters, and the work of Weymann and Sears (**9**) indicates that an increase in opacity causes a deepening of the convection zone. Dr. Baker has recently pointed out (private communication) that ionization of heavy elements can also make a small but non-negligible correction to the depth of the convection zone, and this factor is not always carefully taken into account. And finally the use of mixing-length theory makes our estimates imprecise because it determines the bottom of a convection zone on the basis of linear stability, whereas the indications from finite amplitude stability theory are that the actual boundary of a convective zone is somewhat displaced when one studies the effects of finite amplitude perturbations (**10**).

Now in the pre-main-sequence mixing these small effects probably are not crucial since the bottom of the convection zone passes well into regions where the lithium can be destroyed (**5**) and no qualitative changes are likely to be brought about by small changes in the thickness of the convection zone. Likewise, in post-main-sequence lithium depletion where the main mechanism seems to be convective dilution of lithium (**11**), the qualitative picture does not depend on such small corrections. But in main-sequence lithium depletion the location of the bottom of the convection zone is crucial, for the following reasons.

First, there is the difficulty that if the temperature at the bottom of the convection zone approaches 3×10^6 °K, lithium is destroyed in a time rather less than the main-sequence lifetime of late stars. A mechanism of this kind is not what is called for by the evidence for slow depletion or the main sequence. On the other hand, if the convection zone is too shallow the (presumably) slow mixing process which takes material below the convection zone will have too long a time-scale since the time for the mixing is probably either linear or quadratic in the geometrical distance between the bottom of the convection zone and depth at which lithium is destroyed. This factor makes the depth of the convection zone an important parameter.

At present, estimates of the depth of the convection zone in the Sun range from $1–3 \times 10^5$ km, if we include rumors about unpublished determinations. Not all of these possibilities have been used in complete solar models so that they may not all give agreement with the observed solar radius. Nevertheless, the spread in estimated depths is large, and we may ask whether solar observations can help to settle the problem.

One possible check is that of the observed motions in the solar convective zone. The granulation and supergranulation are generally believed to be convective motions and their horizontal scales are 2×10^3 and 3×10^4 km respectively (12, 13, 14). The granule scale is probably determined by the thickness of the transition layer between the stable photosphere and the adiabatic layers in the deep convection zone, or at least by the scale-height in the upper convection zone, which is closely related. But what determines the scale of the supergranules? One possibility is that it is related to the thickness of the convection zone (13, 15). Though the nature of the convection remains unclear, a discrepancy of a factor 2π is not overly surprising. The picture proposed by Simon and Weiss (16) of motions extending over several scale-heights seems quite reasonable and may explain the supergranule size. On the other hand, it may be that the supergranulation scale is a manifestation of a boundary layer at the bottom of the convection zone analogous to the upper transition layer. Standard mixing-length theory does not show such a boundary layer, since it does not take into account the slowing down of eddies as they approach the bottom of the convection zone. But Böhm and Stückl (17) have attempted to correct this deficiency by setting the mixing length equal to the distance from the edge of the convection zone when this distance is less than the local scale-height. This probably overestimates the effect of the boundary a bit, but it gives a reasonable indication of the existence of a boundary layer at the bottom of the convection zone. (A similar but weaker effect is produced when account is taken of deviation from radiative equilibrium (18).) The calculation of Böhm and Stückl leaves the correlation between vertical velocity and temperature fluctuation as a free parameter. When this parameter is $\frac{1}{2}$, the depth of the convection zone is $1 \cdot 55 \times 10^5$ km and the bottom boundary layer has a thickness of 3×10^4 km. This suggests a rough agreement between current estimates of the convection-zone depth and the observed scale of supergranules, if it is true that the lower boundary layer does set the horizontal scale of supergranulation.

Another possible check on the calculation of the thickness of convection zones comes from theoretical and experimental results on the motion of a spherical shell of rotating fluid. When the shell is contained between rigid, concentric spheres rotating at slightly different speeds rather complicated motions develop. These are not completely understood even for the homogeneous fluid in which viscosity is important only at the boundary layers. However, it is known that in this relatively simple case two important regions may be distinguished (19, 20, 21) by constructing a cylinder coaxial with the rotation axis and tangent to the inner sphere. The cylinder divides the spherical shell into polar and equational regions, as shown in Figure 1. In the polar regions a circulation as indicated in the figure is found; in the equatorial regions no motions occur in the steady state.

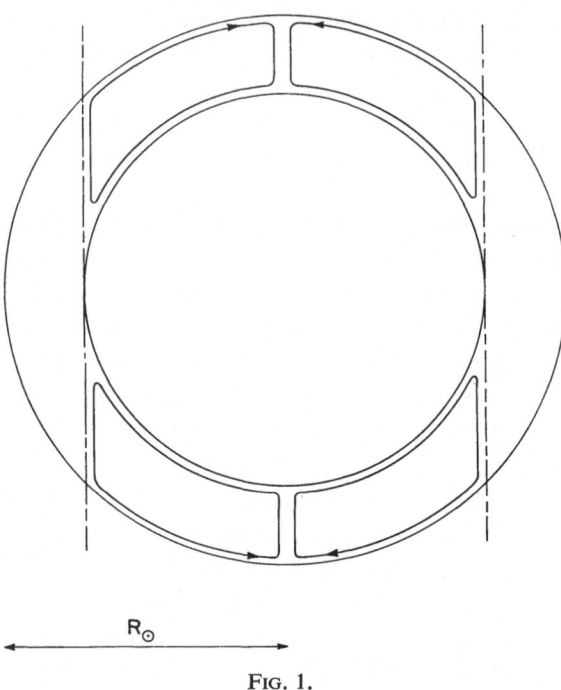

$$R_{\odot}$$

FIG. 1.

Now the solar convection shows a crude analogy to this configuration. The inner stable layers respond to penetrative motions more slowly than the convective zone responds to imposed disturbances such as those on its outer edge resulting from the torque of the solar wind. Moreover, since the bulk of the convection zone is adiabatically stratified, some of the approximations required in the theory of the homogeneous incompressible fluid are applicable. On the other hand, because of turbulent viscosity and the slight degree of instability, we cannot expect the analogy to be really good. Nevertheless, it has been remarked (**22**) that the extreme latitude achieved by sunspots does seem to coincide with the latitude at which the cylinder of Figure 1 cuts the outer sphere if solar parameters are used. That is if d is the thickness of the convection zone, it is related to this critical latitude by

$$d = R_{\odot}(1 - \cos\phi),$$

and with $\phi \simeq 45°$, $d \simeq 2 \times 10^5$ km. This would seem to provide support for the present estimates of the thickness of the convection zones, but until we go farther with the theory of circulation in convection zones, and in particular the effects of stellar wind torques, we can draw only slender comfort from this test.

In sum, the mixing-length theory seems to give qualitatively reasonable but not sufficiently precise estimates of the depths of convection zones.

3. Penetration of Convective Motions into Radiative Zones

The foregoing discussion of the depths of convection zones assumed that the motions stop where the temperature gradient switches over from super- to sub-adiabatic. But now we must confront the possibility that descending motions originating in the unstable layers do not stop at the bottoms of convection zones but continue an unknown distance farther. In effect this displaces the bottom of the convection, or mixed, zone inward and makes it fuzzy. We have as yet no reliable way to deal with this phenomenon and thus far two main approaches have been considered, one using blobs the other convective cells as the agents of convective transport.

In the blob approach, one proceeds as in mixing-length theory with a blob of fluid starting out in near equilibrium with its surroundings. The blob arrives at the edge of the convection zone with a calculable kinetic energy, and the question can then be asked, What is the distance to which the blob can penetrate into the stable layers before it is brought to rest by the buoyancy forces? In this estimate one temporarily suspends the restriction that an element can be destroyed by turbulence, presumably after it travels a distance comparable to its own diameter. This kind of estimation of the effects of penetration was used by Mrs. Böhm-Vitense (23) to discuss granulation in her original mixing-length paper. The same kind of treatment was used by Weymann and Sears (9) in relation to the lithium problem. In effect, the results in this way are equivalent to the argument in which the blob is assumed to penetrate to the point where the local entropy is the same as that in the region where the blob originated (24). Moore (25) has pointed out that the entropy argument is unreliable because it omits entrainment of fluid by the blob and because it neglects the momentum of the blob. The former causes an underestimate of penetration, the latter an overestimate. A phenomenological theory taking account of these effects has been given by Morton et al. (26) and Scorer (27), but only when the diameters of the blobs are much less than the scale-height.

For the present, though, the best available estimate seems to be that of Weymann and Sears. They find that in the solar case only those blobs which form in the upper, highly unstable, convective layers can penetrate deeply enough to cause appreciable mean destruction of lithium. However, if the diameters of the blobs are set equal to their free path for turbulent disruption, and if the blobs have initial diameters comparable to the scale-height in their region of origin, the probability that they can reach the bottom of the convection zone is much too small for appreciable lithium destruction.

It must be stressed, however, that Weymann and Sears used a model in which there is no lower unstable boundary layer such as that found by Böhm and Stückl (17). On the other hand, the lower unstable layer of Böhm and Stückl is calculated on the basis that no penetration is possible, and it would not be correct to use their model directly for an estimation of penetration, though one could iterate their model for this purpose as follows.

Böhm and Stückl used the distance from the bottom of the convection zone as a mixing length when this distance is less than a scale-height. If we now use the entropy argument (or an improved version according to Morton *et al.* (**26**)) to estimate a new penetration distance arising in the Böhm-Stückl model, we can define a new effective bottom of the convection zone. We could then recalculate the structure of the convection zone using the distance to this new bottom as a mixing length. In this way, it is possible to iterate till a self-consistent penetrative model is achieved. But until some such calculation is performed, we cannot draw any conclusions from the present blob calculations.

In the other approach to the penetration problem, one applies the equations of motion, treating small perturbations about the average conditions calculated on the basis of mixing-length theory (**28, 29, 30, 15**). The resulting equations can be solved numerically and the relative amplitude of the velocity field at different depths can be found for various modes. The absolute amplitudes can be established in the solar case by adjusting to the observed surface velocities. Böhm (**29**) has found that this approach would lead to the conclusion that lithium must have been wholly destroyed in the Sun. Of course, this is contrary to observation if we assume that lithium is not presently being produced in the outer solar layers, and the implication of the laminar mode approach would be that all observable lithium in stars as late as the Sun has been recently produced. This conclusion seems unacceptable and the fault must lie in the linear theory.

The reason that linear theory gives an overestimate of the penetration is that it takes full advantage of the upper unstable layer without suffering the turbulent losses inherent in the blob picture. Laminar modes can extend throughout the convection zone and below the zone their amplitude drops roughly exponentially in depth with a penetration distance comparable to their horizontal scale. Thus they are continually driven by the upper unstable layer. It is possible to remedy this failing by means of eddy coefficients, and Professor Böhm is attempting to do this. For the moment though, we have no conclusive results from linear theory which relate to lithium depletion.

In summary, it is not at all clear whether penetrative convection can be the main mechanism involved in lithium mixing. The estimates that do exist make it either too effective or negligible while what is needed is a slow but definite process. Before we can decide finally on the prospects for penetrative mixing we need to know not only the depth of penetration but also the sharpness of the cut-off at the lower limit reached by penetrative motions. It would seem that only if there is a spread in the depths to which motions from a given convection zone can penetrate can we attribute lithium destruction to penetrative mixing; otherwise the penetrative mechanism is an all or nothing proposition and cannot explain the observations.

It may also be that penetrative convection plays an important part in lithium depletion even though it is not the ultimate motion that carries lithium to its destruction. Thus, it is important to know at each spectral type the lowest depth to which penetrative motions extend.

4. Rotational Processes

If we allow for the effects of rotation in the lithium-mixing process, we can find a variety of mechanisms which may be relevant. The most natural of these to consider is probably the Eddington-Sweet current (**31**), which results from the stellar deformation due to rotation. Crudely speaking, this deformation produces a pole to equator-temperature variation, hence a horizontal pressure gradient. Motions are thereby established and these permeate the star, if no gradient of molecular weight intervenes. The circulation time for the Eddington-Sweet current is

$$t_{\text{E.S.}} \cong S t_{\text{K.H.}},$$

where $t_{\text{K.H.}}$ is the thermal (Kelvin-Helmholtz) time of the star and

$$S = \frac{G\bar{\rho}}{\Omega^2},$$

where $\bar{\rho}$ is the mean density, Ω is angular velocity and G is the constant of gravity. In the Sun, $\rho \approx 1$ gm/cm^3, $\Omega = 3 \times 10^{-6}$ sec so that $S \approx 10^4$ and $t_{\text{E.S.}} \approx 10^{11}$ years. Thus it would appear that the Eddington-Sweet currents play no role in lithium depletion in the Sun at present.

However, it is now reasonably well established that the angular velocity of the Sun is at present decreasing with a half-life of about 5×10^9 years as a result of the solar-wind torque (**32, 33, 34**), though this value may have once been lower (see Section 5) and may at present be higher (**35**). In any case, it would seem that the Sun arrived on the main sequence with a somewhat higher angular velocity than at present. If, moreover, the mass-loss rate were initially higher than at present, Ω may have been higher by as much as a factor of 10 and $t_{\text{E.S.}}$ could then have been as low as $\sim 10^9$ years, which means that Eddington-Sweet currents may have played a role in lithium depletion in G stars, at any rate. Unfortunately, we cannot specify precisely what the rate of rotational mixing is without knowing the history of Ω in time, the location of the bottom of the convection zone, and the interaction of the Eddington-Sweet currents with the convection zone. Indeed, these uncertainties are present in the other mechanisms discussed in this section, so that only an indication of the possibly important mechanisms can be given.

In the Eddington-Sweet theory, rotation itself causes a thermal imbalance which gives rise to a slow circulation. But in rotating stars the loss of angular momentum through stellar winds sets up surface stresses which can also give rise to internal circulations. The problem is quite complicated and a number of points of view have been put forward lately. However, there is general agreement that the surface stresses resulting from stellar winds are rapidly transmitted through the outer convection zones, as a result of eddy viscosity. The time-scales on which the radiative cores respond to these transmitted stresses are thus presumed to be longer than those of the convection zone so that the interesting circulations produced in the convection zones need not be considered in detail here.

One suggestion about the response of radiative cores to the external stresses is that the time-scale is so long that the entire radiative core is in relatively rapid rotation compared to the surface layers, in the case of the sun (36, 37, 38). For example, Dicke (36) proposes a rotation rate in the core which is ten times that in the surface layers. In that case the Eddington-Sweet time is $\sim 10^9$ years, and we would expect appreciable mixing during the Sun's main-sequence lifetime.

However, it has been suggested that the Dicke model is not valid because it is unstable (39) and because the surface stresses drive a circulation which would wipe out sharp variations in angular velocity (40). Even the Eddington-Sweet currents resulting from the Dicke model would seem to prevent the model from being self-consistent, as Professor Mestel has privately remarked. Hence, what seems to be important here is that the motions involved in the instability and the circulations induced by the solar-wind torque may be important for the mixing problem, though only suggestions of their importance have been published as yet.

The instability of differential rotation as the mixing mechanism has been favored by Goldreich and Schubert (39). The instability is essentially that first discussed by Lord Rayleigh (41, 42), who showed that a differential rotation is unstable if the angular velocity as a function of distance from the axis, r, drops off more rapidly than r^{-2}. Taylor (43) later showed how viscosity could impede the instability. A stabilizing density gradient can also suppress the instability, but as Lieber and Rintel (44, 45) noted, conductivity can reduce the effect of the density gradient so that for large conductivity, and small viscosity, the Rayleigh instability, in a modified form, can occur in the presence of stabilizing density gradients. Similar reductions of the effect of stabilizing density gradients by radiative transfer apply in the case of shear flow instability (46, 47), Goldreich and Schubert (39) have discussed this instability more fully with the solar problem in mind and conclude that the criterion of Rayleigh is not strongly altered in the solar conditions.* They also appear to be the first to attempt the problem in spherical coordinates, which leads them to an additional instability criterion.

Let Ω be the angular velocity and z be parallel to the axis of rotation. Then Goldreich and Schubert find that instability occurs whenever $d\Omega/dz \neq 0$. (Actually, this is their criterion in the limit of small viscosity.) This result is rather important since it implies that a very slight variation in Ω can produce instability, whereas the Rayleigh criterion demands a great Ω variation for instability. However, I have some misgivings about the interpretation of this result since it is not based on a precise solution of the steady state of motion whose stability is being tested. Recently Barcilon and Pedlosky

* A related stability study has been recently carried out by Fricke (65).

(48, 49) have studied the steady state of a stratified fluid to which a $d\Omega/dz$ is being applied at boundaries. A complicated circulation is set up by the applied $d\Omega/dz$ and it seems clear that, especially for large conductivity and small viscosity, the fluid tries to achieve $d\Omega/dz \approx 0$. The proper stability question would seem to be whether the kind of flow found by Barcilon and Pedlosky is unstable. I am not clear whether the instability found by Goldreich and Schubert is in part simply a manifestation of the fact that there is no solution with $d\Omega/dz = 0$, which does not try to have a circulation, so that if the circulation is left out of the steady state, the perturbations are bound to grow. But in any case, it seems clear that when $d\Omega/dz$ is far enough from zero, motions must be set up, and one possibility to consider is the instability discussed by Goldreich and Schubert.

Presumably, the applied stress requires a slight excess of $d\Omega/dz$ over its value for the onset of instability. This excess should be just enough to permit the resulting weak turbulence to transport angular momentum out of the star at a rate demanded by the stellar wind. Thus, in principle we should be able to estimate an eddy diffusivity for angular momentum transport if we know the rotation and mass loss. The diffusivity can then be used to estimate the turbulent diffusion of lithium. This has not yet been done, partly because the whole picture is still controversial (50).

I have already mentioned that the applied stresses on the boundary of a fluid can set up an internal circulation. Howard, Moore and I (40) have proposed that this circulation is an important feature in the problem of lithium depletion. We discussed the problem in terms of the transient process called 'spin down'. In this process, when a rigidly contained rotating fluid is suddenly subjected to a change in the angular velocity of its container, a boundary layer forms and in turn drives a circulation through the body of the fluid. For a homogeneous fluid, this motion was already discussed qualitatively by Einstein (51), and it has been shown that the effects at the boundary alter the internal angular velocity in a spin-down time which is the geometric mean of the viscous time and the rotation period (52, 53, 54). If the boundary layer, or Ekman layer, is turbulent the spin-down time for a homogeneous fluid is approximately 160 rotation periods (55).

In the stellar case, the fluid is not homogeneous and is not rigidly contained, so that the problem is quite complicated. However, it appears that the replacement of the rigid container by a convection zone only enhances the effects of spin-down. In the first place the exchange of fluid between the convection zone and the interior at the bottom of the convection zone gives a strong coupling between the two regions (40). But more important, it appears that because of turbulent viscosity the whole convection zone can play the role of the Ekman layer in pumping the fluid through the radiative core (56). The effect of stable stratification in the core, on the other hand, works against the spin-down process.

In a simple model, with a stably stratified adiabatic fluid, Holton (57) showed that the effect of spin-down is confined to a layer of thickness $R/S^{1/2}$, where for our purposes R is the radius of the radiative core. Since $S \sim 10^4$ for the Sun, the Holton layer is probably too shallow to directly influence the mixing of lithium. But the Holton layer sets up an imbalance which in turn drives a slow circulation in the interior if the effects of radiative conduction or viscosity are admitted. The spin-down time for this process has been studied by Pedlosky (58) in the case of a cylindrically contained fluid. When the side walls are thermally insulating, Pedlosky's results indicate that the time-scale for stellar conditions would be just the Eddington-Sweet time. The problem with conducting side walls is more difficult to treat, but Pedlosky found a simple special case which, curiously, gives the same spin-down time as for a homogeneous fluid.

In the spherically contained fluid, the Ekman layer encompasses the whole interior and sidewalls do not really enter, while in the Sun, where the convection zone seems to close the interior circulation, this is even more forcefully the case. Hence the actual stellar spin-down time remains in doubt. Moore and I have been looking at simple models for this process and with Newton's law of cooling have found that the Eddington-Sweet time applies for spin-down. However, Newton's law does not take proper account of the small scales set up in the temperature perturbation in the Holton layer, so that the result with a proper diffusive law may be different. At present we are attempting to study this more difficult case.

The possibility that the Holton layer is unstable must also be considered, since this would seem to give a spin-down time close to that for the homogeneous fluid (40). I have already mentioned the difficulty of this stability problem and though we have discussed the problem with experts in this field (especially W. H. Reid) we have not yet been able to reach a definite conclusion. The rather sharp shear layer or Holton layer that is central to the problem may well become unstable, but it may be some time before we can be sure. If the instability does occur we can expect a weak turbulence in late stars and once the instability condition is established, a mixing-length theory can be constructed, as I mentioned earlier. We have already made a crude version of such a theory and it seems clear that the turbulent diffusion of lithium is, on this basis, very dependent on the angular velocity of the star.

From all this discussion we can abstract very little that is definite, but the main point is that the rotational mixing mechanisms certainly give a slow overall circulation and possibly a weak turbulence. Very likely, when late stars first arrive on the main sequence, they have somewhat higher rotational velocities than at the end of their lifetime on the main sequence. It is quite possible that even the Eddington-Sweet time, which characterizes the usual types of circulation and seems to be an upper bound to the time scale for spin-down circulation, is sufficiently short to be of interest in the mixing problem. Much remains to be done on the theory and we are in need of an accurate knowledge of the mass-loss rate in order to evaluate the surface stresses which can drive circulation.

In spite of these uncertainties, the rotational mechanisms seem attractive. In the first place, they suggest a correlation between rotation and lithium abundance and thus suggest an origin for the large scatter in lithium abundance for stars of the same spectral type. In this connection it is interesting that for stars earlier than G2 the dispersion in lithium abundance is much larger than for later stars (1). It is also likely that these stars have a greater dispersion in angular velocities than the later ones since they have weaker convection zones, and less rotational braking. By the same token they will suffer weaker surface stresses and their internal circulations will be more weakly driven, hence the observation that their lithium abundances are higher than in later stars also seems reasonable.

Secondly, Conti (59) has just reported observational evidence of a correlation between lithium abundance and angular velocity in G stars. Though this does not demonstrate the relevance of rotational mixing, it does suggest that we inquire further into this possibility.

And finally, the spin-down mechanism itself permits a qualitative resolution of the following difficulty. If we require a mechanism that brings material from the bottom of the convection zone to deeper layers, we must be aware that the same motions may drag magnetic-field lines into deeper layers too. Indeed, there is no indication of how deep the mixing will penetrate and we demand only that it go deep enough. On the other hand, it now seems probable that the mean solar magnetic field varies with the solar cycle, and Babcock (60) has, on this basis, concluded that the mean solar magnetic field must be confined to shallow layers, since the time-scale for magnetic variation would otherwise be too long (61). But with spin-down, another possibility arises.

In the spin-down of a homogeneous fluid, the time for a complete circulation of the material is generally very much longer than the spin-down time, i.e. than the time required to alter the internal angular velocity of the fluid appreciably. Let us suppose that this discrepancy in the two times can also occur in a stratified fluid. Moreover, the alteration of a magnetic field proceeds by a process similar to the vortex stretching, which alters the angular velocity. Hence, even if the solar magnetic field were to extend deep into the Sun, if the solar spin-down time were 11 years, there would be no dilemma, so long as the circulation time which is relevant to lithium depletion is $\sim 10^9$ years. A theory of this whole process is very difficult, as Moore and I have been finding, since the internal motions can enhance the general field, which in turn alters the co-rotation distance of the solar wind and thus alters the circulation in the interior. Whether a feed back oscillation can thus result is not at all clear, but at least there does seem to be the hint of a resolution of the 'lithium-magnetic field paradox' along these lines.

5. Mass Loss

The last possible process I want to discuss is not strictly a mixing process, though it is the most elegant of all. It is the possibility that late stars lose enough mass on the main sequence so that, after a time, the material we see in the surface convection zones has an appreciable admixture of matter that was once well below the convection zone. The mechanism was considered by Weymann and Sears (9) following a suggestion of Woolf and by Herbig and Wolff (1).

If the structure of the star is not noticeably affected by the mass loss and if the abundance of lithium is sensibly constant down to the depth at which it burns quickly, then the efficacy of this process is governed by the rate of loss of mass and, in effect, the thickness of the convection zone.

Let M_c be the mass in the convection zone and M_1 be the mass in the shell between the bottom of the convection zone (in the sense of Sections 2 and 3) and the depth at which lithium burns. If the lithium abundance is approximately constant down to the depth at which lithium burns, then the mass-loss mechanism does not alter the observations of the lithium abundance until a mass M_1 has been lost, that is until a time t_1 defined by

$$\int_0^{t_1} |\dot{M}| \, dt = M_1 .$$

Here $\dot{M}$ is the rate of mass loss and I have assumed that the process begins when the star arrives on the main sequence ($t=0$). (Though pre-main-sequence effects could clearly be important, not enough appears to be known to estimate them.) If $\dot{M}$ is constant in time, and we consider the solar case where $\dot{M} \sim -10^{-14} \, M_\odot/\text{yr}$ and $M_1 \sim 10^{-2} \, M_\odot$, then $t_1 \sim 10^{12}$ yr. Thus, the process would not even have gotten started in the Sun. However, M_1 is highly uncertain and perhaps it is now being overestimated. But it is more likely that, as has been considered (1, 9), $\dot{M}$ is variable and was larger when the Sun first arrived on the main sequence.

This latter effect can be crudely estimated if we make use of Kraft's (62) recent study of rotation in main-sequence stars. At the time of this writing I have available only his results for the average rotational velocities in the Pleiades and Hyades G stars, namely $V = 19$ and $7 \cdot 9$ km/sec. In the Sun, however, $V = 2$ km/sec. The ages for these three samples of G stars are $\sim 5 \times 10^7$, 5×10^8, and 6×10^9 yr respectively. As Conti has remarked (59), if we treat these data as representative of the time dependence of G-star rotation, we see the half-life of the rotational velocity is time-dependent. Nor is this surprising.

Solar wind theory indicates that the rotational velocity (assuming a fixed structure) varies according to

$$\frac{dV}{dt} = \frac{\dot{M} R_A^2}{M R^2} V,$$

where R_A is the radius to the Alfvén point, M is the stellar mass, and R is the radius of the star. The Alfvén point occurs essentially where the magnetic-field strength has decreased so that it no longer forces the escaping gas to co-rotate with the star. In this formula, I have omitted a factor of order unity which depends on the stellar model, and another factor discussed by Mestel (35). Mestel points out that though an increase in magnetic field increases R_A, it likewise tends to inhibit escape from the equatorial regions, and that the braking is not indefinitely increased as the magnetic field increases. But there is a further effect to consider and that is the role of magnetic fields in the generation and propagation of the waves heating the corona. If the magnetic field causes enhanced heating, then the increase in mass loss from this effect may compensate for Mestel's effect and we could still expect magnetic braking to increase with increasing field.

As to the field itself, it seems very likely that, at least in the solar case it originates in a dynamo mechanism (63). In that case, the field strength will depend on rotation and probably like Ω^2, as Cowling (64) has pointed out. Thus, for a given spectral type both $\dot{M}$ and R_A^2 should depend on rotational velocity. Let us try to parameterize this dependence with the formula

$$\frac{\dot{M}R_A^2}{MR^2} = -\alpha\left(\frac{V}{V_0}\right)^n$$

where V_0 is the rotational velocity when the star arrives on the main sequence and α and n are constants. Then, we find,

$$V^n = \frac{V_0^n}{1 + \alpha n t}.$$

Current estimates give $MR^2/(\dot{M}R_A^2) \sim 5 \times 10^9$ yrs, as the present half-life of solar angular velocity (34) – remarkably coincident with the present age. The above formulae give an instantaneous half-life $(1 + \alpha n t)/\alpha$, which for $\alpha n t \gg 1$ is nt. Hence, if $\alpha^{-1} \ll$ age of the Sun and $n \approx 1$, we recover this coincidence. In fact with $n = 1$, $\alpha^{-1} \approx 5 \times 10^9$ yrs, $V_0 \approx 21$ km/sec, we obtain a passable representation of the data for G stars.

We are now in a position to reconsider the mass-loss mechanism for lithium destruction. To get an upper bound let us assume that all the variation in $\dot{M}R_A^2$ is due to variation in $\dot{M}$. With this assumption we readily find

$$t_1 = \frac{1}{\alpha}\left[\exp\left(\frac{M_1 R_A^2}{MR^2}\right) - 1\right].$$

With $R_A \sim 10R$, and $M_1 \sim 10^{-2}M$, $t_1 = 10^9$ yrs. Thus solar lithium depletion by mass loss could be detectable with the extreme assumption that $\dot{M}$ depends on magnetic fields and thus rotation, while R_A does not. But I think that for G stars we must conclude, that as far as one can tell, the possibility that the mass loss by itself depletes lithium is marginal. On the other hand, the indications from Kraft's observations are that in G stars newly arrived on the main sequence, rotational mixing can be quite important and this, combined with mass loss, may give a reasonable depletion rate. In any case, as Dr. Woolf has pointed out to me, the detection of lithium in certain stars may permit us to place upper bounds on the mass loss. In this connection it would be useful to know how α varies with spectral type.

If t_1 is as low as the extreme estimate suggests, then during the main-sequence lifetime of a G star the lithium abundance will diminish. That is, for $t > t_1$, the mass coming into the convection zone to replace that lost in the wind will be lithium-poor. The lithium abundance, A, will then be governed by

$$\frac{\mathrm{d}A}{\mathrm{d}t} = \frac{\dot{M}}{M} A.$$

Since we used the empirical extreme for $\dot{M}$ to get t_1 above, let us continue the illustration by using it again, though I would not wish to put great weight on this kind of treatment. We then obtain

$$A = A_0 \left(\frac{1 + \alpha t_1}{1 + \alpha t} \right)^{MR^2/M_c R_A^2} \qquad \text{for} \quad t \geqslant t_1,$$

where A_0 is the initial abundance. With $MR^2/(M_c R_A^2) \sim 1$ (probably an overestimate) we find that at $t \sim 5 \times 10^9$ yrs, $A \sim \frac{1}{3} A_0$, which is not sufficient to explain the solar lithium abundance. Nevertheless, there is clearly interest in the mass-loss mechanism and as relevant observations such as those of rotation are increased, the picture should be clarified. For example, if the magnetic field does vary like V^2 we would expect $H = H_0 (1 + \alpha t)^{-2}$, and thus G stars in the Pleiades should have mean magnetic fields 10^2 greater than the Sun. The determination of such data will certainly be very helpful in refining these estimates. And if the mass-loss mechanism does work, the correlation between rotation and lithium abundance is to be expected while the conflict with the magnetic problem at the end of Section 4 does not arise.

References

1. Herbig, G.H., Wolff, R.J. (1966)   *Ann. Astrophys.*, **29**, 593.
2. Böhm, K.H. (1966)   *Zts. f. Naturforsch.*, **21**, 1107.
3. Hayashi, C., Nakano, T. (1963)   *Prog. Theor. Phys.*, **30**, 460.
4. Ezer, D., Cameron, A.G.W. (1965)   *Can. J. Phys.*, **43**, 1497.
5. Bodenheimer, P. (1965)   *Astrophys. J.*, **142**, 451.
6. Böhm-Vitense, E. (1958)   *Z. Astroph.*, **46**, 108.
7. Baker, N,H., Temesvary, S. (1966)   Tables of Convective Stellar Envelope Models, 2nd ed., Report from Goddard Inst. for Space Studies.
8. Schwarzschild, M. (1958)   *Structure and Evolution of the Stars*, Princeton University Press, Princeton.
9. Weymann, R., Sears, R.L., (1965)   *Astrophys. J.*, **142**, 174.
10. Veronis, G. (1963)   *Astrophys. J.*, **137**, 641.
11. Iben, I. (1965)   *Astrophys. J.*, **142**, 1447.
12. Noyes, R.W. (1967)   in *Aerodynamic Phenomena in Stellar Atmospheres*, Ed. by R.N. Thomas, Academic Press, New York, p. 293.
13. Leighton, R. (1963)   *A. Rev. Astr. Astrophys.*, **1**.
14. Bray, R.J., Loughead, R.E. (1967)   *The Solar Granulation*, Chapman and Hall, Ltd., London.
15. Spiegel, E.A. (1966)   *Trans. I.A.U.*, **12B**, 539.
16. Simon, G., Weiss, N.O.   preprint.
17. Böhm, K.H., Stückl, E. (1967)   *Z. Astroph.*, **66**, 487.

18. Baker, N.H., Spiegel, E.A.   unpublished.
19. Proudman, I. (1956)   *J. Fluid Mech.*, **1**, 505.
20. Stewartson, K. (1957)   *J. Fluid Mech.*, **3**, 17.
21. Moore, D.W., Fultz, D.   unpublished film.
22. Hide, R. (1962)   'Some Thoughts on Rotating Fluids'. M.I.T. mimeographed notes.
23. Vitense, E. (1953)   *Z. Astroph.*, **32**, 135.
24. Saslaw, W.C., Schwarzschild, M. (1965)   *Astrophys. J.*, **142**, 1468.
25. Moore, D.W. (1967)   in *Aerodynamic Phenomena in Stellar Atmospheres*, Ed. by R.N. Thomas, p. 405.
26. Morton, B.R., Taylor, G.I., Turner, J.S. (1956)   *Proc. Roy. Soc.*, **A234**, 1.
27. Scorer, R.S. (1957)   *J. Fluid Mech.*, **2**, 583.
28. Böhm, K.H. (1963)   *Astrophys. J.*, **137**, 881.
29. Böhm, K.H. (1963)   *Astrophys. J.*, **138**, 298.
30. Böhm, K.H. (1967)   in *Aerodynamic Phenomena in Stellar Astmopheres*, Ed. by R.N. Thomas, Academic Press, New York, p. 367.
31. Mestel, L. (1965)   in *Stars and Stellar Systems*, Ed. by Aller and McLaughlin, Univ. of Chicago Press, Chicago 8.
32. Schatzman, E. (1962)   *Ann. Astrophys.*, **25**, 18.
33. Brandt, J.C. (1966)   *Astrophys. J.*, **144**, 1221.
34. Weber, E.J., Davis, L., Jr. (1967)   *Astrophys. J.*, **148**, 217.
35. Mestel, L.   *Observatory* (in press).
36. Dicke, R.H. (1964)   *Nature*, **202**, 432.
37. Roxburgh, I.W. (1965)   *Icarus*, **4**, 549.
38. Deutch, A.J. (1967)   *Science*, **156**, 236.
39. Goldreich, P., Schubert, G.   *Astrophys. J.* (in press).
40. Howard, L.N., Moore, D.W., Spiegel, E.A. (1967)   *Nature*, **214**, 5095.
41. Rayleigh, Lord (1920)   *Scientific Papers*, vol. VI, Cambridge, p. 447.
42. Chandrasekhar, S. (1961)   *Hydrodynamic and Hydromagnetic Stability*, Clarendon Press, Oxford.
43. Taylor, G.I. (1923)   *Phil. Trans. Roy. Soc.* (London), **A223**, 289.
44. Rintel, L. (1961)   Thesis, Technion.
45. Lieber, R., Rintel, L. (1964)   *Trans. A.G.U.*
46. Townsend, A.A. (1958)   *J. Fluid Mech.*, **4**, 361.
47. Moore, D.W., Spiegel, E.A. (1964)   *Astrophys. J.*, **139**, 48.
48. Barcilon, V., Pedlosky, J. (1967)   *J. Fluid Mech.*, **29**, 1.
49. Barcilon, V., Pedlosky, J. (1967)   *J. Fluid Mech.*, **29**, 609.
50. Dicke, R.H. (1967)   *Astrophys. J.*, **149**, L121.
51. Einstein, A.   Essay on meandering of rivers in *The World As I See It*.
52. Bondi, H., Lyttleton, R.A. (1948)   *Proc. Camb. Phil. Soc.*, **44**, 345.
53. Charney, J.G., Eliassen, A. (1949)   *Tellus*, **1**, 38.
54. Greenspan, H., Howard, L.N. (1963)   *J. Fluid Mech.*, **17**, 385.
55. Prandtl, L. (1952)   *Essentials of Fluid Dynamics*, Blackie and Son, Ltd., London, p. 356.
56. Bretherton, F., Spiegel, E.A.   to be published.
57. Holton, J.R. (1965)   *J. Atmos. Sci.*, **22**, 402.
58. Pedlosky, J. (1967)   *J. Fluid Mech.*, **28**, 463.
59. Conti, P.S.   preprint.
60. Babcock, H.W. (1961)   *Astrophys. J.*, **133**, 572.
61. Cowling, T.G. (1953)   in *The Sun*, Ed. by G.P. Kuiper, Univ. of Chicago Press, Chicago.
62. Kraft, R.P.   *Astrophys. J.* (in press).
63. Mestel, L.   *Plasma Astrophysics*, 39th Enrico Fermi School (to be published).
64. Cowling, T.G. (1965)   in *Stellar and Solar Magnetic Fields*, Ed. by R. Lüst, North-Holland Publ. Co., Amsterdam.
65. Fricke, K. (1967)   Dissertation, Göttingen.

D. JOINT DISCUSSION OF COMMISSIONS 7, 8, 19, 20, AND 33 MODERN PROBLEMS OF FUNDAMENTAL ASTROMETRY

Wednesday, August 30, 1967

Organising Committee: R. H. Stoy (Chairman), A. Blaauw, G. M. Clemence, J. Delhaye, W. Fricke, F. P. Scott, M. S. Zverev
Meeting Secretaries: W. Gliese, C. A. Murray, G. van Herk

Contents:

INTRODUCTION

As is no doubt the case for all Joint Discussions, the Organising Committee was faced with an embarrassing array of potential subjects and speakers from which to choose combined with the knowledge that any choice it might make could not possibly do justice to more than a small fragment of the subject in the relatively short time available. Some compromise was necessary and it was finally decided to divide the allotted time between three main topics avoiding as far as possible duplication of material already included in the Draft Reports or presented at the various Commission meetings here in Prague.

The papers in the first group summarise the progress that has been made with current cooperative astrometric programmes, and review the evidence available as to the accuracy of the FK4 system. The second group of papers refer to the problems of fundamental proper motions and their applications, while the third group deals with the method of overlapping plates.

This latter topic was deliberately chosen with the intention of focussing attention on what promises to be a most powerful method for solving the most critical modern problem of fundamental astrometry. This is the shortage of astronomers willing to devote themselves to the long hard grind of traditional meridian astronomy with its apparently small and unspectacular returns. While it is true that electronic computers have removed much of the drudgery from the reductions and that the conventional transit circle appears to be entering a phase of rapid development, as was amply demonstrated during the third meeting of Commission 8, it is also true that the new automatic micrometers are not yielding so spectacular an increase in intrinsic accuracy as they are in convenience. Moreover, the indications are that the number of these new transit circles and observers is never likely to be sufficient to cope with more than the semi-continuous observation of strictly limited lists of fundamental stars. In these circumstances the development of powerful photographic methods which promise increased accuracy and a great reduction in the number of fundamental stars required for calibration is a matter of great importance.

From its introduction into astrometry over 80 years ago, photography has tended to be used for mass-production work of rather limited accuracy. In the early years of this century, some of those engaged on the 'Carte du Ciel' project did realise the possibilities offered by overlapping plates but were in no position to exploit them. The situation has been completely revolutionised recently, however, by the advent of high-speed, high-capacity electronic computers combined with automatic measuring machines, while specially designed modern lenses give images of far higher quality and over wider fields than did those available in 1900.

Perek (ed.), Highlights of Astronomy, 278. © I.A.U.

THE AGK3, SRS AND RELATED PROJECTS

Francis P. Scott

1. Introduction

This report consists of a brief summary of the status of a number of programs of an international nature which are now in progress at various observatories of the world. These programs have received strong endorsements from Commission 8 of this Union, as well as from a number of international organizations interested in accurate positions and motions of the stars to the 9th magnitude of brightness. Among the organizations and conferences which have explicitly endorsed the programs are:

(1) The Symposium on Coordination of Galactic Research, Groningen, 1953.

(2) Conference on the Problems of Astrometry, Evanston, 1953.

(3) Astrometric Conference, Pulkovo, 1954.

(4) Conference on Astrometry, Brussels, 1955.

(5) International Union of Geodesy and Geophysics, 1954 and 1963.

(6) International Astronomical Union, 1955.

(7) Second Astrometric Conference, Cincinnati, 1959.

(8) First Inter-American Conference on Astronomy, La Plata and Cordoba, 1959.

(9) Meeting on Problems of Astrometry and Celestial Mechanics, La Plata, 1960.

(10) Committee on Space Research, COSPAR, 1966.

The principal purpose of the astrometric programs now in progress in both hemispheres of the sky is the determination of accurate positions and proper motions on a fundamental system for all stars to the 9th magnitude of brightness. The survey of each hemisphere of the sky comprises a number of interrelated programs designed to accomplish the following objectives:

(1) The improvement of the positions and motions of the fundamental stars.

(2) The relation of the positions of 20 000 reference stars in each hemisphere of the sky to the fundamental system by means of differential observations with meridian circles.

(3) The determination of photographic positions of modern epoch on a fundamental system for all stars to the 9th magnitude through the use of the reference stars in the reduction of the plates.

(4) The derivation of fundamental proper motions for all stars to the 9th magnitude through a comparison of the new photographic positions with similar positions obtained in previous photographic surveys.

The task of improving the positions of the fundamental stars and relating the

Perek (ed.), Highlights of Astronomy, 279–285. © I.A.U.

reference stars to the fundamental system is being carried out in two major programs, the AGK3R for the Northern sky and the Southern Reference Star Program (SRS) for the Southern sky.

2. Northern Reference Star Program, the AGK3R

The AGK3R program resulted directly from resolutions adopted at the Conference on Problems in Astrometry held at Evanston, Ill., September 3–5, 1953. Decisions of an organizational nature, the adoption of the star list, and the division of the star list among the observers were accomplished at a Conference on Astrometry held at Brussels, March 28–30, 1955.

The coordination of the AGK3R Work was carried out by a Committee appointed by Commission 8 during the Dublin Meeting of the IAU in 1955.

Both the Evanston and Brussels Conferences, as well as the Meetings of Commission 8 at Dublin, strongly recommended the revision of the FK3 so as to bring the fundamental coordinate system up to date for the purpose of the AGK3R program. This task was accomplished in 1963 with the publication of the FK4. All work on the revision of the fundamental catalog was done at the 'Astronomisches Rechen-Institut' of Heidelberg under the supervision of the late Dr. Kopff and the present Director, Dr. Fricke.

All results of the AGK3R program were reduced to the FK4 either by the observers themselves or at the U.S. Naval Observatory during the course of compiling the final catalogue.

Observations of the AGK3R stars began in 1956 and were essentially completed by 1963. By the end of 1964 all results had been received at the U.S. Naval Observatory for use in the compilation of a final catalogue of positions of the northern reference stars.

The observations of the AGK3R stars were carried out with meridian circles at Bergedorf, Bordeaux, Heidelberg, Herstmonceux, Nicolaiev, Ottawa, Paris, Pulkovo, Strasbourg and two instruments at the U.S. Naval Observatory. During the course of the program well over 300000 apparent places were computed at the U.S. Naval Observatory for the various observatories taking part in the program.

The final catalog has just been completed at the U.S. Naval Observatory. It contains the positions of 21499 reference stars between $-5°$ and $+90°$ declination. Since this work was completed so shortly before the beginning of the present meeting of the IAU, there has been no opportunity to compare it with other catalogs or to make other investigations of the characteristics of the final results. It has been possible, however, to develop the mean errors in both right ascension and declination, which are shown in Tables 1 and 2.

To understand these tables it should be recalled that the star list adopted at Brussels consisted of a combination of two star lists; one was the KSZ list selected at the

Table 1

AGK3R: mean errors in right ascension

Unit: m.e. cos $\delta = 0\overset{s}{.}0001$

	(C + S) All Stars				Z Stars, $M \leqslant 8\cdot8$				Z Stars, $M \geqslant 8\cdot9$			
Declination	No. Stars	Avg.	50%	95%	No. Stars	Avg.	50%	95%	No. Stars	Avg.	50%	95%
− 5° to + 5°	2154	54	52	93	734	61	56	124	475	71	62	144
+ 5° +15°	2168	50	48	88	703	56	51	110	501	63	58	128
+15° +20°	1089	49	46	90	349	47	44	88	210	58	54	119
+20° +25°	1010	48	46	89	342	51	49	95	258	56	50	113
+25° +40°	2763	44	42	80	1187	47	44	91	507	48	45	99
+40° +50°	1573	49	46	87	629	53	50	105	235	54	51	103
+50° +70°	2252	45	43	78	852	47	44	94	317	43	41	87
+70° +90°	802	49	47	85	261	51	48	99	132	48	43	96

All Stars: Avg. = 50; 50% = 47; 95% = 95.

Table 2

AGK3R: mean errors in declination

Unit: m.e. = $0\overset{''}{.}001$

	(C + S) All Stars				Z Stars, $M \leqslant 8\cdot8$				Z Stars, $M \geqslant 8\cdot9$			
Declination	No. Stars	Avg.	50%	95%	No. Stars	Avg.	50%	95%	No. Stars	Avg.	50%	95%
− 5° to + 5°	2154	115	112	194	734	121	108	252	475	139	126	290
+ 5° +15°	2168	108	103	186	703	108	102	204	501	127	116	264
+15° +20°	1089	112	109	194	349	113	107	217	210	137	126	264
+20° +25°	1010	113	107	204	342	102	094	201	258	125	117	250
+25° +40°	2763	117	111	210	1187	103	095	196	507	121	114	237
+40° +50°	1573	121	116	204	629	118	108	240	235	142	130	282
+50° +70°	2252	116	111	201	852	105	096	204	317	116	105	240
+70° +90°	802	134	129	225	261	124	112	259	132	133	124	272

All Stars: Avg. = 116; 50% = 109; 95% = 214.

Pulkovo and Sternberg Observatories for use in the U.S.S.R. program for relating the proper motions of the stars to the background galaxies, and the other was a list selected at the U.S. Naval Observatory for use in the reduction of photographic plates exposed with an objective grating which would produce a 3·5 magnitude difference between the central image of a star and its first order spectrum. The C stars indicated in the tables are those that were common to both lists. The Z stars are those that remained in the KSZ list after the removal of the C stars, and the S stars are those that remained in the U.S. Naval Observatory list after the removal of the C stars.

The C and S stars were observed 10 times each during the program, whereas, in general, the Z stars were observed only 8 times each, except for the bright Z stars in the zone $+5°$ to $+40°$.

The divisions of the tables of mean errors according to declination were made at points where changes took place in the combination of observatories committed to the program.

3. Southern Reference Star Program, SRS

The Southern Reference Star Program was organized by a Committee appointed at the Moscow 1958 Meeting of the IAU. This Committee held or participated in three Conferences to explore the possibility of organizing a program in the Southern Hemisphere similar to the AGK3R. These Conferences were held at Cincinnati, May 17–21, 1959, La Plata and Cordoba, October 30 – November 3, 1959, and at La Plata, November 7–11, 1960. In addition to the Conferences, considerable correspondence with many observatories was conducted in an effort to solicit commitments to the program.

The SRS program is not as elegantly organized as the AGK3R, nor are all the observations being made simultaneously at all observatories as they were during the AGK3R. It is hoped, however, that these shortcomings will not be detrimental to the results.

It is very fortunate that a number of Northern observatories, working from their home stations and at very inconvenient zenith distances, have agreed to observe as

Table 3

Status of the SRS Program, June 1, 1967

Observatory	Zone	Commitment Stars observations		Date Started mo. yr.		Observations Completed
Abbadia	+ 5° to − 15°	1560	4	4	62	90%
Bordeaux	+ 5 to − 15	1560	4	6	62	72
Bucharest	+ 5 to − 10	1176	4	7	62	100
Nicolaiev	0 to − 20	5984	2	3	64	100
San Fernando	− 10 to − 30	3709	4	3	63	45
Tokyo	− 10 to − 30	3560	4	3	63	95
6-in. N.O.	+ 5 to − 20	{6450 2} {1233 4}		9	66	15
Cape	{− 30 to − 40} {− 40 to − 52} {− 52 to − 90}	10082	4	4	61	{45} {100} {0}
Santiago-Pulkovo	{− 25 to − 47} {− 47 to − 90}	11496	4	1	63	{82} {100}
Bergedorf-Perth	0 to − 90	To start late 1967				
7-in. N.O. El Leoncito	+ 5 to − 90	To start mid 1967				

far South as $-20°$ and in two cases to $-30°$ declination. It is also very fortunate that the Pulkovo, Bergedorf, and U.S. Naval Observatories found it possible to undertake expensive expeditions to the Southern hemisphere. We should also be very thankful that the Cape Observatory threw its full weight into the program. Without its help it would have been difficult to have had a program at all.

The status of the Southern Reference Star observations as of June 1, 1967, is shown in Table 3.

4. Reference Star Lists

As mentioned earlier the Northern Reference Star List consisted of a combination of a star list selected at the U.S. Naval Observatory and the KSZ list.

The list between $0°$ and $-30°$ was compiled at the U.S. Naval Observatory by adding stars in accordance with the AGK3R criteria to the KSZ list, which extended to $-30°$ declination, to bring it up to the density of the AGK3R. The star list South

Table 4

Distribution of the reference stars by magnitude and declination

Magnitude	AGK3R list		SRS list	
	$+90°$ to $+30°$	$+30°$ to $0°$	$0°$ to $-30°$	$-30°$ to $-90°$
$m \leqslant 6·9$	8·1 %	6·5 %	3·8 %	0·4 %
$7·0 \leqslant m \leqslant 7·4$	13·0	10·3	8·6	3·9
$7·5 \leqslant m \leqslant 7·9$	19·3	17·2	15·0	13·5
$8·0 \leqslant m \leqslant 8·4$	25·0	26·7	30·6	23·6
$8·5 \leqslant m \leqslant 8·8$	22·7	23·2	34·3	27·3
$m = 8·9$	4·5	5·7	3·8	6·4
$m = 9·0$	4·2	5·1	3·1	9·6
$m = 9·1$	2·0	3·5	0·4	5·6
$m \geqslant 9·2$	1·2	1·8	0·2	9·8
No. stars	9891	9963	8586	10210

Table 5

Distribution of reference stars by spectral type and declination

Spectral type	AGK3R list		SRS list	
	$+90°$ to $+30°$	$+30°$ to $0°$	$0°$ to $-30°$	$-30°$ to $-90°$
B	3·4 %	3·2 %	3·4 %	0·2 %
A	16·1	16·2	14·6	3·1
F	15·0	17·0	15·5	8·2
G	15·8	16·4	15·7	19·9
K	42·7	40·0	48·3	64·0
M	3·3	2·6	2·3	4·3
Other	3·7	4·6	0·2	0·3
No. stars	9891	9963	8586	10210

of −30° was selected at the Cape Observatory by use of the KSZ criteria and then adding brighter stars until the density of the AGK3R was reached.

The distributions of the resulting star lists in each quarter of the sky, according to magnitude and spectral type, are shown in Tables 4 and 5.

5. Fundamental Star Catalog

It was mentioned earlier that the FK4 had been adopted as a reference coordinate system for the AGK3R program.

There is some doubt as to whether or not it will serve as an adequate system for the Southern Reference Stars.

In relating reference stars to a fundamental system one is interested in:

(1) The precision of the fundamental stars when updated to the epoch of observation of the reference stars.

(2) The possible systematic errors of the updated fundamental stars.

During the past few days Commission 8 has heard several reports that modern observations of the Southern fundamental stars do, indeed, indicate rather large systematic errors in the FK4. It may, therefore, be necessary to consider a revision of the positions of the Southern FK4 stars before compiling the catalog of final positions of the Southern Reference Stars. Material for such revision may be on hand at the completion of the SRS observing program. All SRS observers are, of course, generating relative corrections to the positions of the fundamental stars observed each night in conjunction with the reference stars. In addition, the Cape Observatory is continuing its traditional fundamental work and the Russian and Chilean observers at Santiago are carrying out fundamental series of observations of the southern FK4 stars by use of methods developed at Pulkovo. It is also planned that the Bergedorf and U.S. Naval Observatory expeditions to the Southern hemisphere will do limited programs of a fundamental nature. This material, along with accumulated astrolabe results, may be sufficient to effect the required improvement of the positions of the Southern Fundamental Stars.

6. Photographic Programs

AGK3: The principal purpose in observing the Northern Reference Stars was, of course, to provide reference stars for the reduction of the AGK3 photographic plates. These plates were taken at the Hamburg-Bergedorf Observatory during the years the reference stars were being observed. In all 1939, 5° × 5° plates were taken to cover the sky from the North pole to −5° declination, the last set of plates having been centered at −2°.5 declination.

All AGK2 stars appearing on these plates have been measured. The new AGK3 positions of mean epoch 1958 are being compared with the earlier AGK2 positions of mean epoch 1930 for the derivation of proper motions. For this purpose, the original AGK2 positions have been re-reduced to the system of the FK4. The final solutions may now be started through the use of the AGK3R catalog just completed. It is expected that the resulting positions will have mean errors of approximately $\pm 0''.14$ or $\pm 0''.15$. The mean errors of the proper motions derived from a comparison of the AGK2 and AGK3 positions are expected to be about $\pm 0''.8$/century. This work has advanced to the point where Professor Dieckvoss feels that the first volume of the results, covering the sky from $+90°$ to $+55°$, may be ready for the printer by the end of the year.

Southern sky: The photography of the Southern sky is being done by the Cape and Sydney Observatories. The Cape Observatory has completed the photographs of the zone $-40°$ to $-52°$ with a mean epoch of 1962. The measurement and a preliminary solution of these plates are in progress. The photography of the zone $-30°$ to $-40°$ commenced in March 1966 and is still in progress. The next zone to be undertaken at the Cape will be in the region $0°$ to $-30°$.

The Sydney Observatory is now engaged in the photography of the zone $-51°$ to $-65°$. Earlier, it exposed a set of plates extending from $-48°$ to $-54°$. The latter plates are now being measured at the University of South Florida.

In addition to the current work in the Southern hemisphere, the Cape and Yale Observatories are now finishing the measurement and discussion of earlier series of photographs. The Cape Observatory has but one more volume to publish of its first series of photographs of the Southern sky which started about 35 years ago. The remaining volume is for the South polar cap and includes the stars South of $-80°$ declination.

During 1941, while its instrument was in South Africa, the Yale Observatory photographed the zones $-40°$ to $-50°$ and $-60°$ to $-70°$. Later, about 1953, after its instrument had been moved to Australia, a set of plates was taken for the zones $-30°$ to $-40°$ and $-70°$ to $-90°$. These plates are now being reduced. The results for the region $-30°$ to $-35°$ are in press and those for $-35°$ to $-40°$ will be ready for the printer sometime next year. The remaining zones may require 2 or 3 more years before their reductions are completed.

DISCUSSION

W. Dieckvoss: The revision of the AGK2, which has been carried out at Bergedorf, included the application of systematic corrections to the original coordinates measured at Bonn and Bergedorf before making new plate solutions. The AGK2-revised is on the FK4 system and when published will render the present printed volumes of the AGK2 obsolete.

RECENT WORK ON FUNDAMENTAL ASTROMETRY IN THE U.S.S.R.

M. S. ZVEREV

Absolute determinations of star positions using transit and vertical circles according to the methods developed at Pulkovo are being made at Pulkovo, at Nikolaiev, at the Golossejevo Observatory of the Ukraine Academy of Sciences, and at Santiago de Chile. At this latter observatory Soviet and Chilean astronomers have completed a series of observations of both bright and faint fundamental stars with a photographic vertical circle, and these observations are now being reduced at Pulkovo. Absolute determinations of right ascension will be made at the Cerro Calan Observatory using a large transit instrument recently constructed at Pulkovo. Further details of some of the Southern observations are given in the following paper to be presented by Dr. Anguita.

Independent determinations of right ascension have been made since the time of the International Geophysical Year by the Time Service of the U.S.S.R. using photo-electrically recording instruments and the methods proposed by Professor N. N. Pavlov (1958) and Professor A. A. Nemiro (1957). In these programmes a special effort is made to obtain observations of the zenith stars right through the night in order to derive an independent system of $\Delta\alpha_\alpha$.

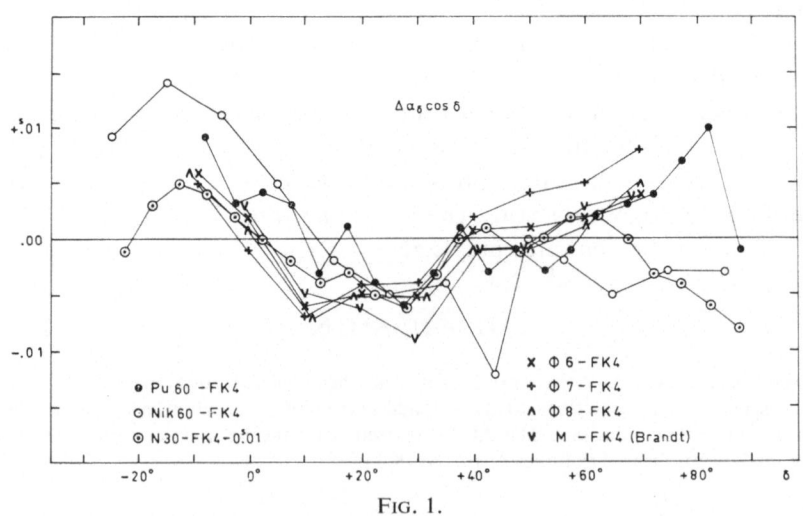

FIG. 1.

Perek (ed.), Highlights of Astronomy, 286–291. © *I.U.A.*

The azimuth is determined both by combining observations of zenith stars at upper and lower culminations and by combining observations of zenith and equatorial stars. (Pavlov's method is only suitable for latitudes higher than 50°.) The catalogues resulting from these observations of the Time Services seem to be very accurate both as regards systematic and accidental errors (Pulkovo, 1966; Brandt, 1963).

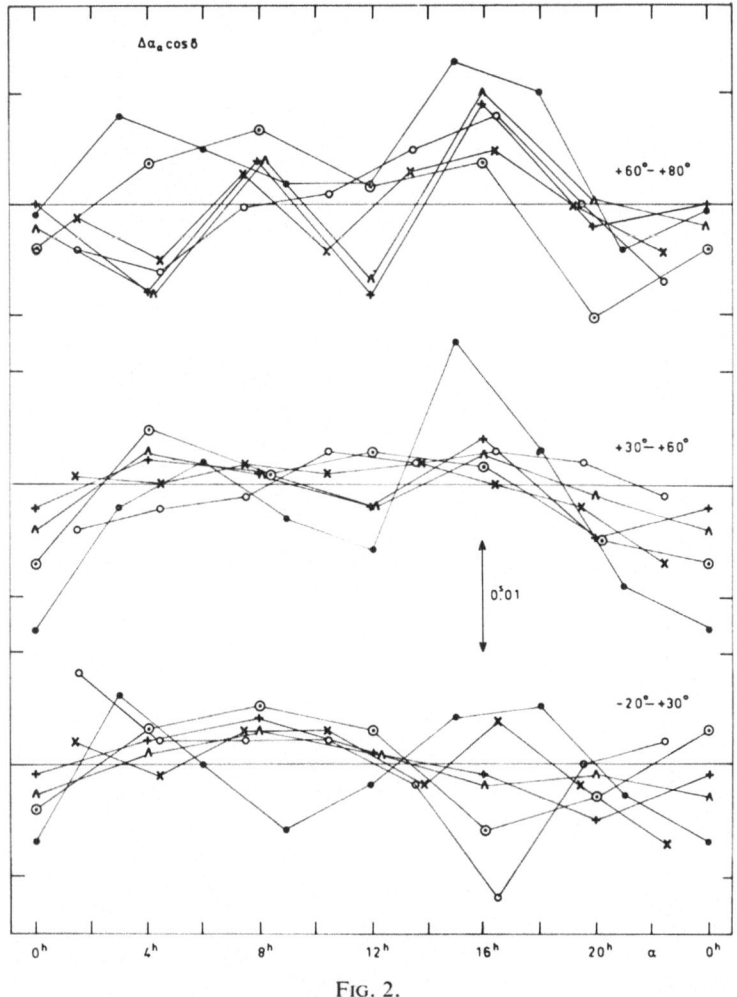

FIG. 2.

Systematic differences at 1960·0 between some of these recent absolute catalogues and FK4 are shown in Figures 1–4, in which the data for Pu 60 must be regarded as preliminary. It will be seen that the new catalogues are in fairly good agreement with one another and that they agree rather better with N30 than with FK4, a fact which

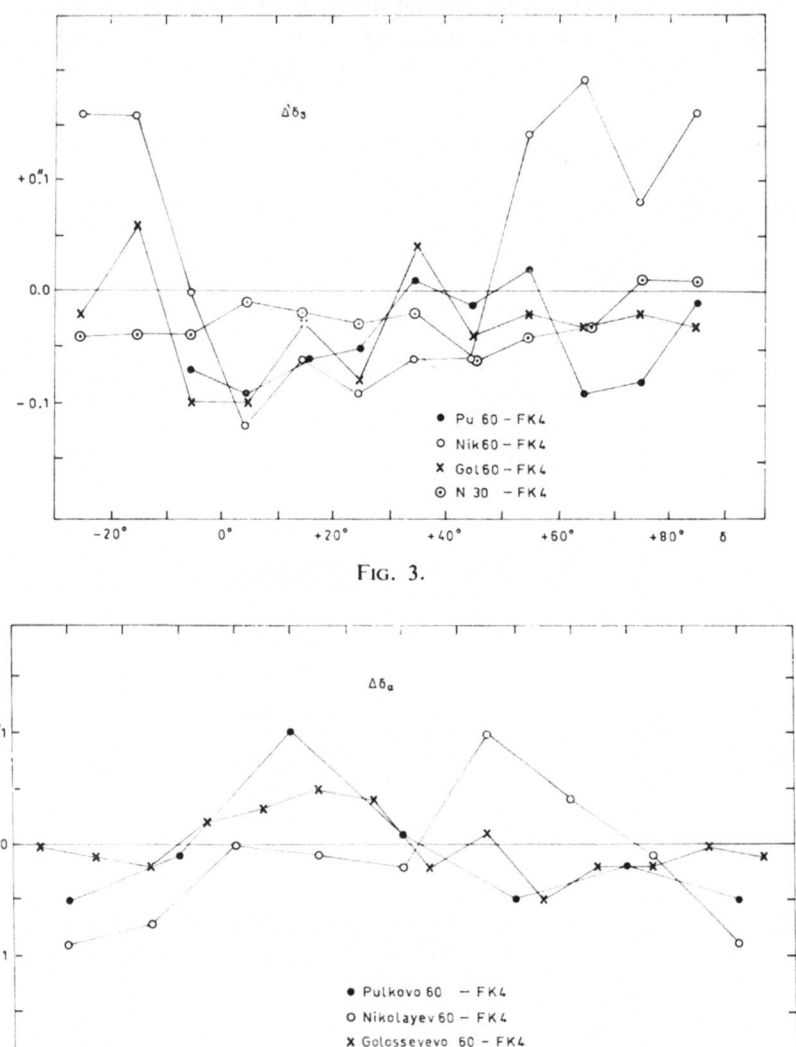

FIG. 3.

FIG. 4.

seems to justify the methods used by Dr. H. R. Morgan for the determination of
proper motions during the compilation of N30.

As may be seen from Professor Nemiro's report to Commission 8, the Russian
observatories are taking their full share in all the international programmes. Observations are also being made of special groups of stars such as those in the areas surrounding the galaxies selected by Professor Deutsch (Bugoslavskaja *et al.*, 1955).

Stars in the North polar area are receiving special attention since they are relatively few in number and the systematic errors in their observed positions often differ considerably from those for other zones. About ten catalogues of the right ascensions of North circumpolar stars have recently become available, some of them as a result of the reductions of observations made many years ago, for instance by W. Fabritius at Kiev about 1880. Most of these catalogues can be regarded as absolute so that they should greatly strengthen the right ascension system in the Northern sky.

Observations and investigations in connection with the Catalogue of Faint Stars (KSZ) have been continued along the lines reported by myself to Commission 8 at

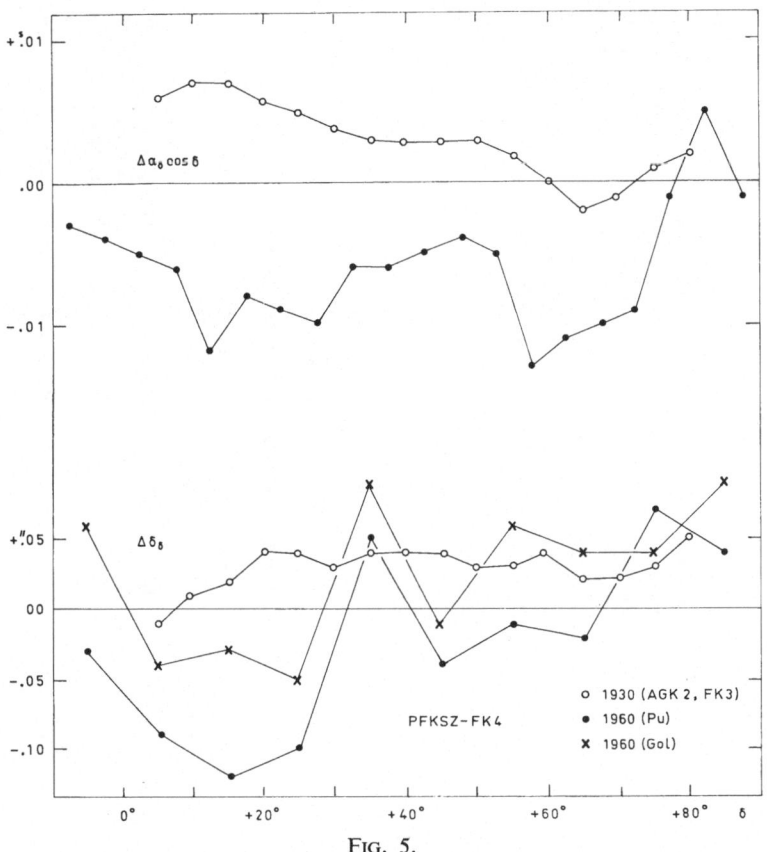

FIG. 5.

Hamburg (IAU, 1966) and by Dr. V.I. Orelskaja and Professor A.N. Deutsch here in Prague. Later today Dr. N.V. Fatčihin will be telling us about the first Pulkovo results from the programme for obtaining absolute proper motions from observations of galaxies.

Bright stars can be related to the galaxies or to minor planet orbits only through the intermediary of faint stars. The determination of the relationship between the

Table 1

Catalogues of the FKSZ Stars ($\delta > -20°$)

No.	Observatory	Coordi-nates	Zone	Number of stars	obs.	Instrument	Observer	Years	Notes
1	Golossejevo	δ	$-26° + 90°$	611	7	VC Wanschaff	Korol	1954–62	Abs.
2	Moscow	α, δ	$-21° + 90°$	589	3	MC Repsold	Oborneva et al.	1953–58	
3	Pulkovo	α	$-10° + 90°$	513	12	GTI Oertel	Nemiro et al.	1954–61	Abs.
4	Pulkovo	δ	$-10° + 90°$	531	9	VC Oertel	Kossin et al.	1955–62	Abs.
5	Pulkovo	α	$-20° + 90°$	587	3	MC Toepfer	Polojentsev et al.	1965–66	
6	Taškent	α	$-30° - 90°$	623	13	MC Repsold	Bykov	1946–52	Abs.
7	Kharkov	δ	$-20° + 90°$	600	4	MC Repsold	Mikhailov	1949–55	

8–21 14 catalogues (1940–58) used for the PFKSZ ($n_\alpha = 29$; $n_\delta = 37$).

Table 2

Catalogues of the KSZ Stars ($\delta > -20°$)

No.	Observatory	Coordi-nates	Zone	Number of stars	obs.	Instrument	Observer	Years
1	Bucharest	α, δ	$-10° + 10°$	3859	4	MC Gautier	Marcus et al.	1955–62
2	Kazan	δ	$+50° + 90°$	2288	2, 5	MC Repsold	Nefedieva	1955–61
3	Kiev	α, δ	$+10° + 30°$	3347	2, 5	MC Repsold	Drofa, Chernega	1956–61
4	Moscow	α, δ	$+30° + 90°$	5257	2	MC Repsold	Guliaev et al.	1957–65
5	Odessa	α	$- 5° - 25°$	2963	4	MC Repsold	Novopashenny	1956–63
6	Odessa	δ	$- 5° - 20°$	2500	4	MC Repsold	Cherniev et al.	1954–63
7	Taškent	α	$- 5° - 20°$	2488	4	MC Repsold	Bykov et al.	1957–61
8	Nikolajev	α, δ	$- 5° - 20°$	2500	4	MC Repsold	Gordon et al.	1956–63

9–19 11 catalogues of the AGK3R stars (1956–62) $n \approx 10$.

position and proper motion systems of the bright and faint stars is thus one of the most important problems of modern astrometry. As FKSZ and FK4 have no stars in common, their systems must be compared either by direct observation or by using an intermediary catalogue.

We have tried both these methods for a preliminary comparison of the PFKSZ and FK4 systems, but, as may be seen from Figure 5, we did not obtain good results. Figure 5 shows the $\Delta\alpha_\delta$ and $\Delta\delta_\delta$ systematic differences between PFKSZ and FK4, as determined from new meridian observations made at Pulkovo and Golossejevo, and as determined by using AGK2 (epoch 1930·0) and applying the differences FK3–FK4. Part of the lack of agreement between the results obtained from the two methods may possibly be attributed to the assumed proper motions, but not all.

The improvement of the Preliminary Fundamental Catalogue of Faint Stars (PFKSZ), which was compiled about 10 years ago (Pulkovo, 1958), is an important and an urgent task. Tables 1 and 2 indicate what catalogues are available for the revision of the Northern section of the FKSZ and the KSZ. The first includes seven new catalogues, of which four are fundamental, and the second – eight. As the KSZ catalogues include FKSZ stars, there are available for the recompilation of the Northern part of the FKSZ 40 catalogues from 19 observatories with about 90 observations of each star in each coordinate. For the Southern part of the FKSZ, however, the only observations available are those from Cape, La Plata, and Santiago.

The Southern KSZ and FKSZ stars are included in the SRS programme, but observations made for this latter programme will not be sufficient for the compilation of a definitive FKSZ catalogue. I would strongly urge all Southern meridian observers to include the FKSZ stars in all their programmes, particularly in programmes of absolute observations.

References

Brandt, W.E. (1963)   *Rotation of the Earth*, Kiev, p. 64.
Bugoslavskaja, E.J., Karimova, D.K., Podobed, V.V., Jakhontov, K.N. (1955)   *Publ. of the 11th Astrometric Conference of the U.S.S.R.*, Leningrad, p. 42.
IAU (1966)   *Trans. IAU*, **XII-B**, p. 124–125.
Nemiro, A.A. (1957)   *Pulkovo Bull.*, **157**.
Pavlov, N.N. (1958)   *Pulkovo Bull.*, **161**.
Pulkovo (1958)   *Publ.*, **72**.
Pulkovo (1966)   *Publ.*, **75**.

MERIDIAN OBSERVATIONS AT SANTIAGO, CHILE

C. Anguita, G. Carrasco, P. Loyola, V. N. Šiškina and M. S. Zverev

Observations of the Southern Reference Star (SRS) and Bright Star (BS) programmes were started in January 1963 with the Repsold meridian circle of the Santiago Observatory. Up to July 1, 1967, a total of 86725 star observations arranged in 1520 series had been observed.

In order to investigate the instrumental system, series of fundamental stars between $+41°$ and $-69°$ SP were included in the observing programme. A total of 214 such series were observed, 123 being with 'clamp East' and 91 with 'clamp West'. The number of individual star observations in these 214 series was 7344. These observations were reduced using Bessel's formula

$$\alpha = T + (u + m) + n \tan\delta + c' \sec\delta,$$

where u denotes the clock correction and c' includes the collimation error, diurnal aberration and the contact width. The value of n which best fits the observations between declinations $+30°$ and $-30°$ was calculated and denoted by n_e. For stars outside this zone, n was calculated from the expression

$$n = \{(\alpha - T) - \overline{(\alpha - T)_e}\}/(\tan\delta - \overline{\tan\delta_e})$$

and $\Delta n = (n - n_e)$ formed.

The differences Δn were grouped by zones of declination $5°$ wide and the mean value $\overline{\Delta n}$ calculated for each zone. The dependence of $\overline{\Delta n}$ on declination was determined graphically. In order to reduce the scatter, the sliding means between two neighbouring groups were plotted. The results for 'clamp East' and 'clamp West' are shown separately in Figure 1, the first curve depending on 2068 values and the second on 1762.

In order to investigate the dependence of $\overline{\Delta n}$ on right ascension, the observations were divided into four zones, each 6 hours wide and centered at 0, 6, 12 and 18 hours. The dependence of $\overline{\Delta n}$ on declination is shown, separately for each of these four right ascension zones, in Table 1. The systematic nature of the values given in this table suggest the need of corrections to the FK4 of the type $\Delta\alpha_\delta$ and $\Delta\alpha_\alpha$. The values of $\frac{1}{2}(\Delta n + \Delta n_{sp})$ and $\frac{1}{2}(\Delta n - \Delta n_{sp})$ are given in Table 2. The former may be used to estimate the $\Delta\alpha_\alpha$ errors in FK4 and the latter errors of the type $\Delta\alpha_\delta$ (Zverev, 1965).

The fact that the implied systematic errors are practically the same for both positions of the instrument makes it appear unlikely that we are dealing with an instrumental

Perek (ed.), Highlights of Astronomy, 292–296. © *I.A.U.*

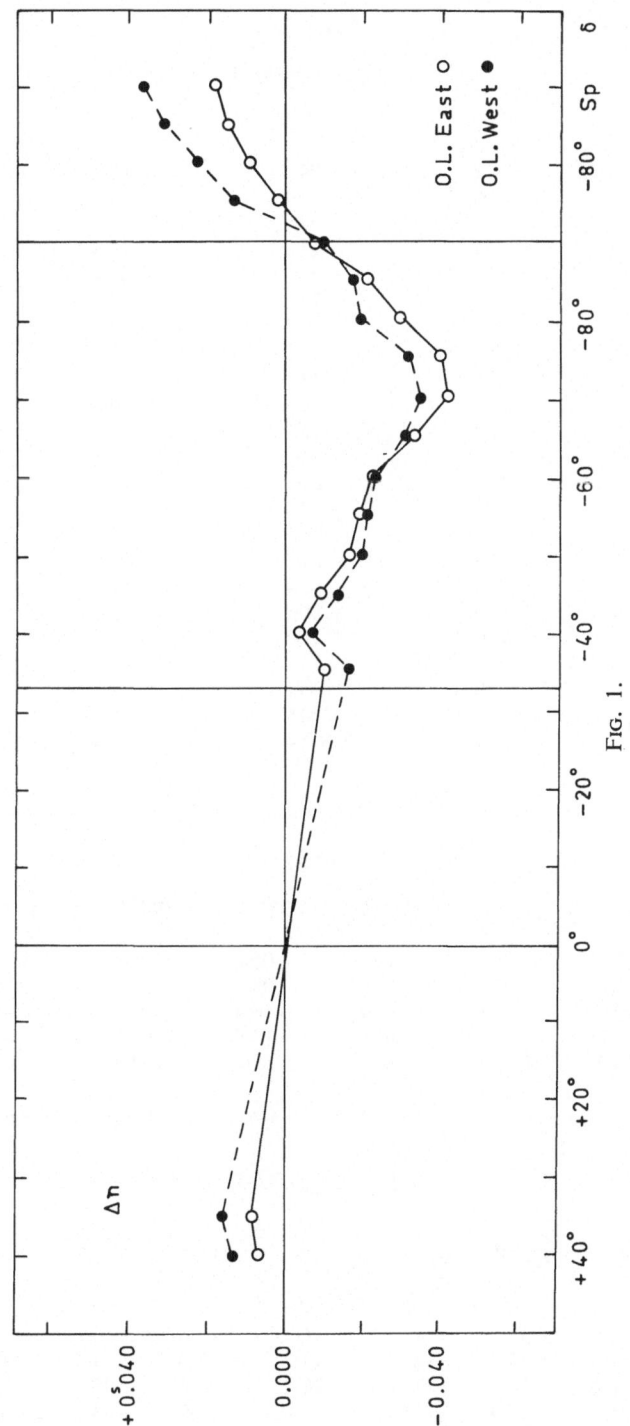

Fig. 1.

Table 1

$\Delta n_{\overline{(E+W)}}$ (0ˢ001)

δ \ α	0ʰ	6ʰ	12ʰ	18ʰ
+40°	+20·8	+04·4	+07·7	+06·3
+35°	+27·0	+06·1	+13·1	+08·7
−35°	−18·1	−12·7	−14·4	−07·9
−40°	−02·8	−08·7	−03·4	−04·0
−45°	−13·2	−15·2	−08·5	−04·5
−50°	−21·9	−16·7	−23·2	−09·2
−55°	−20·3	−22·0	−21·0	−19·5
−60°	−32·0	−22·6	−18·6	−33·2
−65°	−37·5	−29·8	−30·0	−41·1
−70°	−41·2	−36·7	−38·4	−42·8
−75°	−49·6	−37·7	−32·1	−29·5
−80°	−39·5	−24·0	−19·0	−20·6
−85°	−20·8	−18·9	−20·5	−23·3
−90°	−05·6	−04·9	−10·8	−09·6
−85°sp	+03·4	+13·6	+04·8	−01·8
−80°sp	+06·8	+18·4	+17·1	+08·9
−75°sp	+22·6	+21·7	+23·2	+22·3
−70°sp	+31·0	+27·0	+25·7	+43·8
Numbers of Series	30	75	65	27
Numbers of Δn	593	1413	1272	546
α_M	00ʰ1	06ʰ4	11ʰ4	17ʰ7

Table 2

$\frac{1}{2}(\Delta n + \Delta n_{sp})$ 0ˢ001

δ	E	W	$\overline{E+W}$	00ʰ1	06ʰ4	11ʰ4	17ʰ7
90°	−07·4	−08·6	−08·0	−05·6	−04·9	−10·8	−09·6
85°	−09·8	−02·7	−06·6	−08·7	−02·6	−07·8	−12·6
80°	−10·0	+01·6	−04·8	−16·4	−02·8	−01·0	−05·8
75°	−13·0	−00·2	−07·0	−13·5	−08·0	−04·4	−03·6
70°	−12·0	+00·1	−06·2	−05·1	−04·8	−06·4	+00·5
Mean	−10·4	−02·0	−06·5	−09·9	−04·6	−06·1	−06·2

$\frac{1}{2}(\Delta n - \Delta n_{sp})$ 0ˢ001

δ	E	W	$\overline{E+W}$	α: 00ʰ1	06ʰ4	11ʰ4	17ʰ7	Mean	$\Delta\alpha_\delta \cos\delta$
90°	0	0	0	0	0	0	0	0	0
85°	−11·5	−15·4	−13·3	−12·1	−16·2	−12·6	−10·8	−12·9	−12·9
80°	−18·8	−21·2	−19·2	−23·2	−21·2	−18·0	−14·8	−19·3	−19·0
75°	−27·6	−31·2	−29·2	−36·1	−29·7	−27·6	−25·9	−29·8	−28·8
70°	−29·5	−35·4	−32·4	−36·1	−31·8	−32·0	−43·3	−35·8	−33·6

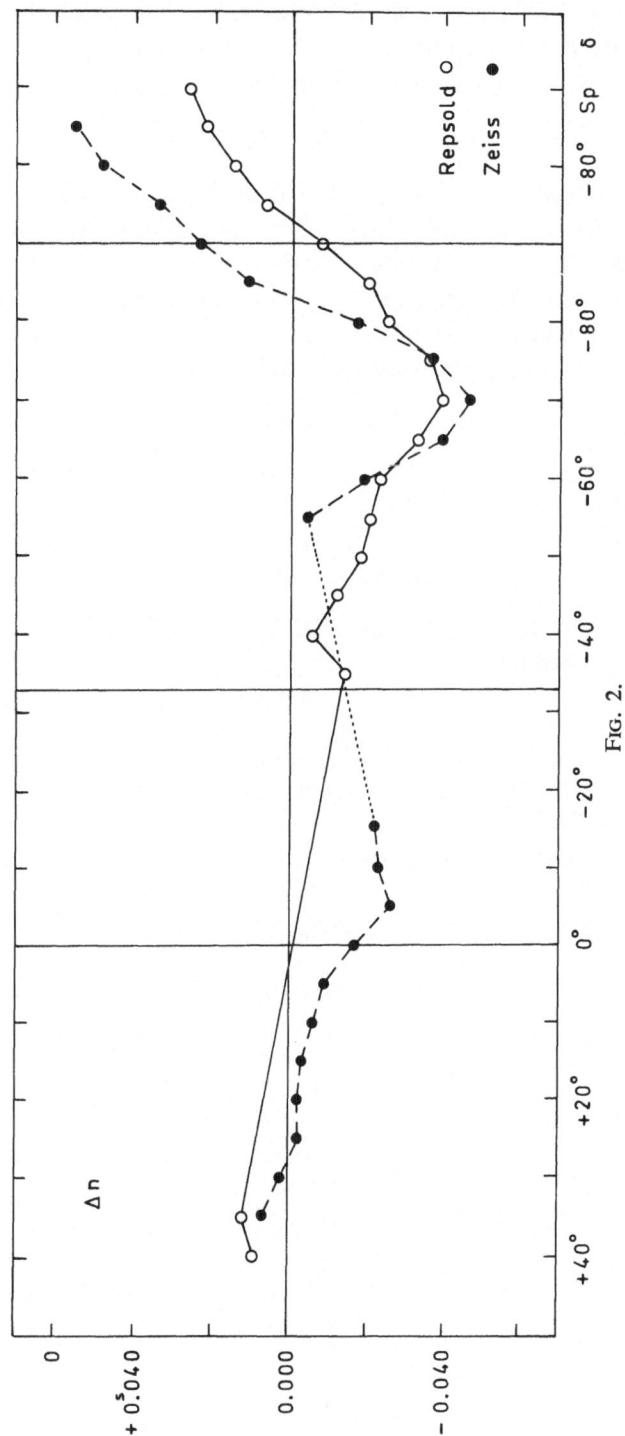

Fig. 2.

peculiarity. This can be further checked, however, by 41 series of observations made between April 1963 and September 1964 with a Zeiss broken transit.

The azimuth of this instrument was determined from star i by the expression

$$a_i = \{(\overline{\alpha - T})_{\bar{z}} - (\alpha - T)_i\}/(\bar{A}_{\bar{z}} - A_i),$$

where $(\overline{\alpha - T})_{\bar{z}}$ is the mean difference between the right ascension and transit time for stars near the zenith, the time being already corrected for contact width, level and diurnal aberration. $A_{\bar{z}}$ and A_i denote the azimuth coefficients in Mayer's formula.

A total of 1372 azimuth determinations were obtained for this instrument. The mean azimuth a_N found from stars near declination $+30°$ was used as a reference azimuth and individual differences Δa were calculated from

$$\Delta a_i = a_N - a_i.$$

As in the case of the meridian circle observations, mean values of Δa were formed for zones of declination $5°$ in width. These mean values were transformed into Δn using the relation

$$\Delta n = b \sin \phi - \Delta a \cos \phi,$$

where b is the level error and ϕ the latitude. But as the level error has already been applied, the relation reduces to

$$\Delta n = - 0\overset{s}{.}835\, \Delta a.$$

The resulting values of Δn are plotted in Figure 2, in which the $\frac{1}{2}(\overline{E+W})$ values of Δn obtained from the Repsold meridian circle are also plotted. There is very good agreement between the values of Δn obtained with the two instruments which seems to indicate that there is a real systematic error of the type $\Delta \alpha_\delta$ in FK4.

Reference

Zverev, M.S. (1965)   *Astr. Zu.*, **42**, No. 4.

PRELIMINARY RESULTS OF THE DETERMINATION
OF ABSOLUTE PROPER MOTIONS OF STARS
REFERRED TO GALAXIES

N. V. FATČIHIN

Absolute proper motions have been determined for 14000 stars in the 82 areas indicated in Figure 1. These proper motions result from measures made on 105 plate pairs taken with the Pulkovo normal astrograph which has a focal length of 3·5 m. The plates, which had an average epoch difference of 22·4 years, were measured film-to-film in a Repsold measuring machine. The number of galaxies used for defining zero proper motion was 271, and among the stars measured were 1283 from the AGK3. The probable error of a determination of proper motion from one plate pair was

$$\pm 0\rlap{.}{''}0052 \text{ (reference stars)}$$
$$\pm 0\rlap{.}{''}007 \text{ (AGK3 stars)}$$
$$\pm 0\rlap{.}{''}0062 \text{ (galaxies).}$$

The images of the galaxies were estimated as having an average quality of 6 on a scale of 10. (Deutsch *et al.*, 1955).

The coordinates of the solar apex and the secular parallax as determined by the Kovalsky-Airy method using the derived proper motion for stars between photographic magnitudes 14·5 and 15·0, but omitting all those for which $\mu > 0\rlap{.}{''}05$, were

$$A = + 302\rlap{.}{°}5 \pm 4\rlap{.}{°}5$$
$$D = + 32\rlap{.}{°}6 \pm 2\rlap{.}{°}7$$
$$(\hbar/\rho) = 0\rlap{.}{''}0096 \pm 0\rlap{.}{''}0006.$$

The variation of the secular parallax with galactic latitude is set out in Table 1. Since only stars within a narrow range of magnitude were considered, this variation is cosmical in origin, e.g. due to a larger percentage of giants in low galactic latitude than in high.

The proper motions of the AGK3 stars on the FK3 and FK4 systems were very kindly provided by Professor Haffner and Dr. Dieckvoss before publication. This permitted a comparison to be made between these systems and that defined by the galaxies. This comparison was limited to those AGK3 stars which are located within 40′ of the optical centre of the measured plates. There were altogether 742 such stars, an average of 9 stars per area.

Perek (ed.), Highlights of Astronomy, 297–300. © I.A.U.

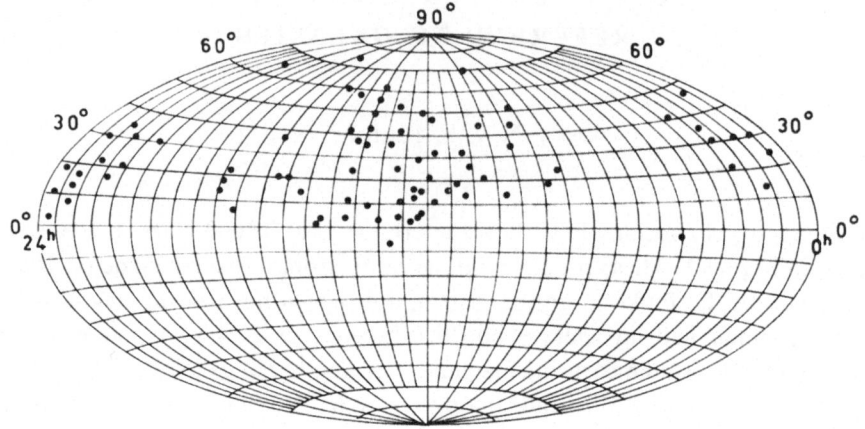

FIG. 1.   *The distribution of the investigated areas.*

Table 1

Variations of the secular parallax with galactic latitude

Galactic latitude	$\left(\dfrac{h}{\rho}\right)$	Probable error	Number of areas
From $\pm 11°$ to $\pm 30°$	$+0\rlap{.}''0047$	$\pm 0\rlap{.}''0013$	21
From $\pm 30°$ to $\pm 50°$	$+0\rlap{.}''0083$	$\pm 0\rlap{.}''0007$	23
From $\pm 50°$ to $\pm 90°$	$+0\rlap{.}''0112$	$\pm 0\rlap{.}''0013$	38

For the determination of the correction to Newcomb's constant of precession the differences $\Delta\mu = \mu_{AGK3} - \mu_G$ were represented by the formulae

$$\Delta\mu_x = \Delta m \cos\delta + \Delta n \sin\alpha \sin\delta$$
$$\Delta\mu_y = \Delta n \cos\alpha.$$

It was found that

$$\Delta m = +0\rlap{.}''0053 \pm 0\rlap{.}''0011$$
$$\Delta n = +0\rlap{.}''0044 \pm 0\rlap{.}''0011,$$

or

$$\Delta p_1 = +0\rlap{.}''0111 \pm 0\rlap{.}''0011$$
$$\Delta E = +0\rlap{.}''0049 \pm 0\rlap{.}''0004,$$

where $\Delta E = \Delta\lambda + \Delta e$.

The derived value of Δp_1 is in good agreement with other determinations. The value of ΔE, on the other hand, is smaller by a factor of about 2. The correction Δe obtained by previous authors must therefore be explained mainly (but not completely) by either the existence of unknown systematic motions of the stars in addition to

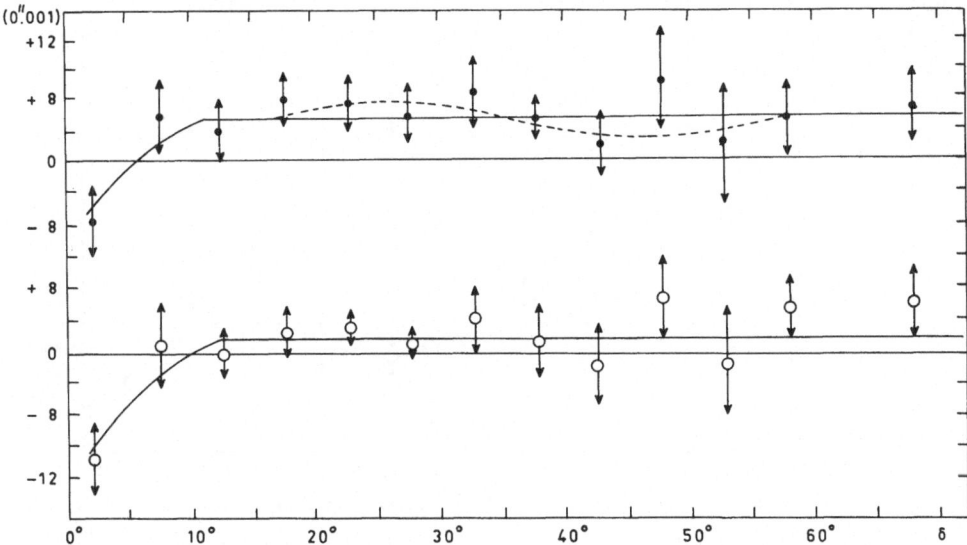

FIG. 2.   *The systematic differences* $(\Delta\mu_x)_\delta$. *Top and bottom – without and with allowance for the correction to Newcomb's precessional constant.*

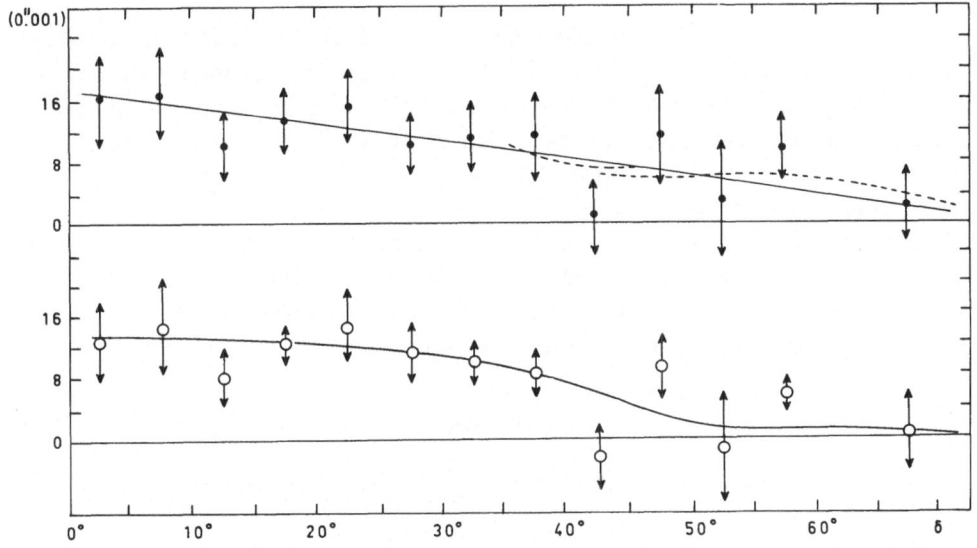

FIG. 3.   *The systematic differences* $(\Delta\mu_y)_\delta$. *Top and bottom – without and with allowance for the correction to Newcomb's precessional constant.*

those due to solar motion and galactic rotation, or by an incorrect allowance for these effects. In our computations the effects of systematic motions of the stars are automatically eliminated since we are working with the differences $\mu_{AGK3} - \mu_G$.

If it is adopted from theoretical considerations that $\Delta\lambda = 0''.0008$, we will have that

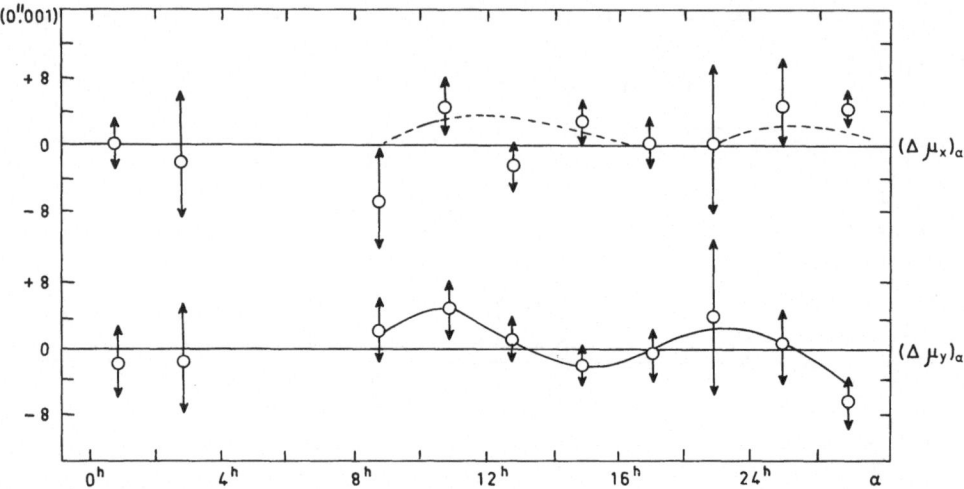

FIG. 4.   *The systematic differences $(\varDelta\mu_x)_\alpha$ and $(\varDelta\mu_y)_\alpha$ with allowance for the correction to Newcomb's precessional constant.*

$\varDelta e = +0''.0041$. This residual is due to errors in the meridian observations of the stars, planets, and the Sun.

The variations in the systematic differences $(\varDelta\mu_x)_\delta$, $(\varDelta\mu_y)_\delta$, $(\varDelta\mu_x)_\alpha$ and $(\varDelta\mu_y)_\alpha$ without and with an allowance for the correction to Newcomb's precessional constant, are set out in Figures 2, 3 and 4. The allowance for the correction does not noticeably change the differences $(\varDelta\mu_x)_\alpha$ and $(\varDelta\mu_y)_\alpha$. The curves of the differences $(\varDelta\mu_x)_\alpha$ in the declination zones $0° - +25°$ and $+25° - +50°$ are practically parallel and similar to the curves in Figure 4.

The systematic differences shown in Figures 2, 3 and 4 are to be attributed to errors in the system of AGK3 since there should be no systematic errors in the proper motions referred to galaxies. It must be emphasised, however, that the present investigation is a preliminary one and its conclusions tentative.

Reference

Deutsch, A. N., Lavdovskij, V. V., Fatčihin, N. V. (1955)   *Izv. Pulkovo*, **154**, 14.

DESIDERATA FOR FK5

W. Fricke and W. Gliese

1. Purpose of a Fundamental Catalogue

A fundamental catalogue compiled from independent and differential observations of stellar positions at various epochs provides the fundamental system of positions and proper motions of the stars. The system ought to be compiled in such a way that it represents the inertial frame of reference as accurately as practicable. Its direct purpose is to serve as the reference system of positional astronomy. It should fulfil the requirements of astronomical research, in particular in celestial mechanics and galactic research, as well as the demands of the astronomical determinations of time and of latitudes. It has also to fulfil the requirements of geodesy and navigation. For all these purposes the system must be uniform over the whole sky (free from regional systematic errors), and it has to be uniform over a large range of stellar magnitudes (free from errors depending on the magnitude of the stars).

2. Desired Stellar Composition of a New Fundamental Catalogue

The FK4 contains 1535 stars brighter than magnitude 7·5. The limitation to bright stars was essentially caused by the optical limit of older meridian circles. The fundamental stars were chosen on the basis of the number of accurate independent observations available per star, and they were selected such that an approximately uniform distribution of fundamental stars exists over the sky.

In further improvements of the fundamental catalogue the number of available independent observations at different instruments will remain one of the important criteria for the selection of more stars. The principle of uniform distribution may, however, no longer be used to exclude stars with observations of high quality. Furthermore, the number of fundamental stars may not be limited owing to the amount of work involved in the compilation of a fundamental catalogue. The data processing may be done by means of electronic computers.

Observations available at the present time suggest, first, a considerable increase of the number of fundamental stars and, second, the extension of the system down to the magnitude limit of about 9·0. Only with the completion of the processing of the observational data will one be able to decide which stars can be included in the fundamental catalogue from the view of systematic accuracy. It is fairly certain that the total number of suitable stars in the whole sky will scarcely exceed about 5000 objects.

Perek (ed.), Highlights of Astronomy, 301–305. © *I.A.U.*

There will remain many stars observed by differential techniques at three or more epochs with respect to some system which can be reduced satisfactorily to the fundamental system. These stars, however, will not attain the accuracy of the fundamental stars. The number of these stars may be of the order of 30000, and their positions and proper motions may form a new general catalogue. In a new general catalogue the requirements of galactic research have to be carefully taken into account. The principle of a uniform distribution of the catalogue stars over the sky can no longer be followed, since it would not be wise to exclude stars with accurate data in the galactic belt. More attention has also to be given to objects of special interest in galactic research. Valuable impetus was given in this direction by such lists as that compiled by Blaauw.

For both catalogues, the fundamental and the general, the determination of the system free from magnitude errors down to magnitude 9·0 is the main problem.

3. Recent Observations Relevant to the Improvement of the FK4 System

A. LIST OF NEW OBSERVATIONS

The communication of observations which observers regard as relevant to the improvement of the Fourth Fundamental Catalogue is of great importance. A list of publications is presented here containing recent observations which will have to be taken into account in improving the FK4 system. (The list does not contain observations made by differential techniques.) Should the list be incomplete, we should be glad to receive more information.

(a) Astrolabe observations:

Neuchâtel: *Publ. Obs. Neuchâtel*, **9**, 1959.

Paris: *Bull. Astron.*, **23**, 1961, 343.

Curaçao: *Netherlands Geod. Comm.*, **1**, 1963, No. 4.

Zi-Ka-Wei: *Acta Astr. Sinica*, **11**, 1963, 73.

Potsdam: *Arbeiten Geod. Inst. Potsdam*, **12**, 1966.

Tananarive: *J. d. Obs.*, **49**, 1966, 387.

Herstmonceux:⎫
Cape: ⎬ private comm., D. V. Thomas, 1967.

Some of these papers contain data of little more than one hundred FK4 stars; in these cases the systematic relation between the observations and the FK4 cannot be determined very reliably.

(b) Observations of the Time Service in U.S.S.R.:

Catalogue $\phi2$:   *Astr. Zu.*, **34**, 1957, 613.

        $\phi6'$:   *Izv. Glav. Astr. Obs. Pulkovo*, **22**, 1961, 2.

        $\phi678$:   *Trud. Glav. Astr. Obs. Pulkovo* (II), **75**, 1966, 29.

(c) Catalogues observed with classical meridian instruments:

W3$_{50}$: *Publ. U.S.N.O.*, **19**, 1964, 1.

Pu$_{60}$: *Izv. Glav. Astr. Obs. Pulkovo*, No. **179**, 1966, 3.

Pu$_{50}$F: *Trud. Glav. Astr. Obs. Pulkovo* (II), **76**, 1966, 5.

(d) Dynamical coordinate system derived from minor planet observations: *Ann. Tokyo Obs.* (II), **10**, 1967, 113.

(e) Right ascension system of the southern polar cap:

Cerro Calan: *Astr. Zu.*, **42**, 1965, 823.

Cape Cat. of Circumpolar Stars: *Roy. Obs. Bull.*, **106**, 1966.

B. ERRORS OF THE FK4 SYSTEMS INDICATED BY RECENT OBSERVATIONS

(a) Right ascension system:

$\Delta\alpha_\delta$: The mean system of the more recent observations indicates that the FK4 right ascensions between $+5°$ and $+50°$ require a negative correction which is $-0^s\!.007$ at $+25°$.

In the Southern circumpolar region ($-70°$ to $-85°$) both the rigid absolute Cape observations of 1936 and Zverev's observations at Cerro Calan show a deviation from FK4 of about $-0^s\!.02 \sec\delta$. According to provisional Cape astrolabe results the fundamental α_δ-system south of $-40°$ requires a negative correction which becomes more pronounced with decreasing declination reaching $-0^s\!.011 \sec\delta$ at $-60°$. The sign agrees with that of the Southern observations just mentioned.

$\Delta\alpha_\alpha$: In the zone $+50°$ to $+80°$ between 11^h and 19^h the α_α-system of FK4 obviously requires a positive correction, the maximum being about $+0^s\!.008 \sec\delta$ at 16^h.

(b) Declination system:

Recent observations do not yet permit of any conclusion on deficiencies of the declination system of FK4. The dispersion among the observations is considerable.

C. UNCONFIRMED ERRORS OF THE FK4 SYSTEM

In the equatorial zone the dynamical right ascension system derived by Yasuda from minor-planet observations deviates from FK4 between $+0^s\!.010$ (12^h) and $-0^s\!.015$ ($20/21^h$). The suggested $\Delta\alpha_\alpha$-curve is not confirmed by the catalogues Pu$_{60}$, ϕ678, and W3$_{50}$.

In the whole region from $-30°$ to $+75°$ the δ_α-system of W3$_{50}$ shows a wave against FK4, its maximum being $+0''\!.08$ at 8^h, its minimum $-0''\!.09$ at 16^h. While Pu$_{50}$F and Herstmonceux astrolabe observations appear to confirm these differences, the averaged results from the astrolabe observations at Neuchâtel, Paris, and Potsdam show no remarkable deviations from the FK4 system.

Magnitude equation: The right ascensions of the brightest stars ($m < 2·7$) in W3$_{50}$ differ from FK4 by about $-0^s\!.002$ while the photoelectrically observed right ascensions in $\phi678$ differ from those in FK4 systematically; the differences increase continuously from $-0^s\!.006$ for stars of magnitude one to $+0^s\!.003$ for stars of magnitude 6.

D. V. Thomas calls attention to an apparent systematic error in the FK4 declinations of bright blue stars (m $< 4 \cdot 0$): $\Delta \delta_{m,c}$ (Herstm. – FK4) $= -0''\!.09$, derived from 12 stars.

4. Suggestions for Further Observations and the Form of their Presentation

Suggestions for further observations which may contribute to an improvement of the present fundamental system arise mainly from known deficiencies and suspected shortcomings of the FK4. There are the following main deficiencies:

(1) The foundation of the system South of $-30°$ on observations with the Cape meridian circle alone (other instruments in the Southern hemisphere play a minor role in FK4);

(2) the inferior accuracy of the right ascension system compared with that of the declinations; this holds for the whole sky;

(3) regional inhomogeneities in the system over the sky;

(4) the limitation to stars brighter than visual magnitude 7·5.

From these deficiencies we conclude that high priority must be given to the continuation of fundamental observations which are suited to eliminate zonal and regional inhomogeneities. Fundamental observations at more than one instrument in the Southern sky are of eminent importance. For the determination of the zero point in right ascension and for the equator point, observations of the Sun and of major and minor planets are necessary. More attention has to be given to absolute azimuth determinations with meridian circles in order to improve the right ascension system.

Concerning the choice of stars for a fundamental program, we suggest that, in programs covering the sky from one of the poles to about $30°$ South or North of the equator, not less than 750 FK4 stars should be included. The desirability of such a large number arises from still existing regional systematic errors. All the other stars in the program should cover the whole magnitude range down to magnitude 9·5, and all necessary care has to be taken for the purpose of eliminating magnitude errors. The inclusion of stars of FK4 Sup, N30 and of Zverev's catalogue of faint fundamental stars is desirable.

So far observations obtained with PZTs and Astrolabes have not contributed to the formation of the fundamental system. In view of the suspected regional systematic errors of the right ascension system in positions and proper motions, the observations with PZTs and Astrolabes may contribute essentially to the compilation of FK5. It is suggested that observers with these instruments present all findings relevant to an improvement of the system in such a way that the observations may be incorporated in FK5.

Moreover it is suggested, quite generally, that in the publication of observations the methods of observing – in particular the determination and elimination of instrumental errors – should be clearly described, even if no change in the methods has been made during several decades.

5. Questions Arising with the Improvement of the FK4

There are some well-known errors in the proper motions of FK4 preventing the immediate applicability of these data in galactic research. These are the errors caused by incorrect precession and by the unprecessional motion of Newcomb's equinox. At present, these errors are known to an accuracy of about 15% of their values. If one contemplates a change of the conventional values of precession in the near future, the most favorable opportunity for the change appears to be with the compilation of a new fundamental system. It will be a matter for discussion within the IAU whether the change shall be made with FK5.

Another question concerns the continuation of the traditional technique applied within the FK series of utilizing instrument series alone for the construction of the μ_α-system. The continuation of this procedure cannot be recommended. Due account will have to be given to the existence of isolated observations with new instruments. While there is no doubt about the great utility of observations at several epochs with one and the same instrument, 'first epoch' observations with new instruments can play an important part in the improvement of proper motions. In addition, the result of the observations of the time service must be incorporated.

Finally, one may ask, 'When will the time be ripe for the compilation of a new fundamental system?' The answer depends of course on the availability of observations which are not yet incorporated in FK4 and which give sufficient evidence of errors in FK4. At Heidelberg, new methods for comparing catalogues with FK4 have been worked out which make use of electronic computers. Preparations have been made to apply these methods in the computation of the systematic differences between FK4 and important observational catalogues already included in FK4 and between FK4 and new observations. In this way information will be gathered continuously on the quality of FK4, and a decision will be facilitated at what time in the future the new system can be formed with profit. It is not unlikely that after the investigation of all observations made within the AGK3R program and after the completion of the SRS program, the new material might justify the revision of the present system and its extension to more stars and to a fainter magnitude limit.

DISCUSSION

G. van Herk: The Quito Observatory should be included in the list of those at which astrolabe observations useful for the correction of FK4 have been made. The desirability of observers publishing the apparent magnitude at which a star is actually observed (i.e. what screen, if any, is used) should be stressed in Section 4, 'Suggestions for Further Observations'. Otherwise the determination of errors dependent on magnitude by later investigators is made unnecessarily difficult.

PRECESSION AND GALACTIC ROTATION ON THE BASIS OF VARIOUS PROPER MOTION SYSTEMS

W. FRICKE

1. Previous Results

Previous determinations of precession on the basis of proper motions resulted in corrections to Newcomb's value of lunisolar precession in the interval $+0\overset{''}{.}75 \leqslant \Delta p_1$ $\leqslant +1\overset{''}{.}50$ per century. From a compilation of several determinations and their reduction to the average of the systems of FK3 and N30 Morgan and Oort (1951) obtained:

$$\Delta p_1 = +0\overset{''}{.}75, \quad P = \frac{A}{4 \cdot 74} = +0\overset{''}{.}43, \quad Q = \frac{B}{4 \cdot 74} = -0\overset{''}{.}15$$

per century, where A and B are Oort's constants. Values in the system of FK4 were derived by Bošniakovich (1965), who applied differential corrections to previous results on the basis of the systematic differences FK4–FK3.

For the following reasons it was found appropriate to investigate in detail whether, on the basis of FK4, some improvement can be made in the determination of precession and galactic rotation:

(1) The differential method of reducing results from one proper motion system to another yields reliable results only on the condition that all important characteristics of the primary solution are taken into account. In most of the previous differential determinations such characteristics of primary solutions as the distribution of the stars in the sky and the different weights assigned to proper motions in different parts of the sky were disregarded.

(2) Different results obtained on the basis of different fundamental systems and from different basic material were usually regarded as caused by the deviations of the systems from each other. Identical basic material and identical methods of treatment must be applied in order to reveal the effects of differences between the proper motion systems.

(3) Some determinations were based on proper motions of faint stars after the motions had been reduced to a fundamental system. The limiting magnitude in FK3 and FK4 is about 7·5. The reduction of proper motions of fainter stars to the fundamental system may involve systematic errors depending on the magnitude unless the motions of the faint stars were measured with respect to brighter stars by techniques which exclude magnitude errors.

Perek (ed.), Highlights of Astronomy, 306–310. © *I.A.U.*

(4) In determinations of precession and galactic rotation the errors of the fundamental proper motions cannot be regarded as negligibly small. In FK4, the averaged standard errors describing the *internal* accuracy are

$$\varepsilon_{\mu_\alpha} = \pm\, 0\rlap{.}''16, \quad \varepsilon_{\mu_\delta} = \pm\, 0\rlap{.}''17 \text{ per century},$$

and the averaged standard errors of the *system* are

$$\varepsilon_{\mu_\alpha} = \pm\, 0\rlap{.}''17, \quad \varepsilon_{\mu_\delta} = \pm\, 0\rlap{.}''07 \text{ per century}.$$

These numbers represent linear averages over the whole sky. The average values resulting from internal and systematic errors combined are

$$\varepsilon_{\mu_\alpha} = \pm\, 0\rlap{.}''24, \quad \varepsilon_{\mu_\delta} = \pm\, 0\rlap{.}''18 \text{ per century}.$$

Although corresponding values for previous fundamental systems are not known, it is certain that they are not smaller. Due to these errors, and in consideration of the greater effect of the *systematic* errors, the weight of the component μ_α is about one half the weight of μ_δ. Furthermore, it has to be taken into account that the errors are different in different parts of the sky.

(5) Williams and Vyssotsky (1947) found that the precessional corrections determined from μ_α and μ_δ of FK3 separately agree with each other, while Morgan (1952) obtained discordant results from both components in the N30 system. It appeared desirable to verify these results by carrying out solutions from μ_α and μ_δ separately on the basis of various proper-motion systems including the FK4.

Guided by the considerations outlined here two different approaches were made to determine the precessional corrections and the parameters of galactic rotation on the basis of the proper motion systems of GC, FK3, N30, and FK4.

2. Results of a Reinvestigation of the McCormick and Cape Proper Motions of Faint Stars

Basic material for this investigation were the McCormick and Cape proper motions of stars of average photovisual magnitude 11·1 presented and analysed by Williams and Vyssotsky (1947) in the FK3 system. The group averages of the motions as they were given by these authors were individually reduced to the systems of N30 and FK4. Under identical conditions, this means that with the weights and parallax factors used by Williams and Vyssotsky, solutions were carried out in the systems of FK3, N30, and FK4. It turned out that the correction to Newcomb's lunisolar precession derived from combined solutions of μ_α and μ_δ is identical in all three systems. The numerical value of the correction, however, depends sensitively on the relative weight of the McCormick and Cape proper motions. Williams and Vyssotsky had assigned to the Cape proper motions twice the weight of the McCormick motions.

Other solutions were carried out in which equal weight was assigned to McCormick and Cape motions. Finally, one may consider solutions as well founded which are based on the McCormick motions alone, since the fundamental proper motions are less accurately known in the Southern sky than in the Northern sky, and since the reduction of the Cape motions to FK3 may be affected by serious errors. The results for Δp_1 are

$+10''.8 \pm 0''.12$ from McCormick motions alone,

$+1''.38 \pm 0''.08$ from McCormick and Cape motions with equal weight.

These results are valid for each of the systems of FK3, N30, and FK4. Since it is not possible to exclude systematic errors from the McCormick and Cape material satisfactorily, there remains the interval

$$+ 1''.08 \pm 0''.12 \leqslant \Delta p_1 \leqslant + 1''.38 \pm 0''.08$$

for the centennial value of the correction. The results of separate solutions from μ_α and μ_δ deviate from each other in all three systems FK3, N30, and FK4. In view of the noteworthy uncertainty of the μ_α-systems, this fact is not surprising.

Combined solutions from μ_α and μ_δ give the same value $Q = -0''.16 \pm 0''.03$ of Oort's constant corresponding to $B = -7 \cdot 6 \pm 1 \cdot 4$ km sec^{-1} kpc^{-1} in all systems. Oort's constant P differs slightly in the three systems and depends on the weight of the proper motions in different parts of the sky. The average value of P obtained from the three systems is $P = +0''.33 \pm 0''.04$ corresponding to $A = +15 \cdot 5 \pm 1 \cdot 9$ km sec^{-1} kpc^{-1}.

All the results demonstrated that differential solutions in N30 and FK4 with respect to a primary solution in FK3 give erroneous results, if the particularities of the primary solution are disregarded. The determinations made by Morgan and Oort (1951) and by Bošniakovich (1965) were affected by such errors.

3. Results from Fundamental Proper Motions of Distant Stars

Basic material for this investigation were the proper motions of 512 FK4/FK4 Sup stars with distances greater than about 100 parsec. Obvious advantages of this material for the determination of precession and galactic rotation are:

(1) All proper motions are individually known in the systems GC, FK3, N30, and FK4 so that errors in the reduction from one system to the other are excluded.

(2) The motions in FK3, N30, and FK4 are presumably much less affected by errors depending on the magnitude than the motions of faint stars.

(3) For most of these stars radial velocities are known so that peculiarities of the stellar motions can be investigated and a check of the photometric distance scale made.

Solutions for the solar motion, the precessional corrections, and the parameters of galactic rotation were performed under strictly identical conditions on the basis of the proper motions directly taken from GC, FK3, N30, and FK4. Effects arising from anomalies in the distance distribution of the stars, from fluctuations in the velocity field due to star streams and from errors in the μ_α and μ_δ systems, were investigated by carrying out a number of alternative solutions.

The most essential result is that the correction to Newcomb's lunisolar precession has the same value, within 0".01 per century, in the systems FK3, N30, and FK4,

$$\Delta p_1 = + 1".10 \pm 0".10 \text{ (p.e.) per century}.$$

The value $\Delta\lambda + \Delta e$ describing the combined correction due to incorrect planetary precession and an unprecessional motion of Newcomb's equinox is

$$\Delta\lambda + \Delta e = + 1".20 \pm 0".11 \text{ per century}$$

in both FK4 and N30, while the value is slightly different in FK3 and GC. The parameters of galactic rotation in the systems FK4 and N30 are (in units of $km \, sec^{-1} \, kpc^{-1}$)

$$\text{FK4:} \quad A = + 14\cdot2 \pm 1\cdot9, \quad B = - 11\cdot8 \pm 1\cdot9$$
$$\text{N30:} \quad A = + 17\cdot1 \pm 1\cdot9, \quad B = - 10\cdot0 \pm 1\cdot9.$$

GC and FK3 reveal deviations, which are not greater than may be expected from the lower accuracy of these systems.

In the material the Scorpio-Centaurus association is represented by twenty members, and some other well-known associations by a smaller number of stars. The complete exclusion of the Scorpio-Centaurus stars caused no appreciable alteration in the results.

By dividing the material into stars nearer than 250 parsec and those farther away it was found that the local group of stars shows no indication of a rotation deviating from differential galactic rotation.

4. Recommended Values of the Precessional Corrections

For investigations to be based on proper motions freed from the effects of incorrect precession and of the zero point error in μ_α, it is recommended that the corrections

$$\Delta n = + 0".44 \text{ corresponding to } \Delta p_1 = + 1".10 \text{ per century},$$
$$\Delta\lambda + \Delta e = + 1".20 \text{ per century}$$

should be applied to motions in the systems FK4 and N30. Proper motions in the systems GC or FK3 should in all cases be reduced to the system of FK4, and then the precessional corrections applied. It is not recommended that, for investigations of stellar motions, averages be taken of proper motions in N30 and FK4. In case of doubt it is preferable to base investigations on each system, separately considered, in order to make apparent the variational width of results due to the differences between the best proper motion systems presently available.

References

Bošniakovich, N.A. (1965)   *Bull. astr.*, **25**, 119. (IAU-Symposium No. 21.)
Morgan, H.R. (1952)   *Astr. Pap. Wash.,* **13**, Part III. (Introduction to N30.)
Morgan, H.R., Oort, J.H. (1951)   *Bull. astr. Inst. Netherl.,* **11**, 379.
Williams, E., Vyssotsky, A.N. (1947)   *Astr. J.,* **53**, 63.

DISCUSSION

W. Dieckvoss: Proper motions for more than 160000 stars of the AGK3 preliminary catalogue have been studied at Bergedorf. The resulting values for Oort's constants are almost identical with those found by Professor Fricke, but the mean errors, about $\pm 0 \cdot 7$ km sec^{-1} kpc^{-1}, are smaller. The distances of these stars range between 100 and 1000 parsec.

G. van Herk: I cannot agree that the motions in FK3, N30 and FK4 are unaffected by errors depending on magnitude. The differences in μ_α between the FK4 and N30 systems show a marked dependence on magnitude in some parts of the sky. For example, in the zone between declinations $-80°$ and $-28°$ the centennial $\Delta\mu_\alpha$ show a total average rate of $-0\rlap{.}''160 \pm 0\rlap{.}''026$ (m.e.) per magnitude. As the brightest stars are unevenly distributed over the sky this must have some influence on the solutions, but I cannot tell how much.

THE RELATIONSHIPS BETWEEN VARIOUS TECHNIQUES
FOR OBTAINING PROPER MOTIONS

C. A. MURRAY

1. Introduction

In many problems of galactic kinematics we need to know the proper motions of faint stars, as well as bright stars, within some well-defined reference frame. Generally speaking, the accuracy required is greater for faint stars on account of their larger distances. Techniques at present used for determining proper motions vary according to apparent magnitude, and it is clearly desirable that all methods should give results which are capable of reduction to a common frame.

2. Fundamental Proper Motions

Approximations to an absolute frame of reference are given by proper motions derived ultimately from long series of meridian observations, as e.g. the FK4 and its extension to intermediate magnitudes, the AGK3. It is important to remember, however, that, apart from their own peculiar systematic errors, all catalogues of fundamental proper motions depend on a conventional value of the constant of precession which, with the motion of the equinox, can at present only be derived by making assumptions about the systematic motions of stars. This may not have much importance in discussion of stellar motions in the immediate solar neighbourhood (~ 100 parsec, say), but it becomes increasingly significant at larger distances.

3. Extragalactic Reference Frame

We have already seen the preliminary results from the KSZ programme, and within the next few years we can hope to see also results from the Lick programme, for referring fundamental proper motions to extragalactic nebulae.

In principle the Lick programme (Wright, 1950) should enable the systematic errors in the AGK3 proper motions outside the galactic zone, and also the precessional motions, to be determined directly without any appeal to stellar kinematics. However, it will be several decades before the Southern extension of the programme, now being carried out by the Yale-Columbia Southern Observatory at El Leoncito, will enable the whole sky to be covered.

In the KSZ programme (Dejč, 1954) the systematic motions of reference stars at

Perek (ed.), Highlights of Astronomy, 311–315. © *I.A.U.*

m_{pg}14–15 will be measured relative to galaxies at the same apparent magnitude, and also relative to KSZ stars for which fundamental proper motions on the FKSZ system will be known. Interpolation between fields containing galaxies and fields centered on KSZ stars will be achieved by making statistical assumptions about the average proper motions of the reference stars. While the KSZ programme is not as fundamental as the Lick programme for determining precession and errors in the fundamental system of proper motions, it will give valuable information about the motions of the reference stars themselves relative to the frame of reference defined by the galaxies.

4. 'Kinematic' Reduction to Absolute Proper Motion

An alternative method for deriving absolute proper motions is to assume a model of galactic kinematics from which the systematic motion of any particular selection of reference stars can be predicted. There is a natural tendency among fundamental astronomers to regard this 'kinematic' method as a regrettable second-best. Its importance, however, lies in the fact that, for magnitudes fainter than about $m_{pg} = 13$, it is the only generally practicable method. Furthermore, it will still have to be used at brighter magnitudes over much of the sky, until the results of the current large-scale photographic and meridian programmes become available. The characteristics of the kinematic model have to be derived from assumptions about stellar motions, and in this sense the kinematic method is no less 'fundamental' than the fundamental method itself.

It is in principle possible to measure absolute proper motions in any field directly with respect to extragalactic objects, provided that a sufficient number of these can be identified and measured. But at $m_{pg} = 16$, the average apparent magnitude of galaxies to be measured in the Lick programme (Vasilevskis, 1953), there will be far too few measurable galaxies in fields of the order $1° \times 1°$, even outside the zone of avoidance. In larger fields, and at fainter magnitudes, such as in Luyten's repetition of the Palomar Sky Survey plates (Luyten, 1963), it may well be possible to use galaxies directly as reference objects; but in general, for faint objects photographed with long-focus telescopes, we shall always need to rely on kinematic methods of reduction. It is, therefore, an important task for fundamental astrometry to provide the data necessary for application of the kinematic method.

The parameters required for the kinematic method can be obtained either from fundamental proper motions alone (solar apex, secular parallaxes), or from an assumed solar motion, together with mean parallaxes derived from a comparison between relative proper motion dispersions and a velocity model derived from radial velocity observations, as first proposed by Oort (1936), and later improved by Binnendijk (1943). As Clube has recently pointed out (1966), this latter approach can in principle lead to a proper motion system which is virtually independent of any fundamental system, provided that sufficient radial velocity and spectroscopic data are available for th? construction of the velocity model. At the present time, radial velocity data for stars fainter than about $m_{pg} = 9$ are very scarce, and the model has to be built up from an extrapolation from bright star observations, with assumptions about the average population of stars at a given apparent magnitude. The model may well be adequate down to $m_{pg} \approx 12$, but at fainter magnitudes there is likely to be an increasing number of distant halo stars, such as was found in the field of M 67 (Murray, 1968), which will affect the mean proper motions, and dispersions. For this reason, any further extrapolation is dangerous.

In addition to the parallactic motion, we also need to know systematic motion of the centroid of any selection of reference stars due to differential galactic rotation, and to the total rotation of the co-ordinate system arising from the circular velocity of the 'local standard of rest'. The differential rotation is usually allowed for by adopting a value of Oort's constant, A, which can be determined satisfactorily from radial velocities; higher-order approximations can also be obtained, in principle, from radial velocity observations of very distant objects. The total rotation on the other hand, corresponding to Oort's A-B, is more difficult to determine. In theory it can only be obtained from absolute proper motions, with assumptions about the systematic motions of stars, but Clube (1966) has argued that a value can be assigned to it, within tolerable limits, from the radial velocity of the Sun relative to very old objects in the Galaxy, and relative to the galaxies in the local group.

Oort's constants are frequently obtained from the motions of early-type stars, and A-B derived in this way must represent the rotation of the extreme Population-I system. On the average, however, reference stars in any particular field will be older, and it is therefore important to determine the appropriate parameters representing rotation and differential rotation for these older stars.

5. Accuracy of Fundamental and Kinematic Reduction to Absolute Proper Motion

With the publication of AGK3, the Northern hemisphere will have a network of stars down to $m_{pg} \approx 11$ whose fundamental proper motions will be known with a mean error of $\pm''.008/a$ (Dieckvoss, 1960). This will be entirely adequate for deriving proper motions from Astrographic Catalogue plates. However, when we consider fields photographed with long-focus telescopes, of the order of $1° \times 1°$ say, we can only expect an average of 9 AGK3 stars; in some fields there will be rather fewer. The reduction to the FK4 system at the centre of a $1° \times 1°$ field will thus have a statistical error of about $\pm''.003/a$ or worse.

The accuracy attainable in the kinematic method depends partly on the cosmical dispersion of the proper motion of the reference stars which, at $m_{pg} = 11$, is roughly the same as the mean error of AGK3 stars (e.g. Vyssotsky, 1954) over much of the sky. At this magnitude, the formal accuracy of the two methods is therefore similar, but in the kinematic method it may always be possible to measure more stars in any particular field. The real errors may, however, be much larger than this on account of imperfections in the assumed kinematic model. The usual assumption of a unique solar apex and velocity, applicable to all stars, is, of course, over-simplified; but as long as we have no independent spectroscopic or photometric information for the reference stars, there seems to be no real justification for a more complicated model.

Vyssotsky and Williams (1948) have shown clearly that the direction to the solar apex varies over more than $20°$ for different spectral types, at both bright and inter-mediate magnitudes, due to systematic variation of mean orbital eccentricity (*ibid.*, p. 51) or age. This uncertainty in the apex can introduce a systematic error of up to $\overset{\cdot}{.}007/a$ in proper motions referred to high-latitude stars with a centennial secular parallax of $2''$.

It has been claimed by Clube (1966) that the mean error of an absolute proper motion derived by the kinematic method from about 20 reference stars with $10 < m_{pg} < 12$, is about $\pm\overset{\cdot}{.}005/a$. For some problems, e.g. study of the kinematics of RR Lyrae stars at distances of the order of 1 kpc, this accuracy is sufficient, but there are many problems, such as the motions of distant clusters and Population-I objects, for which it is quite inadequate.

It goes without saying that if one requires higher accuracy one must be prepared to pay the price by measuring more stars. However, with modern techniques of measurement and computation this is quite feasible. The systematic errors introduced by imperfections in the kinematic model cannot of course be eliminated merely by increasing the number of reference stars, but they may be appreciably reduced if accurate two-colour or spectroscopic data are available. A recent study in the field of M 67, carried out at Herstmonceux, has shown that an accuracy of at least $\pm\overset{\cdot}{.}002/a$ can be obtained without any *a priori* assumption about the solar apex (Murray, 1968).

6. Future Requirements and Programmes

A task for the immediate future is the improvement of our knowledge of the kinematics of stars so that a more satisfactory kinematic model can be built up.

In the first place, the publication of AGK3 will make available the proper motions of 180000 stars, albeit only in the Northern hemisphere. This is six times the number of stars in the two McCormick catalogues, discussed by Vyssotsky and Williams, on which so much of our present knowledge of kinematics of stars down to $m_{pg} \sim 12$ is based. It will provide an unrivalled source of data for kinematic studies, which will be even more valuable when the data from current Southern hemisphere programmes become available.

The KSZ and Lick programmes will give important information on the systematic motions of fainter stars, relative to the extragalactic frame of reference. On account of the relatively small scale and short baseline, however, the accuracy of some of the individual proper motions relative to nebulae measured at Lick will only be about

$\pm\, ''.007/a$ (m.e.) after 20 years (Vasilevskis, 1953). This is greater than the intrinsic proper motion dispersions of many classes of faint stars, so that for more detailed studies of the dispersions it will be necessary to supplement the Lick data with more accurate relative proper motions.

Much observational material is available for deriving very accurate relative proper motions in the Northern Kapteyn Selected Areas. The mean error of an individual proper motion from a combination of the Radcliffe (Knox-Shaw and Barrett, 1934) and Pulkovo (Dejč, 1940) catalogues is about $''.004/a$. But even the second epoch of the Radcliffe catalogue is now about 40 years old, and a repetition of the whole Radcliffe material now would increase the weight of the proper motions by at least a factor of 20. The Radcliffe refractor, and plates, are at present at the University of London Observatory, but it is unfortunately impracticable to repeat the whole programme on this telescope. However, the 26-inch refractor at Herstmonceux is a very similar instrument, and recent tests have shown that plates taken simultaneously on the two telescopes give systematically the same results over at least four magnitudes. It is planned to repeat the bulk of the Radcliffe plates at Herstmonceux, but to have a selection taken at London; in this way it should be possible to control any systematic errors existing now between the two telescopes. The repetition of the Radcliffe material is being supplemented from plates obtained over 50 years ago on the 26-inch refractor at Greenwich, for photometry; most of the Northern areas were covered, and repetition of these plates is nearly complete.

It will be several years before we have the results of all these observational programmes, but we can expect that with the increasing use of automatic and digitized measuring equipment, it will be possible to complete fairly rapidly the large amount of measurement and computation involved.

We can confidently hope that, within the next decade, we shall have far superior data on stellar motions than has hitherto been available, for the determination of absolute proper motions of faint stars.

References

Binnendijk, L. (1943)   *Bull. astr. Inst. Neth.*, **10**, 9.
Clube, S. V. M. (1966)   *Q.J.R. astr. Soc.*, **7**, 257.
Dejč, A. N. (1940)   *Pulkovo Pubs.*, **2**, 50.
Dejč, A. N. (1954)   *Trans. I.A.U.*, **VIII**, 789.
Dieckvoss, W. (1960)   *Astron. J.*, **65**, 171.
Knox-Shaw, H., Scott Barrett, H. G. (1934)   The Radcliffe Catalogue of Proper Motions in the Selected Areas 1 to 115. O.U.P.
Luyten, W. J. (1963)   P.M. Survey with 48-inch Schmidt, I. University of Minnesota.
Murray, C. A. (1968)   *R.O. Bull.* No. 141.
Oort, J. H. (1936)   *Bull. astr. Inst. Neth.*, **8**, 75.
Vasilevskis, S. (1953)   *Astron. J.*, **58**, 126.
Vyssotsky, A. N. (1954)   *Astron. J.*, **59**, 52.
Vyssotsky, A. N., Williams, E. T. R. (1948)   *Pubs. Leander McCormick Obs.*, **10**, 43.
Wright, W. H. (1950)   *Proc. Amer. Phil. Soc.*, **94**, 1.

THE PLACE OF ACCURATE PROPER MOTIONS
IN GALACTIC RESEARCH

A. BLAAUW

Problems in galactic research for the solution of which highly accurate *absolute* proper motions are required are manyfold. The following selection seems of particular interest at the present moment. Problems for which only accurate *relative* proper motions are required are not considered here.

(a) *Objects exhibiting spiral structure* (OB stars, early-type supergiants, luminous Cepheids, youngest open clusters).

Particular reference is made to the problem of the local deviations from circular motion; 'local' implying regions with dimensions of 100–1000 parsecs. Their study is of great interest for the problem of the origin of the spiral structure. Assuming that the youngest stars, ages up to several million years, still possess the kinematical properties of the interstellar medium, these stars may be expected to reveal the regional kinematical characteristics of the gas. 21-cm observations have clearly indicated the existence of deviations from circular motions up to about -30 km/sec. These, however, refer to the line-of-sight component only and the distances of the objects concerned are uncertain.

Velocities of the order of 10–30 km/sec correspond to $0''.002$ to $0''.006$ at a distance of 1000 parsec. Hence, if an absolute accuracy can be obtained of about $0''.002$, important information on regional kinematics may be obtained up to 2 or 3 kpc.

For somewhat older stars, whose motions may not any more exhibit the characteristics of the interstellar medium, accurate proper-motion measurements may throw light on the possible existence of a systematic lag of the mean gas motion with respect to the circular motion in a gravitational field.

For these reasons, continued observations of the list of O and B stars North of $-20°$ declination and brighter than 8·5 visual magnitude, proposed in the IAU Symposium Report No. 1 (1953) for meridian observations appear highly desirable. Exploratory use of recent observations of these stars shows that new proper motions of high accuracy may be obtained. Extension of this list is strongly recommended.

High-precision proper motions of the stars mentioned may also throw important light on possible systematic errors in the fundamental system through the analysis of the proper-motion components perpendicular to the galactic plane.

Perek (ed.), Highlights of Astronomy, 316–318. © I.A.U.

(b) *Bright stars of spectral types A and F for tracing of past orbit and location of origin.*

Strömgren and associates have demonstrated the possibility of locating the region of origin of stars of ages up to a few hundred million years. Application of the method depends to a large extent on the accuracy of the proper motions. For a 6 magnitude A-type star, at about 100 parsecs, a probable error of $0''.002$ corresponds to 1 km/sec; this accuracy is just about sufficient. For fainter, more distant stars the permissible probable errors must be correspondingly reduced.

(c) *Bright stars that served as reference stars in the McCormick faint star programmes.*

The McCormick Observatory has published large numbers of proper motions of faint stars in about 700 regions North of $-30°$ declination, centered on bright stars. The first epoch plates were taken for the determination of the trigonometric parallaxes of these stars.

Meridian observations and, thereby, high-precision fundamental proper motions of these stars are desirable for two reasons:

(1) These stars are used to reduce the proper motions of the faint stars to the absolute system. Studies of the systematic motions of the faint stars have proved very useful in recent analyses for the redetermination of the constants of precession and galactic rotation.

(2) Studies of the absolute proper motions of the faint stars in the McCormick fields. Of particular interest will be the systematic and peculiar motions of the low metal content late-type giants, discriminated photometrically (disk population), which may be assumed to have a smooth density distribution and smoothed-out distribution of velocities. The solar motion with respect to these stars should provide important and accurate information on the component of the Sun's motion with respect to the galactic centre, and hence on this component of the motion of the very young population in the solar neighbourhood.

This programme of bright stars should be carried out in conjunction with the extensive current programme of photographic redetermination of the proper motions of the faint stars in the McCormick fields.

(d) *The Hyades.*

Recent reconsideration of the distance scale has emphasized the importance of the role of the Hyades. Continued observations of the Hyades stars is therefore most desirable, with particular attention to the outlying members.

DISCUSSION

W. Dieckvoss: The AGK3 preliminary catalogue contains about 3500 M-type stars fainter than magnitude 10 and with proper motions known with a mean error of $\pm 0''.008$ per year. Are these stars likely to be useful in the study of the old disk population?

A. Blaauw: Yes, especially if accurate spectroscopic data become available, for then they could be used to determine the solar motion with respect to the galactic centre.

C.A. Murray: The McCormick material discussed by Williams and Vyssotsky does not cover the complete sphere. I feel that the McCormick work should be extended to the Southern hemisphere by utilising the magnitude compensated parallax fields at Southern observatories like Yale and Cape.

ÉTUDES SUR L'EMPLOI DE RECOUVREMENTS DE PLAQUES POUR L'ÉTABLISSEMENT DE CATALOGUES PHOTOGRAPHIQUES

P. LACROUTE

Le but de cette étude est la recherche des possibilités et des limites de l'emploi des recouvrements de clichés.

Le travail n'est pas terminé. Un travail plus complet sera publié dans quelques mois.

1. Nature des résultats obtenus avec solution par clichés séparés

Les méthodes de résolution généralement adoptées consistent à calculer pour chaque cliché à l'aide des coordonnées équatoriales des étoiles de repère et des coordonnées α_0, δ_0, du point tangentiel les 'coordonnées normales' des étoiles de repère, puis à établir des formules reliant les coordonnées normales X, Y aux coordonnées mesurées x, y sur la plaque; $X = x + f(x, y)$; $Y = y + \varphi(x, y)$. On emploiera pour les X, Y et x, y des unités correspondantes. Les paramètres de ces formules sont les 'constantes' des clichés. On applique ces formules aux étoiles de catalogues pour calculer leurs coordonnées normales, puis leurs coordonnées équatoriales. La méthode de calcul établit des relations linéaires entre les coordonnées normales d'une étoile K et les coordonnées normales des étoiles de référence. On a:

$$X_K - x_K = \Sigma_i \left[\beta_{iK}(X_i - x_i) + \gamma_{iK}(Y_i - y_i) \right].$$

Les β_{iK} et γ_{iK} sont les 'dépendances' de la coordonnée de l'étoile K par rapport aux étoiles de repère i. Les β_{iK} et γ_{iK} sont des fonctions des coordonnées mesurées sur la plaque pour les étoiles de repère et l'étoile K. On a $\Sigma_i \beta_{iK} = 1$ et $\Sigma_i \gamma_{iK} = 0$. Pour étudier la précision obtenue sur les coordonnées équatoriales, on peut admettre la substitution de $(\alpha - \alpha_0) \cos \delta_0$ à X et $\delta - \delta_0$ à Y. En fait, l'étude est faite sur la précision des X, Y.

Soient ε_1 et ε_2 respectivement les incertitudes aléatoires des coordonnées des étoiles de repère et celles des mesures de coordonnées sur les plaques. Le catalogue de repère est en outre caractérisé par des erreurs systématiques.

$$\Delta\alpha \cos \delta = F(\alpha, \delta) \quad \Delta\delta = \Phi(\alpha, \delta).$$

L'erreur sur X_K calculée est la somme de l'erreur sur x_K mesuré et des erreurs sur les expressions de la forme $\beta_{iK}(X_i - x_i)$. Dans cette expression β_{iK} est déterminé avec

Perek (ed.), Highlights of Astronomy, 319–337. © *I.A.U.*

une très bonne précision relative, tandis que $X_i - x_i$ qui est en général faible est déterminé avec une précision relative médiocre. On peut donc négliger les erreurs sur β_{iK} et ne considérer que les erreurs sur $X_i - x_i$.

On peut décomposer les erreurs suivant leur origine.

Il y a d'abord des erreurs systématiques provenant des *erreurs systématiques* du catalogue de repère:

$$\Delta\alpha \cos\delta = \Sigma_i \left[\beta_{iK} F(\alpha_i, \delta_i) + \gamma_{iK} \Phi(\alpha_i, \delta_i) \right].$$

On a une sorte de lissage des erreurs systématiques du catalogue de repère, lissage défini par les β_{iK} et γ_{iK}. Ce lissage n'est pas le même en un même point de la sphère obtenu sur deux clichés différents.

On a des erreurs d'*origine aléatoire* qui viennent de la mesure de x_K, des estimations de coordonnées de repère X_i et des mesures des étoiles repères x_i. On peut chiffrer l'amplitude E de ces erreurs d'origine aléatoire:

$$E^2 = \varepsilon_2^2 + \Sigma_i(\beta_{iK}^2 + \gamma_{iK}^2)(\varepsilon_1^2 + \varepsilon_2^2).$$

Dans cette évaluation, le premier terme ε_2^2 est purement aléatoire.

Le terme en Σ_i représente des erreurs qui varient continuement sur la plaque comme les dépendances β_{iK} et γ_{iK}. En fait, ce terme Σ_i représente des erreurs d'origine aléatoire, mais à variation systématique sur la plaque. Ces erreurs sont beaucoup plus fâcheuses que des erreurs purement aléatoires. L'ordre de grandeur de ces erreurs peut être évalué en calculant $\Sigma_i(\beta^2 + \gamma^2)$.

En admettant 16 étoiles de repère réparties à densité uniforme sur la plaque, les formules les plus simples pour relier X à x et y ($X = x + ax + by + c$), la valeur quadratique moyenne de $\Sigma_i(\beta^2 + \gamma^2)$ sur la plaque est 0·247.

Ce sont ces erreurs, assez importantes devant l'erreur aléatoire pure inévitable, qu'on pourra réduire en utilisant les recouvrements de clichés.

2. Moyenne de clichés se recouvrant

En pratique, on fait souvent la moyenne des résultats obtenus sur des plaques se recouvrant mais résolus séparément.

Dans ce cas, l'étoile K étant obtenue sur deux clichés j et l on a pour l'erreur systématique:

$$\Delta\alpha \cos\delta = \tfrac{1}{2}\Sigma_i \left[(\beta_{iKj} + \beta_{iKl}) F(\alpha_i \delta_i) + (\gamma_{iKj} + \gamma_{iKl}) \Phi(\alpha_i \delta_i) \right]$$

c'est une moyenne des erreurs systématiques du catalogue de repère un peu plus lissée que par clichés isolés.

Pour les erreurs d'origine aléatoire on a:

$$E^2 = \frac{\varepsilon_2^2}{2} + \Sigma_i \left[(\beta_{iKj} + \beta_{iKl})^2 + (\gamma_{iKj} + \gamma_{iKl})^2 \right] \frac{\varepsilon_1^2}{4} + \Sigma_i \left[\beta_{iKj}^2 + \beta_{iKl}^2 + \gamma_{iKj}^2 + \gamma_{iKl}^2 \right] \frac{\varepsilon_2^2}{4}.$$

En admettant des clichés qui se recouvrent par quart les mêmes formules et les mêmes étoiles de repère que plus haut, il vient:

$$E^2 = \frac{\varepsilon_2^2}{2} + 0{\cdot}188\,\varepsilon_1^2 + 0{\cdot}247\,\frac{\varepsilon_2^2}{2}.$$

La part systématique d'origine aléatoire est moins réduite que l'erreur purement aléatoire par l'opération de moyenne.

3. Calculs en utilisant les recouvrements de clichés

Les inconnues à adopter sont les α, δ de toutes les étoiles et les constantes des clichés. Suivant une remarque de W. Googe (*Astr. J.*, **72**, 1967, 623) la matrice du système de résolution par les moindres carrés peut être aisément réduite à une matrice ne comportant plus que les constantes de clichés comme inconnues.

En fait, on peut même établir directement cette matrice.

On supposera qu'on part d'une solution approchée par clichés isolés basée sur les coordonnées des étoiles de repère.

On dispose donc de coordonnées de première approximation: α_c et δ_c sur chaque cliché. Les α_c et δ_c du premier calcul peuvent être exprimés comme des combinaisons linéaires des α_i et δ_i des étoiles de repère; on a vu leurs 'dépendances' dans la solution par clichés séparés.

On considèrera, pour donner un exemple, que les formules de liaison entre X, Y et x, y sont du type le plus simple.

$$X = x + ax + by + c$$
$$Y = y + a'x + b'y + c'.$$

Les inconnues seront les α, δ des étoiles et des corrections δ_a, δ_b, δ_c, $\delta_{a'}$ $\delta_{b'}$, $\delta_{c'}$, aux constantes primitives de clichés.

Les étoiles de repère i, de coordonnées α_{iR} et δ_{iR} fournissent des équations de conditions de type $\alpha_i = \alpha_{iR}$ de poids $\sim \cos^2 \delta/\varepsilon_1^2$.

Chaque étoile, de repère ou non, donne sur chaque cliché une équation de condition contribuant à déterminer les retouches à apporter aux constantes primitives pour obtenir la coordonnée inconnue de l'étoile.

$$\alpha = \alpha_c + \left(\frac{\partial \alpha}{\partial X}\right)[\delta_a x + \delta_b y + \delta c] + \left(\frac{\partial \alpha}{\partial Y}\right)[\delta_{a'} x + \delta_{b'} y + \delta c'].$$

Le poids à donner à ces équations est $\sim \cos^2 \delta/\varepsilon_2^2$. C'est le calcul qu'on fera sur un cas concret. Mais on obtient une matrice suffisamment précise pour les évaluations théoriques de précision seulement en négligeant $\partial \alpha/\partial Y$ et en admettant $\partial \alpha/\partial X = 1/\cos \delta$. Il n'y a pas de problèmes polaires, car on peut traiter la question de précision avec d'autres coordonnées que les coordonnées équatoriales.

On peut établir avec toutes les équations de condition et en réduisant la matrice par élimination des coordonnées inconnues des étoiles, un système de résolution par les moindres carrés, avec au premier membre une matrice A reliant les inconnues X que sont les corrections aux constantes de clichés et au deuxième membre B des combinaisons linéaires connues des α_c et δ_c donc des α_i et δ_i des étoiles du catalogue de repère.

On doit résoudre $AX = B$.

On peut étudier la structure de la solution de ce système en le résolvant par approximations successives. Les termes les plus grands de A sont ceux qui relient les constantes d'un même cliché; soit N la matrice qui groupe ces termes.

On peut inverser aisément N.

On résoud $X_1 = BN^{-1}$, puis on calcule $B_1 = B - AX_1$.

On réitère $X_2 = B_1 N^{-1}$, puis $B_2 = B - A(X_1 + X_2)$, etc. …

Ces calculs peuvent être effectués de façon littérale et à chaque pas on peut calculer les dépendances des termes du vecteur X par rapport aux α_i et δ_i, puis les dépendances entre les α et δ calculés après chaque itération par rapport aux α_i et δ_i.

A partir de ces dépendances, on peut étudier l'incertitude à craindre sur les constantes et l'incertitude à craindre sur les α, δ calculés et ceci à chaque itération. Seule l'étude sur les α, δ calculés a été avancée jusqu'ici et on n'a pas tenu compte des incertitudes sur les dépendances elles-mêmes, c'est-à-dire que les résultats seront probablement à retoucher quand ils sont obtenus avec beaucoup d'itérations. On a travaillé jusqu'ici exclusivement en admettant des recouvrements de clichés par quarts; chaque étoile étant obtenue 2 fois. On a considéré ce que sont les dépendances lorsqu'on fait après achèvement du calcul la moyenne des déterminations sur 2 clichés.

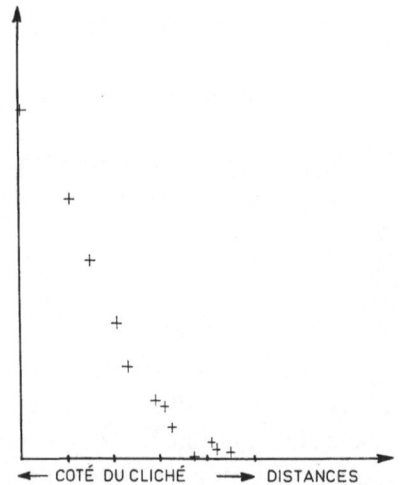

FIG. 1. *Dépendances formules linéaires. $K = 3\cdot7$.*

4. Système du catalogue photographique

Les cartes de dépendances, après quelques itérations, montrent en moyenne à partir du point pour lequel on les calcule, une diminution à peu près linéaire des dépendances en fonction de la distance. Ce caractère est indépendant du type des formules reliant les X, Y aux x, y, du nombre des étoiles repères et du rapport K du nombre total des étoiles au nombre des étoiles de repère. Seule l'ouverture de ce cône de dépendance varie; elle est liée à $\Sigma(\beta^2 + \gamma^2)$ et à la densité en étoiles repères. (Voir Figure 1.)

Ce caractère approximatif, très général, montre comment le système du catalogue de repère est lissé quand on établit un catalogue photographique en utilisant les recouvrements de clichés.

Il est ainsi intéressant de pouvoir calculer correctement l'effet sur le catalogue photographique de changements apportés au catalogue de repère.

5. Erreurs systématiques d'origine aléatoire

L'importance de ces erreurs peut être chiffrée par la valeur de $\Sigma(\beta^2 + \gamma^2)$. On a étudié un certain nombre de facteurs:

A. INFLUENCE DES FORMULES DE RÉDUCTION

Il est évident *a priori* que plus on emploie des formules de réduction souples, avec de nombreux paramètres, moins la propagation des dépendances d'un cliché à un autre sera étendue.

On a comparé les cas suivants:

I. Formules linéaires:

$$X = ax + by + c$$
$$Y = a'x + b'y + c'.$$

II. Formules linéaires avec correction du point tangentiel:

$$X = ax + by + c + px^2 + qxy$$
$$Y = a'x + b'y + c' + pxy + qy^2.$$

III. Formules linéaires avec correction du point tangentiel et distorsion inconnue de centre connu:

$$X = ax + by + c + px^2 + qxy + rx(x^2 + y^2)$$
$$Y = a'x + b'y + c' + pxy + qy^2 + ry(x^2 + y^2).$$

IV. Formules linéaires avec correction du point tangentiel et des termes pour déplacer le centre d'une correction de distorsion préalablement appliquée au centre:

$$X = ax + by + c + px^2 + qxy + s(3x^2 + y^2) + 2txy$$
$$Y = a'x + b'y + c' + pxy + qy^2 + 2sxy + t(x^2 + 3y^2).$$

V. Formules quadratiques:

$$X = ax + by + c + py^2 + qxy + ry^2$$
$$Y = a'x + b'y + c' + p'x^2 + q'xy + r'y^2.$$

Pour chacune de ces formules on a admis les recouvrements des clichés par quart, 16 étoiles de repère bien réparties, au total 60 étoiles par cliché ($K = 3\cdot7$). On a calculé les sommes des carrés des dépendances, pour les moyennes obtenues sur 2 clichés après différents nombres d'itérations. Ce calcul a été fait pour deux positions types sur les plaques; position A, position d'une des étoiles de repère près d'un coin sur un cliché et près du centre sur l'autre, position B, position d'une étoile repère près du bord sur les 2 clichés. (Voire Figure 2 et Tableau 1.) On a en outre admis $\Sigma_1 = \Sigma_2$.

On constate que la convergence est bonne.

On voit en outre, ce qui était prévu, que les formules les plus simples, les moins souples, sont les plus efficaces. La diminution du nombre des constantes inconnues joue le rôle principal pour diminuer les incertitudes, mais la forme sous laquelle ces inconnues interviennent compte également.

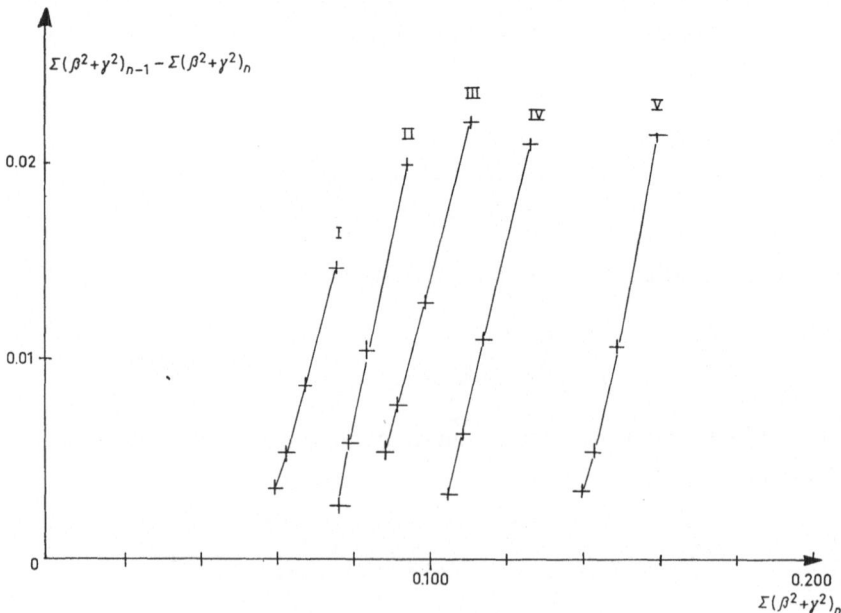

FIG. 2. *Convergence, 16 repères. $K = 3\cdot7$.*

Tableau 1

$$10^4 \, \Sigma(\beta^2 + \gamma^2)$$

	I A	I B	II A	II B	III A	III B	IV A	IV B	V A	V B
1er calcul	1397	1461	2001	1753	2386	1879	2868	1823	3078	2578
1ère itération	872	912	1181	1110	1396	1259	1715	1199	1959	1630
2e itération	726	758	967	914	1137	1062	1463	1024	1720	1431
3e itération	644	671	854	812	997	945	1330	932	1603	1335
4e itération	598	622	795	755	920	981	1258	884	1548	1288
5e itération	566	589	773	719	874	837	1209	852	1512	1258

Tableau 2

$$10^4 \frac{\Sigma(\beta^2 + \gamma^2)}{m}$$

cas	I	II	III	IV	V
m inconnues	6	8	9	10	12
1er calcul	237	235	237	234	235
1ère itération	149	143	147	146	149
2e itération	124	118	122	124	132
3e itération	110	104	108	113	122
4e itération	102	97	118	100	108
5e itération	96	93	95	103	115
limite	87	90	88	100	110

Dans le Tableau 2, on donne les $\Sigma(\beta^2 + \gamma^2)$ moyens des types A et B divisés par le nombre m des inconnues dans chacun des cas du Tableau 1.

Pour un même nombre d'inconnues, les formules qui établissent des liens entre les X, Y et les x, y avec des inconnues ayant un sens physique sont un peu plus favorables.

Pour pouvoir utiliser des formules moins souples avec un nombre faible d'inconnues, il y a grand intérêt à effectuer avant comparaison des plaques toutes les corrections qu'on peut appliquer de façon automatique au lieu de les faire représenter empiriquement par des termes introduits dans les formules de réduction. Ce sont par exemple, des corrections différentielles de réfraction bien calculables et des corrections de champ, dans la mesure où elles sont constantes.

B. INFLUENCE DE LA PROPORTION DES ÉTOILES DE REPÈRE

Il y a deux problèmes à examiner; le nombre des étoiles de repère par clichés et la proportion des étoiles de repère et des étoiles de catalogue.

1. *Nombre des étoiles de repère*

On a fait des calculs de $\Sigma(\beta^2 + \gamma^2)$ pour les mêmes formules, type II, la même valeur, $K = 3 \cdot 7$ la moyenne de deux clichés et des nombres d'étoiles de repère, n, différents.

Le Tableau 3 donne le produit par n de la valeur moyenne de $\Sigma(\beta^2 + \gamma^2)$ sur la surface du cliché.

Tableau 3

$$n \times \Sigma(\beta^2 + \gamma^2)$$

	$n = 16$ repères	$n = 64$ repères
1er calcul	3·002	2·806
1ère itération	1·832	1·645
2e itération	1·504	1·344
3e itération	1·333	1·184
4e itération	1·240	1·101
5e itération	1·194	1·043

On constate que le produit $n \times \overline{\Sigma(\beta^2 + \gamma^2)}$ est à peu près indépendant de n. On peut retenir cette règle pour effectuer des études ultérieures.

2. *Proportion des étoiles de repère et des étoiles de catalogue*

Un autre élément important est le rapport du poids de l'ensemble des équations de condition définissant les coordonnées des étoiles au poids total des équations de condition dues aux repères.

Pour discerner l'importance de ce facteur, on a repris le calcul avec 16 étoiles repères, formules linéaires avec correction du point tangentiel type II, moyenne de clichés, mais avec une série de valeurs de K; depuis $K = 1$, qui correspond à des clichés où il n'y aurait que les étoiles de repère, mais ou on tiendrait compte des comparaisons entre clichés pour ces étoiles, jusqu'à $K = 67$ où le nombre total des étoiles est 67 fois plus grand que celui des étoiles de repère. On a admis $\Sigma_1 = \Sigma_2$.

Le Tableau 4 donne seulement les moyennes de types A et B après chaque itération.

Tableau 4

$$\Sigma(\beta^2 + \gamma^2) \text{ moyens}$$

K	1	3·7	13·7	67
1er calcul	0·1876	0·1876	0·1876	0·1876
1ère itération	0·1363	0·1145	0·1091	0·1076
2e itération	0·1279	0·0940	0·0841	0·0812
3e itération	0·1247	0·0833	0·0707	0·0674
4e itération	0·1237	0·0775	0·0626	0·0585
5e itération	0·1233	0·0746	0·0569	0·0524

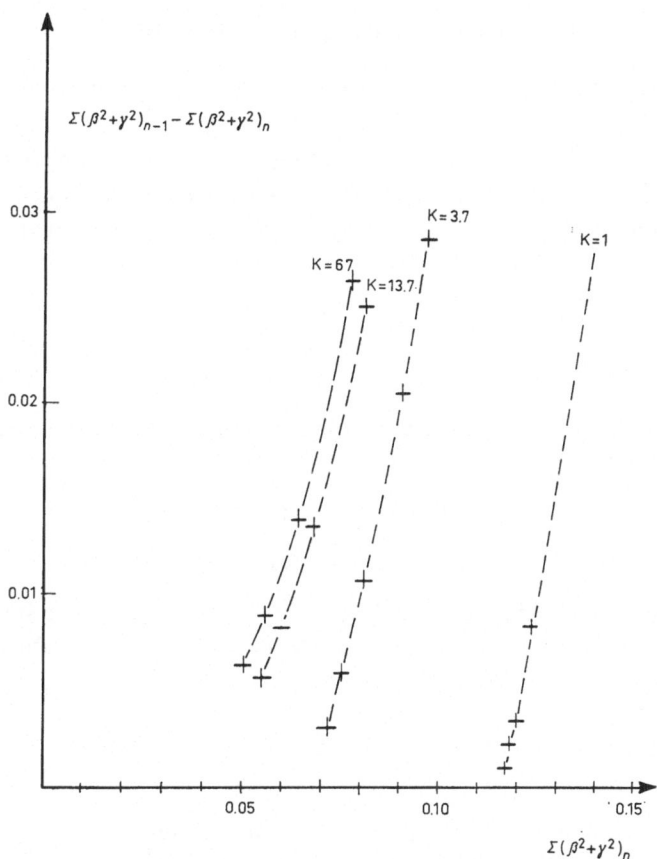

FIG. 3.   *Convergence. Formules linéaires avec point tangentiel. 16 étoiles de repères.*

On a reporté sur un graphique (Figure 3) les différences de valeurs de $\Sigma(\beta^2+\gamma^2)$ en fonction de $\Sigma(\beta^2+\gamma^2)_n$ pour les itérations successives et les différentes valeurs de K.

On constate que lorsque K augmente, la convergence est moins rapide, mais l'extension des dépendances est plus large.

Pour $K=1$ et $K=3\cdot7$ après 5 itérations, on voit bien vers quelle limite on tend. Pour $K=13\cdot7$ et $K=67$, il faudrait quelques itérations de plus pour mieux préciser la limite.

On a cherché une relation entre K et la limite. En calculant limite $\Sigma(\beta^2+\gamma^2)\times\sqrt[3]{K}$ on constate que cette expression n'est pas très variable (v. Tableau 5).

Tableau 5

	$K=1$	$K=3\cdot7$	$K=13\cdot7$	$K=67$
limite $\Sigma(\beta^2+\gamma^2)$	0·121	0·070	0·040 ?	0·030 ?
limite $\Sigma(\beta^2+\gamma^2)\times\sqrt[3]{K}$	0·121	0·108	0·096 ?	0·122 ?

6. Problèmes du catalogue de repère

On peut interpréter les résultats ci-dessus au-delà des hypothèses posées pour les établir.

En effet, quand l'expression $K(2\varepsilon_1^2/\varepsilon_2^2 + 1)$ reprend la même valeur la matrice est la même, on aboutit au même calcul de dépendances, et aux mêmes valeurs de $\Sigma(\beta^2 + \gamma^2)$. Etant donnée la variation de $\Sigma(\beta^2 + \gamma^2)$ constatée en fonction de K, on a:

$$\Sigma(\beta^2 + \gamma^2) \sim \frac{1}{\sqrt[3]{K\left(2\dfrac{\varepsilon_1^2}{\varepsilon_2^2} + 1\right)}}.$$

En désignant par E'^2 la part systématique d'origine aléatoire des erreurs E^2, on a $E'^2 = \Sigma(\beta^2 + \gamma^2)(\varepsilon_1^2 + \varepsilon_2^2/2)$.

Il vient:

$$\frac{E'^2}{\varepsilon_2^2} \sim \left(1 + 2\frac{\varepsilon_1^2}{\varepsilon_2^2}\right)^{2/3}.$$

Cette formule montre l'influence de la qualité du catalogue repère sur les résultats.

En introduisant la notion de travail T exigé pour établir le catalogue de repère; pour des plaques de dimensions données on a:

$$T \sim \frac{n}{\varepsilon_1^2}.$$

En utilisant des résultats antérieurs, notamment le fait que

$$n(\beta^2 + \gamma^2)\sqrt[3]{K\left(2\frac{\varepsilon_1^2}{\varepsilon_2^2} + 1\right)}$$

est à peu près constant pour un type de formule donnée, on peut établir le Tableau 6 et la Figure 4 pour des clichés de 200 étoiles avec des formules de type II, linéaire avec correction du point tangentiel.

Ce tableau et la Figure 4 montrent ce qu'on peut obtenir en utilisant les recouvrements de clichés:

(1) On peut se contenter de catalogues de repère beaucoup moins bon et demandant moins de travail pour obtenir une valeur de E'^2/ε_2^2 donnée.

(2) Dans la mesure où le catalogue est bon, $\varepsilon_1 < \varepsilon_2$, on a intérêt à répartir l'effort sur plus d'étoiles de repère.

Le catalogue AGK2-AGK3 extrapolé des époques 1930–58 vers 1900 sera un catalogue bon en tant que système, mais médiocre au point de vue des erreurs aléatoires; $\varepsilon_1 \sim 0''.4$ et $\varepsilon_1/\varepsilon_2 = 2$.

En utilisant ce catalogue pour la réduction du 'catalogue photographique', si on

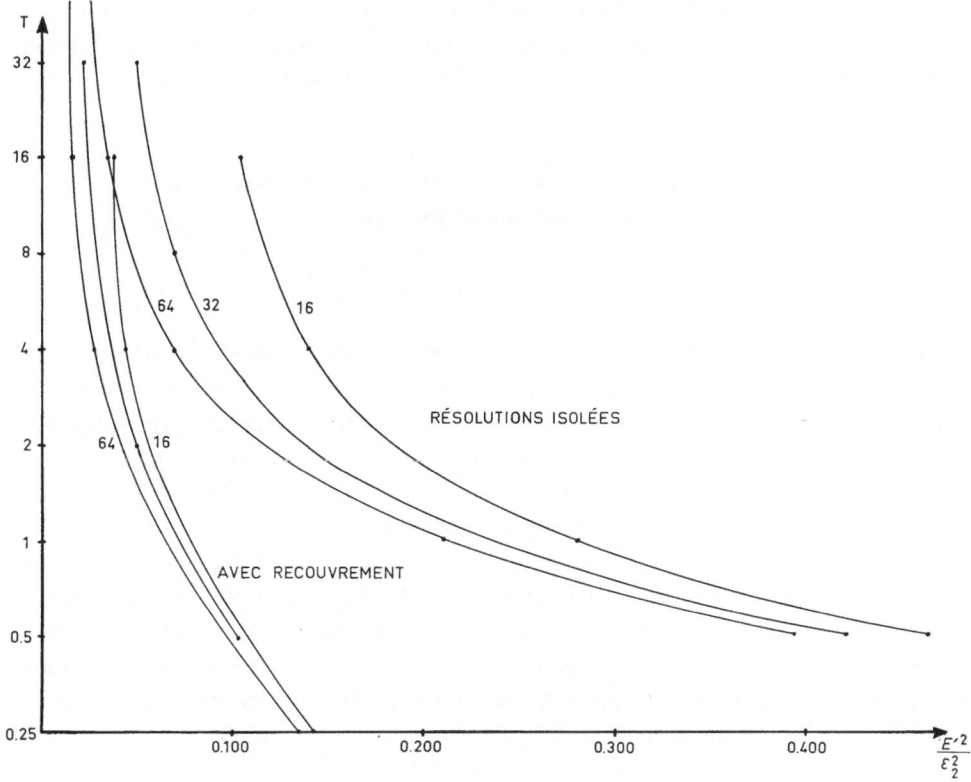

FIG. 4.   *Cliché de 200 étoiles. 16, 32, 64 étoiles de repères.*

Tableau 6

Clichés de 200 étoiles

	$\dfrac{\varepsilon_1}{\varepsilon_2}$	T catalogue de repère	E'^2/ε_2^2 ensemble	E'^2/ε_2^2 isolés	Facteur de poids dû au recouvrement
16 repères	$\frac{1}{4}$	16	0·038	0·105	2·8
	$\frac{1}{2}$	4	0·044	0·141	3·2
	1	1	0·070	0·281	4·0
	2	$\frac{1}{4}$	0·146	0·842	5·8
32 repères	$\frac{1}{4}$	32	0·0239	0·052	2·2
	$\frac{1}{2}$	8	0·0278	0·070	2·5
	1	2	0·0442	0·140	3·2
	2	$\frac{1}{2}$	0·092	0·421	4·6
64 repères	$\frac{1}{4}$	64	0·0150	0·026	1·7
	$\frac{1}{2}$	16	0·0175	0·035	2·0
	1	4	0·0278	0·070	2·5
	2	1	0·0578	0·211	3·6
	4	$\frac{1}{4}$	0·137	0·778	5·7

exploite les recouvrements de clichés, on diminuera par un facteur 5 environ le carré moyen des erreurs systématiques d'origine aléatoire qui sans cela seraient du même ordre de grandeur que les erreurs aléatoires dues aux mesures individuelles.

7. Choix des astrographes et méthodes d'établissement de catalogues photographiques

A. CONSIDÉRATIONS GÉNÉRALES

Ces travaux sont lourds et doivent être automatisés dans toute la mesure actuellement possible. La détermination des coordonnées des étoiles de repère au méridien constitue une grande partie du travail; il apparait que le nombre d'heures d'observation à consacrer à ce travail est difficilement compressible. On devra donc, malgré les calculs plus lourds, utiliser les recouvrements de clichés pour réaliser des économies sur le travail d'établissement du catalogue de repère.

Pour automatiser les calculs sans recherches fréquentes individuelles des erreurs, il faut disposer de trois images à mesurer par étoile. Les recouvrements devront être plus larges que ce qu'on fait d'ordinaire. Ceci sera en outre certainement favorable à l'étalement des dépendances, à la convergence du calcul et à la diminution des erreurs aléatoires. Mais on n'a pas encore exactement chiffré tous ces effets.

On devra automatiser la mesure des plaques dans toute la mesure possible; sûrement l'enregistrement des mesures, peut-être aussi les pointés.

B. ASTROGRAPHE

On a vu que l'étalement des dépendances en surface s'évalue en une unité qui est la dimension du cliché et que cet étalement dépend des formules de résolution utilisées. Pour diminuer le travail d'établissement du catalogue de repère, on devra employer des astrographes au champ le plus grand possible dans la mesure où l'augmentation de champ ne diminuerait pas la précision des mesures et n'imposerait pas des formules plus complexes faisant perdre les bénéfices d'augmentation du champ.

Soit S la surface d'un cliché, n le nombre moyen d'étoiles de repère, N le nombre moyen du total des étoiles mesurées, σ et Σ les densités, en étoiles, E' l'incertitude aléatoire sur le système en un point:

$$E'^2 = \Sigma(\beta^2 + \gamma^2)\left(\varepsilon_1^2 + \frac{\varepsilon_2^2}{3}\right).$$

En utilisant les règles de variations de $\Sigma(\beta^2 + \gamma^2)$ qui ont été établies en fonction du nombre m des constantes, de n, et de K dont on a donné plus haut la valeur, en admettant pour le recouvrement par tiers les mêmes réductions de $\Sigma(\beta^2 + \gamma^2)$ que pour le recouvrement par quart, ce qui est encore à étudier, il vient:

$$\frac{E'^2}{\varepsilon_2^2} = 0{\cdot}355 \times \frac{m}{S} \times \frac{1}{\sigma^{\frac{4}{3}} \times \Sigma^{1/3}} \times \frac{\dfrac{1}{3} + \left(\dfrac{\varepsilon_1}{\varepsilon_2}\right)^2}{\sqrt[3]{1 + \dfrac{2\varepsilon_1^2}{\varepsilon_2^2}}}.$$

La solution la plus avantageuse pour l'astrographe consistera à prendre la plus faible valeur de $(m/S)\,\varepsilon_2^2$. Les clichés des télescopes de Schmitt pourraient bien être les plus avantageux.

Pour réduire le nombre de constantes à introduire on devra corriger les clichés avant de les comparer entre eux de toutes les erreurs systématiques connues.

A égalité d'étalement des dépendances, la solution à grand champ serait la plus avantageuse car il y aurait moins de clichés à prendre, on pourrait les prendre plus vite et on risquerait moins l'influence des mouvements propres dans la comparaison.

C. CATALOGUE REPÈRE

Après le choix de l'astrographe, on devra rechercher la densité en étoiles de repère et l'incertitude ε_1 qui permettent d'obtenir la valeur de E' désirée avec le minimum de travail sur le catalogue de repère. Or le travail est: $T = \sigma/\varepsilon_1^2$.

En utilisant la formule précédente qui donne E'^2/ε_2^2 il vient:

$$T = \frac{\varepsilon_2}{E'^3} \left(\frac{0{\cdot}355\,m}{\Sigma^{1/3} \cdot S}\right)^{3/2} \left(\frac{\varepsilon_2}{\varepsilon_1}\right)^2 \frac{\left[\dfrac{1}{3} + \left(\dfrac{\varepsilon_1}{\varepsilon_2}\right)^2\right]^{3/2}}{\left[1 + 2\left(\dfrac{\varepsilon_1}{\varepsilon_2}\right)^2\right]^{1/2}}.$$

On constate que le travail diminue toujours quand on augmente $\varepsilon_1/\varepsilon_2$, c'est-à-dire quand on répartit l'effort sur un plus grand nombre d'étoiles de repère:

$(\varepsilon_1/\varepsilon_2)$	1/4	1/2	1	2	∞
Facteur sur T	1·46	1·07	0·90	0·80	0·71

Pour des raisons pratiques, on ne devrait pas descendre en-dessous de trois observations méridiennes par étoile, c'est ce qui fixera $\varepsilon_1/\varepsilon_2$ ensuite on calculera σ pour obtenir la valeur E'^2 désirée.

L'établissement d'un catalogue de repères comportant peu d'observations sur un grand nombre d'étoiles peut sembler difficilement compatible avec le désir de faire contribuer au travail plusieurs instruments et de pouvoir comparer leurs résultats. Le plus pratique serait que chaque observatoire ait un programme différent avec des étoiles bien mélangées sur la sphère. La comparaison des systèmes aurait lieu après résolution du catalogue photographique qui donnerait une référence commune précise.

Quand on résoud des clichés séparément, on doit disposer des positions précises des étoiles de repère avant de faire les solutions. Si on dispose seulement de positions approchées, les calculs seront à reprendre complètement sur chaque cliché.

En résolvant un ensemble de clichés avec l'utilisation des recouvrements, on peut terminer tout le travail à partir des positions approchées des repères. Il restera seulement, quand on aura les positions repères précises, à déterminer des corrections systématiques dont le calcul se fera avec sécurité et qui seront lentement variables en α et δ; on se basera pour cela sur les liens établis et étudiés entre le système photographique et le système de repère.

8. Calculs effectués sur l'AGK2-AGK3

Monsieur Dieckvoss, de l'Observatoire de Hambourg, nous a communiqué sous forme accessible mécanographiquement:

(1) Des positions α et δ, AGK2 par cliché, ramenées du FK3 au FK4.

(2) Des positions α et δ, AGK3, déduites d'une représentation approximative des clichés, très peu précise.

(3) Des positions d'étoiles de repère α et δ; AGK2 et AGK3 ramenées au FK4.

Les positions AGK3 sont des positions provisoires déduites des observations de quelques observatoires par F. P. Scott.

Nous avons cherché à apporter des corrections aux constantes de clichés utilisées par Dieckvoss pour les α, δ qu'il nous a communiqués en utilisant en plus des informations apportées par les étoiles de repère, les informations déduites de la comparaison entre les coordonnées d'une même étoile sur des clichés différents.

Pour que la comparaison entre clichés soit plus correcte, nous avons, à partir des positions de Dieckvoss pour époques voisines de 1930 et de 1958, calculé des positions pour les époques exactes 1930·0 et 1958·0, et c'est sur ces positions que nous avons opéré. On a en outre calculé à partir des α, δ, les X, Y et les dérivées partielles

$$\frac{\partial\alpha}{\partial X}\;;\;\frac{\partial\alpha}{\partial Y}\;;\;\frac{\partial\delta}{\partial X}\;;\;\frac{\partial\delta}{\partial Y}.$$

Toute la préparation des données antérieures à l'établissement de la matrice a été effectuée en collaboration avec Monsieur Bacchus.

Les formules prévues par Dieckvoss pour ses résolutions comportaient des termes p et q pour corriger le point tangentiel. Nous avons donc employé les mêmes formules. Cependant, nous avons aussi ajouté un terme fonction de la magnitude, ce qui nous était suggéré comme utile par des expériences antérieures.

Les formules qui relient les coordonnées mesurées x, y et les coordonnées normales

X, Y sont donc:

$$X = ax + by + c + \lambda m + px^2 + qxy$$
$$Y = a'x + b'y + c' + \lambda'm + pxy + qy^2.$$

Soit 10 inconnues.

On a considéré les corrections aux constantes de clichés comme étant les inconnues afin d'éviter le trop grand nombre d'inconnues qu'on aurait obtenu en considérant aussi les α, δ de toutes les étoiles comme inconnus. Il a été fait ainsi parce qu'on n'avait pas vu la possibilité signalée par Goole de réduire la matrice. Les résultats ne sont donc pas les meilleurs possibles, mais l'étude théorique montre qu'ils sont très proches.

Les équations de conditions ont donc été des équations de comparaison entre clichés.

$$\alpha_{Kj} - \alpha_{Kl} = \left[\begin{matrix} \left(\dfrac{\partial \alpha}{\partial X}\right)(\delta_a x + \delta_b y + \delta_c + \lambda m + px^2 - qxy) \\ + \left(\dfrac{\partial \alpha}{\partial Y}\right)(\delta_{a'} x + \delta_{b'} y + \delta_{c'} + \lambda'm + pxy + qy^2) \end{matrix} \right]_{Kl} - \left[\right]_{Kj}.$$

En outre, pour chaque étoile de repère i sur cliché l, on a ajouté une équation de condition du type suivant:

$$\alpha_{iR} = \alpha_{il} + \left[\begin{matrix} \left(\dfrac{\partial \alpha}{\partial X}\right)(\delta_a x + \delta_b y + \delta_c + \lambda m + px^2 + qxy) \\ + \left(\dfrac{\partial \alpha}{\partial Y}\right)(\delta_{a'} x + \delta_{b'} y + \delta_{c'} + \lambda'm + pxy + qy^2) \end{matrix} \right]_{il}.$$

On a donné à toutes les équations le même poids car le catalogue de repère utilisé pour l'AGK3 était le catalogue provisoire basé sur un petit nombre d'observations. Ceci revenait à admettre que l'incertitude aléatoire sur la position admise pour une étoile repère est à peu près égale à l'incertitude aléatoire sur la mesure d'une position sur les plaques, de l'ordre de 0″.2.

Les plaques sont toutes liées, mais chacune directement seulement avec ses voisines immédiates. On établit le système d'équations normales de la résolution par la méthode des moindres carrés; il y a beaucoup d'inconnues, mais beaucoup de coefficients sont nuls. La solution de ce système $AX = B$, où le vecteur X représente l'ensemble des inconnues, donne un résultat correct au sens des moindres carrés pour les équations admises.

Pour établir le système, on a éliminé toutes les équations qui aboutissaient à des différences $\Delta\alpha$ ou $\Delta\delta$ abusivement grandes, supérieures à 1″.2, dues évidemment à des erreurs de mesures, ou d'identification, sans examiner les problèmes posés, n'ayant pas les moyens de le faire. Le système $AX = B$ englobe ainsi 1219 clichés, soit 12190 inconnues.

Le système a été résolu par approximations successives comme il est dit plus haut ; on détache de la matrice A la matrice N qu'on obtient en supprimant tous les termes qui relient les constantes appartenant à des clichés différents. La matrice N se décompose en autant de matrices que de clichés et elles sont aisément inversables.

On calcule une première approximation de X, X ; $X = N^{-1}B$ puis $B_1 = B - AX_1$; on réitère $X_i = N^{-1}B_{i-1}$ suivi de $B_i = B - A(X_1 + X_2 + \cdots)$.

On a ainsi des approximations successives de X :

$$X_1 ; X_1 + X_2 ; \text{etc.} \ldots \quad X_1 + X_2 + \cdots + X_n .$$

Faute de capacités des machines on a limité la résolution à un premier groupe de 691 clichés soit 6910 inconnues entre le pôle et $+42°$ de déclinaison en détachant dans le système ce qui concernait cet ensemble de clichés. Le bord est donc mal résolu et devra être repris dans l'ensemble suivant.

On constate que la convergence est très bonne. On donne dans le Tableau 7 les sommes des carrés des termes du 2e membre divisés chacun par les termes diagonaux correspondants de la matrice.

Tableau 7

$$\Sigma_i \left(\frac{b_i^2}{a_{ii}} \right)$$

	1930	1958
Début	38965	110330
1ère itération	10035	46877
2e itération	1419	5268
3e itération	283	273
4e itération	70	159
5e itération	7·0	16·8
6e itération	1·0	4·48
7e itération	0·25	0·50
8e itération	0·058	

En fait, après quelques itérations, les retouches au vecteur X deviennent négligeables.

En se basant sur les comparaisons initiales des α et δ sur les clichés, on peut déterminer une *limite supérieure* de l'incertitude aléatoire des coordonnés des *premières réductions*. On a trouvé pour l'ensemble les chiffres donnés dans le Tableau 8.

Tableau 8

	1930		1958	
	en $\alpha \cos \delta$	en δ	en $\alpha \cos \delta$	en δ
ε^2	402	447	957	847 en $(0''01)^2$
ε	$0''20$	$0''21$	$0''31$	$0''29$

Il est certain que les évaluations pour 1958 donnent des nombres trop grands, car les solutions utilisées comme point de départ pour établir la matrice sont des approximations très peu précises. On devrait reprendre ces estimations après application des corrections.

9. Précision des solutions

Il est difficile d'évaluer la précision des solutions obtenues parce qu'on ne peut pas inverser la grande matrice.

On est tenté de penser que l'incertitude sur les estimations successives des termes du vecteur X mesure l'incertitude des constantes de cliché. On peut aisément calculer ces incertitudes en inversant la matrice N.

En fait, ce calcul donne bien l'incertitude des constantes de clichés qui appliquent chaque cliché sur le système d'ensemble. Mais il convient de se rappeler que le système d'ensemble est lui-même soumis à des incertitudes. Si bien que les incertitudes réelles sont sûrement plus grandes que celles ainsi calculées.

Pour les constantes appliquant ainsi les clichés sur le système, le gain en poids est à peu près un facteur $1 + K$, par rapport à la solution par clichés isolés.

On peut faire une étude théorique de l'imprécision des constantes en calculant les dépendances entre chaque constante et les coordonnées admises pour les étoiles de repère. Le calcul est du même genre que celui qui a été fait pour obtenir les dépendances entre les α et δ calculés et les coordonnées admises pour les étoiles de repères. Mais les dépendances ne sont pas les mêmes que pour les α et δ et sont différentes pour chaque constante. Les programmes sont prêts, malheureusement les calculs n'ont pas encore pu être exécutés.

10. Remarques sur les constantes

Les corrections aux constantes qui ont été obtenues appellent peu de remarques.

Cependant, comme on est parti pour l'époque de l'AGK3 de représentations empiriques très peu précises, les corrections aux constantes qui sont calculées sont plus grandes que pour l'AGK2. Dans le cas particulier des constantes p et q, on a obtenu ce que donne le Tableau 9.

Tableau 9

	$\overline{p_2^2}$	ε_p^2	$\overline{q^2}$	ε_q^2
AGK2	257	70	304	70
AGK3	829	160	770	160

Les valeurs de ε_p^2 et ε_q^2 données ici résultent de l'hypothèse très pessimiste d'un gain en poids par un facteur 3 sur les solutions par clichés isolés.

11. Cas des λ et λ'

Les coefficients de m que nous avons introduits sont un peu moins significatifs, ce qui n'est pas surprenant. L'existence de ces termes qui n'ont pas une valeur moyenne significativement différente de zéro traduit probablement des erreurs de guidage qui pourraient ne pas exister.

On avait évalué les magnitudes en prenant une unité telle que pour $\lambda = 50$ on a un déplacement de 0″,05 par magnitude. Dans le Tableau 10 on voit ce qui a été obtenu.

Tableau 10

	$\overline{\lambda^2}$	$\overline{\lambda'^2}$
AGK2	702	779
AGK3	2460	3140

Si on admettait que l'incertitude des λ et λ' est celle qui correspond à l'application des clichés sur le système on aurait:

AGK2	AGK3
$\varepsilon_{\lambda,\lambda'}^2 \sim 300$	~ 640

Les λ et λ' seraient dans l'ensemble nettement significatifs. L'incertitude qui sera calculée est certainement un peu plus grande, mais pas beaucoup.

L'application des corrections aux constantes que nous avons obtenues s'effectue aisément.

On peut comparer l'accord entre clichés avant et après. Cette comparaison n'a de

Tableau 11

Clichés		Nbre de comparaison	$\overline{\Delta\alpha\cos\delta}(0″01)$	Nbre de comparaison	$\overline{\Delta\delta}(0″01)$
55–03	avant	263	30·4	276	27·3
	après	273	23·2	276	23·6
facteur sur le poids			$> 1·70$		1·34
60–29	avant	205	24·1	205	25·2
	après	205	20·0	205	21·5
facteur sur le poids			1·45		1·37
62–28	avant	205	22·7	205	22·2
	après	210	20·2	210	19·9
facteur sur le poids			$> 1·25$		$> 1·25$
75–23	avant	234	21·9	234	20·3
	après	234	18·7	234	19·7
facteur sur le poids			1·37		1·06

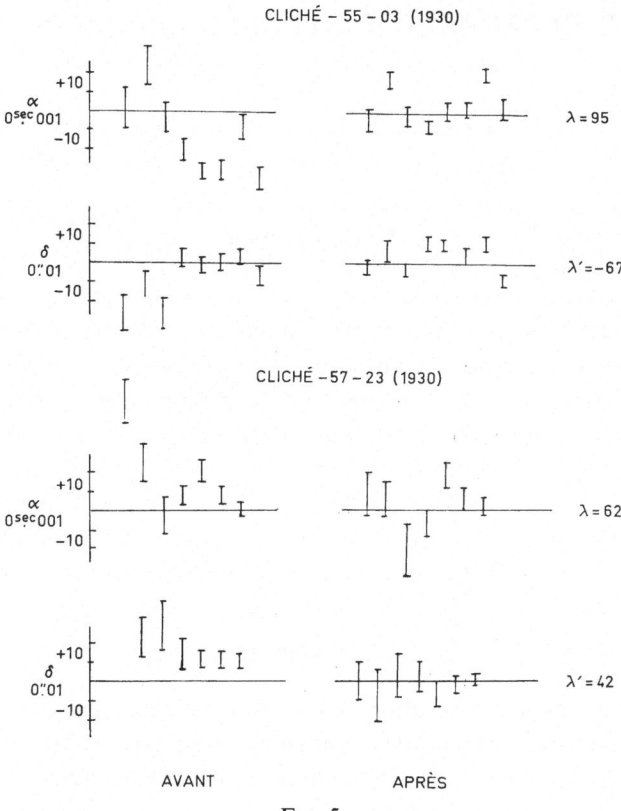

FIG. 5.

sens que pour l'AGK2 dont les premières solutions étaient bonnes en tant que solutions par clichés isolés.

L'étude complète n'a pas encore été faite et on doit se rappeler que les résultats seraient encore meilleurs si le problème avait été posé tout à fait correctement.

On donne (Tableau 11) les valeurs moyennes des écarts avec les clichés voisins après élimination des différences supérieures à 1″.2.

La Figure 5 montre par ailleurs l'utilité des λ. On y a reporté les différences de 2 clichés avec leurs voisins en classant les différences par groupes de demi-magnitude.

Conclusion

L'ajustement des clichés de l'AGK2-AGK3 en utilisant les recouvrements est indubitablement utile.

THE OVERLAP APPROACH TOWARD THE DERIVATION
OF PHOTOGRAPHIC STELLAR COORDINATES

W. D. Googe, C. F. Lukac, and H. Eichhorn

1. Introduction

The use of the conditions imposed by the appearance of one star on more than one photographic plate to strengthen the determination of the relationship between the measurement coordinate systems and the system of celestial right ascension and declination, was first advocated by Eichhorn (1960), and since then various reduction procedures have been proposed to incorporate these conditions in the plate-reduction process (Eichhorn, 1963; Henriksen, 1964; Lacroute, 1964; Googe, 1967; Clube, 1967). During the last several years at the Army Map Service the overlap method has been developed and used in several astrometric projects, and the purpose of this paper is to summarize the main features of the algorithm and some of the projects to which it has been applied.

2. The Algorithm

Two different types of observations enter into the reductions, the celestial coordinates of the reference stars and the rectangular coordinates of all the stars measured on the plate, and for each type observation equations are formed. For the plate measurements (x, y) the equations are:

$$x = a + b\xi + c\eta + \cdots$$
$$y = a' + b'\xi + c'\eta + \cdots,$$

where a, b, c, a', b', c', ... are the usual plate constants and

$$\xi = \xi(\alpha, \delta)$$
$$\eta = \eta(\alpha, \delta)$$

are the star's standard coordinates and thus functions of its celestial coordinates (α, δ). The number and form of the higher-order terms in the expansion depend, of course, on the attributes of the particular reduction being attempted, reflecting as they must the various physical phenomena that enter into the image formation and mensuration processes. The unknowns in these equations are the plate constants and the stellar coordinates.

Perek (ed.), Highlights of Astronomy, 338–342. © *I.A.U.*

A system of equations of the measurement type alone would be singular, that is, it would possess an infinity of least-squares solutions. To remove this singularity as well as to introduce the information contained in the observations of the reference-star coordinates, additional equations are used:

$$\alpha = \alpha$$
$$\delta = \delta.$$

These are interpreted to mean that the quantities on the left are observations of the unknown true coordinates on the right. In principle, there is no reason why these equations could not be made more elaborate. E.g., if several catalogues of reference material were to be used and there were doubts about the uniformity of systems, additional unknowns could be introduced to parameterize the systematic differences.

It should be noted explicitly that there will be a pair of equations for every measured image of a star and another pair for every reference-star observation. Thus the algorithm applies regardless of the pattern of the photography – a star may appear on one plate or many.

Once the observation equations are formulated, the standard non-linear least-squares methods apply. The equations are linearized, weighted, formed into a system of normal equations and a solution calculated. In practice, it is wise to eliminate algebraically the star unknowns in order to produce a system of reduced normal equations containing only plate constant unknowns (Googe, 1967).

If many plates are to be included, the system of equations will be large and some care must be exercised to obtain an acceptable solution. Several iterative methods have been discussed (Eichhorn, 1960; Lacroute, 1964; Henriksen, 1964, 1967) and there are many others available (Varga, 1962; Faddeev and Faddeeva, 1963). In particular, the results of Lacroute (1967) using block Jacobi iteration are most encouraging. It should be remembered, however, that the efficiency of these methods depends on the 'condition' of the coefficient matrix which is a delicate function of the number and type of terms used in the plate constant expansion, on the number and distribution of the reference stars, on the overlap pattern, and on the weighting scheme used. Thus one must be cautious about drawing general conclusions concerning speeds of convergence. At the Army Map Service we have preferred exact methods, that is, some form of direct Gaussian elimination. These not only avoid the problem of convergence but supply useful information about the propagation of errors from one plate to another, e.g. the inverse of the normal equations.

Another problem is the error introduced by the linearization of the non-linear observation equations. Our experience has shown that this is non-existent if care is taken to choose sufficiently accurate initial estimates of the unknowns. In fact, we have iterated many overlap adjustments and in every case the changes computed after the second linearization were negligible.

3. Zonal Reductions

At the Army Map Service we are currently at work preparing several zones of photographic material for reduction with the overlap algorithm. One of these zones, consisting of a series of 90 plates exposed for us at the Sydney Observatory by Dr. H. Wood, covers the Southern sky from $-48°$ to $-54°$. These plates have been measured at the University of South Florida. The region from $-60°$ to $-70°$ covered by Yale plates will also be reduced in the rigorous manner as will be the South polar cap ($-70°$ to $-90°$).

As preparation for this work we have developed programs to reduce all or part of the plates for a zone taken in the usual edge to centre overlap pattern. In order to test the process we have used measurements supplied by Dr. Stoy for the Cape Photographic Zone from $-52°$ to $-56°$. By making use of the greater flexibility allowed in the expansion by the overlap redundancy, it was possible to reduce the residual root mean square considerably. Table 1 gives some results from typical overlap solutions performed to evaluate various models.

Table 1

Solution	Terms in x	Residual Root Mean Square
1	$1, \xi$	$''253$
2	$1, \xi, \eta, \xi^2, \xi\eta$	$''234$
3	$1, \xi, \eta, \xi^2, \xi\eta, \eta^2$	$''229$
4	$1, \xi, \eta, \xi^2, \xi\eta, \xi(\xi^2 + \eta^2)$	$''195$
5	$1, \xi, \eta, \xi^2, \xi\eta, \eta^2, (\xi^2 + \eta^2)$	$''194$

The first three solutions showed considerable systematic error as well as a high r.m.s. value. In particular, the first adjustment produced plate constants which in the extreme gave systematic errors of 4/10 of a second of arc.

It is perhaps of interest that for a complete zone with about 40000 equations in 20000 unknowns the overlap adjustment takes less than 2 hours on the IBM 7094 computer. This includes the combination of direct and reverse raw measurements, application of comparator calibration corrections, classical solution to determine initial values for the plate constants together with the formation and solution of the linearized observation and normal equations.

4. Pleiades catalogue

Another application made of the overlap technique at the Army Map Service has been the reduction of a series of exposures made of the region of the Pleiades cluster

(Lukac, 1967). Thirteen plates were taken with the 26-inch McCormick telescope at the University of Virginia, and one additional plate was exposed on the 20-inch refractor of the Van Vleck Observatory. All together there were 65 exposures. A diffraction grating was used and the 20000 stellar images produced were measured by Eichhorn. Since only about 500 stars are involved, these abundantly redundant data are being used not only for a thorough statistical evaluation of the overlap algorithm but for the investigation of the diffraction-image formation process. The more complete model allowed by the extra redundancy introduced through the overlap conditions has reduced the systematic errors by a factor of about $\frac{1}{2}$. The accuracy of the resulting stellar coordinates will be of the order of 1/100 of a second of arc.

5. Advantages and Disadvantages of the Overlap

The main reason for using the overlap conditions in the plate reduction equations is to strengthen the determination of the relationship between the rectangular and spherical coordinate systems and practical results have verified that this is indeed accomplished. This means that fewer reference stars are needed, and the results of Eichhorn and Gatewood (1967) working with the Northern Hyderabad Zone of the Astrographic Catalogue show that, for this material at least, a decrease of $\frac{2}{3}$ in the number of reference-star observations gives an accuracy comparable to the classical method. At the Army Map Service we have found it possible to reduce accurately some plates in a complex with no reference stars whatever.

The overlap method, naturally, cannot increase the accuracy of the measurements. Nevertheless, putting more information into the reduction allows a more complete model of the image-formation process. This results in a less disparate adjustment. This is one reason why a correct model is more necessary for an overlap adjustment than for a classical solution. But a much more important reason remains. The overlap algorithm has the effect of fusing the plates into a single unit into which unaccounted for distortions can not only propagate but expand and be magnified. Examples of this are given by Eichhorn et al. (1967) and many more could be related. This need for a completely adequate model can only be met by extremely careful statistical and physical analyses.

References

Clube, M. (1967)   Private Communication.
Eichhorn, H. (1960)   *Astr. Nachr.*, **285**, 233.
Eichhorn, H. (1963)   Private Communication.
Eichhorn, H., Gatewood, G. (1967)   *Astr. J.* (in press).
Eichhorn, H., Googe, W.D., Gatewood, G. (1967)   *Astr. J.*, **72**, 626.
Faddeev, D., Faddeeva, V.N. (1963)   *Computational Methods of Linear Algebra*, Freeman, San Francisco, 621 pp.
Googe, W.D. (1967)   *Astr. J.*, **72**, 623.
Henriksen, S.W. (1964)   Private Communication.

Henriksen, S. W. (1967)   *Astr. J.*, **72**, 603.
Lacroute, P. (1964)   *Ann. Obs. Strasbourg*, **6**.
Lacroute, P. (1967)   Private Communication.
Lukac, C. F. (1967)   *Astr. J.*, **72**, 620.
Varga, R. (1962)   *Matrix Iterative Analysis*, Prentice-Hall, Englewood Cliffs, N.J.

REPORT ON OVERLAP METHODS IN
PHOTOGRAPHIC ASTROMETRY

Chr. de Vegt

The usual procedure for deriving standard coordinates ξ, η from the measured rectangular coordinates x, y of a star is to adopt a suitable model for the relationship between the measured x, y, magnitude m, color index c and the ξ, η. This relationship may, in general, be written as

$$\xi = \sum a_{ijkl} x^i y^j m^k c^l; \quad \eta = \sum b_{ijkl} x^i y^j m^k c^l.$$

From the known standard coordinates ξ_r, η_r of some reference stars equally distributed over the plate area, the unknown plate constants a_{ijkl}, b_{ijkl} are found from a least-squares adjustment. After the adjustment the standard coordinates of each field star can be computed. In most cases a statistical criterion for the choice of the proper form of the above relationship between measured and standard coordinates may be useful because the systematic accuracy, obtained from the adjustment, may be seriously affected by the number of unknown parameters included. (Eichhorn *et al.*, 1967). If the same star occurs on two or more plates, i.e. the plates being partly overlapped, one gets the final position by taking the mean of the individual positions on each plate. In this procedure, no direct account is taken of the 'overlap' in the equations of condition for the desired plate constants.

It was first proposed by Eichhorn (1960) to include the positions of the field stars – according to the condition that $\xi_i^1 = \xi_i^2$ and $\eta_i^1 = \eta_i^2$ for each star appearing in the overlap region of the plate 1 and 2, and referring all plates to a common tangential point – in the equations of condition, and solve for the field stars and plate constants together as unknowns.

Recently, Googe (1967) has modified this procedure by an elimination of the field stars from the normal equations so that the overlap condition is expressed by the plate constants themselves. Practical results of Eichhorn's method are now available (cf. Eichhorn and Gatewood, 1967) from a new reduction of the Northern Hyderabad-zone of the Astrographic Catalogue. A part of the material (103 plates) was subjected to an adjustment according to the plate-overlap method in groups of 6 plates each. It could be shown that, in comparison with the classical reduction of the individual plates, the overlap reduction is most effective in increasing the weight of coma and rotation coefficients, whereas zero point and magnitude effects are affected least. As an important result, it is found that an overlap solution using about $\frac{1}{3}$ of the reference

Perek (ed.), Highlights of Astronomy, 343–346. © I.A.U.

stars yields positions with the same systematic accuracy as a classical solution involving the full number of reference stars.

A somewhat different treatment of the problem has been given by Lacroute (1964). Assuming that an initial set of plate constants has been found, the differences $\Delta\alpha = \alpha_v^i - \alpha_v^j$, $\Delta\delta = \delta_v^i - \delta_v^j$ (v runs over all common stars of the plate-pair (i, j)) from the resulting individual positions on the overlapping plates are used to set up equations of condition for differential corrections of the plate constants. The possibility of directly handling the spherical coordinates themselves instead of using standard coordinates, is due to a first order approximation of the $\Delta\alpha$, $\Delta\delta$, treated as functions of the plate constants and the positions. Although parts of the AGK3 material have been sent to Lacroute by Dieckvoss in 1965, no positions according to the overlap method are available at present, so that there is no possibility of making a comparison with the classical reductions performed by Dieckvoss at the Hamburg Observatory. In general, the direct handling of spherical coordinates in overlap technique seems to be due to Eichhorn (1963).

Using the transformations (König, 1933) between the standard coordinates of two overlapping plates referred to different tangential points, de Vegt (1967) has recently set up a rigorous non-linear system of equations of condition involving as unknowns only the plate constants of all overlapping plates. Since the overlap condition is directly expressed in the equations of condition, no previous transformation to a common tangential point is necessary.

To have a comparison between the classical approach – involving only the reference stars – and the overlap solution, an artificial star field, generated by random numbers and divided into 13 $2° \times 2°$ 'plates', was used to obtain 'true' positions, not affected by any error. These true positions were superimposed by different Gaussian distributions $N(0, \sigma_i)$ to get 'observed' positions on each plate. From the resulting field-star positions of both, the classical and the overlap reduction, one can compare (1) the internal scatter (m.e.) of the field-star positions (from the contributions of the neighbouring plates), and (2) the mean differences in the sense 'true minus computed position'. It was found that for the overlap solution the mean error of the positions is reduced by 10–15%, and a better approximation of the true positions is obtained, that means: the overlap solution gives a better systematic representation of the desired position in comparison with the classical solution.

Furthermore, Clube (1967) has described a method to obtain accurate relative star positions from overlapping photographic plates. The configuration of stars in common to each pair of overlapping plates are adjusted for scale and orientation, so that the relative displacements on the celestial sphere of the stars in common are minimized. A direct non-linear relation between the plate constants and the equatorial direction cosines τ_1, τ_2, τ_3 of the stars is used. Starting with an adopted set of provisional equatorial coordinates, the plate constants are computed to get improved direction cosines. Weighted means of the τ_j are then found from overlapping plates on which each star appears, and these are used as improved estimates in repeating the above procedure, so that an iterative process is set up. By this process, the overlapping plates should get a 'rigid connection' and only a few reference stars or starlike extragalactic objects should be necessary to transform the relative star positions to an inertial reference frame. The question of the convergence of the iterative process is under further investigation. Since the whole program is in the development stage, no practical results are available at present. Recently, Lukac (1967) used Eichhorn's overlap method in an investigation concerning relative star positions in the Pleiades cluster.

Some few remarks may be of interest concerning the mathematical problems arising from overlap solutions. In all cases, the resulting linear system of normal equations involves a large number of unknowns, especially if complete catalogue-zones are reduced as a whole, so that an iterative solution scheme would always be necessary. Since the matrix of the normal system is positive-definite some, *a priori*, convergent iterative methods are available, as the Gauss-Seidel method (Faddeev and Faddeeva, 1964) or gradient-methods. Of special advantage seems to be the method of group-iteration (Zadunaisky, 1957; Jefferys, 1963) because, on account of the special structure of the normal equations in most cases, a splitting up into block-type matrices is possible. A difficulty may further arise from the non-linearity of the equations of condition because the necessary neglecting of higher-order terms in the Taylor approximation may invalidate the linearisation process if the initially adopted approximate values for the unknowns differ too much from the exact values. (In the case of de Vegt's overlap solution it was found that the plate constants performed from the classical reduction gave a fairly good approximation.) If this situation prevails, the 'method of damped least squares' by Levenberg (1944) may give an improvement. A detailed discussion of the topological structure of the matrices resulting from overlap conditions is given by Henricksen (1967), who also gives an estimation of the needed machine hours for the reductions. It is shown that an iterative solution-procedure requires fewer operations than does a direct method in this kind of astrometric problems.

As is shown from the practical results of Eichhorn and de Vegt, an overlap solution seems to give a real improvement compared with the classical treatment. However, the question concerning a possible reduction of the numbers of reference stars in overlap techniques should be a matter of further discussion, although Eichhorn's result seems to give a positive hint. Recent investigations by Dieckvoss with regard to the AGK3 reductions have shown that the overlap method may fail if only a small number of common stars is available. Depending also on the accuracy of the measured x, y, the plates then cannot be so 'rigidly' connected by the overlap that the reduced number of reference stars will provide an accurate transformation to an inertial reference system.

346   CHR. DE VEGT

References

Clube, S. V. M. (1967)   Private Communication.
de Vegt, Chr. (1967)   *Astr. Nachr.*, **290** (in press).
Eichhorn, H. (1960)   *Astr. Nachr.*, **285**, 233.
Eichhorn, H. (1963)   Private Communication.
Eichhorn, H., Googe, W.D., Gatewood, G. (1967a)   *Astr. J.*, **72**, 626.
Eichhorn, H., Gatewood, G.D. (1967b)   *Astr. J.* (in press).
Faddeev, D.K., Faddeeva, V.N. (1964)   *Numerische Methoden der linearen Algebra.*
Googe, W.D. (1967)   *Astr. J.*, **72**, 623.
Henricksen, S.W. (1967)   *Astr. J.*, **72**, 603.
Jefferys, H. III (1963)   *Astr. J.*, **68**, 111.
König, A. (1933)   *Handbuch der Astrophysik*, **1**, 508ff.
Lacroute, P. (1964)   *Ann. Obs. Strasbourg*, **6**.
Levenberg, K. (1944)   *Quart. of Appl. Math.*, **2**.
Lukac, C.F. (1967)   *Astr. J.*, **72**, 620.
Zadunaisky, P. (1957)   *I.B.M. Watson Lab. Report.*

RELATIVE STAR POSITIONS FROM OVERLAPPING PHOTOGRAPHIC PLATES

S. V. M. CLUBE

1. Introduction

An account is given of an investigation to obtain accurate relative star positions from overlapping photographic plates. The method described is quite general, but in the first instance is to be applied to plate material obtained with the new Cape astrometric camera.

The Cape 4-component $f/10$ 20-cm astrometric camera produces a field of $5°$ square with a scale of $100''$ per mm in the focal plane. The telescope is fitted with a yellow filter 12 mm from the focal plane, which in combination with the Ilford-R40 emulsion used, transmits more or less uniformly in the range $5300 < \lambda < 6500$ and little elsewhere. Within this spectral range, the maximum longitudinal aberration is only 0·75 mm and the images sensibly round out to the edge of the plate. The instrument is currently being used to obtain a complete coverage of the Southern hemisphere down to $m_v \sim 11·5$ such that each plate, effectively $4°$ square, has overlapping plates centred at each corner and at the middle of each side. In practice, this system of overlapping is applied to a number of zones in declination each about $10°$ broad. Each plate is taken within 30^m of the meridian and is given two 3^m exposures above and below the 1950 standard positions, the geometrical plate centre being estimated to be within $20''$ of the tangential point. At the time of writing, coverage of the zone $-30°$ to $-52°$ is almost complete, and preparations for the zone $0°$ to $-30°$ well advanced.

2. Outline of Technique

The scheme of analysis proposed is to correct the relative coordinates of stars on each plate for such effects as (a) aberration, (b) refraction, (c) optical aberration, and (d) significant relative proper motions, on the assumption that these effects are either calculable to a first order or constant from plate to plate, and then to adjust the configuration of stars in common to each pair of overlapping plates for scale and orientation, so that the relative displacements on the celestial sphere of the stars in common are minimized. The *relative* displacement in spherical coordinates of each pair of plate centres is thus calculated. A large system of overlaps produces a network of such displacements which are then adjusted subject to the overriding condition that this network lies exactly on the celestial sphere.

Perek (ed.), Highlights of Astronomy, 347–351. © *I.A.U.*

Suppose that the centres of two neighbouring overlapping plates can be fixed relative to each other with an accuracy e (standard error) in either coordinate, then two centres 180° apart on the celestial sphere can be fixed with an accuracy of $e\sqrt{n}$, where n is the number of overlaps joining them along a great circle. There are in the limit, $2n$ independent series of overlaps joining two such centres in a whole sky cover- age, although in practice, there is a degree of correlation near the centres in question due to the finite size of the photographic plates. Two centres 180° apart may thus be fixed relative to each other with an accuracy of the order of $e(\sim e\sqrt{n}/\sqrt{2n})$, equal to the accuracy of a single overlap. The practical realisation of a system of relative star positions from overlapping photographic plates may thus be reduced to one of making a single overlap as precise as possible. If m stars are common to each of two overlapping plates and each relative star position can be determined with an average accuracy of σ, the relative centres can be fixed with an accuracy of $e \sim \sigma(2/m)^{\frac{1}{2}}$. By way of example, with the 20-cm camera, it may be expected that $\sigma \sim 1\mu \approx 0''.1$, so that with 8 stars per half plate we anticipate $e \sim 0''.05$, easily comparable and potential- ly better than the typical systematic positional accuracy associated with other in- struments.

Such an order of accuracy approaches the random dispersion of proper motions associated with the stars used in these overlaps so that it is important to complete the photographic cover of the sky in as short a time as possible, or to regard stellar proper motion as an unknown to be sought in the analysis. In the present work, however, we shall without introducing any great error, consider the simple case where all the plates are assumed to be obtained simultaneously at one epoch. In order to refer such a rigid system of relative star positions to an inertial frame, it is necessary to obtain the positions of only a very few star-like extragalactic objects or suitably dimmed 'fundamental' stars on some overlapping plates. The former alternative seems the most promising and is to be investigated with the Cape instrument.

3. Theory of Analysis of Overlapping Photographic Plates

The plates obtained with the Cape camera have typically some 150 star images per exposure. Available facilities and personnel make it impossible to measure the co- ordinates of all these images, and the preliminary investigations are restricted to measurement of stars in the S.R.S. programme – with a density of about 1 per square degree. For each of these stars, provisional equatorial coordinates are available, and in the following discussion these are used as a first approximation to the star positions. It should be understood, however, that these are used only for convenience and that there is no fundamental need for them.

In the first instance, measured coordinates for each 'plate-star' (p, q) are obtained relative to an arbitrary origin chosen to be as near to the tangent point as possible. Then using an assumed focal length, these coordinates are corrected for such calculable effects as differential aberration, differential refraction, and optical distortion to produce 'observed coordinates' (x_1, x_2). It is recognized that these corrections cannot be calculated exactly since small unknown changes in the focal length (x_3), the atmospheric conditions, and the optical characteristics of the telescope may occur. It may reasonably be expected, however, that our estimates of these corrections differ from the 'real' corrections by second-order amounts which are linearly related to x and y.

We now consider all the plate-stars appearing on one plate and form the quantities

$$c_i = (\Sigma_i x_i^2)^{-1/2} x_i \quad (i = 1 \cdot 3) \tag{1}$$

for each star, where x_3 is regarded as an 'observed focal length' and is in the first instance put equal to x_3^0. The following transformation relates $c_i (i = 1 \cdot 3)$ to the star's equatorial direction cosines $\tau_j (j = 1 \cdot 3)$

$$c_i = \Sigma_j \{\Sigma_k (\Sigma_l \alpha_{kl} \tau_l)^2\}^{-1/2} \alpha_{ij} \tau_j \quad (i, j, k, l = 1 \cdot 3), \tag{2}$$

in which $|\alpha_{ij}|$ represents a transformation peculiar to the plate concerned.

In terms of the familiar plate constants for differential scale, orientation, and displacement of the origin,

$$|\alpha_{ij}| \equiv \begin{vmatrix} 1+a & b & c \\ d & 1+e & f \\ 0 & 0 & 1 \end{vmatrix} \begin{vmatrix} -\sin \alpha_0 & \cos \alpha_0 & 0 \\ -\sin \delta_0 \cos \alpha_0 & -\sin \delta_0 \sin \alpha_0 & \cos \delta_0 \\ \cos \delta_0 \cos \alpha_0 & \cos \delta_0 \sin \alpha_0 & \sin \delta_0 \end{vmatrix},$$

where (α_0, δ_0) are the coordinates of the tangent point.

The procedure followed is to solve Equation (2) for $|\alpha_{ij}|$ iteratively using adopted direction cosines τ_j (e.g. corresponding, at first, to the provisional equatorial coordinates) for each plate-star, subject to the condition that

$$\Sigma_j \alpha_{3j}^2 = 1 \quad (j = 1 \cdot 3). \tag{3}$$

Then, with these values of $|\alpha_{ij}|$ and the final values of c_i in Equation (1), we derive improved values for τ_j for each plate-star with corresponding weight w.

Weighted means of τ_j are then formed from overlapping plates on which each star appears, and these are then used as improved estimates in repeating the above procedure. The convergence of this process has not yet been adequately tested, but in principle, it is possible for star positions at the pole to affect those at the equator, thus necessitating at least 5 iterations. In practice, it is likely that such effects will be submerged by the measuring and plate errors.

The only restriction placed upon this method of analysis is in the first iteration. The provisional equatorial coordinates are probably bad enough to produce unnecessarily inaccurate estimate of $|\alpha_{ij}|$ and by implication, inaccurate estimates of the coordinates of each tangent point, (α_0, δ_0). We regard the telescope system and measuring machine as sufficiently rigid to render the tangent point determinate within about 1′ and therefore apply the initial condition that α_0 and δ_0 are known, and solve Equation (2) for the inverted matrix $|\alpha_{kl}|^{-1}$ using c_i as coefficients of the unknowns, since these are, at first, known with greater precision than τ_j.

A FORTRAN programme for this routine has been prepared for use on the ICT 1900 machine at Herstmonceux. Basically, each iteration involves the processing of an unformatted magnetic tape to produce a revised but otherwise identical tape containing a series of records in which each successive set of records pertains to one plate. Each record contains 3 integer numbers (s = serial number on the tape, p, and q if appropriate) and 9 non-integer numbers ($r_k (k = 1 \cdot 9)$). Each set of records contains (1) an initial master record in which $|r_k| = |\alpha_{ij}|$, the transformation peculiar to the plate; (2) a series of star records in which $r_{k=1 \cdot 3} = x_{i=1 \cdot 3}; r_{k=4 \cdot 6} = \tau_{j=1 \cdot 3}; r_{k=7} = w$; and (3) a terminator record for the plate concerned, in which r_k has no significance. Figure 1 illustrates the nature of the reduction routine followed.

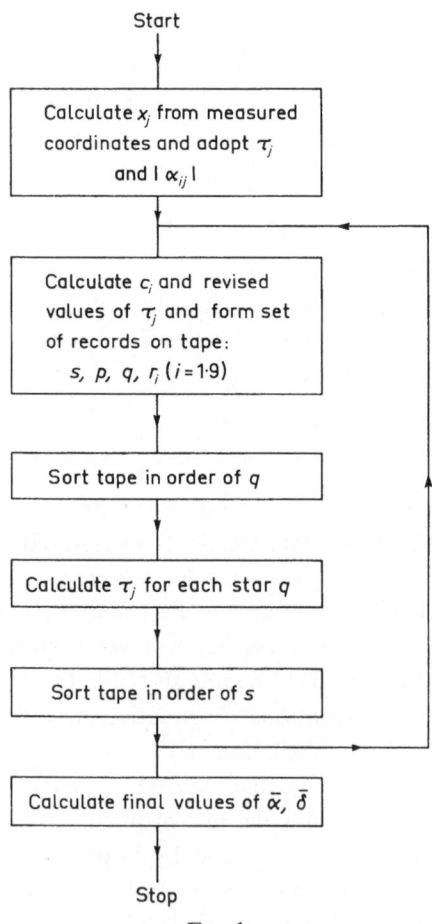

FIG. 1.

4. Current Investigations

At the present, this programme of measurement and reduction is in the development stage, and at the Cape we are engaged on two preliminary investigations. The first is to test both for size and constancy, the effects on the relative star positions induced by the telescope-filter combination and by the star colours and magnitudes and by plate-emulsion irregularities. This is involving an examination of overlapping plates in the Cape E-regions, where accurate photometry is available for fainter stars. The second investigation is to test the predictions of the second section using a series of plates at constant declination in the Cape astrographic zone. Ultimately, it is hoped that this procedure will be extended to plates over the whole of the Southern hemisphere.

Acknowledgements

This investigation was suggested to me by Dr. R. H. Stoy. The organization of the observing programme is arranged by Mr. J. v. B. Lourens and Mr. T. W. Russo, who, with other members of the Cape Staff, are carrying out the plate measurement. I would like to thank Mr. C. A. Murray and Mr. W. Nicholson for discussions which clarified the technique to be used, and for help in writing the reduction programmes.

E. JOINT DISCUSSION OF COMMISSIONS 28, 40, AND 45
EXTRAGALACTIC RADIO SOURCES

Wednesday, August 30, 1967

Organizing Committee: V. A. Ambarcumjan, W. P. Bidelman, R. Minkowski,
M. Ryle, M. Schmidt, I. S. Šklovskij
Meeting Chairmen: W. P. Bidelman, R. Minkowski
Editor: J. R. Shakeshaft

Contents:

OPTICAL STUDIES OF EXTRAGALACTIC RADIO SOURCES

A. R. SANDAGE

*(Mt. Wilson and Palomar Observatories,
Pasadena, Calif., U.S.A.)*

Editorial Note: Dr. Sandage discussed three topics:

  (i) the number of quasi-stellar objects,
 (ii) the optical properties of QSO, Seyfert nuclei, and N-type galaxies,
(iii) the cosmological vs. local hypotheses of QSO distances.

Much of his material was included also in his Invited Discourse. Since this appears elsewhere in this volume he did not wish to repeat it here.

DISCUSSION

Sandage (in answer to question): I have looked for evidence of an underlying stellar population in the QSS near to the N galaxies in the UBV diagram but have not found any.

Rubin: Does the fuzz around 3C 48 affect the colours?

Sandage: No. It is probably a jet as in 3C 273.

Arp: 3C 120 and another compact galaxy which have UBV colours similar to those of QSS are probably intermediate between the QSS and Seyfert objects.

Gratton: Do the different K-corrections for N-type and elliptical galaxies affect the Hubble diagram?

Sandage: Not significantly.

Menon: Has it been established whether QSS do, or do not, exist in clusters of galaxies? If they do not then the 'continuity' argument would be affected.

Sandage: It is not yet established. The two which have been studied carefully do not lie in clusters, but since only 30% of radio galaxies are in clusters this is not yet significant. There are now 13 QSS for which clusters should be detectable if they are there. Some N-type galaxies are in clusters.

Bidelman: Could it not be that QSS turn into clusters of galaxies? This would certainly provide more mass.

Rubin: What is the evidence for the existence of 10^5 quasi-stellar objects?

Sandage: This is based on a small sample which indicates that there are between 1 and 3 per square degree down to 19^m5. The isotropy of the Cambridge radio-source counts, of which about 30% are QSS, suggests that this sample may be typical.

Oke: Photoelectric scans of the N-type galaxy 3C 371 show that it has changed in brightness by over one magnitude during the last 2 years and by 0·10 to 0·15 magnitude in intervals of a few days. The variability is confined to the stellar-like nucleus. The underlying galaxy can be easily distinguished on the scans and shows the usual features; namely, the Mg I lines at $\lambda 5180$, G-band, H and K of Ca II, and the CN-band at $\lambda 3883$. The absolute visual magnitude of the underlying galaxy is -19.9. The non-thermal stellar-like component has the same continuum energy distribution as 3C 48 and in July 1967 was 1/10 as luminous as 3C 48. This object appears to provide the link between Seyfert galaxies and quasi-stellar sources. The rapid variability demonstrates that this property of quasi-stellar sources cannot now be used as an argument against their cosmological nature.

Morgan: I should like to make a short statement on the classification of the optical forms of the radio galaxies. Our system was devised to decrease as far as possible the systematic errors depending

Perek (ed.), Highlights of Astronomy, 354–355. © *I.A.U.*

on the distance of the source. This requires that a minimum number of information elements be used (so that faint sources can be classified), but that there be a sufficient number so that a significant classification can be carried out. We have called attention to a class of supergiant galaxies (our cD class) which have the following interesting properties:

(1) They are the largest single optical complexes of stars and gas known up to now;

(2) They form the 'central bodies' of a considerable number of rich clusters of galaxies;

(3) They have the characteristic appearance of an elliptical-like core (containing peculiarities) surrounded by an extended amorphous envelope;

(4) Many of them are strong radio sources; and

(5) They may include within their boundaries other galaxies of ordinary 'giant' proportions.

It is, of course, of great importance to observe such remarkable objects with the largest possible optical and radio resolution. Their inner regions are complex, and I feel that no two members of this class will be found to be identical in form.

OPTICAL SPECTRA OF EXTRAGALACTIC
RADIO SOURCES

E. Margaret Burbidge

(University of California, San Diego, Calif., U.S.A.)

1. Introductory Remarks

Extragalactic radio sources can be grouped spectroscopically into those that show absorption lines produced by a stellar population (e.g. the nearer radio galaxies, and some more distant ones, usually of D or E type) and those that show only lines (usually in emission) produced by low-density gas (e.g. QSO's and many N-type galaxies). If absorption lines from stars are present, one can estimate a minimum age from the stellar population, and perhaps a mass for the system from the stellar velocity dispersion. If no spectroscopic evidence on the stellar population is available (and this includes evidence from narrow-band filter photometry), then we have much less hold on the problems of age and evolutionary state. Absence of stellar absorption lines does not necessarily mean that stars are absent, but simply that their light, if present, does not contribute appreciably to the overall optical spectrum.

2. QSO's: Emission-Line Spectra

Forbidden and permitted lines are seen, of the general type found in galactic planetary or gaseous nebulae. Usually the lines are broad; a range of widths is sometimes found even in one object. The relative strengths of the emission lines can vary from one QSO to another, but Osterbrock and Parker found that average relative line intensities in 9 QSO's agree quite well with the computed line strengths for a gas with electron density $N_e \approx 3 \times 10^6$ cm^{-3} and electron temperature $T_e \approx 15000$ °K, and 'solar-neighborhood' relative abundances. 3C 273 has a higher electron density than this, but an upper limit is set by the presence of [OIII] in its spectrum.

A single-fluid model with uniform N_e, T_e has been shown, however, not to explain the observations as well as, e.g., a stratified model such as that proposed by Burbidge, Burbidge, Hoyle, and Lynds, or a model with denser filaments embedded in a more tenuous medium (first proposed by Schmidt; later, in more detail, by Woltjer). Perhaps a combination of both would be best. Woltjer's model has $N_e \sim 10^6$ in the background and $N_e \sim 10^8$ in the filaments. The importance of electron scattering was realized by Šklovskij and by Burbidge, Burbidge, Hoyle, and Lynds. The latter authors put it forward as an explanation for the emission-line widths, and particularly for the

Perek (ed.), Highlights of Astronomy, 356–360. © I.A.U.

differences between permitted resonance lines, forbidden lines from higher excitation states, and forbidden lines from low states with fairly large cross-sections for collisional de-excitation.

The relative line strengths can be explained on the basis of abundance ratios of the elements similar to solar-neighborhood values, but, depending on the temperature, some underabundance of the elements heavier than H could also be present. The elements H, He, C, N, O, Ne, Mg, Si, S, Ar, Fe, and possibly Na have been seen in emission. Tentative identifications of Na I by Wampler and Oke and O I by Véron may require more stratification in the model.

The emission-line red-shifts (101 objects) run up to $z = 2 \cdot 22$; there are many around $z = 2$, and it may be noted that the first 9 red-shifts published by Schmidt already included one – 3C 9 – with $z > 2$. The red-shift-apparent-magnitude plot shows enormous scatter and a slope that is a factor of $2 \cdot 5$–3 steeper than the usual Hubble plot for normal or radio galaxies, facts which have led to their abandonment as tools for cosmological research. Some evidence has been found by Strittmatter, Faulkner, and Walmsley for a grouping in the sky of red-shifts according to the size of z.

Variations in the resonance feature Mg II 2798 Å have been observed by Burbidge and Burbidge, by Dibaj and Yesipov, and by Wampler in 3C 345. Šklovskij has explained this in terms of successive excitation and double ionization of moving clouds by bursts of electrons. On the other hand, the Ly-α and C IV lines in 3C 446 were found by Sandage, Strittmatter, and Westphal to fade when the object brightened by 3 mag. in a fashion consistent with the hypothesis that the lines remained at constant absolute flux while the continuum flux brightened. In July 1967, I found the lines to be plainly visible again at their former wavelengths; the object had faded to about $17 \cdot 5$ mag.

A correlation between UBV colors and red-shifts, discovered by McCrea and by Kardašev and Komberg, has been explained by Strittmatter and Burbidge as being due to the effect of the strongest emission lines superposed on a simple smooth underlying energy distribution in the continuum. Several other groups have recently done similar calculations and confirmed the results of Strittmatter and Burbidge.

3. QSO's: Absorption-Line Spectra

By May 1967, about 20 QSO's were known to have absorption lines in their spectra. The published identifications are mostly with resonance lines from the ground states of the cosmically abundant elements, though a few identifications that do not fall into this category have been proposed. Most of the QSO's with absorption lines have large emission-line red-shifts; of the 14 QSO's which have $z_{em} > 1 \cdot 9$, 10 show absorption, while of the remaining 87 with $z_{em} < 1 \cdot 9$, only a further 10 show absorption. This cannot be wholly due to a preponderance of resonance lines in the region 1200–1600 Å among the relevant stages of ionization of the light elements, because Mg II 2796, 2803 Å and Si II 1808, 1817 Å are resonance lines that should appear in QSO's with smaller red-shifts. Mg II absorption has actually been seen in PKS 0812+02, 1229-02, and 1510-08.

The first object found to have many absorption lines was 3C 191; its lines were easily identified by Lynds, the Burbidges, and Stockton with Ly-α, C II, C IV, N V, Si II, Si III, Si IV, and S II transitions at $z_{abs} = 1\cdot95$. A slight shortward shift of the absorption relative to the emission lines was interpreted as being due to an expanding shell of gas around the source of continuum radiation. The absorbing gas cannot be intergalactic, according to Bahcall, because excited fine-structure states of the ground level are populated. The absorption lines are narrow and deep, being only about 3 Å wide in the rest frame. Bahcall, Sargent, and Schmidt found a C/Si abundance ratio of 3·5, as compared with the solar-neighborhood value of 13, but this needs further examination because of saturation curve-of-growth effects.

G. Burbidge noticed a preponderance of values of z_{abs} near 1·95. 3C 191 is one of these cases; PKS 0119-04 is another (it has z_{abs} slightly greater than z_{em}, which means that, if z_{abs} arises in a moving shell, the shell must be *falling in towards* the QSO continuum source and not expanding). Three of the objects with $z_{abs} \approx 1\cdot95$ have two simultaneously present red-shifts; these are PKS 1116+12 (Bahcall, Peterson, and Schmidt), Ton 1530 (Hiltner, Cowley, and Schild; Lynds), and PKS 0237-23 (Arp, Bolton, and Kinman; Burbidge; Greenstein and Schmidt). The latter has $\Delta z_{abs} = 0\cdot25$, corresponding to $\Delta v = 0\cdot08c$ if the shifts are Doppler. It has exceedingly narrow lines ($\leqslant 1$ Å in the rest frame) and many unidentified lines. Other objects are PHL 938 (discovered by Kinman), PHL 5200 (found by Lynds to have a strange spectrum with wide absorption bands adjacent to broad emissions, like a supernova) and PHL 1127 (a radio-quiet QSO studied by Sandage and Luyten).

The conclusion is that absorption lines either have z_{abs} close to z_{em}, or $z_{abs} \approx 1\cdot95$. This suggest a non-Doppler interpretation of the red-shift; Hoyle and Fowler (1967) put forward a model in which a gravitational red-shift is produced in a deep potential well in a configuration of collapsed objects with a pool of gas at the centre. A balance between radiation pressure outwards and gravitational attraction inwards, e.g., might maintain a stable thin shell around a continuum source.

PKS 0237-23 remains the most puzzling object. Arp *et al.* suggested identifications which would require abundances in heavier elements that are decidedly higher than the solar-neighborhood values. More work needs to be done on the identifications.

4. Radio-Galaxy Spectra

We return to the distinction made in Section 1 between objects with a spectroscopically detectable stellar population and those without. There is time to make only a few brief points:

(1) D and E type galaxies with absorption-line spectra of K-giant type (e.g. Vir A, Cen A, For A, and others more distant) must be $\sim 10^{10}$ years old. This fact, and the

time-scale for relaxation of stellar systems, means that one must be wary of postulating that certain objects can evolve into other kinds, by considering the radio properties alone. The optical properties are paramount in considerations of age and evolutionary stages.

(2) Schmidt, in his study of 35 radio galaxies, noted that the absence of absorption features in the continuum is correlated with the strength of the emission lines. Presumably in certain strong radio galaxies the presence of many emission lines of higher stages of ionization and excitation (i.e. lines other than Hα, [O$_{II}$], and [N$_{II}$] which are commonly present in normal galaxies) goes with the presence of continuum radiation that tends to swamp that coming from stars.

(3) There is a strong correlation between red-shift and the presence of strong emission lines, shown by the following tabulation:

> For z <   ·05, 8 out of 44 objects have 'strong emission'
> For ·05 < z < ·10, 13 out of 21 objects have 'strong emission'
> For ·10 < z < ·20, 10 out of 13 objects have 'strong emission'
> For z >   ·20, 9 out of 10 objects have 'strong emission'.

This is presumably a selection effect (one would like to know for how many faint galaxies attempts to measure red-shifts were made but failed because no strong emission lines were present).

(4) K-corrections calculated from a standard elliptical galaxy and applicable for D and E galaxies are dangerous to use for N-type galaxies where an additional UV continuum may be present. Possibly a Seyfert nucleus like that of NGC 1068 would be best, or one of the nearby Zwicky compact galaxies; one of the latter is the best example of a non-radio N-type galaxy that I have seen.

(5) Mg$_{II}$ 2798 Å should have appeared at 4090 Å in 3C 295, but was not seen on Minkowski's spectra. It should also have appeared in 3C 109, with twice the red-shift of 3C 273. This seems to indicate something different in the physical conditions in the two types of objects, even though 3C 109 has an emission-line spectrum otherwise rather similar to those of QSO's.

5. Concluding Remarks

The biggest spectroscopic problem posed by the QSO's at the moment is that of the absorption lines. A 'standard' red-shift of 1·95 raises again the question of gravitational red-shifts. PKS 0237-23 needs more work; there may be additional discrete red-shifts besides the two found, and there may be unusual element abundances.

DISCUSSION

Arp: The identification of the strongest absorption lines in PKS 0237-23 as due to Ti has been confirmed by spectroscopic observations with high resolution.

Oke: There has been no detectable variation of the emission lines in 3C 446 during the past year.

Bidelman: When Fath measured the spectrum of NGC 1068 in 1908 with the Crossley telescope it required a 13½ hour exposure spread over 4 nights.

Arp: A direct photograph of M 87 (Virgo A) with an Hα interference filter has revealed a counter-jet extending 40″ arc from the nucleus in a direction opposite to that of the well-known jet. There are fainter extensions out to 127″ arc. A spectrum shows the [O II] 3727 Å line as well as Hα, in contrast to the spectrum of the original jet which is a blue continuum. If the distribution of galaxies in the neighbourhood of M 87 is plotted they are found to lie along the line of the jet and counter-jet. There is a similar abnormal distribution of galaxies near the radio source Centaurus A (NGC 5128). (Ref. *Astrophys. Lett.*, **1**, 1, 1967.)

Kenderdine: The radio map of M 87 obtained with the 1-mile telescope at Cambridge shows the source extended along the line of the jet and counter-jet to about 30″ arc on either side of the nucleus. This suggests that possibly Virgo A is like many other double sources with two clouds of plasma ejected from the nucleus, but that in this case we can see these clouds optically.

Fomalont: Observations at Cal. Tech. show that M 87 has a large radio halo about 6′ arc in diameter, as do some of the Seyfert galaxies and perhaps 3C 231 (M 82).

RADIO OBSERVATIONS OF ANGULAR STRUCTURE

H. P. PALMER

(Nuffield Radio Astronomy Laboratories, Jodrell Bank)

1. Introduction

Complete information about the angular sizes and structures of discrete radio sources would, in principle, require observations in many position angles and at a large number of resolving powers, of which the highest power must resolve the smallest significant features in the sources. Since some sources are known to have components with differing radio spectra, similar amounts of data would be required at each of several wavelengths. One would then be able to examine statistically the relation between fringe visibility and resolving power at two or three wavelengths, and hence search for structural differences between different types of discrete sources.

2. The Data

The information on the size and structure of radio sources available at present is very incomplete by these ideal standards. What is known could be summarised in many ways. It is considered here in three ranges of angular size (A) $1' \rightarrow 1''$, (B) $1'' \rightarrow 0\cdot1''$, and (C) $<0\cdot1''$.

A. MEASUREMENTS USING RESOLVING POWERS BETWEEN 1′ AND 1″ OF ARC

Almost all the measurements in this range have been made with interferometers using baselines between 2000 and 61100 wavelengths long, at frequencies from 38–2695 MHz. The extent of these measurements has been summarised in columns 3 and 4 of Table 1. For a small number of sources data are available showing resolution over a fair range of baselines and at a number of position angles. In some cases measurements were made of the phases as well as of the amplitudes of the fringe patterns. Fourier inversions were then performed and detailed structures of the sources determined uniquely. Approximately 80 such maps have been published so far of such detailed structures. In a similar number of cases there are not sufficient data for this process, but it is possible to fit to the data plausible source models; normally these have double structures. The results of such analyses are described as 'maps' or 'models' in Table 1, column 3. It will be seen that most of these data come from the observations at Owens Valley **(1, 2)** and a certain amount has already been published

Perek (ed.), Highlights of Astronomy, 361–370. © *I.A.U.*

Table 1

Interferometric measurements with resolving powers between 1′ and 1″ of arc

(1) Wavelength (λ)	(2) Resolving Powers	(3) No. of Structures Reported	(4) No. of Size Estimates	(5) Observatories
7·8 m	≥ 2″	–	18	Jodrell Bank-Malvern (**12**)
1·9 m	47″–1″	6 models	378	Jodrell Bank (**7**)
30 cm	> 40″	62 models	133	Owen's Valley (**1**)
21 cm	> 20″	∼ 60 maps	540	Owen's Valley (**2**)
21 + 70 cm	> 10″	≃ 10 maps	–	Cambridge (**3, 4, 5**)
11 cm	> 3″	≃ 10 models + one map	146	Greenbank (**6**)

from the 1-mile telescope at Cambridge (**3, 4, 5**). The resolving powers of these instruments are less than 10000 wavelengths, so that angular sizes bigger than about 5″ are detected. Much smaller numbers of sources have been studied at higher resolving powers, when some variation of the baseline is often obtained by the fore-shortening that occurs as the sources move across the sky. Data have been published for 10 sources observed at 11 cm at Greenbank (**6**) and for 6 observed at 2 m at Jodrell Bank (**7**). More than 60% of these sources are double.

The second class of data from these observations comprises approximate estimates of the overall angular sizes of sources which have been inferred from limited observations at one or a few baselines. Such estimates are available for much larger numbers of sources as shown in column 4.

A second technique which has been used to give diameter information in this range and somewhat below is that of lunar occultation. After the initial striking success of this method with the first quasar 3C273 later occultations of that quasar have been studied extensively. This series of occultations is now complete, and this quasar will not be occulted again for 12 years. Results from occultations of fewer than a dozen other sources have so far been published, although Hazard (**8**) has recently reported 5 examples of a very large number of occultations observed at frequencies from 41–610 MHz with the 1000-ft. telescope at Arecibo.

B. MEASUREMENTS USING RESOLVING POWERS BETWEEN 1″ AND 0·1″ OF ARC

Two different techniques of observing sources have contributed to the information in the range from 1″ to 0·1″ of arc. Interferometric measurements from Jodrell Bank (**9**) and from the Jodrell Bank-Malvern interferometer (**10, 11, 12**) for about 40 sources are summarised in Table 2.

The second method of obtaining diameter information in this size range uses the scintillation of sources due to the motion of irregularities in the interplanetary solar

Table 2

Measurements with resolving powers between 1″ and 0·1″ of arc

Technique	Wavelength (λ)	No. of Structures reported	No. of Size estimates	Observatories
Interplanetary scintillation	1·7 m	–	45 (140 selected sources observed)	Cambridge (**14**)
Interplanetary scintillation	70 cm	–	70 (138 selected sources observed)	Arecibo (**15**)
Interplanetary scintillation	70 cm	–	10 (Random sample of 27 sources observed)	Arecibo (**15**)
Radio link	70 cm	4 models	12	Jodrell Bank (**9**)
Interferometer	21 cm	4 models	38	J.B.-Malvern (**10, 11, 12**)

wind. This method was discovered by Hewish *et al.* (**13**) and reported to the last IAU. It has since been exploited by the Cambridge group, and by Cohen and his colleagues at Cornell/Arecibo. A measure of the angular diameter of a source is obtained from the ratio of the scale of the diffraction pattern at the Earth to its distance from the diffracting screen. Under simple circumstances this is the ratio of the scale of irregularities in the plasma of the solar wind to one AU, i.e. about 0·3″ of arc. Radio sources smaller than this angular size scintillate while larger ones do not. On this basis Little (**14**) has made a survey at 178 MHz of the scintillation of 140 3CR sources which are of adequate flux density and which pass sufficiently close to the Sun. About 30% have components smaller than 0·3″ of arc. Cohen *et al.* (**15**) have used the Arecibo telescope to observe 140 sources at one or more of the frequencies 195, 430 and 611 MHz. These sources were not confined to the 3CR catalogue, but were selected as likely to show scintillation, which was detected for 70 of them. They also surveyed a smaller but unselected sample of 27 sources at 430 MHz, and found that 10 of these scintillated. They concluded that approximately 40% of sources whose flux at 408 MHz exceeds 2 f.u. emit a detectable fraction of their flux from a region smaller than 0·3″ diameter, and that they could determine the percentage of the total flux which was emitted by these small components.

A greater resolution becomes possible when observations are made closer to the Sun, and at lower frequencies. The diffracting screen becomes phase thick, the scale decreases, and the critical angle decreases. Under these circumstances Cohen *et al.* (**16**) have used the frequency spectrum of the scintillations to determine the angular scale. They show, e.g., that the angular diameter of 3C 138 is about 0·15″, in agreement with Little and Hewish (**17**), while the diameter of 3C 273B is less than 0·02″ (**15**). Complications arise when the phase screen becomes thick and when its Fresnel distance becomes comparable with an astronomical unit. There has been some controversy about the interpretation of measurements made under these conditions, and the analyses of Little and Hewish (17, 18) and Salpeter (19) should be consulted.

Some information about the structure of sources can also be deduced if one knows the properties of the solar wind well enough, and finds variations in the power spectra of the scintillations which correlate with changes in the direction and angular distance between the sources and the Sun. For instance, Rickett has just shown that at 408 MHz the quasar 3C446 (red-shift $z = 1 \cdot 402$) is probably elongated, with a size of about $0 \cdot 4''$ in position angle 94°, by about $0 \cdot 2''$ in p.a. 34°. This agrees with the detailed analysis of the 21-cm interferometric measurements on this source made between Jodrell Bank and Malvern (20, 23).

C. MEASUREMENTS OF RADIO SOURCES SMALLER THAN $0 \cdot 1''$ OF ARC

In 1963 Slish (21) and Williams (22) predicted from spectral considerations that some quasars would be found to have diameters of order $0 \cdot 01''$ of arc, while the discovery of secular changes of radio flux implied that some sources were smaller still. The Jodrell Bank to Malvern 21-cm observations were therefore extended to 11 and 6 cms (12), where 4 sources were still unresolved with a baseline of $2 \times 10^6 \, \lambda$ (Table 3).

Table 3

Interferometric measurements of radio sources smaller than $0 \cdot 1''$ of arc

Wavelength	Baseline Length	Resolving power	No. of size estimates	Observatories
65 cm	183 km	$0 \cdot 3'' (2 \cdot 7 \times 10^5 \, \lambda)$	12	Algonquin-Dominion R.A.O.
65 cm	3074 km	$0 \cdot 02'' (4 \cdot 6 \times 10^6 \, \lambda)$	8	Algonquin-Penticton (23)
50 cm	250 km	$0 \cdot 2'' (4 \cdot 6 \times 10^5 \, \lambda)$	4	Greenbank-Maryland Point
50 cm	2000 km	$0 \cdot 02'' (4 \cdot 0 \times 10^6 \, \lambda)$	1	Greenbank-Arecibo[a]
18 cm	840 km	$0 \cdot 02'' (4 \cdot 7 \times 10^6 \, \lambda)$	13	Greenbank-Haystack[a]
18 cm	4000 km	$0 \cdot 003'' (1 \cdot 95 \times 10^7 \, \lambda)$	1 (3C 273B)	Greenbank-Berkeley[a]
13 cm	1200 km	$0 \cdot 01'' (9 \times 10^6 \, \lambda)$	1 (3C 273B)	Australia (28)
11 cm	127 km	$0 \cdot 05'' (1 \cdot 1 \times 10^6 \, \lambda)$	35 (4 models)	Jodrell Bank-Malvern (12)
6 cm	127 km	$0 \cdot 025'' (2 \cdot 1 \times 10^6 \, \lambda)$	4	Jodrell Bank-Malvern (12)

[a] Private communications at first or second hand.

This equipment used three broadband radio links, and it seemed that for baselines longer than 100 miles an interferometer using rubidium frequency standards and video tape recorders would provide a more satisfactory and economic method of extending the observations. Several systems have been constructed in different parts of the world, and the results known to the author are summarised in Table 3. The first correlation was observed at a wavelength of 65 cm by the Canadians (23) using a baseline 4·6 million wavelengths long. Similar observations were made between Greenbank and several other observatories. It has also been reported that during the observations of OH emissions ($\lambda = 18$ cm) made at 19·5 million wavelengths between Greenbank and Berkeley, correlation was measured for 3C 273B. This gives a size of less than 0.003″ of arc for this quasar. 3C 273 has also been observed in Australia at 9 million wavelengths with a post-detector correlator system (28).

The data from tape-recording interferometers known at the time of writing refer to 3C 273B and 15 other sources. Seven of these are unresolved at one or more of the frequencies of observation. The remaining sources appear to have a fringe visibility significantly less than unity but there are some differences between the results, which may represent actual differences of source size with frequency. The observations reported from tape-recording interferometers have not yet given any indication of the structure of these very small sources.

3. The Distribution of Diameters

The first major diameter surveys at Cal. Tech. at 960 MHz (1) and at Jodrell Bank at 159 MHz (24) were made as complete as possible in order to reduce selection effects. Fifty-five of the sources observed in both surveys had galactic latitudes $> 12°$,

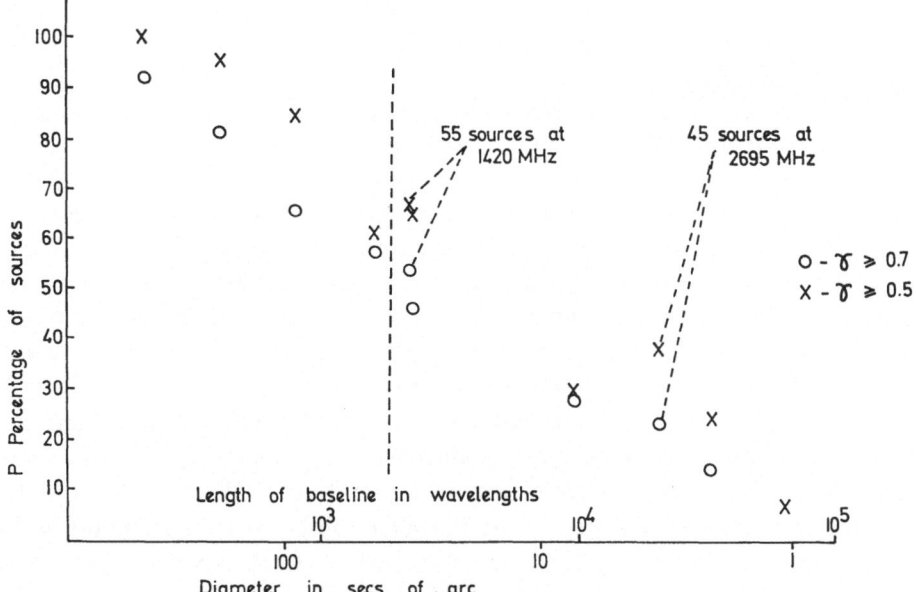

FIG. 1. *Data from Maltby and Moffet* (1) *and Allen et al.* (24) *have been used to find the percentage P of sources with fringe visibility (a)* $\gamma \geqslant 0·7$ *and (b)* $\gamma \geqslant 0·5$. *These observations refer to the restricted group (described in the text) of 55 sources. The points derived from Moffet's data lie to the left of the dashed line.*

and flux densities at 178 MHz > 12 f.u. Figure 1 shows the percentage of sources In this sample having fringe visibilities $\gamma \geq 0\cdot7$ and $\gamma \geq 0\cdot5$ at several baselines (25). it will be seen that more than half of the sources were partially resolved by baselines of 10000 wavelengths or more, and that there is good agreement between the two surveys, in spite of the frequency difference of a factor of 6. The diameter surveys of Fomalont at 1420 MHz (2) and of Clark and Hogg at 2695 MHz (6) included most of these 55 sources. Figure 2 shows that points derived from their new data agree well

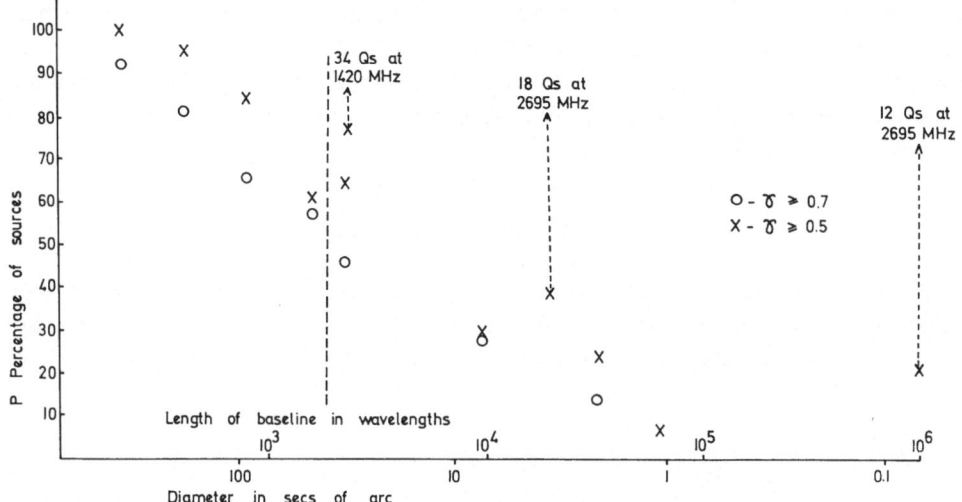

FIG. 2. *This diagram includes values of P (as described for Figure 1) which have been derived from the data of Fomalont (1420 MHz; 2) and Clark and Hogg (2695 MHz; 6).*

with the earlier results, although the frequency range is now increased by a further factor of 3. So it appears that the angular sizes of most radio sources do not change markedly with frequency, at least as far as this range of resolution is concerned.

The data at higher resolving powers are much less complete, for when a new instrument is first set up it is used initially to observe those sources most likely to show small scale structure. No complete surveys unaffected by selection effects have yet been made of sources with sizes smaller than 1″ of arc. The data in Tables 2 and 3 are therefore strongly biased towards sources which have very small diameters, and cannot be used to extend Figure 2 beyond resolving powers of 60000 wavelengths. It is interesting to note, however, that of the 39 sources observed at 11 cm in the Jodrell Bank-Malvern programme, ∼20 were unresolved. A further 6 contained an unresolved component radiating more than 50% of the flux. In several cases it is now clear that these small diameter components have much flatter radio spectra than the remainder of the source, so that at shorter wavelengths they radiate a larger fraction of the total emission (11).

It is difficult to include in Figure 2 diameter data derived from scintillation measurements, partly because the relationship between scintillation index and fringe visibility is not known in detail, and partly because of selection effects. The two sets of data available for unselected samples of sources (140 sources studied by Hewish and Little and 27 sources observed by Cohen *et al.*) both show that approximately 30% of discrete sources contain components smaller than 0·3″ of arc. These components may, however, be contributing only small fractions of the total emission from the sources. The scintillation data should therefore be compared with values of P calculated from interferometric measurements for ranges of fringe visibility going to lower limits of γ such as $\gamma > 0\cdot1$ or $\gamma > 0\cdot3$. The results of the two types of measurements are probably not in contradiction, for we can evidently expect many of the sources which are resolved according to Figures 1 and 2 to contain some structure of much smaller angular size. It is clear that more complete diameter surveys are needed at the higher resolving powers, particularly at wavelengths of 1 m and longer. Scintillation and interferometric data must be presented in the same way before they can usefully be compared, although the present data may already allow an extension of the preliminary examination by Moffet and Palmer (**26**) of the radio spectra of the small and large components known to exist in most radio sources.

4. The Distribution of Diameters for Quasars

Schmidt (**27**) has discussed the optical and radio-luminosity functions of 40 quasars. Thirty-four of these quasars were observed by Fomalont (**2**), 18 by Clark and Hogg (**6**), and 12 by Palmer *et al.* (**12**). Although the complete sample of quasars was not observed, values of P have been calculated on the assumption that the sources which were not observed would have been given values of $\gamma < 0\cdot5$. These values are shown in Figure 3, on which the dashed lines show how P would have changed if all the missed sources had given fringe visibilities $\gamma > 0\cdot5$. The paucity of the data makes these values unreliable, but they suggest that at short wavelengths the radio emission of quasars usually comes from regions whose linear sizes are 10 to 1000 times smaller than the emitting regions associated with the majority of radio sources. The smallest sources also tend to show time variations in their radio flux, and changes in the brightness and other properties of the optical object.

5. Conclusion

The diameter measurements reviewed in this article suggest, in summary, that some quasars, usually those which are optically active, emit radio energy from a very small, isolated region which may be described as a 'core'. Almost all the remaining quasars give radio emission from two separated regions, and there is often a 'core' in one or both of these regions. In several cases the radio emission from these cores is known

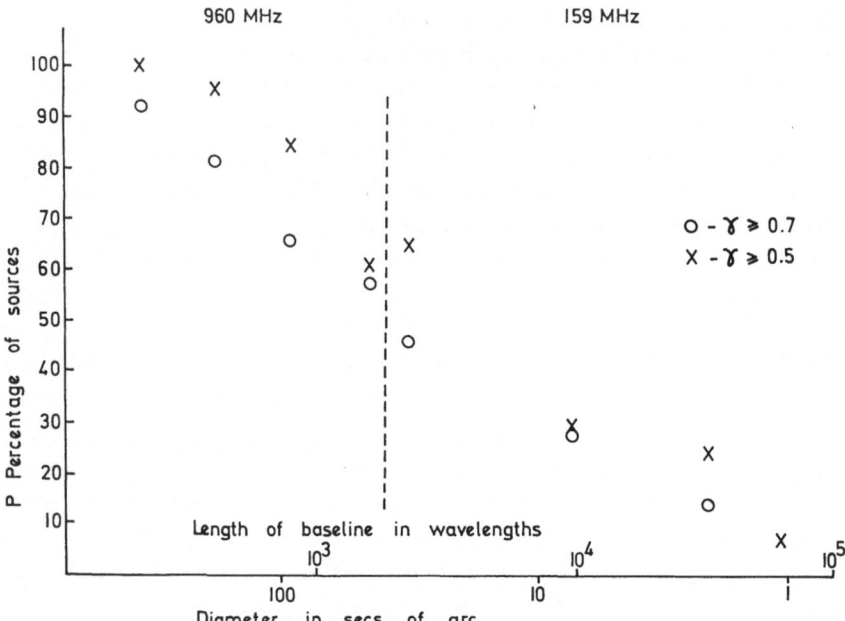

FIG. 3. *Values of P (as defined for Figure 1) derived for those quasars in the list of 40 discussed by Schmidt (27) for which diameter data are available.*

to vary with time, particularly at short wavelengths. The radio spectra of these cores are relatively flat, or the energy emitted by the core may even increase at short wavelengths. Many radio galaxies are also known to have relatively weak radio cores at wavelengths longer than 1 m, but the measurements of radio spectra suggest that, except in special cases such as the Seyfert galaxy NGC 1275, these cores do not usually predominate at shorter wavelengths.

Acknowledgments

I thank many colleagues who have helped with the preparation of this review, by discussion and by advance information about recent work. I also acknowledge gratefully the assistance of Professor F. Graham Smith, who read the paper in Prague, and of Mr. W. Donaldson, who discussed many points at length, and also prepared the tables and figures.

References

1. Maltby, P., Moffet, A. T. (1962)   *Astrophys. J.*, Suppl. **7**, 141.
2. Fomalont, E. B. (1967)   *Astrophys. J.*, Suppl. No. 138.
3. Ryle, M., Elsmore, B., Neville, A. C. (1965)   *Nature*, **205**, 1259.
4. Ryle, M., Elsmore, B., Neville, A. C. (1965)   *Nature*, **207**, 1024.
5. Kenderdine, S., Ryle, M., Pooley, G. G. (1966)   *Mon. Not. Roy. astr. Soc.*, **134**, 189.

6. Clark, B.G., Hogg, D.E. (1966)   *Astrophys. J.*, **145**, 21.
7. Rowson, B. (1963)   *Mon. Not. Roy. astr. Soc.*, **125**, 177.
8. Hazard, C., Gulkis, S., Bray, A.D. (1967)   *Astrophys. J.*, **148**, 669.
9. Anderson, B., Donaldson, W. (1967)   *Mon. Not. Roy. astr. Soc.*, **137**, 81.
10. Adgie, R.L., Gent, H., Slee, O.B., Frost, A.D., Palmer, H.P., Rowson, B. (1965)   *Nature*, **208**, 275.
11. Barber, D., Donaldson, W., Miley, G.K., Smith, H. (1966)   *Nature*, **209**, 753.
12. Palmer, H.P., Rowson, B., Anderson, B., Donaldson, W., Miley, G.K., Gent, H., Adgie, R.L., Slee, O.B., Crowther, J.H. (1967)   *Nature*, **213**, 789.
13. Hewish, A., Scott, P.F., Wills, D. (1964)   *Nature*, **203**, 1214.
14. Little, L.T. (1967)   Ph.D. Thesis, Cambridge University.
15. Cohen, M.H., Gundermann, E.J., Harris, D.E. (1967)   *Astrophys. J.*, **150**, 767.
16. Cohen, M.H., Gundermann, E.J., Hardebeck, H.E., Sharp, L.E. (1967)   *Astrophys. J.*, **147**, 449.
17. Little, L.T., Hewish, A. (1966)   *Mon. Not. Roy. astr. Soc.*, **134**, 221.
18. Little, L.T., Hewish, A.   in press.
19. Salpeter, E.E. (1967)   *Astrophys. J.*, **147**, 433.
20. Gent, H., Miley, G.K., Rickett, B.   in preparation.
21. Slish, V.I. (1963)   *Nature*, **199**, 682.
22. Williams, P.J.S. (1963)   *Nature*, **200**, 56.
23. Broten, N.W., Legg, T.H., Locke, J.L., McLeish, C.W., Richards, R.S., Chisholm, R.M., Gush, H.P., Yen, J.L., Galt, J.A. (1967)   *Nature*, **215**, 38.
24. Allen, L.R., Anderson, B., Conway, R.G., Palmer, H.P., Reddish, V.C., Rowson, B. (1962)   *Mon. Not. Roy. astr. Soc.*, **124**, 477.
25. Allen, L.R., Hanbury Brown, R., Palmer, H.P. (1962)   *Mon. Not. Roy. astr. Soc.*, **125**, 57.
26. Moffet, A.T., Palmer, H.P. (1965)   *Observatory*, **85**, 45.
27. Schmidt, M.   Preprint.
28. Gubbay, J.S., Robertson, D.S. (1967)   *Nature*, **215**, 1157.

DISCUSSION

Hughes: A large percentage of extragalactic radio sources are known to consist of more than one emission region, and it has been suggested that less than about 20% can be described as either simple or 'core and halo' sources. However, most of the observations that lead to this interpretation have been made with interferometers at a few fixed frequencies only, and some caution may be necessary before making a generalization for all frequencies. This may be illustrated by reference to the spectrum of the source 4C 50·11/LHE 100/NRAO 150. The spectrum is similar to those of 3C 279 (a quasi-stellar source) and 3C 84 (the Seyfert Galaxy NGC 1275) but has a double hump at the higher frequencies. Three components can be recognized:

(1) Component A with a spectral index of −0·5 between 78 and 100 MHz.

(2) Component B which has a rising spectrum with an index of about +2·5, a peak at about 300 MHz and, above that, a falling spectrum with an index of −0·5.

(3) Component C which is similar to B but with a peak at about 6000 MHz.

Component C shows no appreciable broadening of the 3′ of arc beam of the 150 ft. telescope at the Algonquin Radio Observatory and the mean position is separated from that of component B, as determined at the U.S. National Radio Astronomy Observatory, by 46″ of arc, with combined errors of about 33″ of arc. The lower frequency 4C position of component A is near to the suggested centre of components B and C, though the error bars overlap the two positions. Hence it appears that the object is at least a double, and quite possibly a triple source. But if measurements were made at frequencies near to the peaks at 300 or 6000 MHz, or below about 100 MHz then the source would only appear as a simple single source; close to the minima in the spectrum at about 100 and 3500 MHz, the source would appear to have two equal components, whilst for small frequency displacements about these values the source would be described as an unequal double. It is quite obvious that before generalizations are made regarding the complexity of a source, interferometer observations are required at a number of frequencies, and the spectrum examined in detail over a very large frequency range.

Locke: A long base-line interferometer has been operating in Canada at a frequency of 488 MHz. Antennas at Penticton and Algonquin Park are used with independent local oscillators. At the spacing of 3074 km only 3 out of the 30 sources observed remain unresolved. They are PKS 1127-14, 3C 345 and CTA 102 which therefore have angular diameters $<0.01''$ of arc. CTA 21 and 3C 273B are partially resolved with $\gamma = 0.5$ and 3C 279, 286, 309·1 have angular diameters of $0.03''$ of arc.

Kellermann: Another long base-line interferometer has been working between Hat Creek, Calif. and N.R.A.O., Green Bank. At a wavelength of 18 cm some sources are unresolved which have been resolved at the longer wavelength in Canada. It is clear that the structure changes with wavelength.

RADIO-SOURCE SPECTRA AND THEIR TIME VARIATIONS

K. I. KELLERMANN and I. I. K. PAULINY-TOTH

(National Radio Astronomy Observatory, Green Bank, W.V., U.S.A.)

During the past few years there has been a large increase in the available data on the spectra of radio sources, particularly at short wavelengths, where a number of sources have shown unexpectedly large time variations, with time-scales of 1 year or less.

The simple power-law spectrum, which is a straight line on a log-log plot of flux density against frequency, is shown by about 30% of sources. Most sources have a spectrum with negative curvature, which steepens at high frequencies. Many have a sharp cut-off, which is almost certainly due to synchrotron self-absorption, at low frequencies. In several of these sources, such as 3C 48, 3C 147 and 3C 295, the spectrum begins to flatten at a considerably higher frequency than the cut-off frequency. This flattening is too sharp to be caused by a change in the energy distribution of the electrons and is probably due to parts of the source becoming optically thick at higher frequencies. Some sources have components which are optically thick even at centimetre wavelengths. These must have angular sizes of $10^{-3''}$ or less. The energy density in relativistic electrons in these compact sources is much larger than the magnetic-energy density, so that the source cannot be stable and variations in the flux density are to be expected.

A total of some 24 variable radio sources are now known, including 13 QSS and 3 galaxies. The QSS for which data are available are known to show variations at optical wavelengths, these being more rapid and more violent than the radio variations. All of the variable sources have radio components of very small angular size and their spectra indicate that these components are optically thick near the wavelength where the flux is increasing and optically thin in the region where the flux is decreasing. This is consistent with the model of a variable component which is initially optically thick, but which becomes optically thin at successively longer wavelengths owing to its expansion. The flux density at a particular frequency thus increases until the source is optically thin at which time it begins to decrease. A characteristic age for this component may be estimated from the observed rate of change of flux, from the maximum flux density reached at different frequencies or from the times at which these maxima are reached. At centimetre wavelengths the characteristic time-scale is about 1 year. The required input of energy in the form of relativistic electrons is about 10^{58} ergs if the QSS are at cosmological distances, and this energy has to be released in a violent event lasting only a few months. Some sources, such as

Perek (ed.), Highlights of Astronomy, 371–372. © I.A.U.

3C 273 and 3C 279, have complex spectra indicating the simultaneous presence of two or three variable components with ages of 1–10 years. Since about 10% of the QSS are currently active, it seems that in the typical QSS there occur periodic explosions every 1–100 years, each with an energy output in particles of about 10^8 Type-II supernovae. Each such explosion gives rise to a temporary increase in the radiated flux and a radio spectrum with positive curvature. During the dormant period, or after the cessation of all activity, the combined remnants of many explosions will give rise to the 'normal' power-law or slightly negatively curved spectrum.

If we consider only that part of the spectrum where the sources are optically thin, then all sources have similar spectral indices. At short wavelengths most radio galaxies are optically thin and the spectral indices are concentrated near $-0\cdot8$. A few galaxies with flatter spectra are Seyfert or Type N galaxies with active nuclei. The QSS have a broader distribution of spectral indices and a larger proportion have flat spectra. This is probably due entirely to self-absorption. QSS selected from low-frequency surveys tend to be optically thin down to these frequencies and have 'normal' spectra. In their linear dimensions and radio properties they tend to be similar to the radio galaxies. QSS selected from surveys at higher frequencies have a greater proportion of sources which are optically thick, which are always very small, often variable and intrinsically more luminous than the sources with steeper spectra.

DISCUSSION

J.A. Roberts: Is there any change of angular size with wavelength in the case of sources with peculiar spectra?

Kellermann: Yes. In 3C 279, for example, the angular diameter is $0\cdot01''$ of arc at centimetric wavelengths but $2''–3''$ of arc at metre wavelengths.

Sturrock: Have you considered that the presence of dense plasma might affect the time-scale of the high-frequency variations?

Kellermann: There is unlikely to be such plasma because of the negligible depolarization of the radio emission.

Scheuer: Another argument against such plasma is that the free-free absorption in it would extend to a frequency of many gigahertz.

INFRARED OBSERVATIONS OF QSS

F. J. Low

(Lunar and Planetary Laboratory, University of Arizona, Tucson, Ariz., U.S.A.)

No abstract has been submitted.

MILLIMETRE-WAVE OBSERVATIONS

E. E. EPSTEIN

(Aerospace Corporation, Los Angeles, Calif., U.S.A.)

No abstract has been submitted since the material contained in the paper is being published in *Astrophys. J.*

THE NATURE OF THE RED-SHIFTS IN
QUASI-STELLAR OBJECTS

L. WOLTJER

(Dept. of Astronomy, Columbia University, New York, N.Y., U.S.A.)

Interpretations of the red-shifts are reviewed.

(1) The gravitational interpretation of red-shifts encounters three difficulties:

(a) If a mass M is surrounded by emitting gas, the volume appears to be too small to explain the forbidden-line emission.

(b) Narrow lines imply a stable, thin, spherical sheet of gas in a strong gravitational field.

(c) The magnitude of the red-shifts is too large.

Fowler and Hoyle proposed a very dense galactic nucleus with 10^{13} neutron stars in a region of 1 parsec diameter. This may get around (b) and (c), but (a) is still difficult.

(2) Kinematical red-shifts can be associated with either large velocities of objects ejected from many galaxies scattered through the universe, or from our galaxy alone. The first possibility leads to a preponderance of blue-shifts, which are not observed; the second, to a total energy far larger than that involved in most strong radio galaxies. Our galaxy is only a quite weak radio source.

(3) The cosmological interpretation of the red-shift meets with the following objections:

(a) There is no clear magnitude-red-shift relation, but large-amplitude variability of individual objects leads us to expect only a weak relation.

(b) Time variations are incompatible with synchrotron radiation, but this is not the case if the relativistic electrons move outward more or less radially. Also, the synchrotron nature of the variable component is not completely certain.

(c) Large energies are involved, but the situation is no worse than that for radio galaxies.

Support for the cosmological interpretation comes from:

(d) NGC 1275, which contains an unresolved radio component with a variable behavior very similar to that of quasi-stellar objects.

(e) The angular separations of the double radio sources associated with some quasi-stellars, which indicate linear separations comparable with those in radio galaxies.

THEORIES PROPOSED TO EXPLAIN QUASI-STELLAR OBJECTS AND RADIO GALAXIES

GEOFFREY BURBIDGE

(University of California, San Diego, Calif., U.S.A.)

In this paper I shall consider briefly two major questions. The first is concerned with whether the QSO's are at cosmological distances or not. The second is concerned with the theoretical models proposed to explain the observed properties of radio galaxies and quasi-stellar objects.

In the Invited Discourses Dr. Sandage and Sir Martin Ryle gave arguments which lead them to believe that the QSO's lie at cosmological distances. They both have strong convictions in this matter. However, Dr. Sandage did remark that he was only giving one side of a 'great debate'. It is presumably up to me to give the other side of this argument. I have no strong beliefs in this matter but it appears that on the basis of the evidence as it is today we cannot reasonably assert that the QSO's are at cosmological distances. If at all I think that the evidence argues against the QSO's being at cosmological distances. The arguments run as follows:

Plots of the red-shift against apparent radio or optical magnitudes show that there is no correlation which can be interpreted as due to a red-shift-distance effect. The diagrams can be interpreted as being due to evolutionary effects if it is assumed that the objects lie at cosmological distances, but in no sense can they be used in evidence for this hypothesis.

The variability of the sources shows us that they are excessively small, and the very high radiation densities which must then be present if the QSO's are at cosmological distances lead to apparent difficulties in the construction of theoretical models. However, models can be constructed, and thus one cannot make a conclusive argument based on apparent contradictions here.

Sandage and Ryle have made much of the continuity arguments but continuity can also be invoked if the QSO's are comparatively nearby. In this case their optical properties are practically identical (apart from the red-shifts) with the nuclei of Seyfert galaxies, and they are weak radio sources similar to those found in the centres of galaxies such as M 82. As a physicist I can make little of the continuity argument one way or the other.

Attempts to detect the presence of intergalactic matter from absorption or scattering of the optical flux from QSO's have failed. This means either that the matter is in a form such that it cannot be detected, or else the objects are not at cosmological distances.

Perek (ed.), Highlights of Astronomy, 376–378. © *I.A.U.*

Arp has suggested that the QSO's and radio galaxies are in some cases associated with comparatively nearby peculiar galaxies. No one is happy about his analysis but it has not been conclusively ruled out.

Strittmatter, Faulkner and Walmsley have found that QSO's with large red-shifts are concentrated in two comparatively small areas in the sky. In my opinion no one has been able to show that this effect is not real. Many people simply hope that it will go away. The discovery that all of the QSO's with enough absorption lines present so that an absorption red-shift can be unambiguously obtained, give the same value of 1·95, with a very small dispersion, is a strong argument in favor of the local hypothesis for QSO's. Šklovskij and Kardashev have attempted to explain this by going back to a Lemaître model, but this is highly improbable. Moreover, if any of the objects with only one line and an emission red-shift very much less then 1·95 are shown conclusively to have an absorption-line red-shift of 1·95, and there are several possible cases, all cosmological interpretations will be ruled out.

The strongest argument in favor of the cosmological hypothesis for the QSO's is the presence of the large red-shifts. However, the recent work of Hoyle and Fowler has shown that the red-shifts may be gravitational in origin. The arguments just made by Woltjer are not strong.

I now turn to the second topic – the present status of theories explaining the outbursts in radio galaxies and quasi-stellar objects.

Before the discovery of the QSO's it had been realized that large-scale violent events take place in the nuclei of galaxies. These give rise to large fluxes of relativistic particles, to highly excited gas, and to large-scale motions of gas and dust. The total amount of energy released is not well known, but minimum estimates reach values as large as $10^7 \, M_\odot \, c^2$ in the strongest radio sources. In the QSO's the minimum total energies required to explain the observations are often less than this but it seems clear that the outbursts in these cases must have a similar origin. The mass energy released is so large that it is clear that very large condensed masses must be present in the nuclear regions of galaxies and in QSO's.

The mechanism by which the energy is converted largely into relativistic particles is not understood. It has usually been assumed that two large clouds of particles are ejected to give rise to a double source. However, the discovery that the sources have fine structure, and that in many cases very small sources are present far from the explosion centre, lead to the conclusion that small fairly massive coherent objects are thrown out in the explosion and that each of these is a secondary source of relativistic particles.

If extremely high mass densities are assumed to be present in the nucleus – at least $10^8 \, M_\odot$ per cubic parsec – then a variety of catastrophic processes will occur. These include multiple supernova explosions, star collisions, and gravitational collapse of a single coherent mass. All of these ideas and others are being explored with a view to understanding the mechanism by which the very large energies are released, largely in the form of relativistic particles. At present we do not have any satisfactory theory which will explain the observed phenomena. Probably the most promising idea is that the energy is released by a large coherent mass in gravitational collapse. If large numbers of small objects are thrown out the large mass must fragment in the collapse. Classical relativity theory shows that it is exceedingly difficult to get out a large fraction of the rest mass in such a process. Some have therefore suggested that, under extreme conditions, modifications of the general theory of relativity are required. Others have concluded that much larger masses are involved and that the energy release takes place at very low efficiency – perhaps 1% to $0 \cdot 1\%$.

There is also no real understanding of the way in which very large masses are condensed into very small volumes in the first place. Some have argued that such large mass concentrations are themselves cosmological in origin. Thus, in the framework of an evolving universe it has been suggested that QSO's are parts of the expanding universe which were 'left behind' and are just now expanding from a high-density state. In the framework of the steady state cosmology it has been argued that matter is created in places where the matter density is highest so that in the nuclei of galaxies very high-density configurations are being continuously formed. Others have attempted to show that large mass concentrations will form through the natural processes of dynamical evolution in the nuclear regions where the star density is high. If one assumes a high enough star density to begin with, then the required conditions can be reached, but these models are also rather *ad hoc* in character.

The discovery of the radio galaxies and then the quasi-stellar objects has led to a revolution in our thinking concerning the evolutionary processes on the scale of galaxies. We know that we are on the verge of major discoveries which relate back to fundamental physics. However, we have as yet no real understanding of many of the fundamental physical processes which are operating in the nuclei of galaxies and in QSO's. Most attention has been paid in recent years to determining the distance scale of the QSO's. When this problem is solved it may be possible to decide unambiguously which models of QSO's are compatible with the observations and proceed to further observational programs which get us closer to an understanding of the physics of explosions on the galactic scale.

DISCUSSION

Oort: What do you think of 3C 371?

G. Burbidge: Because this object has a very small red-shift the apparent difficulties which are present for the QSO's – associated with the high radiation densities – are not present to the same extent. While there is some sort of continuity between QSO's and N-type galaxies, this is not evidence for the cosmological nature of QSO's. The continuity argument also holds if they are close by.

A COSMOLOGICAL INTERPRETATION OF THE ABSORPTION-LINE SPECTRA OF QSO

I.S. ŠKLOVSKIJ

(Sternberg Astronomical Institute, Moscow)

Editorial Note: Dr. Šklovskij presented some work by himself and N.S. Kardašev on the possibility that the concentration of absorption line red-shifts of quasi-stellar objects at $z = 1.95$ points to the Lemaître model of the Universe. No abstract of the paper has been submitted, but see *Astrophys. J.*, **150**, L135, 1967.

DISCUSSION

Gratton: Since Šklovskij bases so much on the redshift $z = 1.95$, what reliability can be placed upon this figure?

E.M. Burbidge: There are seven objects having absorption lines with $z = 1.95$.

Kahn: Would not Šklovskij expect there to be many QSS formed at the time of the 'plateau'?

Šklovskij: I would expect more than have been found but the statistics are not yet good.

McCrea: Does the model give the observed log N – log S relationship for radio sources?

Kardašev: There is no definitive answer yet. It depends upon the luminosity function which has not been worked out for this model.

THE EVOLUTION OF QUASI-STELLAR SOURCES
AND RADIO GALAXIES

M. RYLE

(Mullard Radio Astronomy Observatory, Cavendish Laboratory, Cambridge)

There seem to be two promising ways of investigating the evolution of radio sources: (a) the collection of data on a large number of sources so that attempts may be made to arrange them in an evolutionary sequence, (b) the study of a few sources, for which really detailed information is obtainable, to gain insight into the physical mechanisms occurring.

It is important to distinguish between the most powerful radio sources, the majority of which appear to have a simple double structure at the resolutions now attainable, and the more complex sources. The structure of the latter suggests that energetic particles have been accelerated in multiple 'events', each being of smaller magnitude than the events which give rise to the double sources; the total energy released may, however, be comparable.

It is evident that the evolution of the complex sources presents a more difficult problem, so it will be preferable to try to explain the double sources for a start. A simple model of these has been suggested (Ryle and Longair, 1967), which supposes that two clouds of relativistic plasma are ejected with equal velocities in opposite directions from the parent galaxy. From a consideration of the radiation lifetime and the physical dimensions of the more extensive sources it is evident that the speed of ejection must be comparable with the speed of light. The existence of such velocities implies that the two components will, in general, be observed at significantly different ages and they will have different angular separations from the galaxy. By examining a large number of sources it is then possible on the basis of this model to derive the power and age of each of the components. When the results are combined it appears that the model provides a reasonably consistent picture for the evolution not only of the extensive radio galaxies but also the QSS; the wide dispersion in the radio power and physical dimensions can be attributed mainly to the different ages at which the sources are observed.

From these results it appears that at high frequencies (where the effects of self-absorption do not persist for long) the emission from each plasma cloud remains relatively constant for a period of about 10^5 years despite the considerable increase in the size of the cloud which occurs during this time. At greater ages the further expansion is accompanied by a rapid decrease and the emission falls to a small value after about 3×10^6 years. The second phase corresponds to the expected variation of emis-

Perek (ed.), Highlights of Astronomy, 380–383. © *I.A.U.*

sion from a cloud which is expanding adiabatically, but the first phase seems to require a progressive enhancement of the magnetic field; there is good evidence (Williams, 1966) from the spectra of some of the most compact QSS (3C 147, 286 and 287) that the fields are initially small (10^{-5}–10^{-4} gauss). This model of powerful sources is therefore consistent with the sudden release of at least 10^{61} ergs in the form of relativistic particles in an initially weak magnetic field; it requires a mechanism for the amplification of this field as the plasma clouds move out through the galaxy.

Whilst such an analysis provides some evidence on the physical mechanisms involved, it is likely that further information will be obtained by studying sources which have a complex structure. A number of sources are known in which four main components are distributed along a line, suggesting that two separate periods of particle production have occurred (3C 46, 274·1, 452). In other cases (e.g. 3C 465), which have even more components, estimates may be made (Macdonald *et al.*, 1966) of the variation of magnetic field with distance from the galaxy; it may then be possible to infer how the magnetic field in each component changes with time.

In some sources, such as the Seyfert galaxy NGC 1275, it seems that a more or less continuous acceleration of particles occurs. Early studies of the source (3C 84) revealed a compact component and an extensive halo some 30′ of arc in diameter. The work of Dent (1966) on the variability, of Pauliny-Toth and Kellermann (1966) on the spectrum, and of Barber *et al.* (1966) using long base-line interferometry have indicated that there are components having angular diameters $<0.001''$ of arc and $0.05''$ of arc. These components are coincident with the Seyfert nucleus.

Recent observations (Ryle and Windram, 1968) of this region with the 1-mile telescope at Cambridge enable 6 separate but associated components to be distinguished (Figure 1). In addition to those mentioned already there is one 5′ of arc in diameter, the same size as, and coincident with, the outer envelope of gaseous filaments of NGC 1275. Another (3C 83·1 B) is clearly associated with the neighbouring galaxy NGC 1265 and consists of an intense core with a curved tail extending away from NGC 1275. Yet another coincides with the 14^m galaxy IC 310, and this also shows an extension away from NGC 1275. The remarkable shape of these two additional sources confirms the suggestion by the Burbidges (1965) that the emission from NGC 1265 has its origin in NGC 1275.

We may now examine all the components in order to establish their energy requirements from their spectra, physical size, and radio power. In the case of the most compact component of 3C 84, particles having a total minimum energy of 3×10^{52} ergs are necessary if it is supposed that protons having 100 times the electron energy are present. The particle production has probably occurred within a period of less than 1 year and, as discussed by Van der Laan (1966) and Kellermann (1966), the lifetime of the source is likely to be limited to a few years by their escape from the region. The observed variability of this component is therefore to be expected, and the greater part of the electron energy will escape into the surrounding medium.

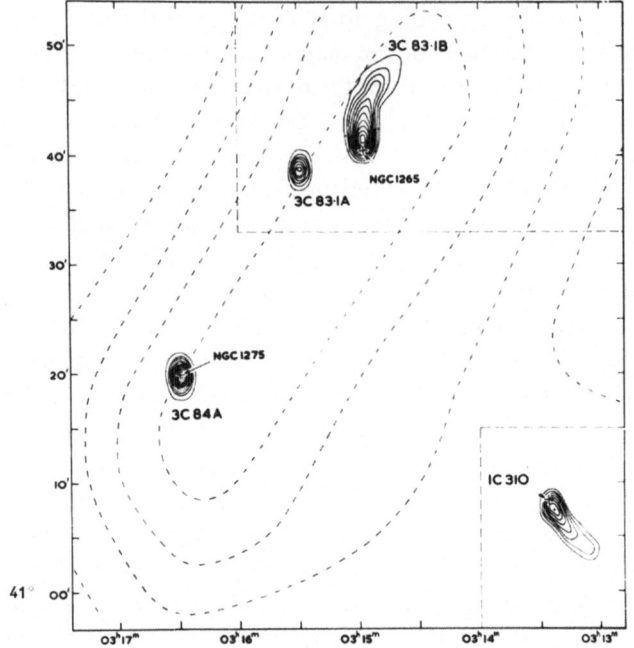

FIG. 1.   *Map obtained at* 408 *MHz of the sources around NGC* 1275.

The 0″.05 of arc component, which is comparable in size with the Seyfert nucleus, requires a total particle energy of $\approx 10^{53}$ ergs; the electron life may be determined by a radiation loss time of ≈ 100 years or by escape from the nucleus in a rather shorter time, but a supply of the order 7×10^{51} ergs per year is necessary in either case. A similar rate is needed to account for the 5′ of arc component.

These results agree well with the conclusions of Dibaj and Pronik (1966) from optical studies. They find two components of the nucleus – one of which is about 0·1 parsec in diameter and has a gaseous mass of $35M_\odot$, while the other has a diameter of 10 parsec and a gaseous mass of $10^4\ M_\odot$. These components are of size comparable with, and may be identified with, the two smallest radio-emitting regions. This supports the suggestion of Dibaj and Pronik that their origin can be attributed to supernova activity within the Seyfert nucleus since the particle energy associated with the most compact source is comparable with that in Cassiopeia A (J. E. Baldwin, private communication).

If we consider the sources associated with NGC 1265 and IC 310, it appears impossible on energetic grounds to explain the emission except on the basis of the interaction of relativistic particles ejected from NGC 1275 with the magnetic fields of these galaxies. The extensive halo component of 3C 84 may similarly be due to the interaction of particles with fields in the intergalactic medium. The total particle energy required is about 3×10^{53} erg per year continuing for a period of about 5×10^6 years.

Provided that the particles are trapped this conclusion is not dependent upon an assumption of equipartition of energy since a reduced rate of emission from each electron leads to a correspondingly longer electron life.

The results therefore suggest that NGC 1275 has been responsible for the continuous production of energetic particles over the past 8×10^6 years at a rate which has decreased from about 3×10^{53} to 2×10^{52} ergs per year. The total particle energy released is then at least 4×10^{60} ergs and is comparable with that of the most powerful QSS and radio galaxies known. NGC 1275 differs in that the energy release takes place over some millions of years rather than over 10^2–10^3 years only. This difference of time-scale may be explicable in terms of energy production by supernova explosions. In the case of NGC 1275 the explosion of one supernova every 10 years could provide the energy to maintain all three radio components and could continue for several million years, whereas in the powerful sources a chain reaction of supernovae, as suggested by G. R. Burbidge (1961), might perhaps release the same total energy within 100 years, a period comparable with the light travel time across the nucleus.

It seems therefore that, besides the age of the source, the rate at which energy is released may be of great importance in determining its character. The intrinsically most powerful are likely to have a brief period of unusual optical emission and will have the simplest radio structure. An increase of the time during which particle acceleration occurs, from say 10^2 to 10^4 years, is unlikely to have much effect on the radio emission except at very high frequencies, but may result in a marked reduction in the optical emission. A further increase in this time may cause the optical peculiarities to be so faint as to be undetectable except for the nearest sources, while still producing intense radio emission. Alternatively, if a series of brief events of moderate power occur, spaced at long intervals, relatively intense optical sources might be produced without appreciable radio emission. Are these the radio quiet QSO's?

References

Barber, D., Donaldson, W., Miley, G.K., Smith, H. (1966)   *Nature*, **209**, 753.
Burbidge, G.R. (1961)   *Nature*, **190**, 1053.
Burbidge, E.M., Burbidge, G.R. (1965)   *Astrophys. J.*, **142**, 1351.
Dent, W.A. (1966)   *Astrophys. J.*, **144**, 843.
Dibaj, E.A., Pronik, V.I. (1966)   *Izv. krym. astrofiz. Obs.*, **35**, 87.
Kellermann, K.I. (1966)   *Astrophys. J.*, **146**, 621.
Macdonald, G.H., Neville, A.C., Ryle, M. (1966)   *Nature*, **211**, 1241.
Pauliny-Toth, I.I.K., Kellermann, K.I. (1966)   *Astrophys. J.*, **146**, 634.
Ryle, M., Longair, M.S. (1967)   *Mon. Not. R. astr. Soc.*, **136**, 123.
Ryle, M., Windram, M.D. (1968)   *Mon. Not. R. astr. Soc.*, **138**, 1.
Van der Laan, H. (1966)   *Nature*, **211**, 1131.
Williams, P.J.S. (1966)   *Nature*, **210**, 285.

THE MAGNETODYNAMIC MODEL OF QUASARS

L. M. OZERNOJ

(P. N. Lebedev Physical Institute, Academy of Sciences of the U.S.S.R., Moscow)

1. Introduction

Although the energy problem is one of the main difficulties of the cosmological interpretation of quasar red-shifts, there is no need to go beyond conventional physical laws in seeking an explanation. Various proposals (**1–10**) have been made which differ in important details. I believe a suitable basis for a theory of quasars to be a supermassive plasma configuration, the matter of which is in mostly regular motion in a magnetic field (**5, 7**). Such a magnetodynamic configuration I shall call a 'magnetoid'. On this model a quasar is a galaxy with a dense non-stellar nucleus, the magnetoid, which gives rise to the characteristic non-thermal luminosity. The mass of the quasar is determined by its stars while the parameters of the magnetoid depend on the stage of evolution or conditions of formation in a similar way to those of radio-galaxies or active nuclei of galaxies.

2. Important Features of Quasars

The most important features of quasars are, firstly, the exceptionally powerful continuum radiation, up to 10^{46}–10^{47} ergs^{-1} and, secondly, the time variations of this radiation the rapid component of which indicates linear dimensions of no more than 10^{15}–10^{17} cm. A statistical investigation of the optical variability of 3C 273 reveals significant differences from a gaussian distribution, and it may be concluded (**11**) that the source must represent what is effectively a single body rather than a large number of independent events such as multiple supernova explosions.

The constancy of mean output of 3C 273 over nearly a century means that the source must be in quasi-stationary equilibrium. At the rate of 10^{47} erg s^{-1} for 3×10^5 years (a likely time-scale for a quasar) the energy required is 10^{60} erg. The mass involved is therefore not less than 10^8–$10^9 M_\odot$. If a plasma cloud of such a mass were initially in *static* thermal equilibrium it would radiate its internal energy within a few years and, without some source of energy, would contract rapidly as predicted by the theory of general relativity. If, however, the equilibrium is *stationary* rather than *static* and the plasma has regular internal motions in a magnetic field, then the time-scale can be sufficiently extended to account for the observations.

Perek (ed.), Highlights of Astronomy, 384–388. © I.A.U.

3. Features of the Magnetodynamic Configuration

In the quasi-stationary state, the main motion of the magnetoid matter takes the form of orderly circulation of the plasma in a regular inhomogeneous magnetic field. There is also likely to be magnetoturbulence on various scales. The equilibrium is quasi-stationary because of continuous outflow of matter from the surface as well as transport of angular momentum and magnetic field.

The magnetoid has two important properties which explain the main peculiarities of quasars:

(i) A powerful energy output from a small volume. On reaching the boundary of stable equilibrium the laminar motion in the magnetoid is apparently destroyed. In (5) and (7) consideration was given to the limiting situation where the 'break-down' of quasi-equilibrium occurs in a state of magnetoturbulence. For the simplest case when the magnetoturbulent energy E_{mt} is small compared with the thermal energy E_t, an analytical solution is possible. Instability occurs at a magnetoid radius of $R_{cr} \approx 3 \cdot 5 \, (3 + \tau^{-1}) \, R_g$, where $\tau = E_{mt}/E_t$ and R_g is a gravitational radius. If, e.g., $\tau = 2/9$ (the maximum turbulent velocity being limited by the velocity of sound so that the damping of magnetoturbulence is small) then $R_{cr} \approx 26 R_g$, which is much less than in the static case. The energy corresponding to the critical state equals $\mathscr{E} \approx -0 \cdot 11$ $(3 + \tau^{-1})^{-2} \, Mc^2$, and putting $\tau = 2/9$ we obtain $|\mathscr{E}| \approx 2 \times 10^{-3} \, Mc^2 = 4 \times 10^{51} \, M/M_\odot$ erg.

It should be stressed that the total output of energy is proportional to the mass. In the polytropic approximation the magnetoid lifetime of about 10^6 years is determined by the thermal losses and is independent of the mass. If the mass does not exceed the value $M_* \approx 2 \times 10^{10} \, (3 + \tau^{-1})^{-2} \, M_\odot$, i.e. about $4 \times 10^8 \, M_\odot$ if $\tau = 2/9$, then nuclear reactions take place during the quasi-static contraction, providing an extra energy source. For large masses, however, with $M > M_*$, gravitational contraction turns out to be the main energy source.

(ii) Variability of the radiated energy. This is readily understood if the large-scale circular motion is inhomogeneous, anisotropic and, in particular, not axially symmetrical. The time variations will consist of a long-period component due to the circular motion and fast fluctuations associated with the turbulence and active flare processes. A comparison of the theory of these variations (12) with the light curve of 3C 273 over 80 years (13, 14) shows that the observations are in agreement with the idea of large-scale motion in the magnetic field of a quasar nucleus. This field may be thought of as providing 'magnetic rails' on which the circular motion proceeds with little damping.

4. Observational Tests of the Magnetodynamic Model

The model described above explains some of the observations and also predicts certain effects which may serve as observational tests (15). First of all, however, it is important to check the main consequences of the theory, i.e. the existence in quasar nuclei of strong magnetic fields and large-scale motion. The first may be tested by analysis of the synchrotron radiation of quasars. Such an analysis of the infrared radiation of 3C 273 (**7**, **16**) indicates a field strength of several hundred oersted – an unprecedented value for extragalactic sources but easily provided by the magneto-dynamic model. The presence of large-scale motion is confirmed by the statistics of the optical continuum variations, as discussed already. In addition the regular motion will lead (**12**) to periodic alterations of the position angle of linearly polarized radiation. It is of interest that variations of position angle have recently been observed (**17**). The changes are non-monotonic and are, besides, irregular on a short time-scale. Such irregularities are to be expected as a consequence of the turbulent-convective motions. Frequent photoelectric observations of quasars such as 3C 446 are vital for acquiring further information about the magnetoturbulence and important information about explosion processes in a magnetoid can be extracted from analysis of the radio variations (**20**).

5. The Magnetodynamic Model and the Origin of Quasars

Magnetoids may result as a consequence of the contraction of an initially more rarefied medium and be dependent upon a suitable configuration of the magnetic field and rotation (**6, 9**). Formation may occur either in the intergalactic medium (especially in the past) or in the nuclei of existing galaxies.

Magnetoids in intergalactic space are likely to originate in rotating gas clouds of mass up to 10^{12} $M_\odot$ which have a magnetic field roughly perpendicular to the rotation axis. In the central region, plasma will form a magnetoid while elsewhere fragmentation into star clusters will take place. Annihilation of oppositely directed magnetic fields near the rotation axis is expected to release magnetic energy explosively. The consequent shock wave will expand in each direction along the rotation axis and produce relativistic particles and plasma jets. Since the particle acceleration takes place at the shock front, this should be a region of strong radiation.

Magnetoids may also be formed in galactic nuclei from the interstellar gas concentrated there. In this case, however, the total mass does not exceed 10^8 $M_\odot$ and the energy output is less, corresponding to the phenomena in the nuclei of Seyfert galaxies.

6. The Relation of Magnetoids to Star Formation

The explosive release of energy on the formation of a magnetoid, and during its development, leads to the ejection of matter in the form of a plasma. On cooling, this plasma will fragment into stars (**18**), thus providing a new process for the formation of stars and star clusters (see Figure 1).

Before cooling and fragmentation the plasma will contain large numbers of relativistic particles and will radiate strongly, thus explaining the pairs of radio-sources associated with certain peculiar galaxies. Calculations indicate that the mass of plasma ejected in this way may reach that of small galaxies. The fragmented parts are then expected to have masses of the order of those of globular clusters, which may account for the unusual number of globular clusters in NGC 4486. This can be regarded as the physical realization of Ambarcumjan's idea (19) on star formation from

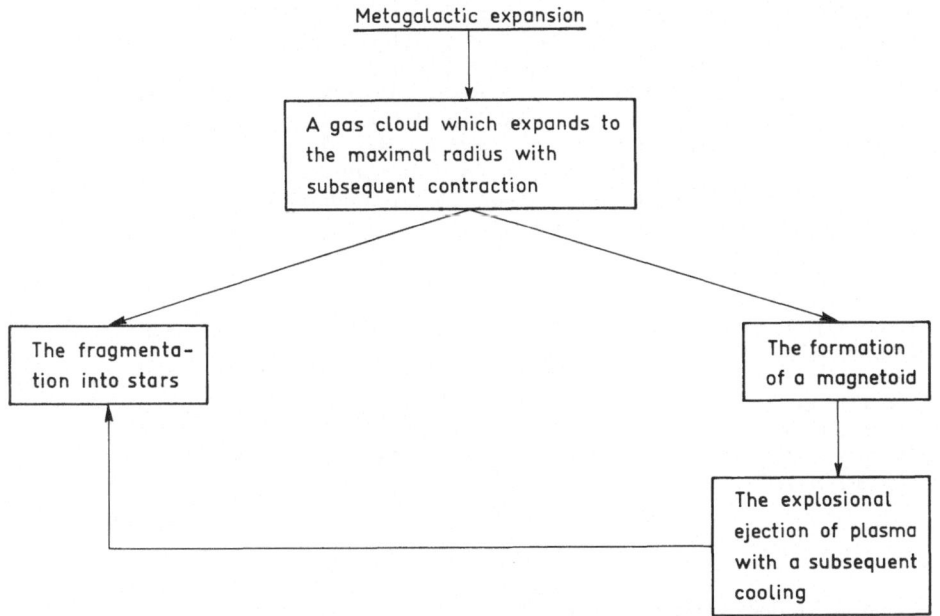

FIG. 1.   *The scheme of a dualistic way of star formation in a galaxy.*

material denser than interstellar matter. The eruptional genesis of stars and stellar systems turns out, however, to be only a part of the common scheme of gravitational condensation, with the matter traversing from the rarefied dispersion phase into the dense structure phase (21) via the intermediate stage of a magnetoid.

7. Conclusions

The magnetodynamic model provides the necessary energy source for quasars and explains the principal observational data. It suggests that such varied objects as quasars, radio-galaxies and active nuclei of galaxies may have similar energy reservoirs, quantitatively different due to the different conditions giving rise to the magnetoids. It is hoped that a more detailed treatment will account for the differences between these phenomena.

References

1. Fowler, W.A. (1964)   *Rev. mod. Phys.*, **36**, 545; (1966)   *Astrophys. J.*, **144**, 180.
2. Ginzburg, V.L., Ozernoj, L.M. (1964)   *Zu. eksp. teor. Fiz.*, **47**, 1030.
3. Ozernoj, L.M. (1966)   *Soviet Phys. Dokl.*, **10**, 581.
4. Layzer, D. (1965)   *Astrophys. J.*, **141**, 837.
5. Ozernoj, L.M. (1967)   in *Proc. Symp. "Variable Stars and Stellar Evolution" (Moscow 1964)*, Nauka Publ. House, Moscow, pp. 140–155.
6. Ozernoj, L.M. (1967)   *ibid.*, pp. 13–39.
7. Ozernoj, L.M. (1966)   *Soviet Astr.*, **10**, 241.
8. Sturrock, P. (1965)   *Nature,* **205,** 861; (1966)   *Nature*, **211**, 697.
9. Piddington, J.H. (1966)   *Mon. Not. R. astr. Soc.*, **133**, 163.
10. Bisnovaty-Kogan, G.S., Zel'dovic, Ja.B., Novikov, I.D. (1967)   *Astr. Zu.*, **44**, 525.
11. Gudzenko, L.I., Ozernoj, L.M., Čertoprud, V.E. (1967)   *Nature*, **215**, 605.
12. Ozernoj, L.M., Čertoprud, V.E. (1967)   *Astr. Zu.*, **44**, 537.
13. Ozernoj, L.M., Čertoprud, V.E. (1966)   *Soviet Astr.*, **10**, 15.
14. Smith, H.J. (1965)   in *Quasi-Stellar Sources and Gravitational Collapse*, Ed. by I. Robinson *et al.*, Univ. of Chicago Press, Chicago, p. 221.
15. Ginzburg, V.L., Ozernoj, L.M. (1967)   in *Proc. of 3rd Texas Symp. on Relativistic Astrophysics, New York* (in press).
16. Šklovskij, I.S. (1965)   *Soviet Astr.*, **9**, 683.
17. Kinman, T.D., Lamla, E., Wirtanen, C.A. (1966)   *Astrophys. J.*, **146**, 964.
18. Ozernoj, L.M.   *Astr. Zu.* (in press).
19. Ambarcumjan, V.A. (1958)   in *La Structure et L'Evolution de L'Univers (Proc. of 11th Solvay Conference)*, Ed. by R. Stoops, Brussels, p. 241.
20. Ozernoj, L.M., Sazonov, V.N. (1968)   *Nature* (in press); *Astr. Zu.* (in press).
21. Ozernoj, L.M., Černin, A.D. (1967)   *Astr. Zu.*, **44**, 1131; *Zu. eksp. teor. Fiz.* (Pisa) (in press); *Astr. Zu.* **45**, (in press).

DISCUSSION

Cowling: What does a magnetoid look like?
Ozernoj: Unfortunately I cannot draw one.

A ROTATING DISK MODEL FOR A QUASAR

F. D. KAHN

(Astronomy Dept., University of Manchester)

It has often been suggested that the energy supply of a quasar is derived from a central body with a mass of 10^9–$10^{10}\ M_\odot$. However, it is known that such bodies have no stable and static equilibrium configurations beyond a relatively low degree of condensation. Only a very limited amount of energy can be extracted before instability sets in. Basically the instability is due to the dominance of radiation pressure in supporting the mass against its self-gravitation.

But if a massive object is in rapid rotation, then it is partly supported by centrifugal force, and is much less liable to an instability of this kind. It is therefore interesting to investigate the properties of a rotating disk-like object with a mass of the order of 10^9–$10^{10}\ M_\odot$, and to see whether its overall properties match those of any quasar. The most important parameters are $2\sigma_0$, the mass per unit area of the disk, and Z, its half-thickness. In the case of the quasar 3C 273 B, assumed to be at a distance of 450 Mpc, the most suitable pair of values are $\sigma_0 = 10^7$ gm/cm^2 and $Z = 10^{16}$ cm.

With these parameters the following properties are predicted for the model. The energy output is derived from the collapse of the disk towards its central plane, but this is slow and proceeds through a sequence of quasi-static states. The outflow of radiation is checked by electron opacity, and radiation pressure is much more important than gas pressure. The matter in the disk is hot enough to be fully ionized but the temperature in its central layer is only about 3×10^5 deg K. No important nuclear reactions can therefore take place, and so the object does not change its composition as it evolves.

The radiative loss per unit mass must equal $4\pi Gc/\kappa = 6 \times 10^4$ erg g^{-1} s^{-1} everywhere, where κ = electron opacity. This necessarily follows when radiation pressure and electron opacity dominate. Thus the outer layers, which are cooler, must release radiation at the same rate as the inner layers, which are much hotter. The outer layers thus overcool, and convection begins.

Now in a structure that is supported by radiation pressure, the temperature everywhere is determined simply in terms of the pressure, and is uniform at any level. The rising and the falling convective streams are therefore equally hot. Because of this there is no heat exchange between them, and the convected bubbles may have a much longer lifetime than they do when gas pressure dominates. Perhaps the fluctuations in brightness of a quasar are due to the irregular arrival of large bubbles at its surface.

Perek (ed.), Highlights of Astronomy, 389–390. © I.A.U.

Finally, the predicted lifetime for the object is 3×10^5 years, the colour temperature in the visible is about 50000 deg K, and the natural period of the disk for dynamical vibrations about 10 years.

DISCUSSION

E.M. Burbidge: Is not most of your energy thermal? How do you explain the observed optical polarization?

Kahn: Something will obviously have to be added to the theory.

G. Burbidge: Most of the observed energy is in polarized infrared.

Bidelman: Have you considered element production?

Kahn: No.

SHORTER CONTRIBUTIONS

P. A. Sturrock: Observational evidence indicates that the same explosion mechanism occurs in both quasars and radio galaxies, giving rise to jets or to single or double radio clouds. An understanding of the nature of quasars and of the structure of radio galaxies is tied inextricably to the understanding of this explosion mechanism.

An explosion is an instability. This instability converts stored energy efficiently into the kinetic energy of high-energy electrons – and probably also high-energy protons. The instability therefore develops strong electric fields. This indicates that the explosion is a plasma instability. Further considerations indicate that the stored energy is magnetic in form. In these respects the explosion is the same as a solar flare, and is therefore interpreted as a 'galactic flare'.

This interpretation of the explosion mechanism suggests the following model of quasars and radio galaxies. The condensation of an object of galactic mass in intergalactic space containing a weak (about 10^{-7} gauss) intergalactic magnetic field leads to the initial formation of a compact object, identified as a quasar or a galactic nucleus, which is coupled magnetically to intergalactic space. The magnetic-field configuration contains current sheets. The tearing-mode resistive instability may occur in the vicinity of these sheets, giving rise to a recoupling of the magnetic field lines, permitting the magnetic field to relax to a lower-energy state. Certain field configurations, which seem likely to arise, may give rise to the formation of a jet, or of a single or a double radio cloud.

In this framework, one is led to interpret a quasar as an evolving form of a galactic nucleus. The magnetic field plays an important role in the formation of a massive compact mass, both by enhancing the accretion process and by providing a mechanism for getting rid of angular momentum. It is believed that, at a certain stage, the mechanism for propagation of angular momentum becomes ineffective so that further accretion leads to the formation of the main body of the galaxy.

S. von Hoerner: It still seems possible to explain quasars by stellar collisions with subsequent explosions, as first mentioned by T. Gold (Texas Symposium 1964). I have made some estimates with the following results.

If we assume all stars to be like the Sun, then a stellar cluster in virial equilibrium (or the dense centre of a larger system) has only two degrees of freedom; its radius R and the number N of stars. If we want stellar collisions to continue for a longer time, then the colliding and exploding stars must be replenished by dynamical interaction, which means that the collision time $t =$ (mean free path)/(r.m.s. velocity) must equal the relaxation time T. The condition $t = T$ then leaves only *one* degree of freedom, which we will use up by demanding a collision rate, $C = N/t$, of $C = 10$ collisions/year to explain the time-scale of the variability.

Insertion of only one number, $C = 10$ col/year, makes the system entirely defined and yields $R = 0.01$ parsec for its radius, $M = 3 \times 10^6$ $M_\odot$ for its mass, and $t = T = 3 \times 10^5$ years for its lifetime; these are just the values required for quasars (at cosmological distance) from completely different arguments. If we further assume that the energy, released by explosions and transformed into radiation, is $\frac{1}{2}\%$ of mc^2, we derive a luminosity of $L = 10^{46}$ erg s^{-1}, again just the right value. Using only *one* free parameter yields a good fit for *five* quantities (C, R, M, T, L), which seems to support the collision hypothesis (if it is not just chance, of course).

But it is impossible to obtain such an extremely high density $(10^{12} M_\odot$ parsec$^{-3} = 10^{-10}$ g cm^{-3}) from a more normal one within 10^{10} years by stellar-dynamical means; an estimate gave a decrease in radius of only a factor of 10. The stellar system thus must have started with already a very high density $(10^{-13}$ g cm^{-3}) provided in the gaseous state before the formation of stars. The question, whether or not a stellar system can finally produce collisions, then mainly depends on the maximum gas density which can be reached, in the centre of a cloud or of a rotating disk, before star formation sets in. For comparison: with a density of 10^{-13} g cm^{-3}, Jeans' criterion for formation of stars of solar mass gives a temperature of $400°$K. (I would like to add that the observed dense centres of spiral and elliptical galaxies, too, must have been provided in the gaseous state.)

F. JOINT DISCUSSION OF COMMISSIONS 29, 35 AND 42
CLOSE BINARIES AND STELLAR EVOLUTION

(Wednesday, August 30, 1967)

Organizing Committee: F. B. Wood (Chairman), R. Kippenhahn,
M. Plavec, J. Sahade
Meeting Chairman: F. B. Wood
Secretaries: J. Grygar, S. Kříž
Editors: M. Plavec and J. Grygar

Contents:

OPENING REMARKS

Frank B. Wood

(Flower and Cook Observatory, Philadelphia, Pa., U.S.A.)

I should like to open on a personal note recalling the developments that led to my own interest in this topic beginning many years ago. I do this in part because it illustrates how interest in one particular problem sometimes leads one into another quite different one. Actually, there were two apparently unrelated problems which led to my interest in evolution of close binaries.

The first interest arose from a paper published by Raymond Smith Dugan and Frances Wright some 30 years ago (Princeton Contribution 15). This was an investigation of the changes of period of a number of eclipsing binary stars. At about this time we had become aware that rotation of the line of apsides in systems with elliptical orbits depended in part on the internal constitution of the stars and that the determination of the period of this, together with other information derived from the light and velocity curves, could permit calculation of the degree of central condensation of the stars. I suspect that this may have been at least in part the motivation for the work, because in the discussion of almost every system, mention is made of whether or not the changes could have been caused by apsidal rotation.

What Dugan and Wright actually found was that in many cases there were erratic and non-periodic period changes which could not be explained by any known cause. At any rate, this paper made a profound impression upon me, and for years afterwards I speculated upon what physical mechanism might be responsible for sudden period changes.

The second development came in 1946 when I was finding some difficulty in interpreting the system R Canis Majoris – if anyone wants a system hard to interpret I recommend this one. Since only one spectrum could be observed, determination of absolute dimensions presented some difficulty. In an effort to find limiting solutions, it suddenly occurred to me that laws of celestial mechanics applied here also and that an upper limit to the sizes of the components could be set by using the particular solution of the three-body problem in which one component had infinitesimal mass. Fortunately, the difficult mathematics had been worked out long previously and were clearly presented by Moulton in his text on celestial mechanics. Thus, into eclipsing star studies could be introduced the zero-velocity surface which has now become familiar and which I am sure will be referred to in this discussion.

Actually, the first use of this concept in eclipsing stars had been made by Kuiper in 1941 in his study of β Lyrae. In defense of being unaware of a paper in my own field

Perek (ed.), Highlights of Astronomy, 394–395. © I.A.U.

published in a professional journal, I can only say that the paper appeared late in 1941. Earlier that year I had gone on active duty with the U.S. Navy and saw no astronomical literature for nearly 5 years. In preparing for publication my dissertation written 5 years earlier, I had reviewed carefully all the stars included, but β Lyrae was not one of these.

These two interests came together in the late 1940's when, still wondering about irregular period variations, I found that, with one exception, and that itself an unusual system, all systems showing such fluctuations had one of the components near the zero-velocity limiting surface and that in systems which for extended intervals had shown no period variation, neither component approached the limiting surface. This led to the suggestion published in 1950 in the *Astrophysical Journal* that we might use this criterium to divide eclipsing systems into two general classes – one in which neither component approached the limiting surface and a second in which at least one component reached it. A few years later a further refinement was suggested by Z. Kopal, who suggested that the second of these be further subdivided and who introduced the terms 'detached' and 'semi-detached' – the term 'contact binary' had, I believe, been used earlier. In the same paper, I suggested mass ejection as the cause of period changes, and noted that, if correct, the period of mass loss would be short compared to the total life of the star.

Time does not permit even a summary of all of the work done in recent years on evolution of close double stars. Su-Shu Huang, one of our speakers today, was the first to compute in some detail the effect of mass loss upon the period, and Crawford first realized what paths of stellar evolution as computed for single stars would mean in a double-star system when one component started from the main sequence to the red giant stage. Many puzzles, of course, still remain. Are the strange departures from the normal mass-luminosity relation caused by conditions existing as the systems form or do they later evolve into this state? What is the final result of continued mass loss? Hopefully, after today's papers and discussions we may be a step or two nearer to the solution of some of the problems involved and will be aware of as yet unasked questions concerned with stellar evolution and close double stars.

ON THE ORIGIN OF THE ALGOL SYSTEMS

M. PLAVEC

(Astronomical Institute, Ondřejov, Czechoslovakia)

1. Properties of the Algol Systems

The 4th edition of the Finding List for Observers of Eclipsing Variables (Koch *et al.*, 1963) contains 145 sufficiently well-observed eclipsing binaries brighter than $8 \cdot 5^m$ at maximum light. Among them, 59 binaries, or 41%, are systems with both components on the main sequence. The second largest group, 52 binaries or 36% of all systems, are systems similar to Algol. These can be characterized as follows:

(1) The primary (more massive) components are main-sequence stars, fitting well into the mass-luminosity relation defined by visual binaries and by eclipsing binaries with both components on the main sequence (detached systems).

(2) The secondary components are of later spectral type than the primaries, and can be best characterized as subgiants. They are overluminous for their mass as well as for their spectral class.

(3) As a rule, the secondary components fill their respective critical Roche lobes (innermost Lagrangian surfaces).

These properties can be demonstrated on the accompanying Figures 1–3, based on our re-discussion of 46 Algol systems with sufficiently reliable data. Full circles refer to binaries in which both spectra are measurable for radial velocities; hence, the determination of their absolute dimensions is free from any assumptions. However, the data are usually not particularly accurate, and an attempt was made in our discussion (Plavec, 1967*b*) to estimate the uncertainty; therefore, in the figures, circles connected by a straight line refer to the same star. Open circles represent binaries in which only the spectrum of the primary was measured. In order to solve the mass function for the masses, an additional assumption is required. In our discussion, (Plavec, 1968*b*), an effort was made to apply two independent assumptions and to compare the results; again, the uncertainty is marked in the figures.

Figure 1 is a H–R diagram for the primary (more massive) components. The broken heavy line marks the locus of single stars of luminosity class V. A few primaries may be already somewhat evolved off the main sequence (as W UMi or SZ Psc), and the primary of RZ Sct is apparently a supergiant, but otherwise the primaries scatter along the luminosity class-V line.

Figure 2 is similarly a H–R diagram for the secondary components. Although a few of them are giants of luminosity class III, and others show practically no overlumi-

Perek (ed.), Highlights of Astronomy, 396–408. © I.A.U.

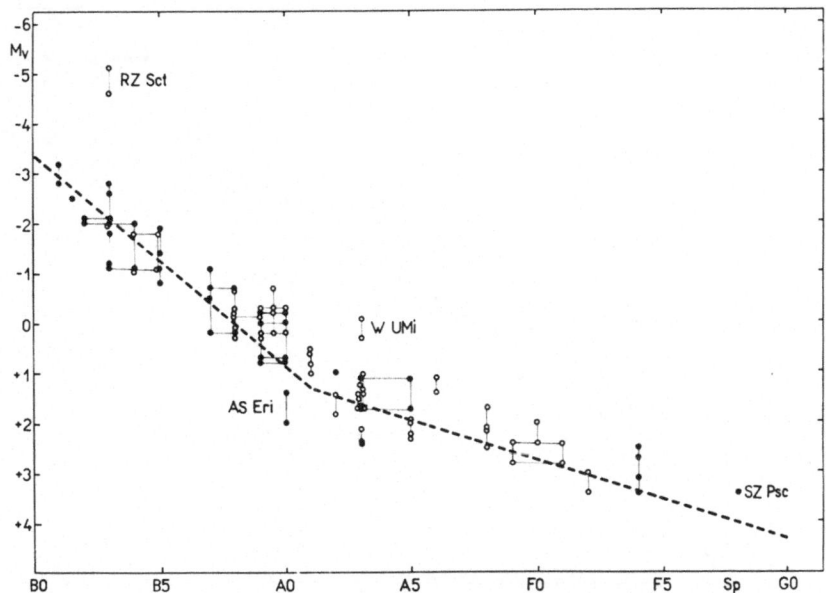

FIG. 1. *H-R diagram for the more massive components.*

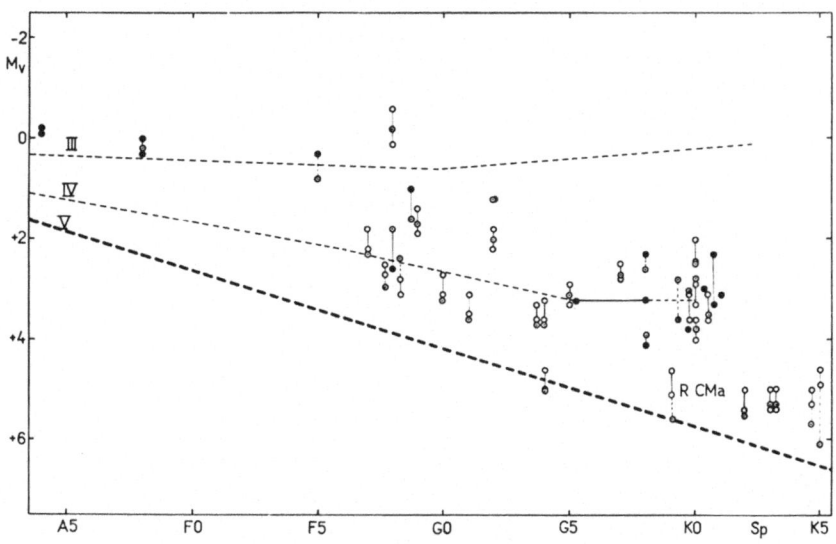

FIG. 2. *H-R diagram for the less massive components.*

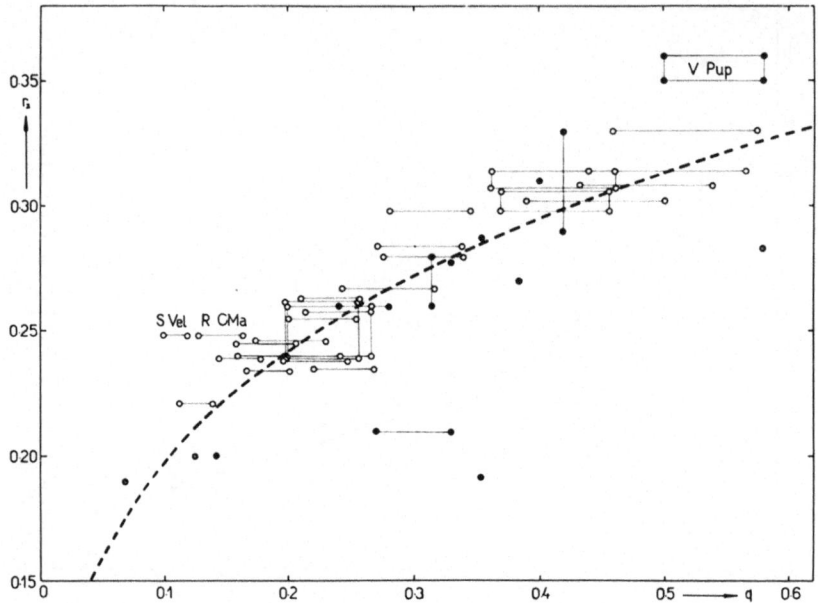

the primary is already on or near the main sequence. Therefore the radius of the se-condary is relatively large, and under favourable conditions it may be comparable with that of the primary.

Such a binary has a relatively very high probability of being discovered as an eclips-ing variable, since the fractional dimensions of the stars are large and comparable with each other, and the eclipses tend to produce deep minima. However, the duration of this phase is quite short. We shall see in Section 4 that the theory of large-scale mass exchange predicts another 'Algol' phase, photometrically almost equally favourable, but with a duration about 50 times longer. Our calculations indicate that the observed number of the young binaries to these transformed systems should be about 1:30.

Nevertheless it is worth while to look for the binaries with young contracting sub-giants. However, the most probable combination is not a genuine semi-detached system with a contact subgiant, but a binary in which the secondary is already signi-ficantly smaller than its Roche limit – this is because towards its end the pre-main-sequence contraction is slower. Roxburgh (1966) suggested that KO Aquilae might be such a binary. However, the absolute dimensions of this system are very poorly known. I think that young contracting subgiants could be best identified spectro-scopically.

3. Phases of Secular Expansion

It is evident that the existence of a great number of the semi-detached binaries similar to Algol can be explained only if we admit that the evolution of each compo-nent of the binary is greatly affected by the presence of the other star. In the theory I am presenting here it is assumed that the dominant role is played by the Roche limit,

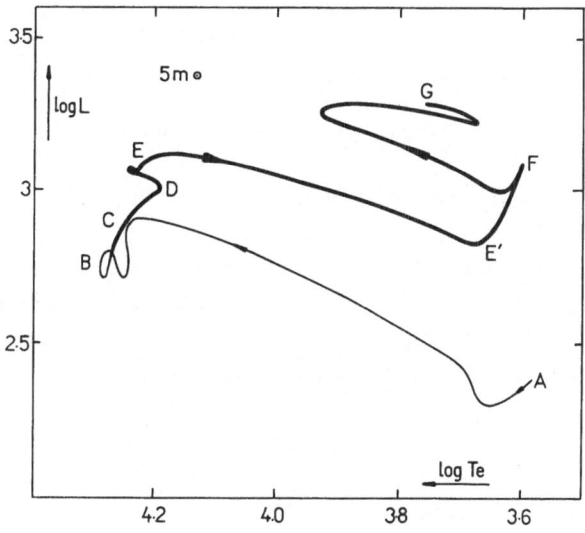

FIG. 4.   *Evolution of a star of 5 $M_\odot$ in the H-R diagram.*

which represents the upper boundary to the permitted size of each component. Each component is expected to follow its normal course of evolution as long as it remains smaller than its Roche limit. When, however, its expansion due to the internal forces brings its surface to the Roche limit, the star begins to lose mass. Because the more massive stars develop faster, it is always the primary component that reaches the Roche limit first.

Figure 4 shows the evolution of a star of 5 $M_\odot$, according to Iben (1965), and Figure 5 shows the corresponding secular change of its radius. We recognize two phases of

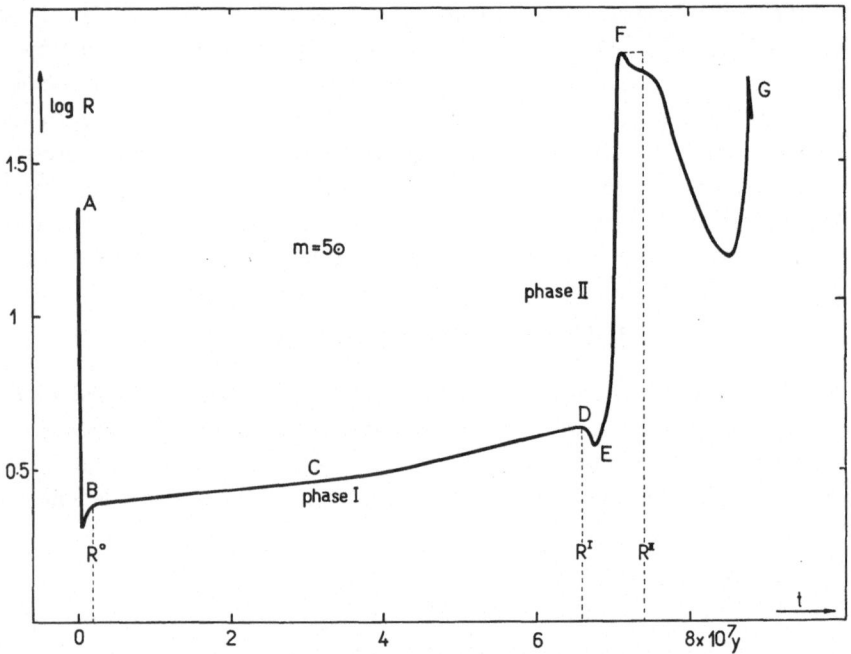

FIG. 5.   *Secular change of radius with time for a 5 $M_\odot$ star.*

expansion, at which the star can attain the Roche limit: Phase I (or A, as Kippenhahn and Weigert call it) of slow expansion when the star burns hydrogen in its core and its envelope expands simultaneously as the convective core contracts; and Phase II (or B) when hydrogen is burning in the shell surrounding the helium-rich core, and the star expands rapidly, changing into a late-type giant. Certainly there are further phases of expansion in the later evolution of this giant, but these will not be considered here.

Whether and when the primary component fills its Roche limit, depends on the mass and distance of the secondary component. A more detailed discussion has been published elsewhere (Plavec, 1967a, 1968a). May it suffice here to say that the primary reaches the Roche limit in Phase I if the period of the system is shorter than P^I, given by the equation

$$\log P^{\mathrm{I}} = 0{\cdot}73 \log M_1 - 0{\cdot}86 - T(q). \qquad (1)$$

Similarly, the primary component attains the Roche limit in Phase II if the period of the system fulfils the inequality

$$P^{\mathrm{I}} < P \leqslant P^{\mathrm{II}}, \qquad (2)$$

where

$$\log P^{\mathrm{II}} = 2{\cdot}20 \log M_1 + 0{\cdot}04 - T(q). \qquad (3)$$

Here the periods are expressed in days, M_1 is the mass of the primary component, and $T(q)$ is a slowly varying function of the mass ratio q; for the sake of a rough estimate, we can take $T(q) = -0{\cdot}45$.

From (1) it follows that in Phase I, the Roche limit is reached by the primary component in short-period systems only, since P^{I} is about $0{\cdot}8$ day for a primary of 3 $M_\odot$, and about $1{\cdot}9$ day for a primary of 9 $M_\odot$. Nevertheless, just these short-period systems appear to be the progenitors of many of the Algol systems.

4. The Mass Exchange

Crawford (1955) was the first to suggest that the semi-detached binaries are products of a large-scale mass transfer between the components. Rough model calculations by Morton (1960) and by Smak (1962) added a great support to this hypothesis, but I think I am correct to say that the definitive theoretical proof that such a process is physically possible was not afforded until in the model calculations by Kippenhahn and Weigert (1967). During the past year, further calculations illuminating the problem have been performed not only in Göttingen (Giannone et al., 1968) but also in Warsaw (Paczyński, 1966, 1967) and by our group at Ondřejov (Plavec et al., 1968a, b).

The work of all the three groups is based on the same principles. Rotational and tidal distortion, affecting only the outer layers of the stars, are neglected, and spherical symmetry is assumed as well as circular orbits.

All the material driven by the expanding star beyond the Roche limit is expected to flow rapidly to the other component and settle down on it. It is assumed that the total mass and the total amount of orbital angular momentum remain preserved. (This may not be the best assumption, but we have as yet little observational evidence contrary to this assumption.)

In this 'conservative case', the distance between the two components varies according to the law

$$A = A_0 \left(\frac{M_1^0 \cdot M_2^0}{M_1 \cdot M_2} \right)^2, \qquad (4)$$

where the noughts refer to the values at the beginning of the mass exchange.

The fractional size of the Roche limit has been tabulated (Kopal, 1954; Plavec, 1968a) or can be represented by the following approximate formula:

$$r_1^* = 0\cdot38 + 0\cdot2 \log(M_1/M_2); \tag{5}$$

the absolute mean radius of the Roche limit is then

$$R_1^* = A\cdot r_1^*. \tag{6}$$

When the mass exchange begins, both factors in (6) decrease, so that the Roche limit of the mass-losing star shrinks rapidly. However, Equation (4) shows that A reaches a minimum when the two stars are equal in mass, and then begins to increase again. Eventually, this increase more than compensates the monotonous decrease of r_1^*, and the absolute radius of the Roche limit begins to increase again.

This has serious consequences for the process of mass loss. Consider the beginning of this process. The expanding star has just reached the Roche limit and has lost a small fraction of its mass. This loss cannot affect perceptibly the deep layers, and the star tends to continue its expansion. If it were free to adjust itself to thermal equilibrium, it would have a larger radius. But the radius is imposed upon it in the form of the

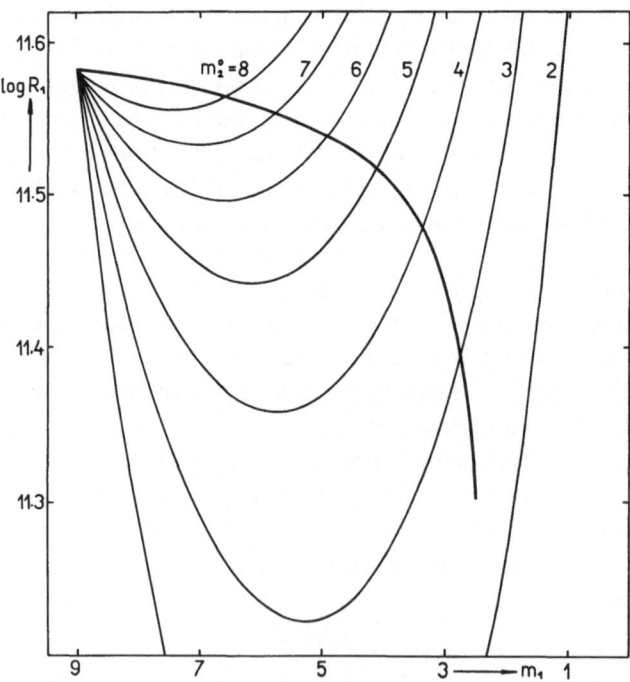

FIG. 6.   Change of the Roche limit during mass loss of a star with $M_1{}^0 = 9\ M_\odot$, for various initial masses of the secondary $M_2{}^0$. Abscissa is the instantaneous mass M_1.

Roche limit, and the Roche limit contracts instead of expanding. Therefore the star deviates strongly from thermal equilibrium, and the time-dependent terms in the equation for energetical balance play an important role. As usual under such conditions, the evolution is rapid – in our case it proceeds roughly on the Kelvin time-scale of the outer layers. The mass loss goes on, and is particularly rapid at the early phases, when the star's radius is forced to decrease so rapidly.

It is not difficult to understand that the rapidity of the mass loss depends also on the initial mass ratio. The greater is the initial disparity in masses, the more pronounced is the variation of the Roche radius. This can be seen from Figure 6, where we see how the size of the Roche limit of a star originally of 9 $M_\odot$ varies during the mass exchange in dependence on the original mass of the other star.

The actual computation of the evolutionary sequence of stellar models deviating from thermal equilibrium is made by means of Henyey's method for integrating the time-dependent equations of stellar structure (Henyey *et al.*, 1964). In our Ondřejov group, we have investigated several cases of mass exchange in Phase I, using this method modified by S. Kříž for this purpose. The necessity of a modification lies in the fact that we have one more unknown – the time interval Δt during which the star

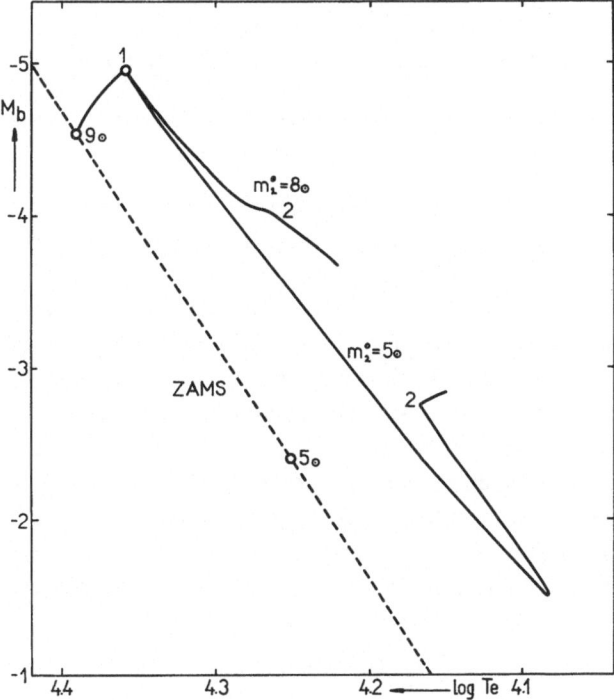

FIG. 7.   *Theoretical H-R diagram shows the evolution of a star of 9 $M_\odot$ during mass loss. Mass exchange begins at the point '1', rapid mass loss ends at '2', slow mass loss follows.*

loses the mass ΔM – but we have also one additional equation of condition, namely Equation (6) which fixes the radius.

Figure 7 shows the evolution of a primary component of 9 $M_\odot$ in a system where the other component has a mass of 5 $M_\odot$ in the first case, and 8 $M_\odot$ in the second case. The primary was supposed to reach the Roche limit at point '1', where its central hydrogen content has dropped from $X_c = 0.602$ to $X_c = 0.25$. As soon as the mass loss begins, the star's luminosity decreases, because a part of the luminous flux is absorbed in the expanding outer layers. This process is much more conspicuous if the original masses are very different, because the radius decreases more rapidly.

The phase of rapid mass loss is relatively very short, 10^4–10^5 years. Figure 8 shows

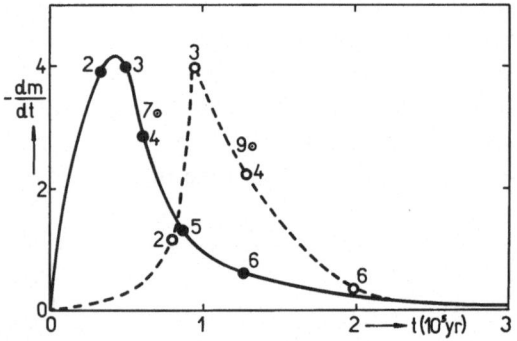

FIG. 8. *The rate of mass loss (in* 10^{-5} *solar masses per year): Full line, initial masses* 7 $M_\odot$ +5 $M_\odot$; *dashed line,* 9 $M_\odot$ +8 $M_\odot$.

the variation with time of the rate of mass loss for two configurations with the two stars initially not much different. The average rate of mass loss is 2×10^{-5} $M_\odot$/yr. For the combination 9 $M_\odot$ +5 $M_\odot$, the rates are ten times higher.

A star of 9 $M_\odot$ loses, during the phase of rapid mass transfer, a total of 2.4 $M_\odot$ if the secondary has initially 8 $M_\odot$, and 5.2 $M_\odot$ if the secondary has 5 $M_\odot$. It has been found by our three groups that the final stage of the rapid mass loss can be determined approximately but much more easily by computing a series (not an evolutionary sequence) of stationary models with continuously decreasing mass. In Figure 6, the heavy line is the locus of such models in thermal equilibrium. Its intersection with the appropriate curve describing the change of the radius of the Roche limit indicates approximately the final mass of the mass-losing component at the end of the rapid mass transfer, when the star's thermal equilibrium is very nearly restored.

5. The Algol Stage

The final stage of the rapid mass loss in Phase I is a subgiant, overluminous for its mass and usually also for its spectral type. This subgiant continues to fill the critical

Roche lobe. Its evolution proceeds now again on the nuclear time-scale. With increasing chemical inhomogeneity in its interior, where hydrogen continues to be converted into helium, the core shrinks and the envelope expands slowly. This phase is analogous to the main-sequence slow expansion (Phase I in Figure 5), and its duration is of the same order. The star fills the Roche lobe but no rapid mass loss occurs, since simultaneously with the expanding star the radius of the Roche limit increases, too. This is because the mass-losing star is now the less massive component, and with continuing mass loss the disparity of masses increases and causes the distance A to increase, as follows from (4). Although the rate of mass loss is very low, the total amount of material lost in this stage is not negligible; e.g., according to Kippenhahn and Weigert (1967), in the system consisting initially of stars of $9\,M_\odot$ and $5\,M_\odot$, the new secondary loses another $0\cdot7\,M_\odot$ within $1\cdot8 \times 10^7$ yr. This phase ends when the mass-losing star begins to contract and detaches itself from the Roche limit.

We have here a long stage when the system is photometrically quite prominent, when the subgiant fills the Roche limit, and when it is losing mass at a rate of about 10^{-8}–$10^{-7}\ M_\odot$/yr. This mass flow is probably sufficient to cause the observed phenomena of period changes and gaseous streams.

This is therefore the stage in which we observe the Algol systems. It is interesting to realize how much effort was devoted to the study of an evolutionary phase which appears to be only a fade-out of a much more violent and conspicuous process.

While this theory is, in my opinion, very promising in its qualitative explanation of the evolution of the Algol systems, I would not dare at this stage to attempt a quantitative comparison with an observed system. First of all, the models are still rather crude; the uncertainty introduced already in the models of the single stars by the assumptions about opacity, nuclear energy rate, composition etc., is augmented here by the simplifications mentioned in Section 4.

But, disregarding this, the number of free parameters is rather large. The final stage is not determined uniquely by the original masses. It is clear enough that the position of the heavy line representing the stationary models in Figure 6 will be different for different internal constitution of the mass-losing star. Since the phase of rapid mass loss is very fast, the chemical composition remains practically the same as at the beginning. Therefore when introducing e.g. X_c (abundance of hydrogen in the centre) as a parameter, we introduce in fact the age of the star. Figure 9 shows how the final product of the rapid mass loss depends on X_c.

The phase of slow mass loss is long and the evolution of the subgiant in it is not negligible – this is another complication. If we admit that part of the material leaves the system, we are introducing two more unknown parameters (Paczyński and Ziółkowski, 1967). But perhaps we shall be forced to make this assumption, since it is difficult to understand how the initial secondary can accommodate the huge amount of material carrying a large amount of angular momentum.

All we can do now is to establish which regions of the H-R and mass-luminosity

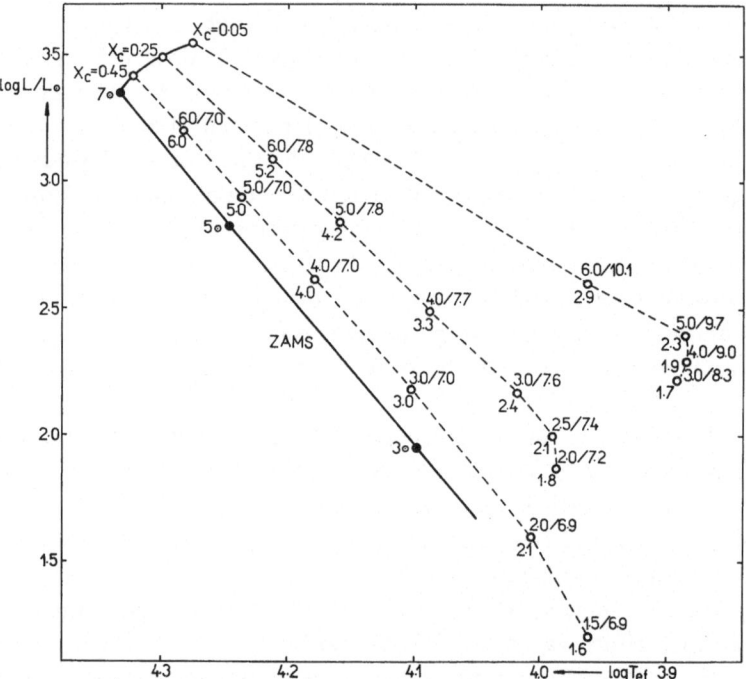

FIG. 9.   *H-R diagram of the mass-losing components at the end of rapid mass transfer. Initial mass 7 $M_\odot$. Dashed lines are loci of stars of the same X_c. Lower numbers are final masses of the mass-losing components. Upper numbers are the corresponding initial and final mass of the mass-gaining component, respectively.*

diagrams are covered by the models based on our simple theory. This has been done by Giannone *et al.* (1968), who introduced the 'generalized main sequences', and in a somewhat different way by our Ondřejov group (Plavec *et al.*, 1968b).

A fact of great importance is that in each single case, the mass ratio is more than reversed, i.e. the roles of the components are always interchanged and the discrepancy in masses increased. As a consequence, the final period is always longer than the original one. Starting from this fact, and from our computations of the type shown in Figure 9, we made rough estimates of the initial parameters of the 46 Algol systems for which we have reliable data. These systems can be explained as products of mass exchange in Phase I, except perhaps for binaries with low masses; however, Kippenhahn *et al.* (1967) have shown that semi-detached systems with subgiants are produced also by primaries of low mass expanding in Phase II.

6. Observational Tests

Although a direct comparison of the theoretical models with observed binaries is evidently difficult, I believe that the theory can be tested by observation.

We have seen (cf. Figure 9) that the final mass of the subgiant is the smaller the smaller was the initial mass ratio. Now in the homogeneous model of a star of 7 $M_\odot$, the convective core contains a mass of 1·7 $M_\odot$. Within this mass, therefore, hydrogen is deficient and also other abundances are affected by the nuclear reactions. Now only this part of the original star is left at the end of the rapid mass exchange in binaries with a final mass ratio about 0·2; this result obtains roughly also for other initial masses of the primaries. This means that in the observed Algol systems with $q \leqslant 0·2$, we should expect anomalous abundances (in particular, deficiency of hydrogen and overabundance of helium) in the atmospheres of both components. Such systems are e.g. AS Eri, DN Ori, AW Peg, XZ Sgr, R CMa, S Equ, and S Vel. The present primary components are naturally much easier to study, but in their atmospheres the effect may be obliterated by strong superficial currents which may develop when the material is falling on this star. It is our plan at Ondřejov to study this problem with our new 79″ telescope.

Another possibility is to catch a binary at the stage of rapid mass loss. This phase is short, but may be spectroscopically rather conspicuous, since the rate of mass loss in favourable cases is as high as 10^{-4} $M_\odot$/yr, which is only one order of magnitude smaller than a nova outburst.

References

Crawford, J.A. (1955)   *Astrophys. J.*, **121**, 71.
Giannone, P., Kohl, K., Weigert, A. (1968)   *Z. Astrophys.*, **68**, 107.
Henyey, L.G., Forbes, J.E., Gould, N.L. (1964)   *Astrophys. J.*, **139**, 306.
Iben, I. (1966)   *Astrophys. J.*, **143**, 483.
Kippenhahn, R., Weigert, A. (1967)   *Z. Astrophys.*, **65**, 251.
Kippenhahn, R., Kohl, K., Weigert, A. (1967)   *Z. Astrophys.*, **66**, 58.
Koch, R.H., Sobieski, S., Wood, F.B. (1963)   *Publ. Univ. Pa. astr. ser.*, **9**.
Kopal, Z. (1954)   *Jodrell Bank Ann.*, **1**, 37.
Morton, D.C. (1960)   *Astrophys. J.*, **132**, 146.
Paczyński, B. (1966)   *Acta astr.*, **16**, 231.
Paczyński, B. (1967)   *Acta astr.*, **17**, 1.
Paczyński, B., Ziólkowski, J. (1967)   *Acta astr.*, **17**, 7.
Plavec, M. (1967a)   *Bull. astr. Inst. Csl.*, **18**, 253.
Plavec, M. (1967b)   *Bull. astr. Inst. Csl.*, **18**, 334.
Plavec, M. (1968a)   *Adv. Astr. Astrophys.*, **6**, 201.
Plavec, M. (1968b)   *Bull. astr. Inst. Csl.*, **19**, in press.
Plavec, M., Kratochvíl, P. (1964)   *Bull. astr. Inst. Csl.*, **15**, 165.
Plavec, M., Kříž, S., Harmanec, P., Horn, J. (1968a)   *Bull. astr. Inst. Csl.*, **19**, 24.
Plavec, M., Kříž, S., Harmanec, P., Horn, J. (1968b)   *Bull. astr. Inst. Csl.*, **19**, in press.
Roxburgh, I.W. (1966)   *Astr. J.*, **71**, 133.
Smak, J. (1962)   *Acta astr.*, **12**, 28.

DISCUSSION

Koch: Have any of the listed binaries with small mass ratio, e.g. S Equulei, been studied for abundance anomalies?

Plavec: Not so far I know. It is a difficult task, because for reliable abundance determinations you need spectra of high dispersion, and our stars are relatively faint.

Masevič: Could you indicate the regions in the H-R diagram where you expect the very fast mass loss to occur?

Plavec: I think I can refer to my Figure 5, which shows that the mass-losing components lie above the main sequence in the general region of the subgiants or giants.

Abhyankar: Have you considered the evolution of the component which is gaining mass?

Plavec: Not yet, but certainly this is a very important problem. We have postulated tacitly that the mass-gaining component is capable of accumulating a huge amount of material, often equal to several solar masses, within about the same time as the primary loses it. It is questionable whether such a 'conservative case' of mass transfer is possible. We must bear in mind that the flowing material carries with it a great amount of angular momentum, which could make the accommodation difficult, since the mass-gaining star may get on the verge of rotational break-up. Personally I consider this problem the weakest point in the whole theory of mass-exchange, and its solution should not be delayed. The result may easily be that we must admit a certain mass loss from the whole system.

Weigert: Dr. Paczyński has really calculated several cases where a mass loss of the whole system was assumed. When we normally do not take into account a mass loss from the system, we do not think that this is the normal case in reality. We only neglect it since we would have to introduce unknown parameters which can change the result to either side. We only wish to see which systems can be explained with the most simple assumptions.

Paczyński: The Warsaw group (B. Paczyński and J. Ziólkowski, *Acta astr.*, **17**, 1967, 7) computed a number of evolutionary tracks for close binaries with arbitrarily chosen rates for the mass- and angular momentum losses. In all these cases the final mass ratio was closer to unity because not all the mass lost by one component was captured by the other star. I think this is in better agreement with the observed mass ratios.

Mammano: The new Bamberg variable BV 412 may be an example of systems in which the more massive component is in contact with its Roche lobe, while the companion is well detached. However, the only anomaly we can see is that the K line is unusually strong for the A0 spectral type, and its intensity varies with phase (A. Mammano, R. Margoni, R. Stagni, *Mem. S.A. It.*, **38**, 1967, in press).

THE EVOLUTION OF CLOSE BINARIES AND
THE FORMATION OF THE WOLF-RAYET STARS

BOHDAN PACZYŃSKI

(Institute of Astronomy, Polish Academy of Sciences, Warsaw, Poland)

I shall discuss here the evolution of massive binaries in the *case B*. In this case the initially more massive component fills up the Roche lobe after the exhaustion of hydrogen in the core. By a massive binary I understand the binary in which the phase of a rapid mass exchange is terminated by helium ignition in the core of the initially more massive star. The helium ignition stops the contraction of the core and the expansion of the envelope of that component. In a small mass star the degeneracy plays the role of helium ignition in this respect. The last case will be discussed by Dr. A. Weigert in his lecture.

The evolution of the primary component of 9 $M_\odot$ in the *case B* was computed by Kippenhahn and Weigert (1967) and is shown in Figure 1. The first part of the evolu-

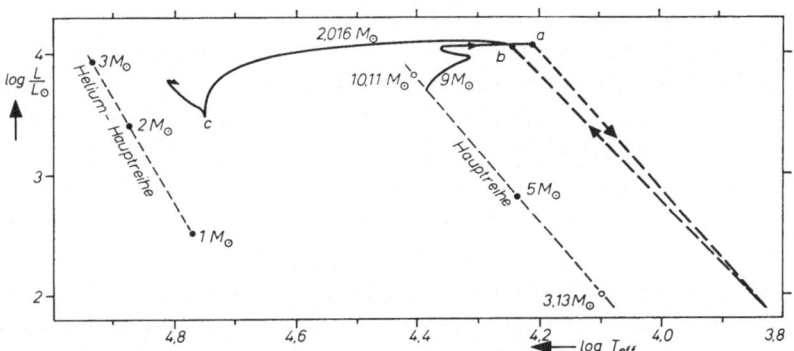

FIG. 1. *Evolutionary track on the theoretical H-R diagram for the primary component of the close binary studied by Kippenhahn and Weigert (1967, case B).*

tionary track is the same as for a single star. The mass exchange begins after the Roche lobe has been filled up by the primary. At the beginning the mass transfer takes place on the thermal time-scale of the envelope of this star and is therefore extremely rapid. When the larger part of the hydrogen-rich envelope has been lost the mass exchange slows down and proceeds on the thermal time-scale of the contracting helium core. The process is terminated by the helium ignition. Later on the star evolves on the local nuclear time-scale of the hydrogen-burning shell source that causes the decrease of

Perek (ed.), Highlights of Astronomy, 409–413. © I.A.U.

stellar radius. The original primary approaches the sequence of homogeneous helium-burning models on the H-R diagram.

Similar computations were carried out by Paczyński (1967b) for a star of 16 $M_\odot$. The evolution was followed through the phase of the most rapid mass exchange, and the results were in agreement with those obtained by Kippenhahn and Weigert. From the Weigert's discussion of the sequences of inhomogeneous models for helium-burning stars (Giannone, 1967) it follows that the evolution of massive binaries in *case B* should practically always look like that shown in Figure 1. This means that as a result of mass exchange we get a detached binary with the initially more massive component resembling closely pure helium star. The other component is now more massive and close to the zero-age main sequence. It was suggested by Paczyński (1967a, b) that we have here a system with the Wolf-Rayet type component.

We shall summarize here some basic observational data concerned with the W–R phenomenon. These stars have spectra with very prominent and wide emission lines of ionized helium and highly ionized nitrogen or carbon. The surface temperatures are very high, probably of the order of 50000 °K (Kuhi, 1966). The average visual absolute magnitude is near −4. The bolometric corrections are certainly large but unknown. Many W–R stars show small intrinsic-light variations. About 50% of the W–R stars are known to be members of the binary system with the other component being a normal OB star. The photometric observations of the eclipsing systems of this type suggest that the surface brightness of the W–R star is higher than that of the OB component (Kron and Gordon, 1950). The basic data for eleven Wolf–Rayet binaries with known orbital periods are given in Table 1. Five of them are eclipsing variables. We see that the typical mass of the W–R component is slightly below 10 $M_\odot$, that of the OB star near 25 $M_\odot$, and the average mass ratio $M_{WR}/M_{OB} = 0.3$.

We think that the high luminosities and surface temperatures of the W–R stars are not compatible with our knowledge of the position occupied on the H–R diagram by the massive stars in the phase of gravitational contraction towards the main sequence (Iben, 1965). It seems that the hypothesis that those stars are in advanced stage of stellar evolution (Burbidge and Burbidge, 1958; Limber, 1964; Divine, 1965), presumably in the helium-burning phase, and that they have lost in some way almost the totality of the hydrogen-rich envelope may account much better for their luminosities and surface temperatures, and perhaps for some anomalies in the abundances of the elements like hydrogen, carbon and nitrogen. The evolution of a massive binary in *case B* may account in the most natural way for most of the properties of those W–R stars that are members of binary systems. The masses, mass ratios and the periods of the observed binaries (Table 1) fall into the range of these quantities derived from the theoretical considerations (Paczyński, 1967b). The hypothetical model binaries with the initial masses of the primary in the range of 15–30 $M_\odot$ appear to be the best candidates for turning into the known W–R binaries.

It is well known that the W–R stars split into WC and WN sequences. It is not clear

Table 1
Eleven Wolf–Rayet Binaries

HD Number	Name	Spectral type	$M_{WR}\sin^3 i$	$M_{OB}\sin^3 i$ ($M_\odot$)	M_{WR}/M_{OB}	Period	References
152270		WC7 + 08	1·9	6·9	0·27	8^d·82	Struve (1944)
168 206[a]	CV Ser	WC7 + O	8·2	24·8	0·33	29·64	Hiltner (1945b); Hjellming and Hiltner (1963)
186943		WN5 + OB	5·8	21·0	0·28	9·55	Hiltner (1945a)
190918		WN5 + B0	0·2	0·8	0·26	85·0	Wilson (1949)
193576[a]	V444 Cyg	WN6 + B1	9·7	24·8	0·39	4·21	Wilson (1942); Kron and Gordon (1950)
193928		WN6	$f(m) = 4·94$		–	21·64	Hiltner (1945a)
197406		WN7	$f(m) = 0·07$		–	4·32	Bracher (1966)
211 853[a]		WN6 + B0	7·6	19·6	0·39	6·69	Hiltner (1945a); Hjellming and Hiltner (1963)
214419[a]	CQ Cep	WN5	$f(m) = 4·38$		–	1·64	Hiltner (1944); Hiltner (1950)
228766		WN7 + OB	4·6	22·3	0·21	10·6	Hiltner (1951)
	CX Cep	WN5	$f(m) = 5·38$		–	2·13	Hiltner (1948)

[a] Eclipsing variable.

whether this is the result of different abundances or of different excitation conditions in the atmospheres. We should like to show that genuine abundance differences are to be expected if our model is correct. In the inner half of the original mass of the primary component almost all carbon and substantial fraction of the oxygen are transformed into nitrogen in the CNO cycle during the phase of core hydrogen burning. It is very likely that after the mass exchange the W–R component will have a nitrogen-rich and carbon-poor atmosphere. Now, as a result of $N^{14}(\alpha, \gamma) F^{18}$ $(\beta^+ \nu) O^{18}$ reactions all nitrogen in the core will be transformed into oxygen, and carbon content will increase following helium burning. In massive helium stars the convective cores are large and have their masses almost constant during the evolution. We may expect that a large change in the C/N ratio will be present at the boundary of the convective core. If the star has a mass above 7–8 $M_\odot$ it will be probably vibrationally unstable and will eventually eject all the excessive mass (Boury and Ledoux, 1965). The convective core has a mass of 8 $M_\odot$ for a helium star of 12 $M_\odot$ (Divine, 1965). Therefore, if the mass of our star after the mass exchange exceeds 12 $M_\odot$ it may, as a result of further mass loss, uncover those layers that were once within the convective helium-burning core. In that case the atmosphere will have no nitrogen and hydrogen, but carbon and oxygen may be abundant.

It is not clear whether in all massive binaries following the evolution corresponding to the *case B* the initially more massive component may, after the mass exchange, show the W–R type characteristics. If not, the duplicity of the corresponding stars might be difficult to discover, as the helium component may not be bright enough in the visible light. This may be expected in the case of the binary studied by Kippenhahn and Weigert (1967). It is possible that vibrational instability of the helium star is required to generate W–R type spectrum. This instability, if present, might lead to light variations with a period of $0\overset{d}{.}01$–$0\overset{d}{.}04$. It would be interesting to see whether this can be observed. It would be also very interesting to know whether there are any single W–R objects that belong to the extreme population I, and how much they are different from the W–R binaries.

References

Boury, A., Ledoux, P. (1965)   *Ann. Ap.*, **28**, 353.
Bracher, K. (1966)   *Astr. J.*, **71**, 156.
Burbidge, E. M., Burbidge, G. (1958)   in *Encyclopedia of Physics*, Vol. **51**, Springer-Verlag, Berlin-Göttingen-Heidelberg, p. 134.
Divine, N. (1965)   *Astrophys. J.*, **142**, 824.
Giannone, P. (1967)   *Z. Astrophys.* **65**, 226.
Hiltner, W. A. (1944)   *Astrophys. J.*, **99**, 274.
Hiltner, W. A. (1945a)   *Astrophys. J.*, **101**, 356.
Hiltner, W. A. (1945b)   *Astrophys. J.*, **102**, 492.
Hiltner, W. A. (1948)   *Astrophys. J.*, **108**, 56.
Hiltner, W. A. (1950)   *Astrophys. J.*, **112**, 477.
Hiltner, W. A. (1951)   *Astrophys. J.*, **113**, 317.
Hjellming, R. M., Hiltner, W. A. (1963)   *Astrophys. J.*, **137**, 1080.

Iben, I. Jr. (1965)   *Astrophys. J.*, **141**, 993.
Kippenhahn, R., Weigert, A. (1967)   *Z. Astrophys.*, **65**, 251.
Kron, G. E., Gordon, K. C. (1950)   *Astrophys. J.*, **111**, 454.
Kuhi, L. V. (1966)   *Astrophys. J.*, **143**, 753.
Limber, D. N. (1964)   *Astrophys. J.*, **139**, 1251.
Paczyński, B. (1967*a*)   I.A.U. Colloquium "On the Evolution of Double Stars", Observatoire Royal de Belgique, *Communications*, Série B, No 17, 111.
Paczyński, B. (1967*b*)   *Acta Astr.*, **17**, 355.
Struve, O. (1944)   *Astrophys. J.*, **100**, 384.
Wilson, O. C. (1942)   *Astrophys. J.*, **95**, 402.
Wilson, O. C. (1949)   *Astrophys. J.*, **109**, 76.

DISCUSSION

Underhill: In considering Wolf-Rayet stars it is important to realize

(i) that they are really very constant stars – light and spectrum variations are the exception, not the rule;

(ii) a great variety of objects is included in the class Wolf-Rayet including central stars of planetary nebulae; thus one model need not explain all objects;

(iii) the evidence concerning the surface temperatures is very weak – we can only say that these temperatures are about the same as those of O stars;

(iv) our theories of spectrum-line formation are too rudimentary to permit any deductions about abundances.

Paczyński: What do you think about estimates of the surface temperature from the depths of eclipses?

Koch: It is not possible to infer (J_{WR}/J_{OB}) from the ratios of the eclipse depths in those eclipsing binaries containing WR stars unless bandpass effects are allowed for.

Mammano: A WR nucleus of a planetary nebula has been recognised by F. Bertola (*Publ. astr. Soc. Pacific*, **76**, 1964, 241) as being of type WN. Can this be taken into account in your scheme?

Paczyński: I know that some nuclei of planetary nebulae show W–R type spectrum. But I do not think that these objects belong to the extreme Population I.

WHITE DWARFS IN CLOSE BINARY SYSTEMS

ALFRED WEIGERT

(Universitäts-Sternwarte, Göttingen, Germany)

1. The Evolution to White-Dwarf Stage

While it has not yet been possible to give a detailed step-by-step treatment of the evolution of a single star from the main sequence to the white-dwarf stage, such a treatment is available for close binary systems. It has been shown that by calculating the evolution including mass exchange in a system of main-sequence stars of small mass and relatively large separation, one can follow the system to its final stage of a white dwarf and a more massive main-sequence star. This type of evolution arises when the original primary has exhausted its central hydrogen content when mass exchange starts, and the mass of its helium core is small enough so that electron degeneracy prevents the ignition of helium.

The evolution of such a system has been calculated by Kippenhahn *et al.* (1967). The results of these calculations illustrate those processes occurring in a star undergoing mass loss. The system considered consisted initially of a pair of main-sequence stars of 1 $M_\odot$ and 2 $M_\odot$ separated by 6·6 $R_\odot$. The 2 $M_\odot$ star first exhausted its central hydrogen content and then filled its critical Roche volume in 5·7 × 10⁸ years, during the phase of core contraction and envelope expansion. By that time, the star had moved in the H-R diagram (Figure 1) from the main sequence (point *A*) to point *D*, at which time a rapid mass loss starts, the time-scale of which is the thermal time-scale of the outer layers. This very rapid change is necessary for the star to keep its radius down to the now steeply decreasing critical radius R_{cr}. Later on, the mass loss becomes gradually slower, especially when (between *E* and *F*) R_{cr} again increases. Now the interior is able to adjust thermally. The interior of the star contracts and, consequently, the envelope expands further mass shifting outside the critical volume. At points *E*, *F*, *G* and *K*, the star's mass is 1·55, 0·96, 0·28 and 0·26 $M_\odot$ respectively. The entire phase of mass exchange lasts for 1·1 × 10⁸ years (points *D–K* in Figure 1). During the last phases of mass loss, the star becomes a red giant, although a most peculiar one: Its helium core contains 96% of the star's mass and is sufficiently condensed to exhibit the typical properties of a white dwarf. The hydrogen-rich envelope which contains 4% of the star's mass and 99·7% of its volume is very extended. Now, as this envelope contracts and in 1·2 × 10⁷ years the star moves over to the left of the main sequence (to point *N*), the system again becomes detached. During this period,

Perek (ed.), Highlights of Astronomy, 414–419. © *I.A.U.*

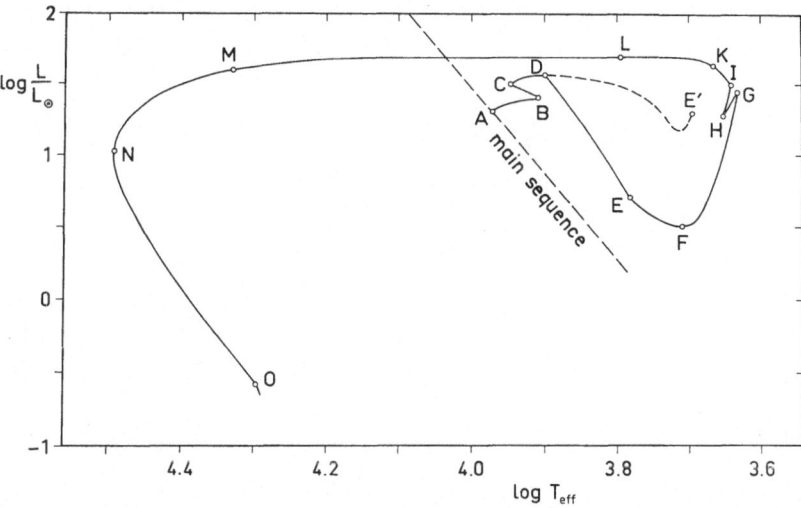

FIG. 1. *H-R diagram showing the evolutionary track of the original primary in a system of 2 $M_\odot$ and 1 $M_\odot$, as described in the text and as calculated by Kippenhahn et al. (1967). The star starts with 2 $M_\odot$ on the main sequence (point A); at point C central hydrogen is exhausted and mass transfer to the companion starts at point D. (Without mass loss, the track would have followed the dashed line.) Mass loss ceases at point K, and at point O the star has become a white dwarf.*

the shell source has burned outwards up to $M_r/M \approx 0.99$, and has supplied most of the luminosity. Finally its energy production decreases and the star moves in the H–R diagram downwards into the white-dwarf region (3×10^6 years to point O). During this entire evolutionary process the companion has not had enough time to evolve appreciably away from the main sequence. So we finally are led to a detached system containing a white dwarf and a main-sequence star, their distance having been increased by the mass transfer to about 50 $R_\odot$.

It is important to note that the final phases of mass loss last for a relatively long time (about 10^8 years) so that there is a reasonable chance for such systems to be observed. In this phase, the system is semi-detached since the mass loss is still going on, and the calculations show the now secondary component to be enormously overluminous (about $\Delta M_{\mathrm{bol}} = 9$ magnitudes above the main sequence M–L relation). This may suggest that many of the observed systems of low mass are in a similar phase of evolution, since it is characteristic for them to have a secondary of unusually high over-luminosity.

The evolution of the interior of the star during mass exchange can be seen from the changes of T and ρ at the centre (Figure 2). At the onset of rapid mass loss at point D, the central core cannot follow thermally since its time-scale is much longer than that of the outer layers. Thus, the interior has to change adiabatically implying here that $T_c \sim \rho_c^{2/3}$. This gives the steep increase of the curve in Figure 2 after point D (parallel to $\psi = $const.). Later on, when R_{cr} increases and the mass loss becomes slower, the

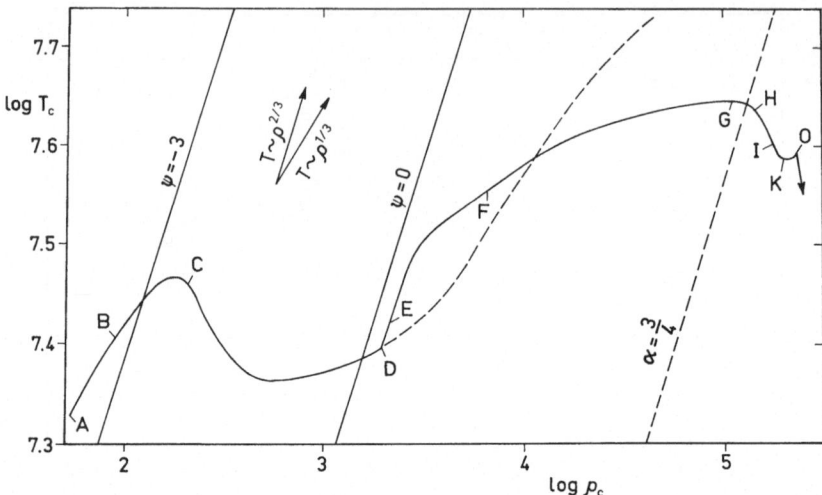

FIG. 2.   *The values of temperature T and density ρ at the centre of the star, the evolutionary track of which is given in Figure 1. The letters along the curve correspond to those along the evolutionary track. (The dashed curve after point D gives the values of T_c and ρ_c as they would have been without mass loss.) Two lines of constant degeneracy parameter ψ are indicated. (Figure taken from Kippenhahn et al., 1967.)*

centre can adjust; stars of lower mass have larger ρ_c and the curve goes to the right in this diagram. If the core would not further contract, or if the outer layers would not expand with further central contraction, then the mass loss would stop after the first very rapid phase. But the 'mirror effect' (where contraction of the core is coupled to expansion of the outer layers) is still operative, and further mass from the envelope is shifted outside the critical volume.

With increasing ρ_c, the centre becomes gradually degenerate. Due to this, ignition of helium is prevented, since T_c no longer increases with ρ_c. For a homologically contracting core, one can easily show that

$$\frac{d \ln T_c}{d \ln \rho_c} = \frac{4\alpha - 3}{3\delta},$$

where the matter obeys an equation of state expressed in the general differential form

$$d \ln \rho = \alpha \times d \ln P - \delta \times d \ln T.$$

With increasing non-relativistic degeneracy of the electron gas, $\alpha \rightarrow 3/5$ and $\delta \rightarrow 0$, and for $\alpha \leqslant 3/4$ (δ still > 0) we have $d \ln T_c/d \ln \rho_c \leqslant 0$, i.e. T_c goes down with further contraction (compare Figure 2, where the line for $\alpha = 3/4$ is indicated).

The mass loss ceases when with negligible residual mass in the envelope, the 'mirror effect' no longer holds, i.e. when all parts of the star contract simultaneously.

To predict when this will happen in a particular star is difficult without specific numerical calculations. Therefore, one cannot say whether or not all possible initial systems of such small mass will evolve in the same manner. A rough estimate shows that all primaries with initially somewhat less than 3 $M_\odot$ *can* become white dwarfs in this way; their helium cores as left after the central hydrogen burning phase are too small to ever achieve the helium-burning temperature.

Numerical calculations for this type of evolution were comparatively easy; the physical processes involved here (mass loss, nuclear burning) are very simple as compared to those which must occur in a highly evolved single star. This simplicity is the only reason why the entire evolutionary sequence of a binary component could be calculated, whereas no such calculations are available for single stars.

If stars behave similarly under a mass loss starting in later phases of evolution, then it is likely that more massive white dwarfs can also be obtained in a very similar evolution. If we assume that the mass loss of the original primary starts *after* central helium burning and again the whole envelope is stripped off, then for stars starting with up to about 5 $M_\odot$, the remaining carbon-oxygen core is not massive enough to achieve carbon burning. D. Lauterborn has started calculations for such a case. He begins with an initial system of 5 $M_\odot$ and 2 $M_\odot$, separated by 300 $R_\odot$. This separation is large enough for the primary to fill its critical volume only after exhaustion of helium in the central core. After the mass exchange, about 1 $M_\odot$ was left of the original primary, which means that again the whole hydrogen envelope was stripped off. His most recent calculations indicate that the remnant starts to move in the H–R diagram from the red giant region to the left, the separation of the components after mass exchange having increased to about 700 $R_\odot$ (Lauterborn, to be published).

2. On the Further Evolution of White Dwarfs

Although at the first glance one might suspect that the further evolution of a white dwarf would be rather boring, such a conclusion would not take into account the complications which can arise from two effects: the influence of the companion, and the fact that the white dwarf is not the idealized configuration (homogeneous, completely degenerate) which one likes to imagine. In fact, different and peculiar phenomena may occur in rather short time-scales. This has been shown in several recent calculations.

Numerical calculations were carried out (Kippenhahn, Thomas, Weigert, to be published) on the further evolution of the white dwarf of 0·26 $M_\odot$ which originated from the 2 $M_\odot$ + 1 $M_\odot$ system described earlier. Near its surface, this white dwarf still possessed a hydrogen-burning shell source since a small hydrogen-rich envelope was left after the mass loss. This shell source, which had decreased very much in its energy production, becomes thermally unstable, the instability being caused by the fact that the shell source has sharply decreased its radial extension. Two thermal pulses resulted, the time interval between them being about 600 years. In each of the pulses, the luminosity and the radius of the star greatly increased so that the envelope again filled the critical volume. However, in both pulses the resulting mass loss was negligibly small. Only after these pulses were finished, did the star settle down again into the white-dwarf region.

Many such white-dwarf models were tested for pulsational instability by D. Lauterborn. The model's eigenperiod was around 1 min, and a very small excitation was found when the model evolved through the extrapolated Cepheid strip. This excitation is too small to expect an appreciable amplitude to be built up in a reasonable time interval. But, of course, it may be that greater excitation occurs when this model is in the midst of a thermal pulse. A pulsation excited under these circumstances may be visible for a long time.

All these phenomena may also be present in a single white dwarf as long as it has a small superficial hydrogen layer. For a white dwarf in a close binary system, additional disturbances must arise as the (now more massive) companion evolves and expands over its critical Roche lobe. Then hydrogen-rich matter is returned to the white dwarf. The resulting rate of mass increase can be rather high (at least much higher than for any reasonable accretion of interstellar matter on a single white dwarf). Numerical calculations were carried out by Giannone and Weigert (1967). A mass increase was assumed with a rate of $10^{-9} M_\odot$/year on a white dwarf of $0.5 M_\odot$, consisting originally of helium. The hydrogen heated up to ignition, and the new shell source turned out to be thermally unstable in all cases which were treated. Depending on the initial model, the instability was due either to degeneracy or to the very small radial extension of the shell source. Thermal runaways resulted increasing the temperature in the shell source to such an extent that even helium may be ignited in the succeeding phase (which is not yet calculated). Violent effects may be expected to occur, but the results are difficult to predict. Only a small fraction of the produced energy penetrated to the surface, most of it was absorbed by expansion of the outer layers. In the evolution calculated so far, the main observable effect at the surface should come from the large amount of kinetic energy of the infalling gases. This has to be dissipated and finally radiated away, and very hot and peculiar photospheric and chromospheric layers may result. An appreciable amount of X-ray radiation, e.g., can be expected to originate from such a star.

Many varieties of rapid changes, caused by quite different effects, are thus theoretically possible for close binaries containing a white dwarf, as shown even by these few calculations. One might feel tempted to relate one or the other of them to some observed activity in such systems (novae, pulsation), but it is as yet much too early for such speculations.

References

Giannone, P., Weigert, A. (1967)   *Z. Astrophys.*, **67**, 41.
Kippenhahn, R., Kohl, K., Weigert, A. (1967)   *Z. Astrophys.*, **66**, 58.

DISCUSSION

Conti: Do you anticipate any difference in evolution for a *smaller* primary mass in your case C calculation?

Weigert: We estimated very roughly a limiting initial mass of 2·8 $M_\odot$ for a star to follow the evolution to white-dwarf stage. All stars of smaller mass have too small helium cores to ignite helium. For case C the corresponding limiting initial mass would be about 5 $M_\odot$. In stars up to that mass the C–O core is too small to achieve $C + C$ burning.

Conti: I understand, then, the 5 $M_\odot$ upper limit for case C is analogous to the 3 $M_\odot$ upper limit for case B. Is the evolution of the primary directly to the white dwarf the same in both cases?

Weigert: It seems to be very similar, as far as one can tell up to now from the calculations by D. Lauterborn.

Mestel: Can you apply your calculation to the Sirius A and Sirius B system?

Weigert: The Sirius system cannot be explained directly by our calculations, since the separations of the calculated systems are too small. But also in the Sirius system some mass exchange must have occurred, maybe, together with mass loss of the whole system.

Tayler: In the example with initial masses 2 $M_\odot$ and 1 $M_\odot$ the masses after the first mass exchange are 0·26 $M_\odot$ and 2·74 $M_\odot$ with a separation of 15 $R_\odot$. This separation is not large, so would you expect a considerable return of mass to the white dwarf when the 2·74 $M_\odot$ becomes a red giant? In general, for favourable masses and separations might there be several exchanges of mass back and forth?

Kippenhahn: The reason why we have not tried to compute several mass losses, one following the other, was that in most cases you encounter contact systems. We do not know well enough up to now how to deal with such systems.

Leung: I would like to draw attention to one of the important observational approaches to the study of evolution of binaries. From eclipsing binary in clusters we have an estimation of the age and the location in the H–R diagram with respect to main sequence. There are many of these systems. We may be able to follow the evolution of binaries through this approach. It is my wish that observers would observe these systems.

Plavec: It would be very interesting to catch a system at the phase of rapid mass loss. We believe that two binaries of this kind are known, namely β Lyrae and V 367 Cygni.

Roxburgh: I would like to emphasise the desirability of investigating evolution of very close systems into contact systems. Have any of the groups investigated this problem?

Weigert: We always avoided to calculate such systems which have to come to a contact. The reason is that we do not know up to now how to treat the contact case. If a mass loss of the whole systems occurs, we would have to introduce completely unknown parameters (specific orbital angular momentum carried away).

ARGUMENTS INDICATING THAT Ap AND Am STARS ARE EVOLVED SPECTROSCOPIC BINARIES

E. P. J. van den Heuvel

(Sterrenwacht 'Sonnenborgh', Utrecht, The Netherlands, and Sterrenkundig Instituut der Vrije Universiteit, Brussels, Belgium)

1. Introduction

Roughly three types of evolution of spectroscopic binaries may occur (**1, 2, 3, 4, 5, 6**):

Case (a): the primary fills its Roche limit already before the end of core hydrogen burning;

Case (b): the primary fills its Roche limit after core hydrogen exhaustion but before the onset of helium burning, and

Case (c): the primary does not fill its Roche limit before the end of core helium burning.

Since long-period variables of Population I as well as Population II may reach radii of 10 AU (**7**), case (c) will occur in systems with periods up to 100 years. Since about 50% of the A and B stars are spectroscopic binaries (**8**), many evolved spectroscopic binaries are expected to occur in the galactic system. Computations (**9**) show that, using the mass-ratio distribution of unevolved systems derived by Kuiper (**10**), the number of systems with one evolved component is expected to be at least as large as the number of unevolved systems, in the spectral regions A and B.

2. Expected Characteristics of Evolved Spectroscopic Binaries

When the primary component has passed through all nuclear burning stages and has become a white dwarf, the remaining secondary will be expected to show the characteristics summarised in Table 1:

(1) Due to its increase in mass (up to 87% of the mass of the primary may be transferred towards the secondary (**2**)), the star will, in the Hertzsprung-Russell diagram, have moved upwards along the main sequence (Figure 1). In clusters it may have become (one of) the bluest and/or brightest main-sequence star(s), beyond the turn-off point (**11**). Since only little of its total amount of hydrogen has been burned at that moment, the new primary will be close to the Zero Age Main Sequence (ZAMS).

(2) Table 2 (after Hack (**12**)) lists some of the many eclipsing binaries in which at

Perek (ed.), Highlights of Astronomy, 420–431. © *I.A.U.*

Table 1

Expected characteristics of evolved spectroscopic binaries compared with observed characteristics of Ap and Am stars

Evolved spectroscopic binaries, expected

Ap, Am stars, observed

(1) In clusters: close to the ZAMS and near or beyond the turn-off point

(1) In clusters: close to the ZAMS and near or beyond the turn-off point

(2) Slow rotation

(2) Slow rotation

(3) Abundance anomalies of heavy elements and iron peak, except in case (b) if M_{prim} smaller than 2·8 $M_{\odot}$

(3) Abundance anomalies of heavy elements and iron peak

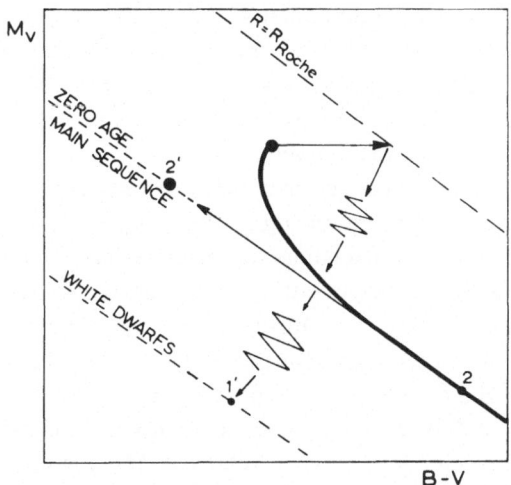

FIG. 1. *The expected positions of the components of an evolved spectroscopic binary in the colour-magnitude diagram of a cluster. The heavily drawn curve indicates the cluster main sequence.* R_{Roche} *is the Roche limit of the primary before the mass exchange.*

Table 2

Some of the many eclipsing systems in which the more massive star is an unevolved A or B star and the less massive star is an evolved subgiant (12). Initial periods are given for three initial mass ratios

Name	Sp_1	Sp_2	P (days)	$\dfrac{m_2}{m_1}$	$P_0(\alpha_0 = 0\cdot5)$	$P_0(\alpha_0 = 0\cdot2)$	$P_0(\alpha_0 = 0\cdot1)$
U Sge	B9	gG6	3·38	0·30	1·72	7·05	33·6
TX UMa	B8	gG4	3·06	0·30	1·56	6·40	30·4
SX Hya	A3	gK6	2·90	0·35	1·87	7·67	36·5
ST Per	A3	gG5	2·65	0·18	2·72	11·30	53·0
S Cnc	A0	gG5	9·49	0·36	6·40	26·2	125·0
RY Per	B4	F2	6·86	0·20	1·69	6·94	33·0
RG Cep	A5	gK3	12·42	0·38	9·05	37·1	176·0
Lib	A0	gG2	2·33	0·44	2·07	8·50	40·4

present the *more massive* primary is an *unevolved* A or B star, while the *less massive* secondary is a probably *evolved* subgiant, filling its Roche limit. In these systems the mass exchange has probably already taken place. Often the unevolved A or B stars in close systems are observed to *rotate slowly* (13); for $P < 10^d$ the rotation is often (and for $P < 4^d$ in practically all cases) synchronised with the orbital motion (14). Computations (15) show that in systems with $P > 10^d$, large mass exchange may also slow down the rotation of the unevolved star (exchange of small amounts of mass may accelerate the rotation of the surface layers, as is probably observed in systems like U Cep (16)). In the computations it is assumed that, at the moment of accretion on the A or B star, the exchanged matter moves in the system with velocities synchronised with the orbital motion of the components. It is found that for rigid-body rotation as well as for conservation of angular momentum in each shell in the stellar interior, the final angular velocity of rotation will not exceed $6 \cdot 8 \times 10^{-5}$ sec^{-1}, the average being less than $4 \cdot 4 \times 10^{-5}$ sec^{-1}. For A0 stars these angular velocities correspond to $v_e \approx 85$ km/sec and 57 km/sec, respectively.

(3) Except in case (b), with primary mass below $2 \cdot 8$ $M_\odot$ (2), heavy-element synthesis and mass ejection may be expected during the last stages of evolution of the primary. Especially, since after the mass exchange the primary has become an evolved star of *low mass*, one expects that a degenerate core will develop, and flash stages like the helium and carbon flash will occur. During flashes many neutrons will be produced and the conditions will be favourable for heavy-element synthesis in the way outlined by Fowler *et al.* (18). Hence, when the primary has become a white dwarf, the surface layers of the remaining A or B star may be covered with the ejected heavy elements. Since in case (b) a primary less massive than $2 \cdot 8$ $M_\odot$ directly evolves into a pure helium white dwarf (2), also some slowly rotating blue stragglers *without* abundance anomalies may be expected in A clusters.

3. The Observed Characteristics of Ap and Am Stars

These are summarised in the right-hand part of Table 1:

(1) As Figure 2a–2g show, in five galactic clusters the bluest and/or brightest stars are Ap or Am, while in all galactic clusters Ap and Am stars scatter about the blue edge of the top of the cluster main sequence. Since Ap and Am stars have masses (19, 20, 21) and surface gravities (22, 23) similar to those of stars on the main sequence that have the same B–V, these bluest stars are probably also the *heaviest* cluster stars.

(2) The rotation of Ap and Am stars is intrinsically slow (24, 25, 26, 27, 28); $v_e \sin i \approx 40$ km/sec for Am stars and ≈ 50 km/sec for Ap stars.

(3) Ap and Am stars show probably real overabundances of heavy elements and the iron peak (29, 30) and underabundances of light elements like He, C and O.

Since in clusters like the Hyades and Coma at least 20 to 30% of the stars are spectroscopic binaries (31), one certainly expects to find a number of evolved spectro-

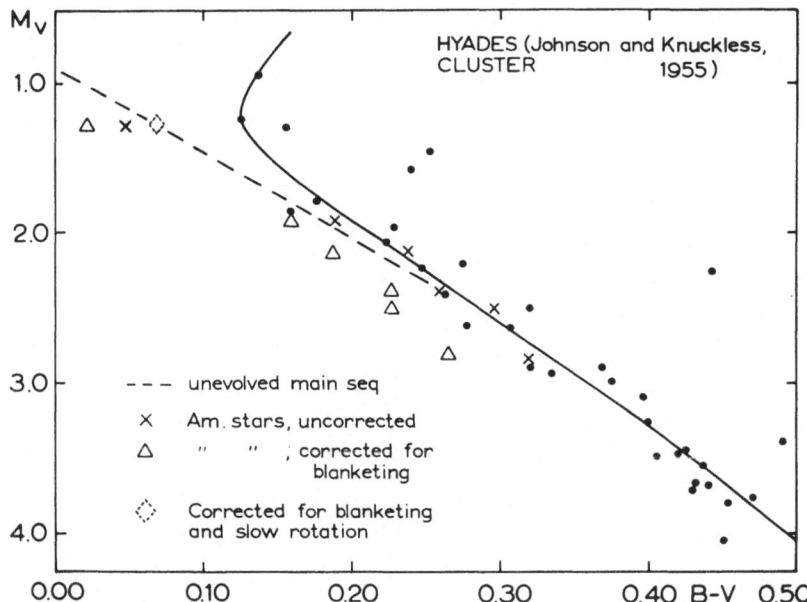

FIG. 2a. *The positions of Ap and/or Am stars in the colour magnitude diagrams of seven galactic clusters (borrowed from sources (**40**), (**41**), (**42**), (**43**), (**44**)).*

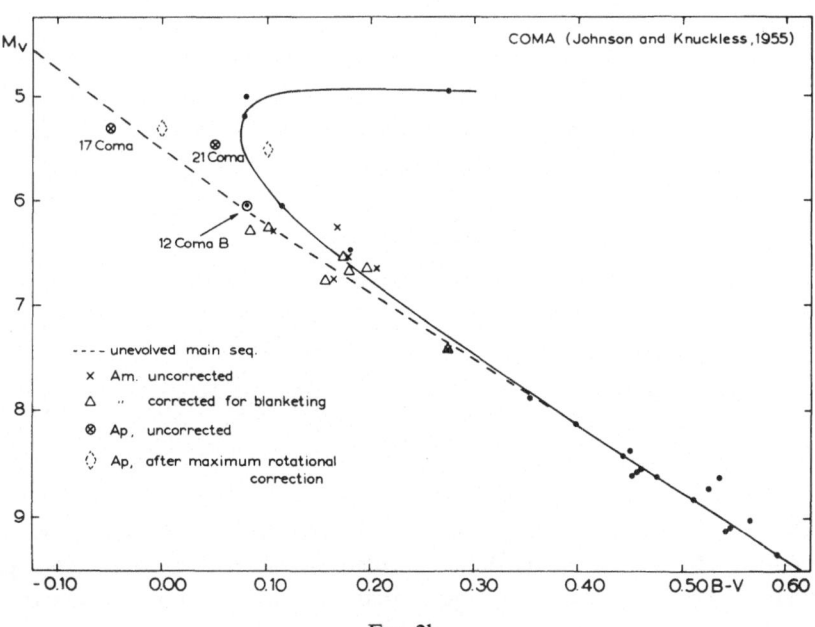

FIG. 2b.

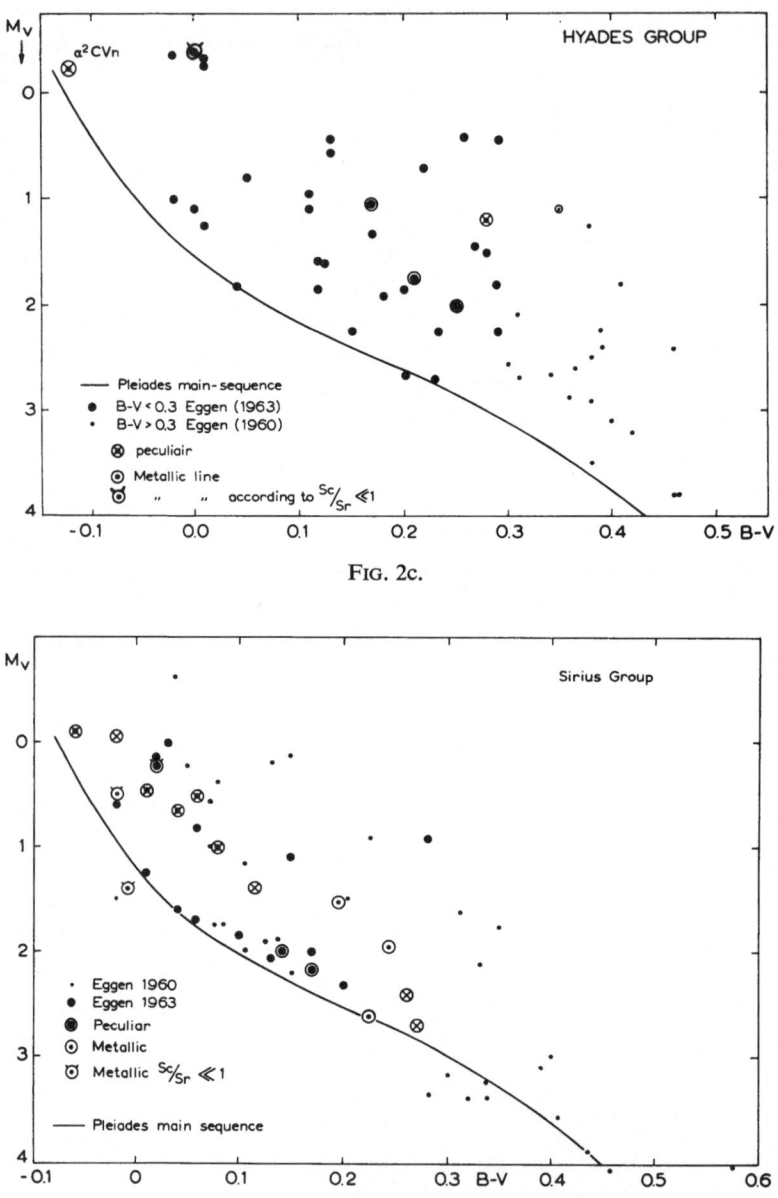

FIG. 2c.

FIG. 2d.

scopic binaries in such clusters. Ap and Am stars are the only stars in these clusters which show the three above-mentioned expected characteristics of evolved spectroscopic binaries. For this reason we suggest that Ap and Am stars *are* spectroscopic binaries with an evolved companion (as was first suggested by Renson (**32**)).

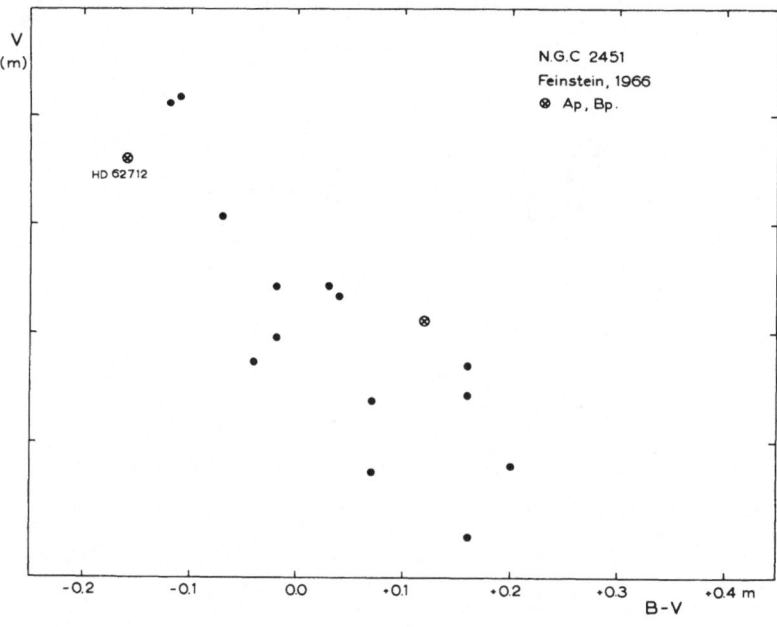

FIG. 2e.

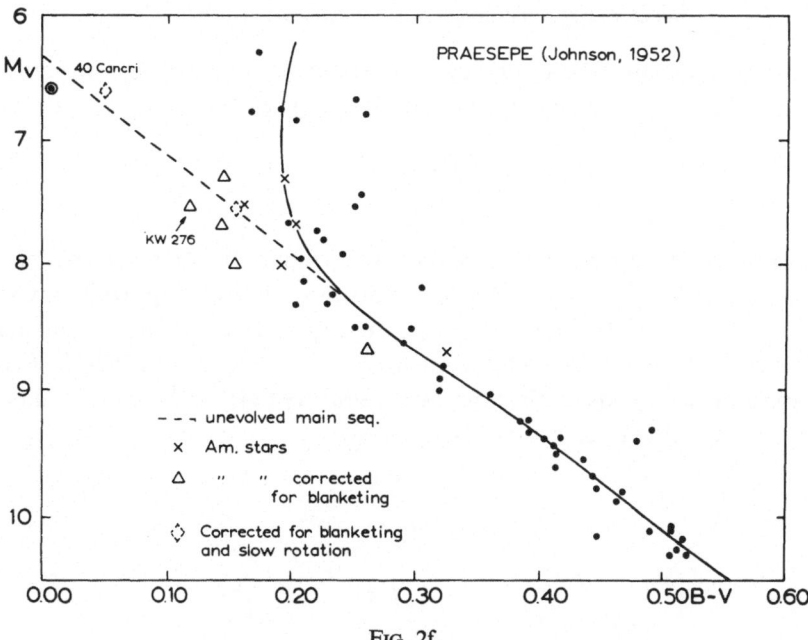

FIG. 2f.

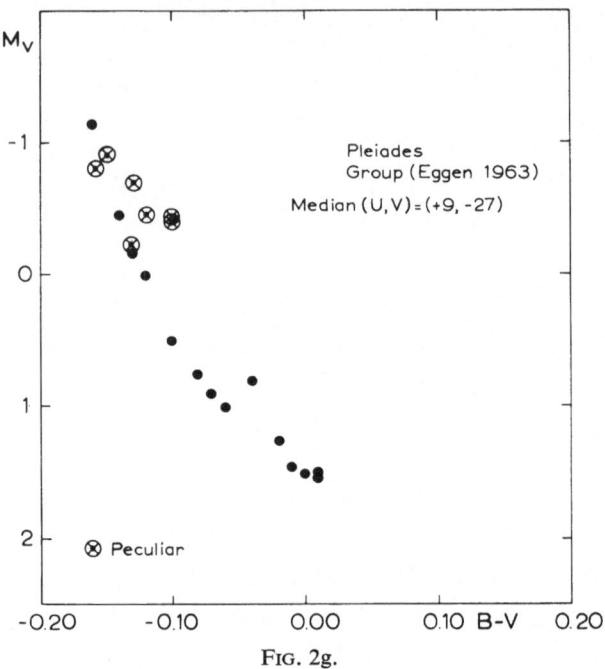

FIG. 2g.

4. Possible Reasons for the Observed Different Percentages of
Spectroscopic Binaries among Ap and Am Stars

From conservation of orbital angular momentum (33) the change in period as a consequence of mass exchange is for circular orbits given by

$$P'/P_0 = \left(\frac{\alpha_0}{(1 + \alpha_0)^2} \cdot \frac{(1 + \alpha')^2}{\alpha'} \right)^3, \tag{1}$$

where α_0 denotes the mass ratio of secondary and primary before the exchange and α' the mass ratio of new secondary and new primary after the exchange (both are smaller than 1). If $\alpha_0 > \alpha'$, the final period P' is larger than P_0. If $\alpha_0 < \alpha'$ then P' is smaller than P_0 (see Figure 3). Assuming the final white dwarf to have a mass of 0·7 $M_\odot$ (which is the average for white dwarfs on the high-mass sequence (34)) and the original primary mass to be M, the final mass ratio α' is equal to

$$\alpha' = \frac{0 \cdot 7}{(\alpha_0 + 1) M - 0 \cdot 7} \tag{2}$$

and

$$P'/P_0 = \left[\frac{\alpha_0 (M)^2}{((\alpha_0 + 1) M - 0 \cdot 7) \, 0 \cdot 7} \right]^3. \tag{3}$$

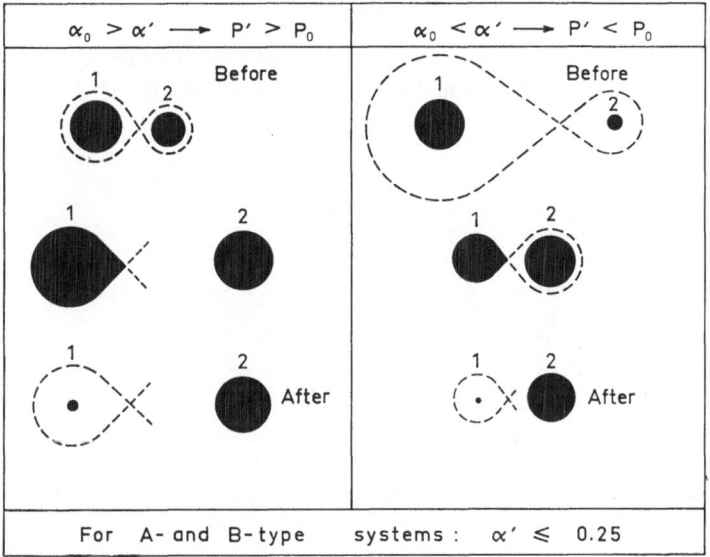

FIG. 3. *The change in separation as a consequence of mass exchange, as a function of the original and final mass ratios α_0 and α'.*

Consequently, if the period is increased by a factor a due to the exchange, α_0 must have been equal to

$$\alpha_0(a, M) = \frac{a^{1/3}(0\cdot7\,M - 0\cdot49)}{M^2 - 0\cdot7\,a^{1/3}\,M}. \tag{4}$$

Assuming the mass-ratio distribution derived by Kuiper (10) and by Van Albada (personal communication)

$$\varphi(\alpha_0) = 2/(1 + \alpha_0)^2, \tag{5}$$

the percentage of spectroscopic binaries of which the period increases by a factor a or more is equal to

$$Q_M(a) = \int_{\alpha_0(a, M)}^{1} \frac{2}{(1 + \alpha_0)^2}\,d\alpha_0. \tag{6}$$

Equation (6) yields $Q_{1\cdot5}\,(1) = 0\cdot36$; $Q_2\,(1) = 0\cdot50$ and $Q_4\,(1) = 0\cdot75$. Hence, for A5–F2 stars ($M = 1\cdot5$–$2\,M_\odot$), only in 36% to 50% of the cases the orbit will be increased, while for $M = 4\,M_\odot$ (B7–A0) in 75% of the cases the period will be increased after the exchange. Since furthermore a $0\cdot7\,M_\odot$ white-dwarf companion, in the same orbit, will induce two times larger variations in radial velocity in a $2\,M_\odot$ primary than in a $4\,M_\odot$ primary, the *observable* percentage of spectroscopic binaries among B8p stars is expected to be much lower than among Am stars (which in general have

E. P. J. VAN DEN HEUVEL

Table 3

The distribution of the periods of B-type spectroscopic binaries in Moore and Neubauer's (35) catalogue

P (d)	B0-B5	B6-B9·5
< 1$^{\rm d}$	1·2%	2·9%
1 – 1·5	8·1	2·9
1·5 – 2	7·0	8·6
2 – 3	20·9	11·4
3 – 9	32·5	40·0
9 – 27	15·1	17·1
27 – 81	7·0	11·4
81 – 243	7·0	5·7
> 243	1·2	

spectral type later than A3). With the distribution of periods of normal B-type spectroscopic binaries (Table 3, after Moore and Neubauer (35)) one computes that after the mass exchange only 35% of the stars with a mass of 4 $M_\odot$ will show variations in radial velocity larger than 20 km/sec, while 65% of the stars of 2 $M_\odot$ mass will show such variations in radial velocity. Since only 6% of the spectroscopic binaries in Moore and Neubauer's catalogue show variations in radial velocity below 20 km/sec, a velocity variation of 20 km/sec is roughly at the limit of detectability for the resolutions used for radial velocity catalogues. In this way, or by the presence of a less massive white-dwarf companion for Ap stars than for Am stars, one may explain the observed apparently low percentage of spectroscopic binaries among Ap stars (36) and the high percentage among Am stars (24). In fact we expect all Ap stars to be binaries, which, however, will only be recognised as such if sufficiently high spectral resolutions are used.

5. Suggested Reason for the Difference between Ap and Am Stars

The solid curve in Figure 4 represents the lower limiting binary period for case (b) evolution computed from Iben's results for $\alpha_0 = 0·5$ (this means: systems below the solid line are in case (b)). For other α_0-values the limiting periods differ by less than 15% from these values. The periods of the bulk of the normal spectroscopic binaries are also indicated. Since most B-type systems have $3^{\rm d} \leqslant P \leqslant 9^{\rm d}$ (35) and practically all normal A5 to F2V spectroscopic binaries have $P > 100^{\rm d}$ (37), most B-type spectroscopic binaries are in case (b) while most later A-type systems are in case (c). Case (a) is relatively unimportant since it occurs only for B-type systems with $P < 2^{\rm d}$ and for A-type systems with $P < 1^{\rm d}$. Since also the transition from the dominance of Ap stars to the dominance of Am stars (among slow rotators with enhanced metal spectrum) occurs in the early A spectral region, we suggest that Ap stars are the products of case (b) evolution of systems with original primary mass above 2·8 $M_\odot$, while Am stars are

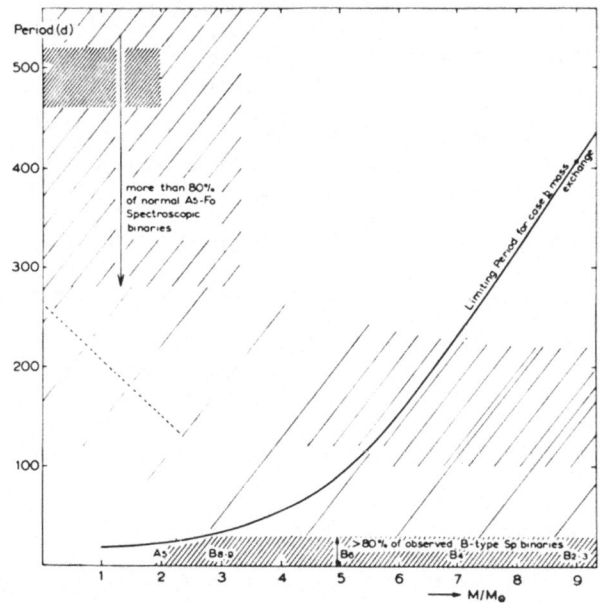

FIG. 4. *The limiting binary period for case (b) mass exchange as a function of the original primary mass, for an original mass ratio of 0·5. Systems below the heavy curve are in case (b). The heavily shaded areas indicate the periods of the bulk of the normal A- and B-type spectroscopic binaries. The dashed curve indicates the limiting period when the occurrence of the helium flash is taken into account.*

the products of case (c) evolution. Since primaries less massive than 2·8 $M_\odot$ will, by case (b) evolution, directly evolve into pure-helium white dwarfs (2), no Ap stars are expected in clusters in which unevolved stars near the turn-off point are less massive than 2·8 $M_\odot$. In these clusters only Am stars or slowly rotating blue stragglers without abundance anomalies – like 40 Cancri in Praesepe (38) – are expected to be found. On the other hand, Am stars are expected, in principle, in any cluster, but will be rare in B-clusters (since most B stars are in case (b)). Indeed, some low Sc/Sr Am stars occur already in the Pleiades (39)).

References

1. Kippenhahn, R., Weigert, A. (1967)   *Z. Astrophys.*, **65**, 251.
2. Kippenhahn, R., Kohl, K., Weigert, A. (1967)   *Z. Astrophys.*, **66**, 58.
3. Paczyński, B. (1966)   *Acta astr.*, **16**, 231.
4. Paczyński, B. (1967)   *Acta astr.*, **17**, 1.
5. Paczyński, B., Ziolkowski, J. (1967)   *Acta astr.*, **17**, 7.
6. Plavec, M. (1967)   *On the evolution of double stars*, ed. J. Dommanget (Obs. Royal, Uccles, Belgium), p. 83.
7. Smak, J. (1964)   *Astrophys. J.*, Suppl. **11**, 141.
8. Batten, A. (1967)   *A. Rev. Astr. Astrophys.*, **5**, 25.

430 E. P. J. VAN DEN HEUVEL

9. Van den Heuvel, E.P.J. (1968)   *Bull. astr. Inst. Netherl.*, **19**, 326.
10. Kuiper, G.P. (1935)   *Publ. astr. Soc. Pacific*, **71**, 451.
11. McCrea, W.H. (1964)   *Mon. Not. R. astr. Soc.*, **128**, 147.
12. Hack, M. (1963)   in *Star Evolution*, Ed. by P. Gratton, Academic Press, New York, p. 452.
13. Huang, Su-Shu (1966)   *A. Rev. Astr. Astrophys.*, **4**, 35.
14. Plaut, L. (1959)   *Publ. astr. Soc. Pacific*, **71**, 167.
15. Van den Heuvel, E.P.J. (1968)   *Bull. astr. Inst. Netherl.* (in press).
16. Struve, O. (1963)   *Publ. astr. Soc. Pacific*, **75**, 207.
17. Struve, O. (1946)   *Astrophys. J.*, **103**, 76.
18. Fowler, W.A., Burbidge, E.M., Burbidge, G.R., Hoyle, F. (1965)   *Astrophys. J.*, **142**, 423.
19. Eggen, O.J. (1965)   *A. Rev. astr. Astrophys.*, **3**, 235.
20. Abt, H.A., Conti, P.S., Deutsch, A.J., Wallerstein, G. (1966)   Paper presented at the 122nd Meeting of the AAS.
21. Hyland, A.R. (1967)   *Proc. astr. Soc. Australia*, **1**, 14.
22. Baschek, B., Oke, J.B. (1965)   *Astrophys. J.*, **141**, 1404.
23. Van 't Veer-Menneret, C. (1966)   in *IAU Symp.* 26, Ed. by H. Hubenet, Academic Press, New York, p. 261.
24. Abt, H.A. (1961)   *Astrophys. J.*, Suppl. **6**, 37.
25. Kraft, R.P. (1965)   *Astrophys. J.*, **142**, 703.
26. Sargent, W.L.W., Strittmatter, P.A. (1966)   *Astrophys. J.*, **145**, 938.
27. Van den Heuvel, E.P.J. (1967)   *Bull. astr. Inst. Netherl.*, **19**, 11.
28. Schöneich, W. (1967)   *Astr. Nachr.*, **61**, 290.
29. Sargent, W.L.W. (1966)   in *IAU Symp.* 26, Ed. by H. Hubenet, Academic Press, New York, p. 247.
30. Conti, P.S. (1965)   *Astrophys. J.*, **142**, 1594.
31. Kraft, R.P. (1965)   *Astrophys. J.*, **142**, 681.
32. Renson, P. (1965)   *Ann. Astrophys.*, **28**, 679.
33. Huang, Su-Shu (1963)   *Astrophys. J.*, **138**, 471.
34. Eggen, O.J., Greenstein, J.L. (1965)   *Astrophys. J.*, **141**, 83.
35. Moore, J.H., Neubauer, F.J. (1948)   *Lick Obs. Bull.*, **20**, 1.
36. Jaschek, C., Jaschek, M. (1958)   *Z. Astrophys.*, **45**, 35.
37. Abt, H.A. (1965)   *Astrophys. J.*, Suppl. **11**, 429.
38. Conti, P.S., Wallerstein, G., Wing, F. (1965)   *Astrophys. J.*, **142**, 423.
39. Conti, P.S., Strom, S.E. (1967)   *Astrophys. J.* (in press).
40. Johnson, H.L., Knuckless, C.F. (1955)   *Astrophys. J.*, **122**, 209.
41. Treanor, P.J. (1960)   *Mon. Not. R. astr. Soc.*, **121**, 503.
42. Eggen, O.J. (1963)   *Astr. J.*, **68**, 483.
43. Eggen, O.J. (1963)   *Astrophys. J.*, Suppl. **76**.
44. Feinstein, A. (1966)   *Publ. astr. Soc. Pacific*, **78**, 301.

DISCUSSION

Roxburgh: How did you determine the rotation velocities of 30 km/sec?

Van den Heuvel: Since A or B stars in close binaries with period of less than 10 days are observed to be slow rotators, I expect that these stars will continue to rotate slowly. This was my argument.

Mestel: I heard last year from Dr. H. Abt that he had begun a study of the proportion of Ap stars in *close* binary systems, and that he was finding a surprisingly high number. Is there any observational astronomer present who can confirm or deny this?

Batten: I also understood from Dr. Abt a few months ago that he was finding a high proportion of spectroscopic binaries amongst Ap stars.

Van den Heuvel: I think all Ap stars are spectroscopic binaries. Only, as shown by the formulas for the change in period, the period is probably increased while the companion has a low mass. Hence, I expect that spectroscopic detection of the presence of the companion will not be easy.

Underhill: Could one paraphrase your remarks to read that membership in a close binary leads to slow rotation, and that slowly rotating stars, however they achieve this state, very frequently have a spectrum which is classified as being peculiar? I refrain at this time from interpreting the peculiarities in term of particular physical conditions and abundances.

Van den Heuvel: I do not think that all slow rotators should have abnormal spectra. There are slow rotators which have normal spectra, like the blue straggler 40 Cnc in the Praesepe cluster. It has a normal spectrum as Dr. Conti has shown. Abnormal spectra occur in my opinion only in evolved spectroscopic binaries in which nucleosynthesis has taken place.

INVESTIGATION OF POSSIBLE ABUNDANCE ANOMALIES IN CLOSE BINARIES OF SPECTRAL TYPES A0-A2 & F5-F6

YOJI KONDO*

(Goddard Space Flight Center, NASA, Greenbelt, Md., and Kitt Peak National Observatory**, U.S.A.)

1. Observations

This investigation was originally conceived as a follow-up on a work by Abt (1965), in which he finds the close binaries, with periods of about 1–100 days, and with spectral types of about A4–F2 (IV–V), to be all Am stars. The stars to be observed were chosen from the 'Fifth Catalogue of the Orbital Elements of Spectroscopic Binary Stars' (Moore and Neubauer, 1948). The criteria in making the selection were that the spectral types are A0–A2 and F5–F6 (IV–V) and that the periods are between about 1–100 days.

Spectra of 11 binary systems were obtained at the Coude focus of the 84-inch reflector of the Kitt Peak National Observatory during three nights in May 1967. In addition to the binary stars, 5 standard spectrum stars, i.e. α Lyr (A0V), θ Leo (A2 V), σ Boo (F2 V), ι Peg (F5 V) and 110 Her (F6 V), were observed to provide comparisons. The dispersion used was 13·5 Å/mm. Typically 2 or 3 spectra were obtained for each object. A spot densitometer was employed to furnish calibrations for the plates.

2. Discussion of the Observations

The spectra obtained were first examined on a spectrocomparator. Tracings of the spectra were obtained using a microdensitometer with a slit width corresponding to about 0·07 Å.

Conti's criterion was adopted in investigating metallicity for the binary stars in the A0–A2 region, although other spectral features were also examined. Conti (1965) has reported that metallicity can be found for early A-type stars using the relative weakness of the Sc II line at 4246 Å and the relative strength of the Sr II line at 4215 Å. This ratio, which is called Sc/Sr, is close to unity for a normal star in this spectral region.

* Visiting Astronomer, 1967, Kitt Peak National Observatory.

** Operated by the Association of Universities for Research in Astronomy, Inc., under contract with National Science Foundation.

In case of the stars in the F5–F6 region, strengths of Balmer lines and metallic lines were compared with those of the spectra of the standard spectrum stars.

The results are tabulated in Table 1.

In the spectrum of δ Lib, both Sc II and Sr II are too weak to be compared meaningfully. In case of ω UMa, the blending of lines are such that it is not possible to evaluate Sc/Sr reliably. The spectra of TX Leo are rather unusual. The ratio Sc/Sr is slightly greater than unity, while, in the spectra of θ Leo, Sc/Sr is never greater than unity.

50 Dra is a double-line system. In two out of the three spectra obtained for this binary system, the spectral lines of the two components are well separated, making it possible to examine the lines of each component separately. Corresponding lines of both components are of similar intensity. The Sc/Sr for both components are distinctly smaller than unity indicating that both primary and secondary components are metallic. In another double-line system, η Vir, the separations of the spectral lines of the two components were very small at the time the two spectra were obtained, and it is not possible to evaluate Sc/Sr for the components separately.

HR 5317, which is classified as F6 IV in the *Catalogue of Bright Stars* (Hoffleit, 1964) has Balmer lines stronger than those of 110 Her (F6 V), while its metallic lines indicate a later spectral type than 110 Her. This is a characteristic of a metallic-line star. The difference in spectral types, as may be determined independently from the Balmer lines and from the metallic lines, appears to be about 3/10 of a spectral type. It was also found that there exists an apparent overabundance of calcium. An apparent overabundance of calcium was reported several years ago by Preston (1961) for an early to middle F type star HD 174 704, which he classifies as a metallic-line star. HR 4536, classified as F5 in the B.S. Catalogue, shows metallic lines which are slightly stronger than those of ι Peg (F5 V). The Balmer lines of HR 4536 are only very slightly stronger than those of ι Peg, and the difference in spectral types, as determined from the Balmer lines and from the metallic lines, appears to be about 1/10 of a spectral type.

In this survey, when metallicity is found in a binary system, it appears to be stronger for a binary with a shorter period, although it is not possible to make any generalization with the limited number of binary systems observed in this study.

It is to be noted that 9 out of 11 close binary systems, if we include ω UMa classified as a metallic-line star by Conti (1965), are found to have metallic characteristics.

3. Concluding Remarks

According to Abt (1961) all metallic-line stars are members of spectroscopic binary systems. Also, according to more recent work by Abt (1965), which was quoted earlier, all short-period binaries in the spectral range about A4 to F2 (IV–V) are found to be Am stars.

The possibility may be considered that metallicity is a norm rather than a peculiarity for close binaries, at least within a certain spectral range.

Table 1

Preliminary results on metallicity in close binaries

Star	BS (= HR) No.	V	B-V	BS Sp	Lick Binary Catalog No.	P (days)	Sc/Sr	Metallicity	Comments
δ Lib	5586	4·92	−0·01	A0	271	2·33	?	?	Eclips. Var.
–	6641	6·28	–	A0	328	2·82	≪ 1	yes	
ω UMa	4248	4·75	–	A1 V	208	15·84	? c*	?	Conti finds Sc/Sr < 1. Double-line system.
50 Dra	7124	5·36	–	A1	359	4·12	≪ 1	yes	Both components show metallicity.
φ Aql	7610	5·23	−0·02	A1 V	384	3·32	≪ 1 c	yes	Eclips. var.
TX Leo	4148	5·63	–	A2	201	2·45	≳ 1	no	
η Vir	4689	3·88	+0·02	A2 V	231	71·9	< 1 c	yes	Double-line system. Var.?
	6917	5·83	+0·06	A2 V	346	9·61	≪ 1	yes	
ι Del	7883	4·42	+0·04	A2 V	409	11·04	≪ 1 c	yes	
–	4536	5·72	–	F5	222	32·86	–	(yes)	Balmer and metallic lines slightly stronger than ι Peg (F5 V).
–	5317	5·93	+0·47	F6 IV	262	2·70	–	yes	Balmer and metallic lines stronger than 110 Her (F6 IV).

*c Indicates stars observed also by Conti (1965).

Further investigations of close binaries in the middle to late F spectral region are desirable. This is especially so in view of the fact that the absence of deep convective zones for stars earlier than the early F type has been suggested by some to have relevance on metallicity.

Some years ago, F.B. Wood suggested, in a somewhat different context, a possibility that secondary components of close binary systems are metal-rich. A systematic spectroscopic investigation, at high dispersion, of double-line systems, particularly in the early A region, may yield useful information on the metallicity in secondary components. For F-type stars, an investigation of metallicity in double-line systems may be a rather difficult undertaking due to the absence of a convenient criterion.

Acknowledgements

The author is indebted to Dr. H.A. Abt for helpful advice and discussions in conducting this investigation. Interesting conversations with Professor F.B. Wood and Professor K. Wurm are appreciated. It is a pleasure to thank Dr. G.E. McCluskey for his assistance in reduction of the data and for useful discussions. Special thanks are due Mr. M. Snowden for his kind assistance in obtaining the observations. Also, various assistance and cooperations received from the staff of the Kitt Peak National Observatory are gratefully acknowledged.

This work has been supported by NASA National Academy of Sciences – National Research Council Postdoctoral Research Associateship.

References

Abt, H.A. (1961) *Astrophys. J.*, Suppl. **6**, 37.
Abt, H.A. (1965) *Astrophys. J.*, Suppl. **11**, 429.
Conti, P.S. (1965) *Astrophys. J.*, **142**, 1594.
Hoffleit, D. (1964) *Catalogue of Bright Stars*, 3rd Rev. Ed., Yale Univ. Observ.
Moore, J.H., Neubauer, F.J. (1948) *Lick Observ. Bull.* **521**.
Preston, G.W. (1961) *Astrophys. J.*, **134**, 797.
Wood, F.B. (1962) in *Vistas in Astronomy*, **5**, Ed. by A. Beer, Pergamon Press, New York, p. 119.

DISCUSSION

Batten: I think this approach is important. There are over 70 known spectroscopic binaries with spectral types A4-F2 that have not yet been classified as Am. Until these are checked, it seems premature to generalize about the percentages of binaries among Am stars and normal A stars.

Koch: The hot component of δ Librae is earlier than A0 (on the evidence of the UBV photometry) so that the conclusions from the systems tested are more unanimous than might appear from the slide shown.

Kondo: The hot component of δ Librae is possibly a star earlier than A0. Perhaps B9·5. I agree with Dr. Koch on this point, and it may more correctly be said that 9 out of 10 systems observed show some measure of metallicity, since δ Librae should perhaps be excluded from the list.

Mestel: I believe, Dr. Abt has stated that A stars in long-period binaries ($P > 100$ days) were invariably rapid rotators and *normal* in their spectra. Does anyone know whether this is universally agreed, and whether Dr. Abt still holds to this?

Kondo: Dr. Abt has suggested a possibility in his paper (1965) that slower rotation in close binary systems have relevance on metallicity. I am not aware if he has revised his views.

Batten: I understand Abt's view, with which I do not agree, is that Am stars may be in either short- or long-period systems but that normal A stars are only in long-period systems.

Fliegel: To my knowledge, no definitive study has appeared concerning meridional circulation in binaries. However, it appears possible that the metallicity in Dr. Kondo's stars may be produced by the enhancement in binaries of such circulation which brings up material from the depths of the stars.

R.C. Smith: I have recently studied meridian circulation in surface layers of early-type stars. In addition to the circulation reversal noted by Öpik (1951)* and Mestel (1966)* I find a second reversal very near the surface. There are therefore three distinct circulation zones, and it is very unlikely that meridional circulation could bring material up from the deep interior and produce abundance anomalies.

* Öpik, E.J. (1951)   *Mon. Not. R. astr. Soc.*, **111**, 278.
  Mestel, L. (1966)   *Z. Astrophys.*, **63**, 196.

THE EVOLUTION OF CLOSE BINARIES
AND THE Am STARS

PETER S. CONTI

(Lick Observatory, University of California, Santa Cruz, Calif., U.S.A.)

In this paper, I shall be concerned with three related questions: (a) what is the evolutionary fate of close binary systems, (b) is this concept consistent with the number of evolved close binary systems, and (c) are these evolved systems the Am stars?

For the first question, it should be kept in mind that as stars evolve to the giant region they greatly expand. The interesting point then arises, what is the value of the critical period, P_c, for a given mass (and mass ratio), such that the primary in a close binary system with any period, $P \leqslant P_c$, will reach its Roche limit during its post-main-sequence evolution? This can be answered numerically by using Kepler's law and writing it in the form:

$$R_0 = ar = 4 \cdot 2 \, r (m_1 + m_2)^{1/3} \, P^{2/3} . \qquad (1)$$

In Equation (1), the masses are in solar units, the period is in days, the mean separation, a, and Roche radius, R_0, are in solar radii, and r is the Roche function tabulated by, say, Plavec and Kratochvil (1964). The Roche function takes on the values $0 \cdot 37 \leqslant r \leqslant 0 \cdot 60$ for the primary star for mass ratios $1 \cdot 0 \leqslant m_2/m_1 \leqslant 0 \cdot 1$, respectively.

To simplify the discussion I adopt the value of $0 \cdot 5$ for r (corresponding to a mass, it tio of $0 \cdot 27$), which is sort of a mean value for this quantity. Then, for a given mass ra is simple to find the period for any radius R_0. Since radii or their corresponding periods are related to the temperature and luminosity, they can be placed on an HR diagram. For this I used the temperature, $B-V$ and bolometric corrections of Johnson (1966). The result is Figure 1. Also shown are the evolutionary tracks of Iben (1966, 1967) transformed in a like manner. It is apparent that binaries of quite appreciable periods will reach their Roche limits during their evolution. The horizontal part of each track corresponds to the time when hydrogen burning in the core has ceased and a very rapid evolution is proceeding. Kippenhahn and Weigert (1967) distinguished three possible cases for close binary evolution. For case *a*, the primary reaches its lobe before the end of core hydrogen burning; for case *b*, the primary reaches the lobe after core hydrogen burning but before core helium burning; and for case *c*, the primary reaches its lobe after core helium burning.

As can be seen from Figure 1, only binaries with periods less than about 1 day will

Perek (ed.), Highlights of Astronomy, 437–442. © I.A.U.

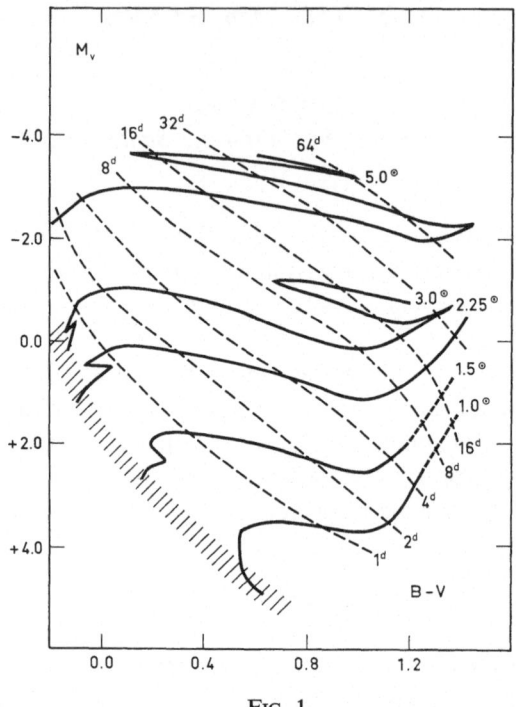

FIG. 1.

be in case *a*. Some binaries will reach case *c*, but a large number of known binaries with periods between 1 day and, say, 10–50 days will be in case *b*. Kippenhahn *et al.* (1967) have followed the detailed evolution of a system in case *b*. They find that if the primary is less massive than 2·8 ☉, it transfers material to the secondary and goes directly to the white-dwarf state. We see this represents masses for the primary corresponding to A and F stars and periods up to *at least* 10 days. We will consider these systems only in what follows.

It is necessary to choose a sample of stars that have been reasonably well studied for duplicity. Suppose one picks the stars in the *Bright Star Catalogue* (BSC). There are 97 known spectroscopic binaries (SB) in this catalogue with periods less than 10 days, spectral type A or F. Unfortunately, the BSC stars have not been uniformly well studied for duplicity. The only complete study of a sample of stars for duplicity has been that of Abt (1961, 1965). Among a sample of late A-early F stars, he finds 5·5% are short period binaries ($P < 10$ days). If one assumes that this percentage would also apply to the early A and late F stars, then the total number of A and F short-period SB in the BSC would be 195.

Eclipsing semi-detached (SD) systems have been supposed to be the evolved close binaries (Crawford, 1955). From Kippenhahn *et al.* (1967) it is possible to estimate the time-scale for the semi-detached mode for the A and F short-period SBs. For their A star with a mass of 2 ⊙, the time on the main sequence (MS) is 5×10^8 years. This is to be compared with the (total) time as a SD system of $8 \cdot 5 \times 10^7$ years. In other words, about 1/6 of the MS lifetime is spent in the SD mode. Then, the number of SD systems of types A and F in the BSC should be 1/6 the number of SBs. By actual count there are four identified SD systems, all eclipsing. This number may be under-estimated by a small factor since there are several more 'Algol' types in the BSC not completely studied, and a few other EBs may have been undetected as yet. Perhaps there are five eclipsing SD systems. But this number is an underestimate of the total since eclipsing systems are preferentially viewed edge on. A factor five (Conti, 1967) more semi-detached systems are viewed non edge on. There are thus about 30 semi-detached A and F systems in the BSC.

This expected total number of SD systems is within close to that predicted from the estimated number of short-period SB systems (195) and the time-scale arguments given by Kippenhahn *et al.* (1967) for case *b* evolution. It is therefore legitimate to assert that the SD systems are the descendants of the short-period SB stars (at least for A and F spectral types). This *must* be so since the short-period SB stars *will* evolve, *will* fill their Roche lobes, *will* transfer material to the secondary, and/or *will* lose

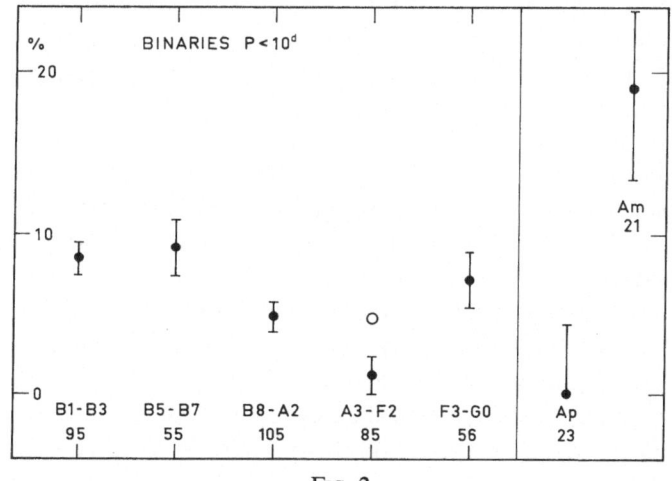

FIG. 2.

mass from the system. Then the observed small mass ratios of the SD systems, the overluminosity of the subgiant secondary, and the number of stars are all readily explained.

How do the Am stars fit in this evolutionary scheme? I will discuss only the short-period Am stars, which, according to Abt (1961, 1965), make up an appreciable fraction of these stars. If the Am stars were evolved systems, then we would expect them to be semi-detached, or at the next stage where they have white-dwarf companions (Conti, 1965; Van den Heuvel, 1967). A number of Am stars are double-lined. Out of the 27 identified Am short-period SB in the BSC, 7 have double lines. In none of these are the reported spectral types of the secondaries very different from the primaries. Therefore, at least 25% of the short-period Am stars are not semi-detached (nor do they have white-dwarf companions). 3 Am stars (RR Lyn, WW Aur, and δ Cap) are eclipsing binaries, but all of these are detached systems. Additionally, none of the eclipsing SD systems known (see, e.g., Hack, 1963) are identified as Am systems. So there is no evidence that the Am stars are presently SD systems.

Is it possible that all Am stars (aside from the 25% with double lines) have white-dwarf companions? Let us suppose they all did have such companions and then exa-mine the consequences. As such, there must be binaries on the main sequence *for the first time* which are their progenitors. Let us now look at the statistics of short-period binaries along the main sequence. It is most important to make a count of only those stars that have accurate and consistent spectral types. The only complete studies made, so far, have been those of Slettebak (1954, 1955) and Slettebak and Howard (1955), who gave MKK classifications and rotational velocities for nearly all stars brighter than $5^m.0$ and North of $-20°$. If we take these spectral types, and the fractional number of binaries known *at that time* (i.e. before Abt's work), we get Figure 2. I have used Slettebak's spectral type divisions (combining his last four groups into two) and listed the Am (and Ap) stars separately. The numbers below the spectral classes are the number of stars in that group; the vertical limits give the change in binary percentage for one more (or less) binary.

We see the result that from spectral classes B1 to G0 there is a sensibly constant percentage of binary stars, *except* in just one group, that of the A3–F2 stars. In fact, the *one* A3–F2 star in Figure 2, HD 118 216, is a non-eclipsing SD system (Conti, 1967). It would seem much more than a coincidence that it is just *this* group with practically *no* binaries that covers the spectral range of the Am stars. We see the well-known result that the percentage of Am stars is very high. It is only when the Am stars are added to the A3–F2 group that the fraction of short-period SB is the same in the A3–F2 group as in all the other spectral types (open circle in Figure 2). Unless one would care to assert that *all* short-period SB stars are evolved, or that there are no *unevolved* A3–F2 short-period SB stars, Figure 2 would force one to the conclusion that the short-period Am stars are not evolved. (Note this argument is not applicable for Ap stars.) If the short-period Am stars are not evolved, then the Am phenomena is not a product of *post-main-sequence* evolution.

One might ask, at this point, Isn't there an incompleteness in the numbers of known SB at all spectral types? This is true, but it should not completely change the argu-ments in the paragraph above since short-period binaries are those most easily detec-ted. In any event, among almost the same group of A3–F2 stars, Abt (1961, 1965) has made a systematic study of duplicity. The results are well known: he finds the short-period binaries are all Am stars and he finds no normal A stars with periods less than 100 days! If we are to assert that the Am stars are not evolved, then the SD systems are *their* descendants. Since the SD systems show no Am anomalies, we conclude that the Am phenomena is confined to the outer layers of the star. This is consistent with arguments about rotation, depth of convective zone, and line strengths in Am stars advanced by Stromgren (1963) and others.

In a spirit of speculation, I would like to suggest one possible reason for the Am phenomenon, assuming it to be largely characterized by abundance anomalies. Going back to Figure 1, we visualize that close binaries pass through regions of contact interaction on the *approach to* the MS when on Hyashi convective tracks. It is known that the approach to the MS is a time of great surface activity for stars (e.g. T Tauri stars). There may be sufficient energy available for nuclear interactions in the region between the contracting stars. Although no detailed mechanisms are offered here, the contraction phase during the evolution of a star is at least as active, on the surface, as at any subsequent time.

This would suggest that Am stars will be found in the youngest clusters, e.g. NGC 2264. Although none have been found as yet, it should be kept in mind that few young clusters have been fully studied this far down on the MS. Also, one would suppose that a binary system is also a necessary condition for an Am star and only those clusters with short-period binaries will contain them. This has been suggested for the Pleiades (Abt *et al.*, 1965), where there is only one marginal (classically defined) Am star, but where there are no short-period binaries among the late A stars, either.

References

Abt, H.A. (1961)   *Astrophys. J., Suppl.*, **6**, 37.
Abt, H.A. (1965)   *Astrophys. J., Suppl.*, **11**, 429.
Abt, H.A., Barnes, R.C., Biggs, E.S., Osmer, P.S. (1965)   *Astrophys. J.*, **142**, 1604.
Conti, P.S. (1965)   *Astrophys. J.*, **142**, 1594.
Conti, P.S. (1967)   *Astrophys. J.*, **149**, 629.
Crawford, J. (1955)   *Astrophys. J.*, **121**, 71.
Hack, M. (1963)   *Stellar Evolution*, Academic Press, New York, p. 452.
Iben, I. (1966*a*)   *Astrophys. J.*, **142**, 1447.
Iben, I. (1966*b*)   *Astrophys. J.*, **143**, 283.
Iben, I. (1967*a*)   *Astrophys. J.*, **147**, 624.
Iben, I. (1967*b*)   *Astrophys. J.*, **147**, 650.
Johnson, H.L. (1966)   *A. Rev. Astr. Astrophys.*, **4**, 193.
Kippenhahn, R., Weigert, A. (1967)   *Z. Astrophys.*, **65**, 251.
Kippenhahn, R., Kohl, K., Weigert, A. (1967)   *Z. Astrophys.*, **66**, 58.
Plavec, M., Kratochvil, P. (1964)   *Bull. Astr. Inst. Csl.*, **15**, 165.
Slettebak, A. (1954)   *Astrophys. J.*, **119**, 146.
Slettebak, A. (1955)   *Astrophys. J.*, **121**, 653.
Slettebak, A., Howard, R. (1955)   *Astrophys. J.*, **121**, 102.
Stromgren, B. (1963)   *Quart. J. R.A.S.*, **4**, 8.
Van den Heuvel, E.P.J. (1967)   *Bull. astr. Inst. Netherl.*, **19**, 11.

DISCUSSION

Van den Heuvel: You say that when the Am star becomes semi-detached, the Am character disappears. Do you agree that blue stragglers in galactic clusters descend from semi-detached systems? How can you then explain the fact that Pesch recently found that two blue stragglers in the old galactic cluster M 67 are Am stars? Then the Am character would only be absent during the s-d stage and reappear after the disappearance of the subgiant companion. Does not this seem strange?

Conti: I think the statement that there exist two Am stars in M 67 is not greeted by universal agreement among spectroscopists.

ORIGIN OF BINARIES FROM A CONSIDERATION OF ANGULAR MOMENTUM

SU-SHU HUANG

(Dept. of Astronomy, Northwestern University, Evanston, Ill., U.S.A.)

According to various studies, more than one half of stars exist as members of binary and multiple systems. Therefore any theory for star formation must account for this fact. In other words, the problem of the origin of binaries has a far broader implication than is implied as a part of binary-star study. For this reason one cannot over-exaggerate its importance.

1. Observed Results

Statistical studies of binary stars have been proven to be useful for understanding stellar evolution (Hynek, 1967). In order to discuss their origin we should similarly examine the statistical results. Two facts which will be briefly reviewed here appear to be crucial:

A. ORIENTATION OF ANGULAR MOMENTUM VECTORS

The early investigations on this problem were confined to visual binaries, because they are the only kind of binaries whose orientation in space may be ascertained by observation. Even so, the actual determination of the orientation of a visual binary is not an easy task, and orbital planes of only a limited number of visual binaries have been determined. Chang (1929), who collected 16 visual binaries with their inclinations definitely established, found a random orientation in space of orbital planes. Later Finsen (1933) undertook a similar study of 28 visual binaries and obtained the same conclusion. Few additional determinations of the orbital orientation of visual binaries have been performed in recent years. Perhaps their conclusion may still be regarded as representing the final result along this line of study.

While both Chang and Finsen used a small number of systems with their orientations definitely ascertained, later investigators were inclined to rely on a large number of binaries for which the sign of the inclination, i, is not determined. A review of their methods and results has been given by Huang (1968). Most investigators obtained the same conclusion of random orientations, although there were some dissensions.

While it is difficult to find the sign of i for visual orbits, it is not even possible to determine the magnitude of i for spectroscopic binaries. It may be argued that the velocity curve of a spectroscopic binary depends upon the orbital inclination. A factor of $\sin i$ enters into the amplitude of the curve. This fact may aid us to detect

Perek (ed.), Highlights of Astronomy, 443–450. © I.A.U.

any preferential orientation of these binaries from the statistics of the velocity-curve amplitudes of stars in different regions of the sky. Actually the problem is complicated by the systematic variation of stellar mass from the galactic equator to both poles, because the velocity-curve amplitude depends also upon the stellar mass (Shajn, 1926; Struve, 1950; Abt, 1966).

Although it is difficult to infer orbital orientations for spectroscopic binaries, a different situation obtains when they show the sign of eclipse. For then the inclination will be close to $\pi/2$. Because of our own special location in the Galaxy, the inclination of the orbital plane of an eclipsing binary at galactic latitude, b, with the galactic plane can assume any value from b to $\pi/2$. Therefore, orbital planes of eclipsing binaries found in high galactic latitude necessarily make large angles with the galactic plane, while those found in low galactic latitude may make either large or small angles.

Because of this fact one can examine whether orbital planes of eclipsing binaries show any preferential orientation parallel to the galactic plane. Based on the data collected by Koch et al. (1963) which cover known eclipsing binaries over the entire sky down to the 13th magnitude at minimum light, a statistical study (Huang and Wade, 1966) shows that there is no such preferential orientation.

B. DISTRIBUTION OF BINARY SEPARATIONS

The histogram in the figure obtained by Kuiper (1935) represents the frequency distribution of binaries in a volume of space around the Sun as against $\log a$, where a denotes the semi-major axis of the relative orbits expressed in AU. It combines the statistical data of eclipsing, spectroscopic and visual binaries as well as common-proper-motion pairs. However, it excludes binaries whose combined brightnesses are below absolute magnitude 6·45, and the differences in brightness of those two components are less than 4 mag. K and M giants are excluded from the statistics because they do not follow the usual mass-luminosity relation which has been used in order to derive a. How much Kuiper's histogram will be modified by Shapley's (1948) conclusion about the high abundance of W UMa systems is not clear at present.

Because of the difficulty involved in collecting statistical data, the histogram is not expected to represent an accurate picture of the distribution of binary separations. However it does give a general trend and serves as a test for any theory or theories that concern formation of binaries.

2. Theories

The random orientation of angular momenta of binaries gives us a clue as regards their origin. They must be the result of random processes. Consider an interstellar medium in which turbulence prevails. One can imagine a velocity field composed of random motions of eddies of different sizes. When a stellar condensation is formed in it, what is the net effect of all these eddy motions on the final stellar angular momentum? It is obvious that angular momenta due to motions of small eddies will cancel off. But the net angular momentum due to large eddies will in general be different from zero. Hence an amount of angular momentum is acquired by simply collecting matter in a medium in turbulent motion. Being a random process such a mode of acquisition inevitably leads to a Maxwellian distribution of stellar angular momenta (Kuiper, 1955; McCrea, 1959). If we write $\Omega = x\Omega_p$ where Ω_p denotes the most probable value of stellar angular momentum, the distribution of x may be written as

$$\phi(x)\,dx = \frac{4}{\sqrt{\pi}}\,x^2\,e^{-x^2}\,dx. \tag{1}$$

The orbital angular momentum of a binary is given by

$$\Omega = M_1 M_2 \left[\frac{Ga(1-e^2)}{M_1 + M_2}\right]^{1/2}, \tag{2}$$

where M_1 and M_2 stand respectively for the masses of two components, e the orbital eccentricity and G the gravitational constant. By neglecting the variations of M_1, M_2 and e, from one binary to another, Kuiper (1955) assumes $x = \alpha_1 a^{1/2}$, where α_1 is a constant. If we further neglect the spin-angular momentum, Ω as given by Equation (2) will be distributed according to Equation (1). From this condition we may write

$$F(\log a)\,d\log a = \phi(x)\,dx, \tag{3}$$

where $F(\log a)$ represents the distribution of $\log a$ as consistent with the histogram in the figure and is given by (Kuiper, 1955)

$$F(\log a) = \frac{x\phi(x)}{2\log e}. \tag{4}$$

The value of α_1 can be determined by fitting the calculated maximum of $F(\log a)$ with the observed maximum shown in the histogram. In this way $F(\log a)$ can be computed according to Equation (4) and is shown as the broken curve in Figure 1.

Kuiper's statistical calculation is incomplete because he has neglected the variation of M_1, M_2 and e. Also there is the question about the distribution function of Ω. We shall first discuss the latter point. Stars are believed to be formed in groups (Roberts, 1957). Each emerges from an interstellar medium which has its own characteristic turbulent-velocity field. Therefore the reasoning that the angular moments are distributed according to the Maxwellian law is correct only for binaries in a single group. For different groups, the values of Ω_p will be different because the turbulent velocity fields of the pre-stellar media are different.

Let us assume that Ω_p itself follows the Maxwellian distribution with the most probable value Ω_0. Perhaps such an assumption overstates the case. However, we will examine its effect as a preliminary study. If we write $\Omega = x\Omega_0$, it can be easily shown that the distribution function now becomes

$$\phi(x)\,dx = \frac{16}{\pi}x^2\,dx\int_0^\infty e^{-(x/y)^2 - y^2}\frac{dy}{y}. \tag{5}$$

When $\phi(x)$ as given by this equation is substituted into Equation (4), we obtain the broken curve interspersed with dots, as is shown in Figure 1. Thus the discrepancy between the observed and calculated distribution is much reduced by assuming star formation in groups.

The variation of a due to e may be regarded as small. This is especially true for

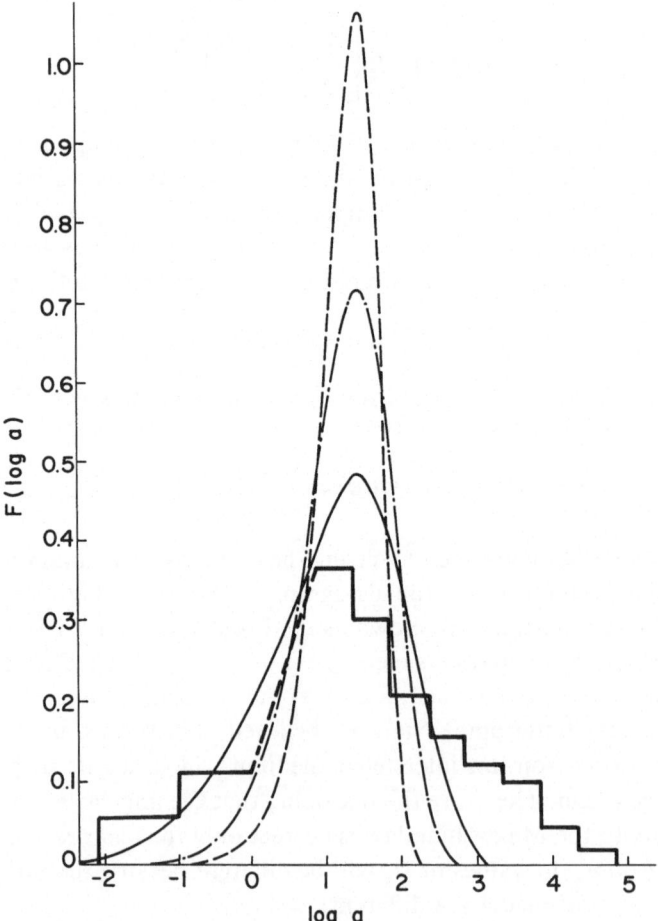

FIG. 1.   *Distribution of binary separations according to observational data (histogram) as well as statistical theories (curves) discussed in the paper.*

close binaries. Therefore we may neglect the effect of e by setting it equal to zero everywhere. At the same time, instead of treating M_1 and M_2 directly we may use y and z, defined by

$$y = \frac{M_2}{M_1 + M_2} \quad \text{and} \quad z = \frac{M_1 + M_2}{M_\odot} \tag{6}$$

as two independent variables. It may be argued that the variation of y in observable binaries is limited to perhaps between 0·2 and 0·5, while the range of variation of z is much wider, say from 0·4 to 6. Therefore the main contribution to the variation of t is due to that of x and z. We may propose to derive the distribution of $\log a$ from the distribution of x, given by Equation (5) and the distribution of z which is assumed to follow a power law, namely

$$\Psi(z)\,dz = \frac{B\,dz}{z^m}, \quad z_0 \leqslant z \leqslant z_1, \tag{7}$$

where B is a normalizing factor and can be easily obtained. As an approximation we may take $m = 2\cdot35$, which was obtained for single stars (Salpeter, 1955).

Instead of looking for the distribution of a, we may equivalently examine the distribution of t, defined by

$$t = \alpha_2 a \quad \text{with} \quad \alpha_2 = y^2 (1 - y)^2 / \lambda^2, \tag{8}$$

where λ represents a dimensionless parameter given by

$$\lambda = \frac{\Omega_0}{(Ga_1 M_\odot^3)^{1/2}}, \tag{9}$$

if a in Equation (8) is understood to be expressed in units of a, which may be taken as 1 AU. Hence t differs from a only by a constant factor since y is not assumed to vary. The distribution function $F(t)$ of t can now be expressed as a double integral (e.g., Chandrasekhar and Münch, 1950), namely

$$f(t)\,dt = \iint_\Delta \phi(x)\,dx\,\Psi(z)\,dz, \tag{10}$$

where the integration domain, Δ, is defined by

$$t \leqslant \frac{x^2}{z^3} < t + dt \tag{11}$$

in the x-z plane.

Substitution of Equations (5) and (7) into Equation (10) and simple reduction of the resulting equation yield the normalized distribution of t:

$$f(t)\,dt = \frac{8B}{\pi}\,t^{1/2}\,dt \int_{z_0}^{z_1} z^{(q - 2m)/2}\,dz \int_0^\infty e^{-z^3 t/\omega^2 - \omega^2}\,\frac{d\omega}{\omega}. \tag{12}$$

The distribution with respect to $\log a$, i.e. $F(\log a)$ is given by Equation (4) only with $x\phi(x)/2$ replaced by $tf(t)$. The double integral in Equation (12) can be roughly evaluated with the aid of the Gaussian quadrature formula (e.g. Chandrasekhar, 1950). The distribution of $F(\log a)$ which was derived by assuming $z_0 = 0\cdot 5$ and $z_1 = 5\cdot 9$ is plotted as a solid curve in Figure 1.

While the present probing study does show that star formation in groups and the mass difference of stars both broaden the distribution function of binaries with respect to $\log a$, the general shape of the calculated curve cannot be said to agree with the observed histogram. This is especially true at large values of a, where the calculated curves decline sharply while the observed one shows a broad shoulder. The cause of this discrepancy is unlikely due to the neglect of variations in mass ratio y and eccentricity e, although their inclusion in the study could somewhat modify the calculated distribution. A more important reason may well be stellar encounters which statistically widen the binary separation (Chandrasekhar, 1944; Takase, 1953). Also for very wide pairs their formation as a result of three-body encounters in interstellar space cannot be ruled out. On the other hand the slope at small values of a appears to show no great divergence from that of the observed histogram. That does not mean the initial distribution of separations has not been modified by some events, such as disintegration of multiple systems, that have occurred after the formation of binaries.

From the present study and the slope of the observed histogram in Figure 1 we may draw a tentative conclusion that the origin of binaries could be dichotomic. Each binary in one group may be found from a single initial condensation in the interstellar medium. Perhaps this group centers the distribution at about $a = 10$ AU. The binary in the other group may be formed from two initial condensations. This means that the binary was formed by capture likely in the very early stage when stars were just emerging from the medium. The separations in the second group may have a maximum at about $a = 1000$ AU.

Acknowledgements

It is my pleasure to thank Mr. Clarence Wade, Jr. for his evaluation of the integrals used in the present investigation on the IBM-360 digital computer at Goddard Space Flight Center. The present investigation has been supported by a research grant from the National Aeronautics and Space Administration.

References

Abt, H. A. (1966)   in *Vistas in Astronomy*, **8**, Ed. by A. Beer and K. Aa. Strand, Pergamon Press, London, p. 75.
Chandrasekhar, S. (1944)   *Astrophys. J.*, **99**, 54.
Chandrasekhar, S. (1950)   *Radiative Transfer*, Clarendon Press, Oxford.
Chandrasekhar, S., Münch, G. (1950)   *Astrophys. J.*, **111**, 142.
Chang, Y. C. (1929)   *Astr. J.*, **40**, 11.
Finsen, W. S. (1933)   *Circ. Union Obs.*, **90**, 397.

ORIGIN OF BINARIES 449

Huang, S.-S. (1968) in *Vistas in Astronomy* (in press).
Huang, S.-S., Wade, C., Jr. (1966) *Astrophys. J.*, **143**, 146.
Hynek, J.A. (1967) in O. Struve memorial Volume (in press).
Koch, R.H., Sobieski, S., and Wood, F.B. (1963) *Pub. Univ. Pennsylvania* (Astr. Series, 9).
Kuiper, G.P. (1935) *Publ. astr. Soc. Pacific*, **47**, 15, 121.
Kuiper, G.P. (1955) *Publ. astr. Soc. Pacific*, **67**, 387.
McCrea, W.H. (1959) Modèles d'Étoiles et Évolution Stellaire, Comm. Ninth Coll. Intern. d'Ap.,
  Liège, p. 332.
Roberts, M.S. (1957) *Publ. astr. Soc. Pacific*, **69**, 59.
Salpeter, E.E. (1955) *Astrophys. J.*, **121**, 161.
Shajn, G. (1926) *Mon. Not. R. astr. Soc.*, **86**, 543.
Shapley, H. (1948) *Centennial Symposia* (Harvard Obs. Monographs, 7), p. 249.
Struve, O. (1950) *Sky and Telescope*, **9**, 216.
Takase, B. (1953) *Ann. Tokyo Astr. Obs.* (2), **3**, 192.

DISCUSSION

J.A. Hynek: Do I understand then, that all common-proper-motion pairs are the result of later capture processes?

S.-S. Huang: According to the present study the majority of common-proper-motion pairs result from captures. However, the captures likely occurred at the time of their formation. If two stars happen to be formed near together, and if their dynamical energy turns out to be negative, the two will become a binary. Hence such a capture does not have its usual sense which requires a three-body collision. Of course for binaries with separations in the order of interstellar distances, they may be captured in the usual sense.

I.W. Roxburgh: I feel I must say a few words in defence of theoreticians. The arguments in which the angular momentum is assumed to come from galactic rotation are only done to show that there exists an angular-momentum problem. The fact that the angular-momentum vectors are distributed at random shows that the turbulence is the dominant effect and there is an even greater angular momentum problem than that obtained on using the galactic rotation.

S.-S. Huang: I agree with you that when you discuss the effect of angular momentum on stars, it does not matter where the angular momentum comes from. However, what I have emphasized in my discussion is the fact that a deep gulf of misunderstanding exists between theoretical and observational astronomers. Each group is preoccupied with its own problem without taking a look at what the other has found. Such a lack of communication perhaps is one of the serious problems in astronomy and the present one about the origin of stellar angular momentum is only one of many examples. For the sake of advancement of astronomy we should do our best to fill up such a communication gap.

L. Mestel: In discussing the formation of the initial protostars, we must certainly include not only the vorticity of the galactic rotation but that of the turbulence. In addition, the magnetic coupling between the star and a stellar wind emitted during the Hayashi phase will in general tend to rotate the angular-momentum vector.

S.-S. Huang: Magnetic coupling between the star and the stellar wind will have its effect on the stellar angular momentum just as the turbulent velocity field of the pre-stellar medium. However, one may take the coupling as another process of random nature that helps make the distribution of stellar-angular momenta Maxwellian-like.

M. Savedoff: Emphasize H field and tidal effect will reduce L parallel to Galaxy, especially as it was emphasized in sessions of Commission 34 that H field is parallel to plane. Has bias of statistics produced by a number of stars near Milky Way plane been reduced from statistics of orientation of orbital planes? This would tend to explain your figures.

S.-S. Huang: Statistical studies of orbital orientation are subject to some limitation mainly because of the difficulty of determining the sense of tilt of the orbital plane with respect to the celestial sphere. For those visual binaries whose orientation in space have been definitely determined by spectroscopic observations, their orientations appear to be at random. This is the result of Chang and of Finsen

I quoted. Subsequently, however, other statistical studies have been performed which use many binaries whether their orientations are determined or not. Some obtained random orientation but others derived different conclusions. Those studies performed prior to 1966 have been reviewed in an article which will appear soon in *Vistas in Astronomy* edited by Dr. A. Beer.

A. Batten: Dommanget presented results to Commission 30 last week that suggest poles of visual binary orbits avoid the poles of the Galaxy. This must pose even worse problems than the random distribution.

THE ORIGIN AND EARLY EVOLUTION
OF CLOSE BINARY STARS

I. W. ROXBURGH
(London, U.K.)

(Manuscript not received by November 6, 1967.)

THE STRUCTURE OF W URSAE MAJORIS SYSTEMS

LEON B. LUCY

(Dept. of Astronomy, Columbia University, New York, U.S.A.)

According to an argument of Kuiper's, contact binaries with unequal components cannot exist at age zero. This result comes from a comparison of the mass-radius relation for single, homogeneous stars with the condition to be fulfilled by the radii if the surface of the binary is to be an equipotential. This conclusion is contradicted, however, by observational evidence that W UMa stars do occur at age zero with unequal components.

The spectral types of W UMa stars are later than F0, so that their outer layers will be convective. The adiabatic constants of the envelopes of the components must then be equal because the two stars are in contact. Then, since the radius of a star with a convective envelope is sensitive to the adiabatic constant, it follows that the mean radii of the components differ, in this case, from the radii of the corresponding single stars. Thus Kuiper's argument does not apply when the common envelope is convective.

It can then be shown that zero-age contact configurations are possible when the common envelope is convective provided that the C–N cycle is dominant in one component and the proton-proton reaction is dominant in the companion. This leads to a model for a W UMa which accounts for some of the properties of these stars.

A paper describing this investigation in detail has been submitted to *Astrophys. J.*

DISCUSSION

Paczyński: I would like to make two remarks on the problem of the W UMa-type binaries. First, many such binaries show very rapid period variations with $P/(dP/dt)$ being of the order of Kelvin-Helmholtz time-scale. In some systems the orbital period increases, in others it decreases. But in any given binary the sign of the period changes seems to be constant during the last 60 years. I wonder whether these variations are secular or not. Second I have computed the time-scale for the orbital angular-momentum loss due to the emission of gravitational waves from many W UMa-type binaries (cf. Paczyński, *Acta Astr.*, **17**, 1967, 00). This time-scale turned out to be very close to the time-scale of nuclear evolution. Perhaps in future, when we shall have a good theory of evolution of the W UMa-type binaries we shall have a chance to check the physical reality of the gravitational radiation.

Lucy: It would be of interest if an observer would tell us what percentage of W UMa have secular changes in orbital period.

COMMENTS

L. MESTEL

(Dept. of Mathematics, University of Manchester, U.K.)

I would like to make a few comments, relevant to different points raised in the Discussion.

(i) When discussing stellar angular momentum, we should remember that substantial magnetic braking is likely during contraction towards the main sequence in both the Hayashi and post-Hayashi phases, for we have every reason to expect a strong stellar wind. Even if the star has earlier lost most of its primeval magnetic flux, it is plausible that there will be a dynamo-generated field, as in the solar atmosphere, able to control the gas flow out to radii well beyond the stellar surface, and leading to a considerable increase in the amount of angular momentum carried away for a given loss of mass. (In fact, numerical computations suggest the possibility that a star which *has* retained a strong primeval field does not necessarily lose more angular momentum than one with only a dynamo-driven field, but rather that the same amount of braking is achieved for a much smaller mass loss.) The point is that I do not think one can strictly describe any star in these phases as 'non-magnetic', following its rotational evolution without feeling any magnetic torque. In particular, it is not clear that stars contract in a state of close balance between centrifugal force and gravity at the photospheric equator, as assumed by Dr. Roxburgh: magnetic braking *may* always be efficient enough to keep the ratio well below unity, in spite of the contraction.

(ii) I had expected that Dr. Huang would discuss the model that he and I have independently put forward for the formation of close binary main-sequence systems, using magnetic braking. Basically, the idea is that if a pair of coupled proto-stars are each losing angular momentum of spin, but if there is strong spin-orbit coupling, tending to synchronize periods of rotation and revolution, then the loss of spin angular momentum will be made up at the expense of the orbital angular momentum, and the two stars will approach each other, remaining a close binary system in spite of their contraction by large factors to their main-sequence states.

Although I still like this picture, I want to point out a possible difficulty. We want to apply the braking theory both to individual stars – explaining at least semi-quantitatively the distribution of angular momentum along the main sequence – and to double stars. The danger is that if the process works at about the right efficiency for single stars, it will not be efficient enough for the double-star problem; whereas if the double-star process works adequately, then the same theory applied to single stars will yield main-sequence rotation periods that are far too long. The difficulty may be avoided, however, if the rate of magnetic braking depends in a *non-linear* way on the instantaneous angular velocity, so that the process works at maximum efficiency only when centrifugal force is close to gravity. This will be the case when the stellar wind is a 'centrifugal' rather than a thermal wind, with the mass and angular-momentum loss decreasing sharply as the centrifugal forces decrease.

(iii) The classical process of spin-orbit synchronization is 'tidal friction' – the continuous destruction of the energy of the tidal motions set up within the components of a non-synchronized binary system. But tidal friction decreases like the sixth power of the mutual separation of the two components, and so may not always be efficient enough, even when the friction is due to eddy viscosity in a largely turbulent star. However, a magnetic star tends to keep its corona co-rotating with the bulk of the star. In a non-synchronized binary system there will result violent shearing motions in the common corona; if these motions become turbulent, there may be enough dissipation to maintain nearly synchronous motion.

(iv) In connection with Dr. Van den Heuvel's contribution: accepting the strong correlation between the Am (and possibly also the Ap) characteristic and membership of close binary systems, perhaps we should look upon binary membership in this context not as a primary phenomenon, but rather as the cause of *slow rotation*, through spin-orbit coupling. (Normal A stars rotate rapidly, and Dr. Abt has remarked that those A stars that are found in wide long-period binaries where spin-orbit coupling is weak do have normal spectra.) If we associate the Am or Ap characteristic with the presence of a photospheric magnetic field, the problem reduces to explaining why slowly rotating stars have strong *surface* magnetic fields. The reason may be that in rapidly rotating stars the Eddington-Sweet circulation flows and drags the field lines beneath the surface; but in a slowly rotating, strongly magnetic star, thermal equilibrium is achieved without any meridian circulation, and the field lines appear above the surface.

(v) Finally, taking up Dr. Huang's and Dr. Batten's remarks on the orientation of stellar angular-momentum vectors, again I would like to emphasize that a lot happens in between formation of a protostar and its arrival on the main sequence. In particular, if the angular-momentum vector and the magnetic dipole are neither parallel nor perpendicular, the net couple on the star due to a stellar wind will be inclined to the angular-momentum vector, and so will not only brake the star, but will also rotate the angular-momentum vector. It is hoped that further computations will show whether the tendency is to align the angular momentum and magnetic dipole, or to make them perpendicular. Meanwhile, it seems premature to draw firm conclusions about the processes of protostar formation, though I agree that the vorticity associated with the turbulence in gas clouds must be included along with that due to the overall galactic rotation.

A MODEL FOR THE NOVA OUTBURST

WILLIAM K. ROSE

(Physics Dept., Massachusetts Institute of Technology, Cambridge, Mass., U.S.A.)

Models for a $0.75\,M_\odot$ hydrogen-exhausted star that is approaching the white-dwarf state have been computed. It has been assumed that the star is in a close binary system and therefore can accrete small amounts of hydrogen-rich material ($\approx 7 \times 10^{-8}$ $M_\odot$/yr). Hydrogen burning commences when the mass of the hydrogen-rich envelope exceeds about $10^{-4}\,M_\odot$. Eventually the mass accretion causes the non-degenerate hydrogen-burning shell to become thermally unstable, and therefore the nuclear reaction rate increases very dramatically. Pulsational instability is found to develop as a consequence of the very high rate of nuclear-energy generation.

A theory of novae must explain the observed fluxes and mass loss. In addition, it must explain the short rise times as well as the observed dispersion in rise times that characterize the nova outburst. The present model for a nova outburst assumes that the energy released during the period of early decline is energy that has been stored in the form of pulsations. The observations of novae indicate that the amount of energy that is released during the period of early decline is about 3×10^{44} ergs. The computed pulsational modes for the models indicate that about 6×10^{44} ergs could be stored in the form of pulsational energy for these models.

On the other hand, the calculations suggest that the rise time observed for a nova outburst might be caused by thermal expansion of the outer layers of a thermally unstable star that is also pulsationally unstable. This is true because the relative amplitude of the adiabatic modes in the vicinity of the surface are greatly increased with respect to those for interior regions of the star as a consequence of thermal expansion. This implies that energy in the form of pulsations has been transported closer to the surface.

The present model for a nova outburst assumes that the energy released during the period of slow decline, which may continue for a number of years, is due to the thermal diffusion of photons that were initially emitted during the period of high nuclear energy generation. The maximum surface luminosity that results in the computed models because of the high rate of nuclear-energy generation is approximately 5–6 magnitudes above that for the initial model, and therefore is in agreement with the schematic light curve for a nova outburst due to McLaughlin.

A more complete discussion of these results as well as references to earlier work will be published in a forthcoming issue of the *Astrophys. J.*

Perek (ed.), Highlights of Astronomy, 455. © *I.A.U.*

CONCLUDING REMARKS

JORGE SAHADE

(Observatorio Astronómico, La Plata, Argentina)

This has been a very stimulating and important meeting. It has given a chance for theoreticians and observers to get together and discuss some of the problems that are worrying the astronomers in the field of close binaries.

In this Joint Discussion we have dealt with two main problems. There were two papers – one by Su-Shu Huang and the other by Roxburgh – on the question of the origin of close binaries. But the main bulk of the papers were on the problem of the evolution of close binaries after they have reached the main sequence. On the latter problem I would like to make a few comments. Let me start by reminding you that the problem of the Algol stars was posed many years ago when the single-lined system of R Canis Majoris was first observed. R Canis Majoris is characterized by a very small mass function. If one wants to play with the mass function in order to get an idea about the two masses, one ends up with the possibilities that either one of the components has a very small mass or that both components must have small masses.

The second question which is posed by the Algol systems is that of the sizes and masses as related to the spectral types and luminosities of the components. It was clear from the study of U Sagittae and U Cephei that the secondary components of the Algol systems are subgiants with masses much smaller than those of the main-sequence primary components, and this conclusion has found confirmation in many cases where direct determination of the masses of the two components has been possible. The problem of the Algol stars was dealt with several times, and I am very glad to see that three groups – there was going to be a fourth group that dropped out because the German group went ahead too fast – have attacked this problem in such a way that we can now understand very nicely how main-sequence systems can evolve into the systems where the primary component is a main-sequence star while the secondary is a subgiant whose mass can be quite small. Actually we have cases such as that of XZ Sagittarii, where the mass of the subgiant is of the order of 0·2 solar mass; and case B studied by Kippenhahn and Weigert shows how we can get an Algol system of this kind from a system originally on the main sequence.

We must not forget, however, that the spectra of the components – sometimes we can photograph the spectra of each component – appear quite normal. Even the spectrum of the subgiant looks very similar to that of a normal single subgiant star. Perhaps this is an observational fact, which we should keep in mind. There are now large telescopes in the Southern and Northern hemispheres, and we should try to

Perek (ed.), Highlights of Astronomy, 456-458. © *I.A.U.*

analyze the spectra of the subgiant components obtained during minima, and compare them with those of the normal single subgiant stars.

One would expect abundance anomalies in such secondary components, and an attempt to try to derive observational evidence on the matter has been mentioned here; another attempt was made by Douglas Hall, while he was at Indiana University, by applying Strömgren's narrow-band photometry. His conclusion, which should be checked, was that perhaps the secondary components of Algol systems are CN-under-abundant. We should try to go on in this direction.

Another application of narrow-band photometry to this type of project has been made by Sisteró, of the Córdoba Observatory, at the Cerro Tololo Interamerican Observatory in Chile. I still do not know what are the results, but I just mention it as another attempt to apply new techniques to Algol systems and try to secure further information.

As I have said, it seems to me that we are going in the right direction to try to explain what happens to a close binary when it starts evolving off the main sequence. But it also seems to me that the general feeling of the people who are making the computations is that now everything or rather every case can be explained by the results of the computations that have been made or are being made. I am afraid we will be put out of business if we could explain everything at once, so we should better not do that....

I think that we should try to consider whether there are other facts that should be taken into account in some cases. Mention has been made e.g., of β Lyrae, and mention has been made of the Wolf-Rayet stars. Let me remind you that although there has been quite a debate about the masses of the components of β Lyrae, at the present time everybody, or almost everybody, agrees that the secondary (less luminous) component is the more massive component. Yet the size of this object is much smaller than the size of the primary star. So we are dealing here with a system where the component from which the stronger stream originates, the component which appears to be more evolved, is more massive and much smaller than the primary component, which is a B8 II object. And there are quite a few systems where the secondary component seems to be underluminous, small and yet the more massive star. We should remember that such objects do exist, and that they may be very important for our understanding of some other type of evolution in close binaries.

The question of the Wolf-Rayet stars is still controversial. I think that Miss Under-hill now favours the same opinion that I have been advocating for several years, namely that the Wolf-Rayet stars are contracting objects. Perhaps we both would like to see another set of computations, which will consider stars that are contracting along the Hayashi tracks. Perhaps we might get a kind of system similar to the Wolf-Rayet stars, or perhaps I may be shown to be wrong.

And there are other interesting systems. This afternoon, when listening to the papers, I felt that perhaps a review paper on the observational aspects, bringing to this audience all the relevant observational facts collected over the years, might have been a very good idea. There is a great number of systems which are very interesting and which show effects that may be important enough to be taken into account in the computations – e.g., in cases when the masses of the components are large. The system HD 47 129, for instance, is a very interesting binary. There is a very large expanding envelope around the system, as it exists also in β Lyrae and in the Wolf-Rayet binaries. Thus the question of the mass loss to the system, as it has been considered by the Polish group, is a very important point in some cases, and the observations, in the case of HD 47 129, seem to indicate that radiation pressure may be a mechanism to pay attention to.

Very important also is the investigation of the German group into the problem of what will happen after one of the components of a close binary has become a white dwarf. Certainly, wherever we have to deal with eruptive phenomena, one of the components happens to be a white dwarf or a star which is becoming a white dwarf.

Before I finish, I should like to say just a few words on the peculiar A and metallic-line stars. It would seem to me that at present we still need some more information before being able to draw more definite conclusions. Some people, for instance, think that the Ap stars have not gone through the giant stage and are still unevolved; and there seems to exist observational evidence in favour of such a conclusion. As to the case of the metallic-line stars, there are many binaries – as Van den Heuvel has already mentioned here – in which both components display metallic-line spectra. Therefore if the metallic-line stars are evolved binaries, then the evolution may have started from a different kind of objects than those we have dealt with this afternoon.

Concluding these remarks, I should like to say that I was surprised to see so many people in this room interested in close binaries when there was another meeting going on, on quasars. I think this is very good, since it shows that astronomers remain interested in close binaries and, therefore, that we can expect much progress in this field in the near future. I think we can look forward either to the next meeting of the IAU, or to an earlier meeting where such progress will be reported. I think the people who have had the idea of holding this meeting here should be congratulated, because it has been a very good, stimulating meeting.

SPECIAL MEETINGS

COMMISSION 17
A. LUNAR PROBES

(Saturday, August 26, 1967)

Chairman: Dr D. Menzel

Out of the programme, published in the I.A.U. *Transactions* **XIIIB**, the following contributions were selected for publication in the present volume:

1. I. Taback: A Description of the Lunar Orbiter Spacecraft.
2. W. E. Brunk: Photographic Results of the Lunar Orbiter Program.

A DESCRIPTION OF THE LUNAR ORBITER SPACECRAFT

ISRAEL TABACK

1. Introduction

The U.S. Lunar Orbiter Spacecraft Program was conceived to search out potential Apollo landing sites, and to return detailed photographic coverage of the lunar surface for scientific study. The first flight was launched in August 1966, and the successful launch of the fifth orbiter was in August 1967. All five were successful and returned to Earth a large amount of photographic data.

Presented in this paper is a short description of the spacecraft system with particular emphasis on the photographic system. The intent is to supply information which will permit a better understanding of the photographic data to be presented by Dr. William Brunk in the following paper.

2. Mission Plan

Figure 1 schematically shows the launch and flight operations which were applied for all of the orbiters. During the launch phase, telemetry and tracking are accomplished by the Eastern Test Range (ETR) stations. At about the time the spacecraft

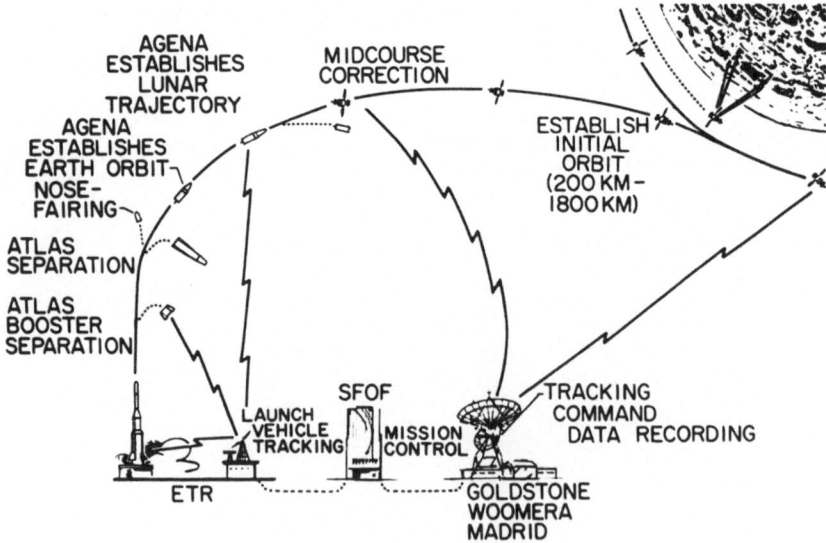

FIG. 1. *Mission operations.*

Perek (ed.), Highlights of Astronomy, 462–470. © *I.A.U.*

is inserted into a translunar trajectory, tracking, telemetry, and command transmission is accomplished by the Goldstone (Nevada), Madrid (Spain), and Woomera (Australia) stations of the NASA Deep Space Instrumentation Facility (DSIF). Centralized control is accomplished from NASA Space Flight Operations Facility (SFOF) in Pasadena, Calif.

3. Spacecraft Configuration

Figure 2 shows the general configuration of the 390 kg (850 pound) spacecraft which is approximately $1\frac{1}{2}$ m (5 feet) high and $1\frac{1}{2}$ m in diameter, excluding the solar panels and antennas. The span across the deployed antenna booms is about 6 m.

With the exception of the rocket engine and its fuel tanks, which are used for midcourse maneuvers and deboost into lunar orbit, essentially all of the major spacecraft components are attached to an equipment mounting plate. That mounting plate, the underside of which remains oriented toward the Sun at all times except when the spacecraft is maneuvering, provides thermal control for the components attached to it.

In the flight configuration, all of the main spacecraft structure above the equipment mounting plate is covered with a highly reflective shroud of aluminum-coated mylar which constitutes part of the thermal control system. The only protrusions through that shroud are the camera lenses and micrometeoroid detectors.

The spacecraft power system is a conventional solar array/storage battery type. A 12 amp-hour nickel cadmium battery is used to supply the spacecraft power requirements during the launch phase prior to solar array deployment and during those periods of the lunar orbit when the spacecraft is in the Moon's shadow. Because of the rather severe weight penalty for battery capacity, it has been imperative to mini-

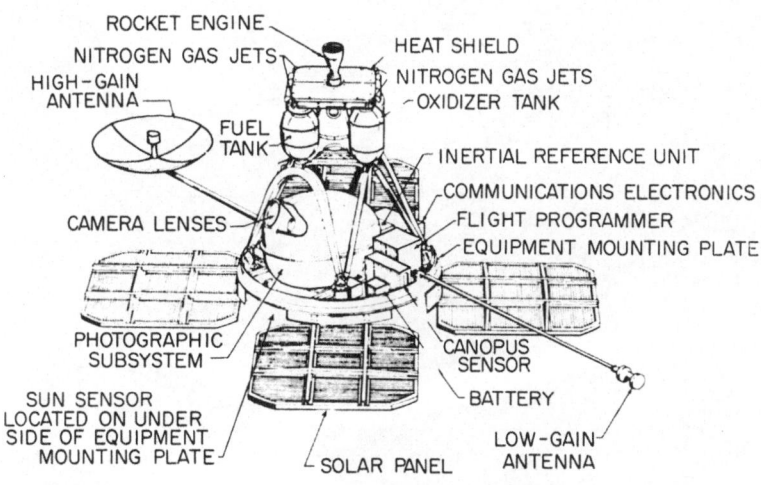

FIG. 2. *Spacecraft configuration.*

mize the power consumption of the residual power load 5, including such communication system items as the transponder, command decoder, etc. When the solar array is in sunlight, a maximum of about 375 W is available to handle all power demands including battery recharging. The voltage of the spacecraft dc-power bus can vary from a minimum of 22 V to a maximum of 31 V when the array is in operation. A shunt regulator is used to prevent the voltage from exceeding 31 V.

The star Canopus and the Sun are the primary references for spacecraft attitude orientation. For maneuvering, or when those references are occulted, a strapped-down gyro system is used. Attitude control is accomplished by a cold-gas system.

A photograph of the camera system is shown in Figure 3. The spacecraft camera system employs two lenses (Figure 4), which take simultaneous pictures on a roll of 70-mm wide aerial film. One of the lenses has a 610 mm (24 inch) focal length and can take pictures from an altitude of 46 km with a resolution of approximately 1 m. The other lens, which has a focal length of about 80 mm (3 inches), takes pictures with a resolution of approximately 8 m. The film is developed onboard using a method which presses the film into contact with a web that contains a single-solution processing chemical. After the film has been dried, it is ready for read-out and transmission to Earth.

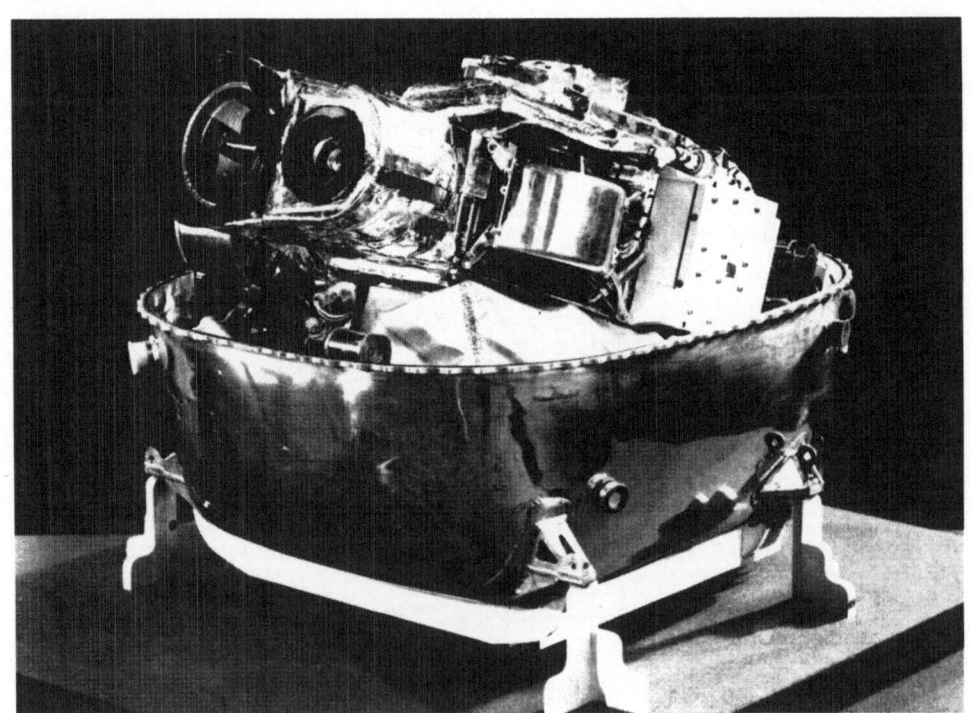

FIG. 3. *Camera subsystem.*

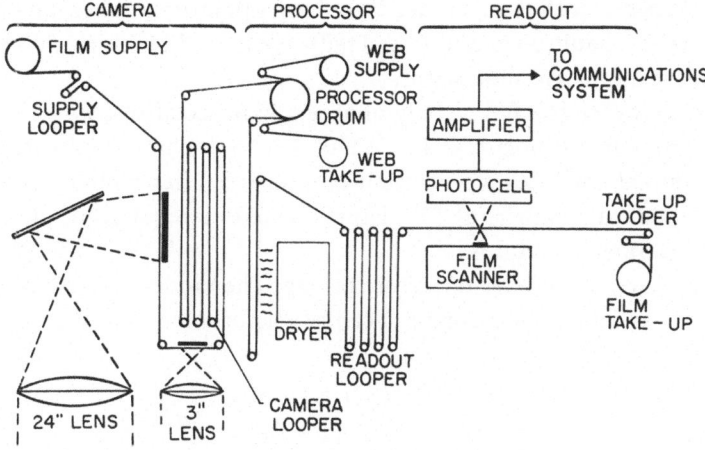

FIG. 4. *Spacecraft photographic system.*

Figure 5 is a drawing of the read-out system which uses a line-scan tube as the light source for film scanning. The line-scan tube itself electronically scans the beam of light through an excursion of 2·5 mm (0·1 inch) in a direction parallel to the film travel. Sweep of the light beam in the other direction (across the width of the film) is accomplished by a mechanically driven optical system, which also serves to reduce the diameter of the beam of light to approximately 5 microns. In read-out, this type of scan produces a series of 'framelets' which are 2·5 mm in one direction and the width of the film in the other direction. As the figure shows, scanning begins at one edge of the film and continues across the film until the opposite edge is reached. At that time the film is advanced and is scanned in the opposite direction. This sequence is repeated, with the scan rates being such that 43 min are required to scan about 28 cm (1 foot) of film which contains one high-resolution and one medium-resolution photograph.

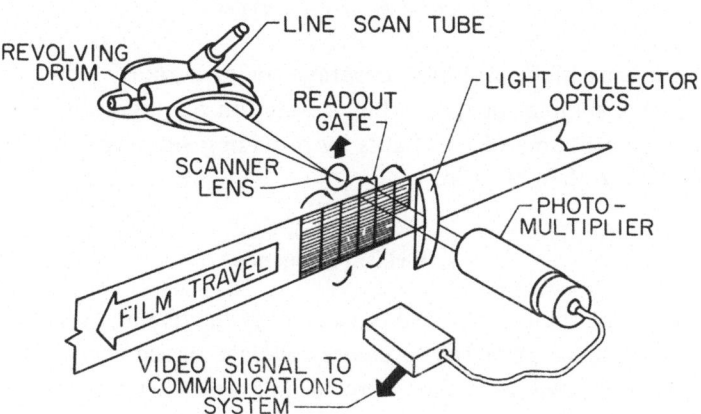

FIG. 5. *Photographic system read-out schematic.*

Collecting optics direct the transmitted light into a photomultiplier, and the resulting electrical signal is conditioned and mixed with synch and blanking pulses and fed to the communication system modulator.

Each spacecraft carries a film supply of about 85 m (260 feet), which is sufficient to photograph, from an altitude of 46 km, 12000 km² of lunar surface with a resolution of 1 m and 200000 km² with a resolution of 8 m. It is interesting to note that about one million standard commercial TV pictures would be required to photograph that area with comparable resolution.

Figure 6 indicates the modes of picture taking. The moderate-resolution and high-resolution fields of view have a common centre. Pictures may be taken at a rate which

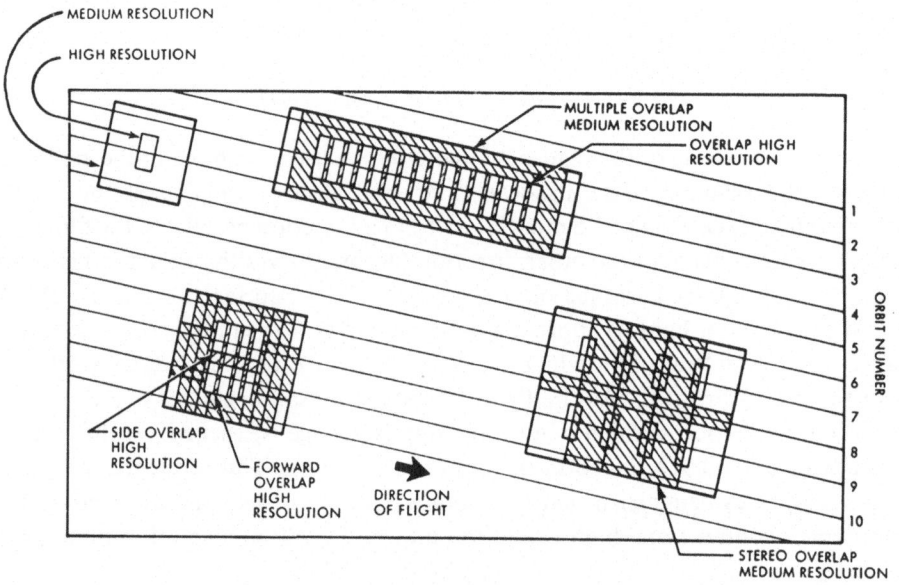

FIG. 6.   *Photographic modes.*

results in contiguous high-resolution coverage, or at a rate which provides stereo coverage with only samples of spaced high-resolution photography. Stereo coverage was also obtained on some of the flights by providing side-overlap of photography with a proper base to height ratio.

4. Orbital Sequence

Figure 7 shows the sequence of orbits secured by the five missions. The first three were at low inclinations, primarily to secure coverage of the near-equatorial Apollo sites. On these three flights, 8 sites were located and vertical stereo and oblique photography obtained. Flight four resulted in photography which essentially covered

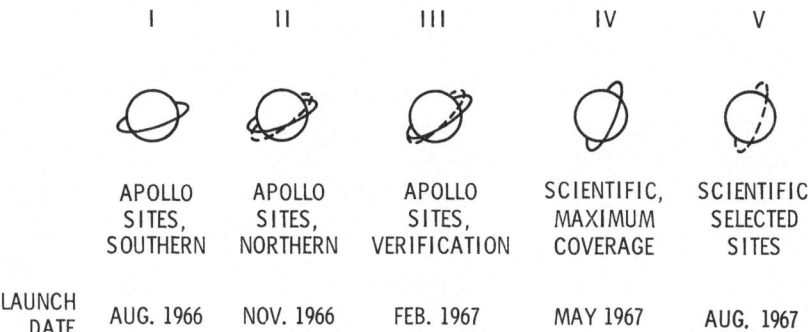

	I	II	III	IV	V
	APOLLO SITES, SOUTHERN	APOLLO SITES, NORTHERN	APOLLO SITES, VERIFICATION	SCIENTIFIC, MAXIMUM COVERAGE	SCIENTIFIC, SELECTED SITES
LAUNCH DATE	AUG. 1966	NOV. 1966	FEB. 1967	MAY 1967	AUG. 1967

FIG. 7.   *Lunar Orbiter missions.*

the entire front face of the Moon at resolutions of at least one order of magnitude greater than previously available from Earth-based photography. The result, a mosaic which covers a large portion of a gymnasium floor, is available for those who wish to walk on the Moon. The fifth flight covered over 40 sites of particular scientific interest at a resolution of a few meters, and completed coverage of the far side of the Moon.

Table 1 is a tabulation of the orbital parameters for each flight and indicates the accuracy with which these orbits were established at the Moon. These photographic orbits were, in general, those secured after first establishing a somewhat coarser orbit, and then making an orbital correction to secure proper perilune altitude, period, and orbit inclination. The apolunes and perilunes are in kilometers and the orbit inclination is given in degrees.

Table 1

Final orbit parameters

	I			II		
	Apo	Per.	Inc.	Apo	Per.	Inc.
Desired	1824·0	40·0	12·03°	1788·7	35·91	11·88°
Actual	1816·7	40·5	12·00°	1787·8	35·91	11·89°
Δ	−7·3	+0·5	−0·03°	−0·9	0	+0·01°

	III			IV		
	Apo	Per	Inc.	Apo	Per	Inc.
Desired	1792·9	35·84	20·87°	6111·0	2701·0	85·50°
Actual	1793·1	35·85	20·85°	6114·4	2706·3	85·48°
Δ	+0·2	0	−0·02°	+3·4	+5·3	−0·02°

	V		
	Apo	Per	Inc.
Desired	6065·6	101·3	84·62°
Actual	6065·4	100·8	84·60°
Δ	−0·2	−0·5	−0·02°

5. Results

The orbiters returned a considerable quantity of data from the Moon, not all of which was photographic. Table 2 lists the data secured.

Continuous tracking of the orbiters has resulted in a definition of many of the higher harmonics of the gravitational field. This work has been reported by Michael and Tolson (1965).

Table 2

Information return

1. Information for Apollo Sites
2. Information on Surface Properties
3. General Geological Information
4. Special Interest Photographs
5. Non-photographic
   (a) Selenodesy
   (b) Micrometeorite Flux
   (c) Radiation Flux
   (d) Surface Height (V/H Sensor)

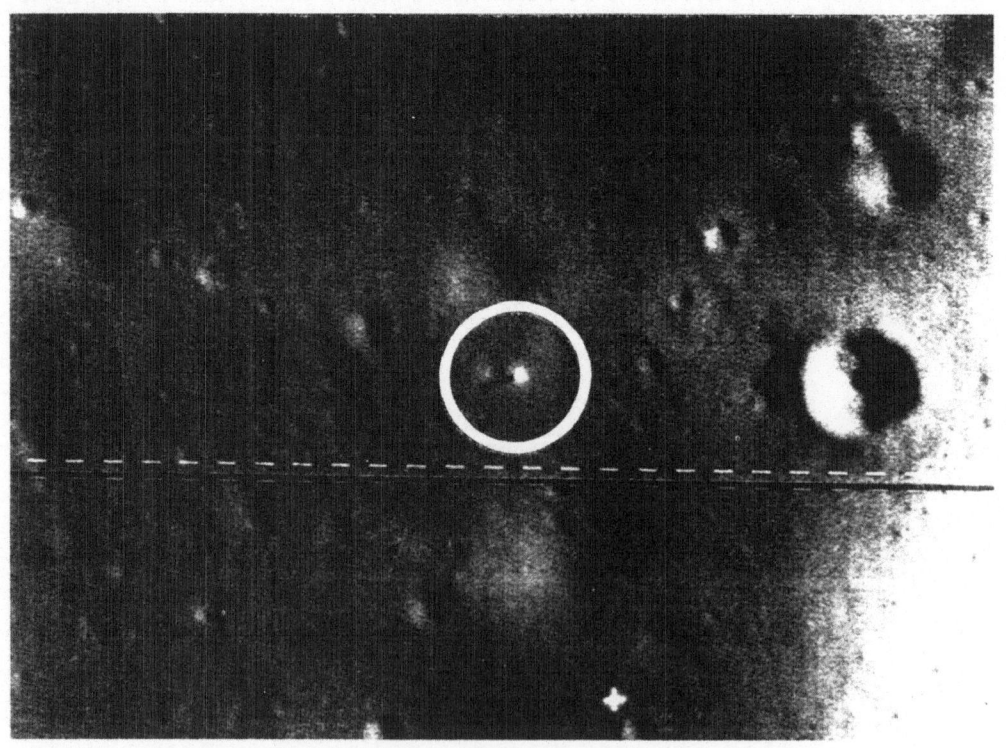

FIG. 8. *Landed Surveyor.*

The micrometeoroid detectors, 20 pressurized cans on each spacecraft, have been punctured at a rate which to date averages approximately $\frac{1}{3}$ of the rate which has been measured in the near-Earth environment. Data are still being collected from the three operating spacecraft which are still in orbit.

The radiation detectors have provided measurements during solar flares indicating that the environment near the Moon is similar to that of deep space.

6. Selected Photographs

Only two pictures will be shown in this paper, which are indicative of the fact that the photography secured provides some information regarding lunar surface properties.

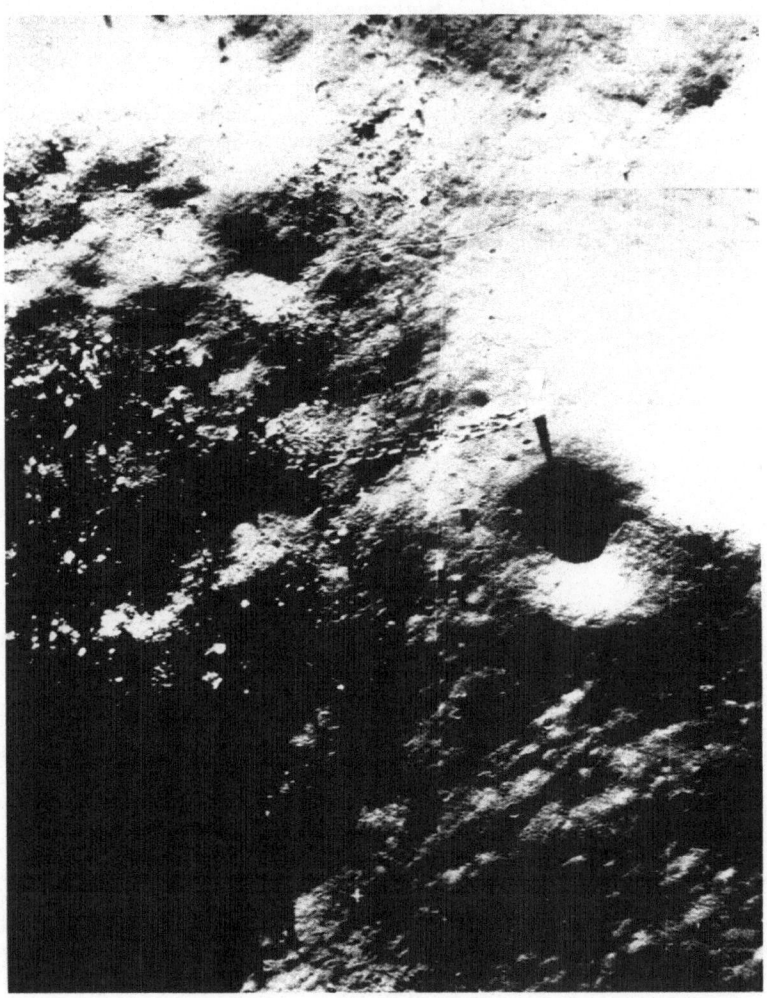

FIG. 9.   'Rolling Stones'.

Figure 8 is a picture taken of Surveyor-I by Lunar Orbiter-II. A positive identification was made by correlating numerous objects in the Surveyor picture with the same objects in the orbiter picture. This identification allowed extrapolation of Surveyor results over a wide area.

Figure 9 shows the tracks of two boulders which have rolled down an incline and are still visible. By making estimates of the boulder density it is possible to draw some conclusions regarding the bearing properties of the lunar surface.

In conclusion it can be stated that the Lunar Orbiters have provided not only the specific information required for manned exploration, but have secured photographic coverage of the entire lunar surface at a resolution never before attained.

References

Kosofsky, Leon J., Broome, G. Calvin (1965)   Lunar Orbiter: A Photographic Satellite. Presented at Society of Motion Picture and Television Engrs. (Los Angeles, Calif.), Mar. 28–Apr. 2.

Michael, William H., Jr., Tolson, Robert H. (1965)   The Lunar Orbiter Project Selenodesy Experiment. NASA paper presented at the Second International Symposium on the Use of Artificial Satellites for Geodesy (Athens, Greece), Apr. 27–May 1.

Taback, Israel (1964)   Lunar Orbiter: Its Mission and Capability. Preprint 64-7, Am. Astronaut. Soc., May 1964.

Taback, I., Brummer, E. A. (1965)   The Lunar Orbiter. Presented at AIAA Unmanned Spacecraft Meeting (Los Angeles, Calif.), Mar. 1–3.

PHOTOGRAPHIC RESULTS OF THE
LUNAR ORBITER PROGRAM

WILLIAM E. BRUNK

(NASA, Washington, D.C., U.S.A.)

The principal goal of the Lunar Orbiter program was to obtain photographic coverage of selected areas of the lunar surface at resolutions of 1 and 8 m. As the success of the program was greater than originally anticipated, the photographic coverage was extended to include the entire front side of the Moon at a resolution one order of magnitude greater than possible from the Earth and almost all of the far side. In this paper, a selection of photographs obtained from all five missions will be presented. These photographs show many interesting and previously unobserved features. However, no attempt will be made herein to analyze or interpret such features.

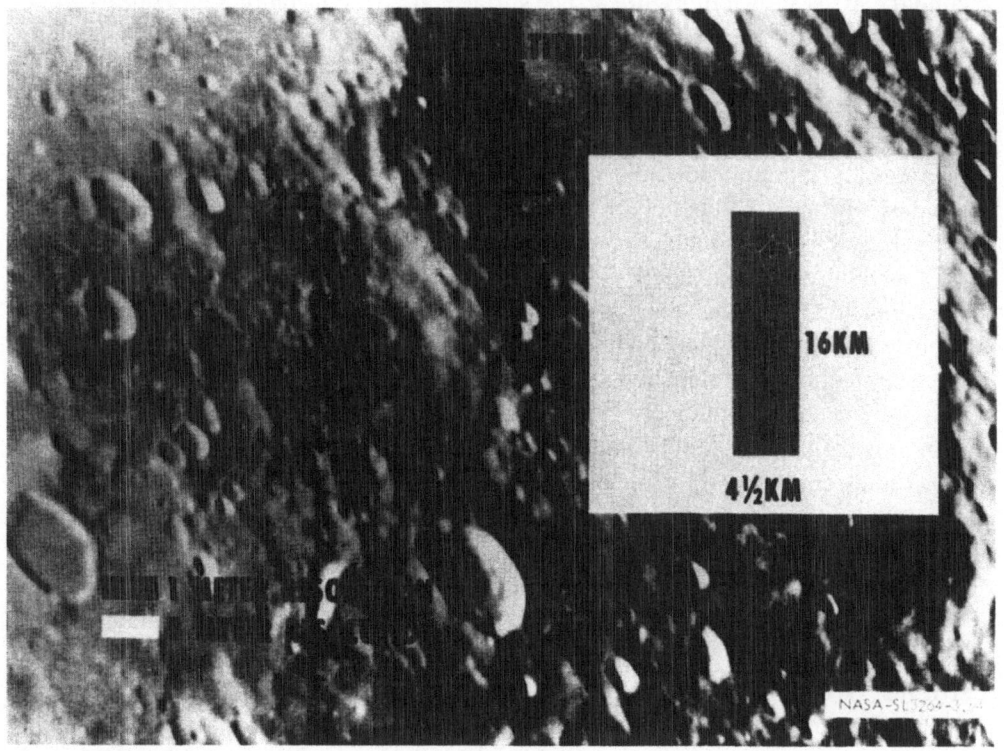

FIG. 1.

Perek (ed.), Highlights of Astronomy, 471–523. © I.A.U.

Analysis and interpretation is being carried out by a number of scientists and will be reported in the journals.

The relative coverage obtained with the two Lunar Orbiter cameras is shown in Figure 1 for a spacecraft altitude of 46 km and with the optical axis of the cameras normal to the lunar surface. The 24-inch focal-length lens produced a high-resolution photograph with 1-m resolution over an area of 4·5 by 16 km as indicated by the dark gray color, while the 3-inch lens produced a moderate-resolution photograph with 8-m resolution over the lighter gray area of 32 by 37 km. Both a high- and moderate-resolution photograph are obtained with each exposure. Portions of the high- and moderate-resolution photographs of a region in the Sea of Tranquility are shown in Figure 2. The photographs are scaled to give the same linear resolution.

The primary sites that were to be photographed during the Lunar Orbiter-I mission are shown in Figure 3. The sites, indicated by the black dashes, were possible Apollo landing sites as determined from Earth-based photography. The rectangle represents the zone of possible Apollo landing sites. A moderate-resolution photograph of a secondary site is shown in Figure 4. The surface, at a resolution of 8 m, appears to be completely pockmarked with craters. The orbit of Lunar Orbiter-I was elliptical with a perilune altitude of 200 km and apolune altitude of 1850 km. The high-resolution photography of the frontside, taken near perilune, was blurred due to a malfunction in the altitude-velocity corrector in the camera system. However, excellent high- and moderate-resolution photographs of the lunar far side were obtained. The resolution was considerably less than on the front side as they were taken when the spacecraft was near apolune. A moderate-resolution photograph of an 1100 by 1300 km region in the western portion of the far side is shown in Figure 5. A similar sized area in the eastern portion of the far side is shown in Figure 6. The dark-floored crater near the bottom is Tsiolkovsky, first detected on the U.S.S.R. Luna-III photography. Many interesting features were noticed on the high-resolution photography of the far side. In Figure 7 is shown a double crater where material from the larger, approximately 52-km diameter, crater has flowed over into the smaller crater. Shown in Figure 8 is 25-km diameter crater with a highly fractured floor. The fractures appear as an irregular pattern like that observed in a dried mud flat. Many examples of craters with fractured floors were seen in later high-resolution photographs on both the near and far sides of the Moon.

The first photographs of the Earth from the vicinity of the Moon were taken on this mission. Figure 9 shows a portion of a high-resolution photograph which shows the Earth just beyond the Eastern limb of the Moon. At the time of the exposure, the American continent was on the limb and Europe was located close to the terminator.

The primary goal of the Lunar Orbiter-II mission was further photography of possible Apollo sites. Representative samples of high-resolution photography of these sites are shown in Figure 10. Although all of the primary sites looked smooth on Earth-based photography, many appeared filled with craters when examined in detail. The

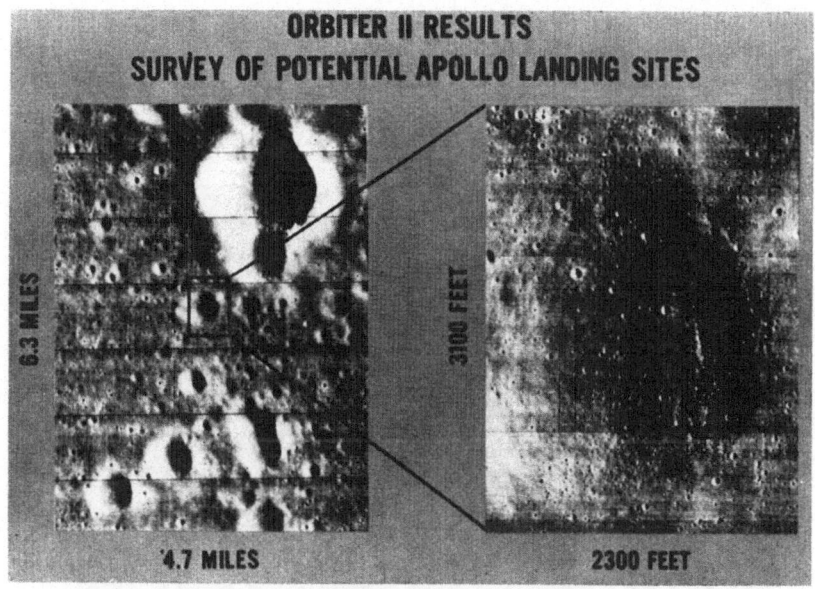

FIG. 2.

FIG. 3.

Fig. 4.

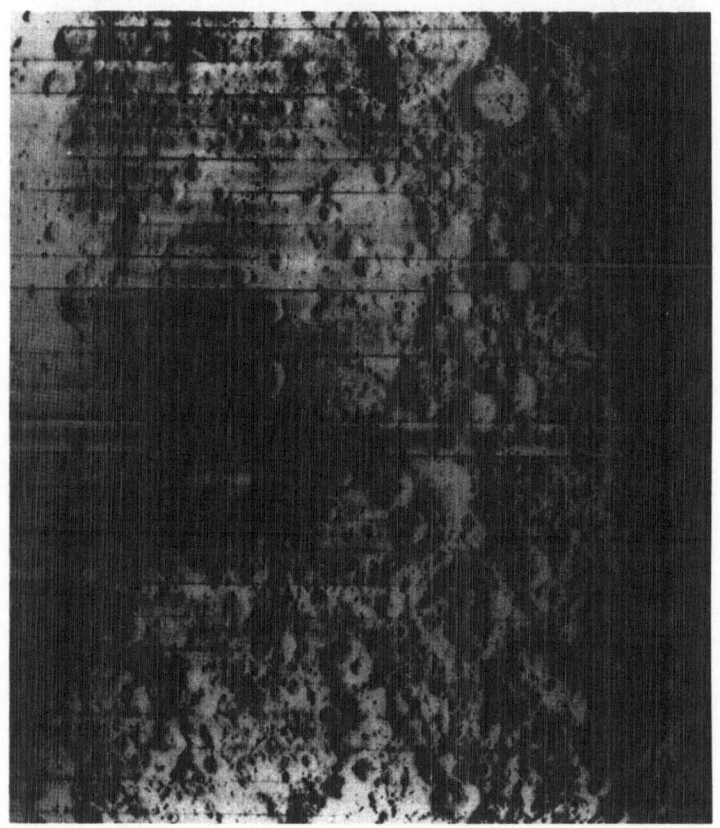

FIG. 5.

FIG. 6.

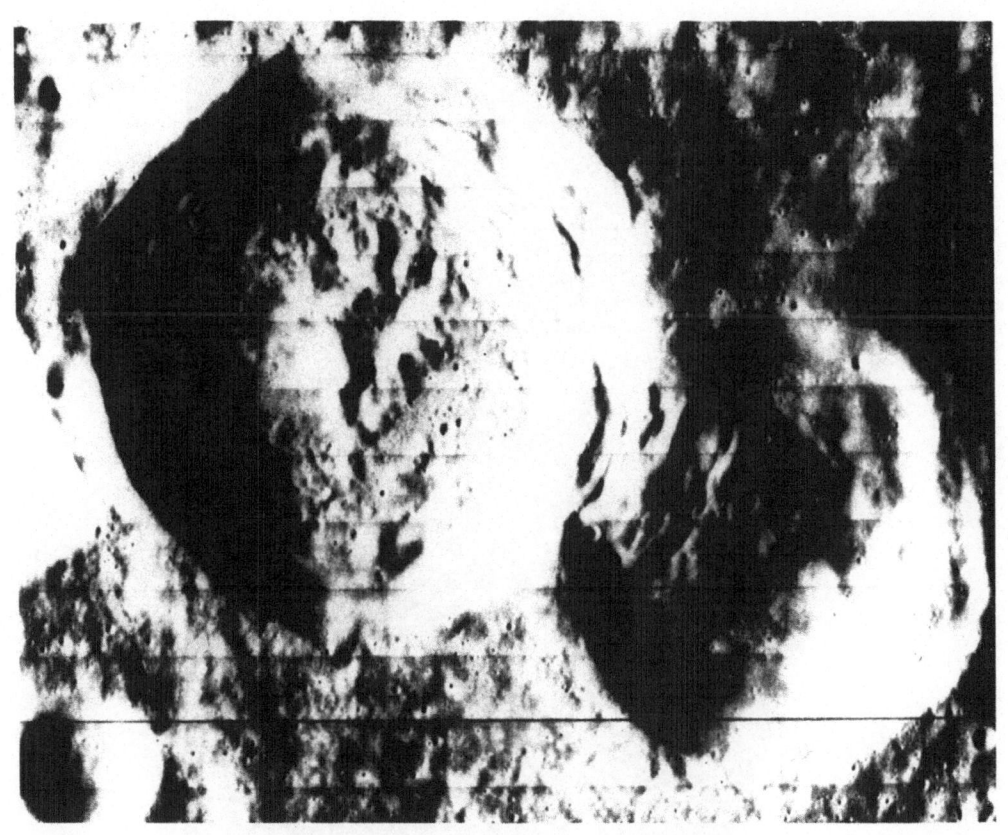

FIG. 7.

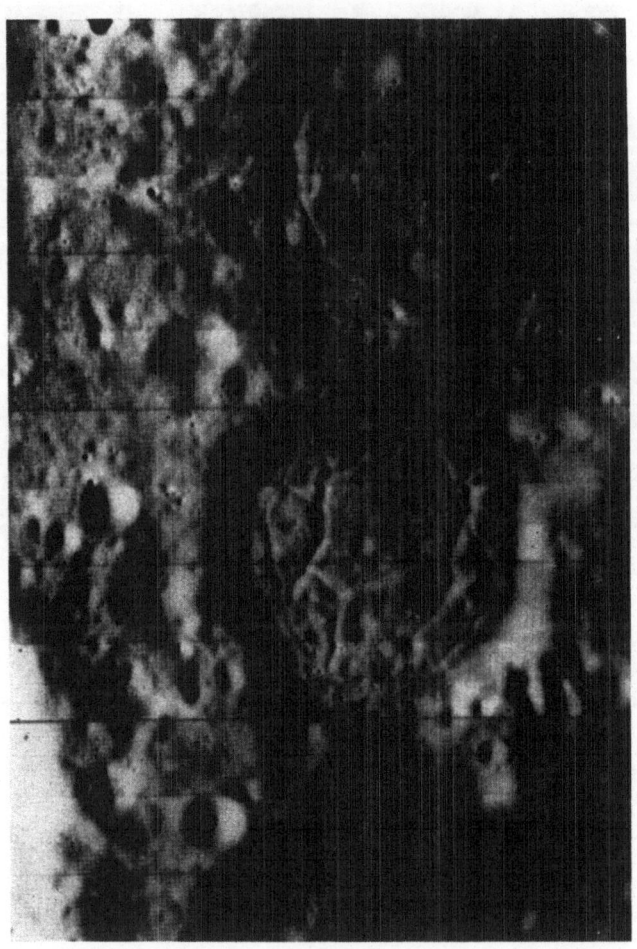

FIG. 8.

FIG. 9.

resolution in these pictures is approximately 1 m. Interesting features were found in many of the photographs. A rock-strewn area in the southeastern part of Mare Tranquillitatis is shown in Figure 11. The area covered by the photograph is approximately 400 by 500 m. The rock formations are part of a linear feature appearing as a lighter colored ray on the mare surface. A portion of another high-resolution photograph is shown in Figure 12. The large crater is approximately 180 m across and the area is in the western part of Mare Tranquillitatis.

With the exception of the Earth–Moon picture, all photographs shown thus far were taken with the camera's optical axis nearly normal to the lunar surface. Starting with the Lunar Orbiter-II mission, oblique photography was undertaken. The first oblique photograph was that of the crater Copernicus. An Earth-based photograph of the Copernicus region to illustrate the geometry of the situation is shown in Figure 13. The spacecraft was traveling from right to left along the dashed line and the photograph was taken when the spacecraft was at the point indicated by the ×. The fields of view of the moderate- and high-resolution photographs are indicated by the solid white lines. The moderate-resolution photograph is shown in Figure 14. Copernicus is seen near the limb with the double crater Fauth just below it. A section of the high-resolution photograph is seen in Figure 15 with the southern rim of Copernicus in the foreground and the northern rim in the background. The features seen on the floor of the crater are 300-m high mountains. Oblique photography was subsequently used to study other features of interest. A portion of the

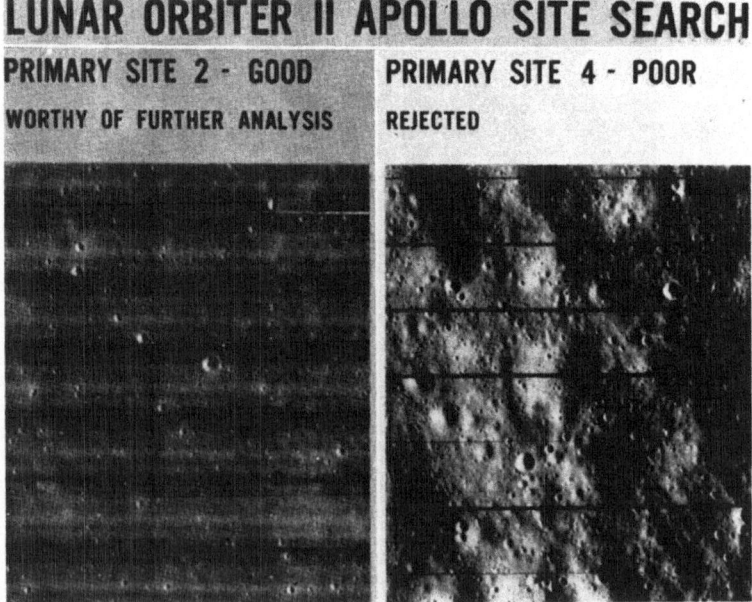

FIG. 10.

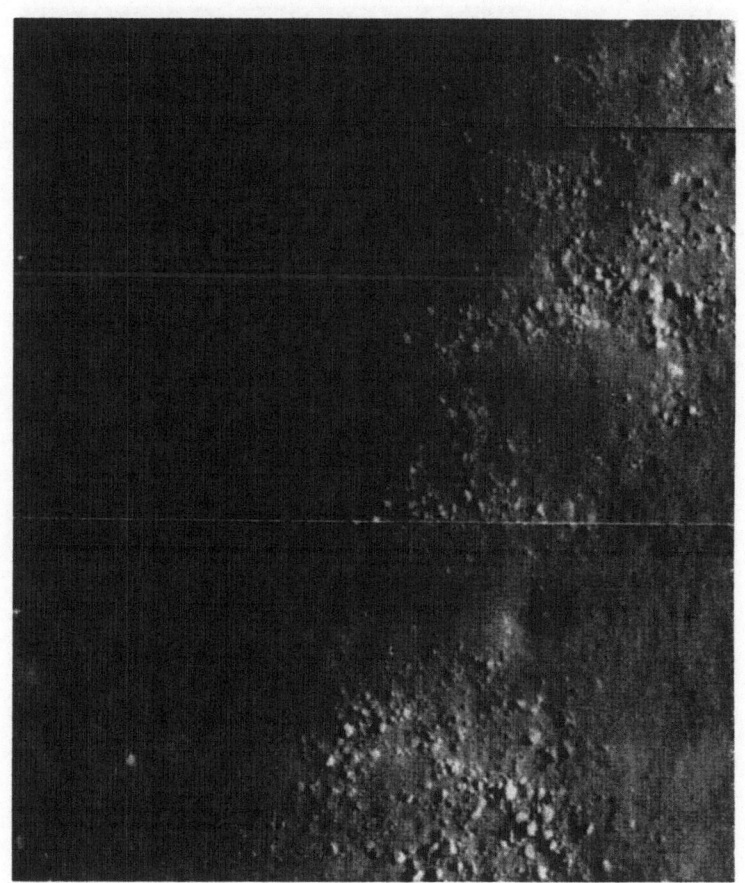

FIG. 11.

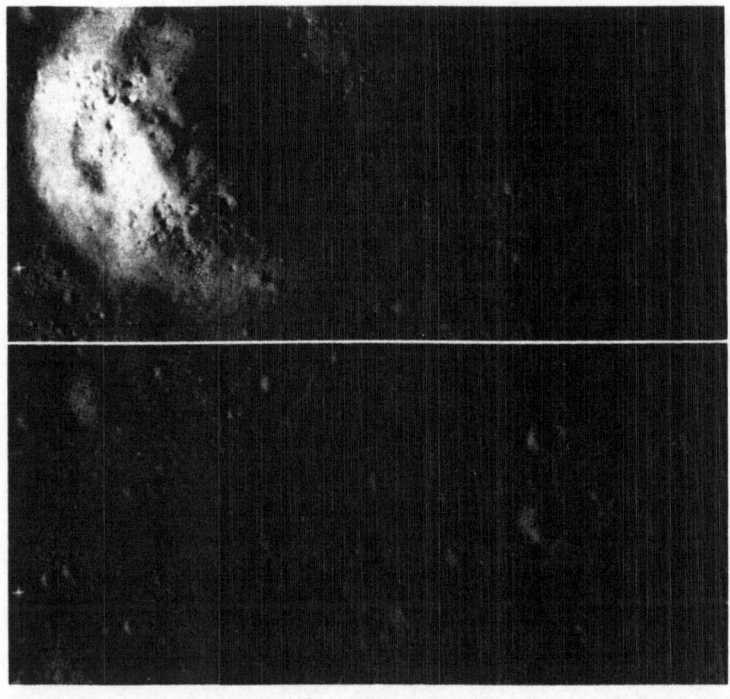

FIG. 12.

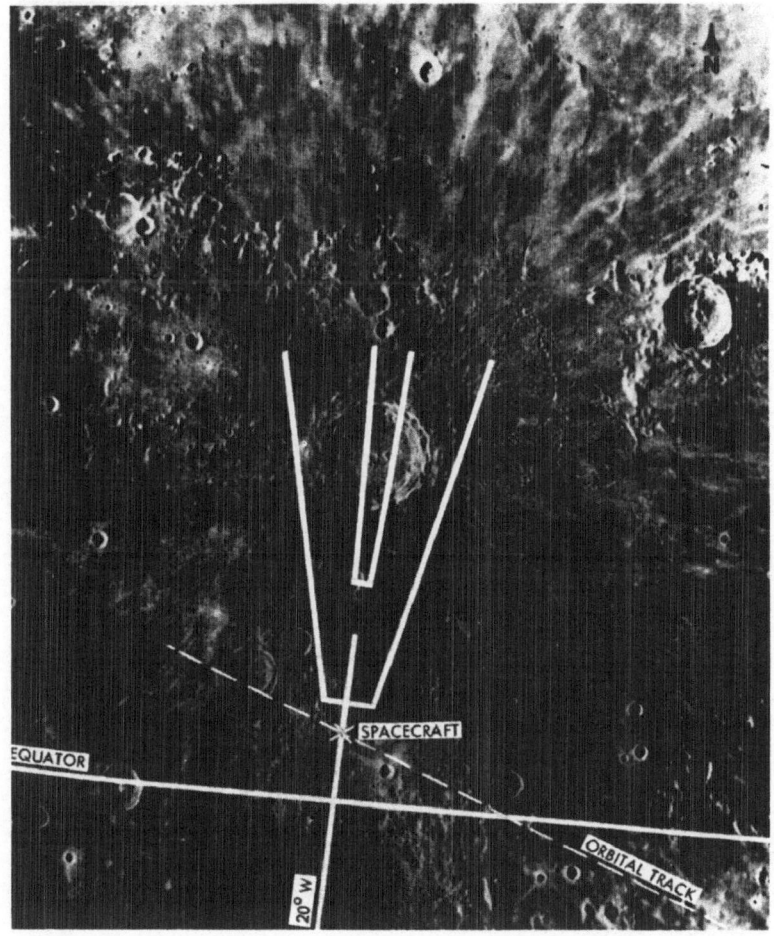

Fig. 13.

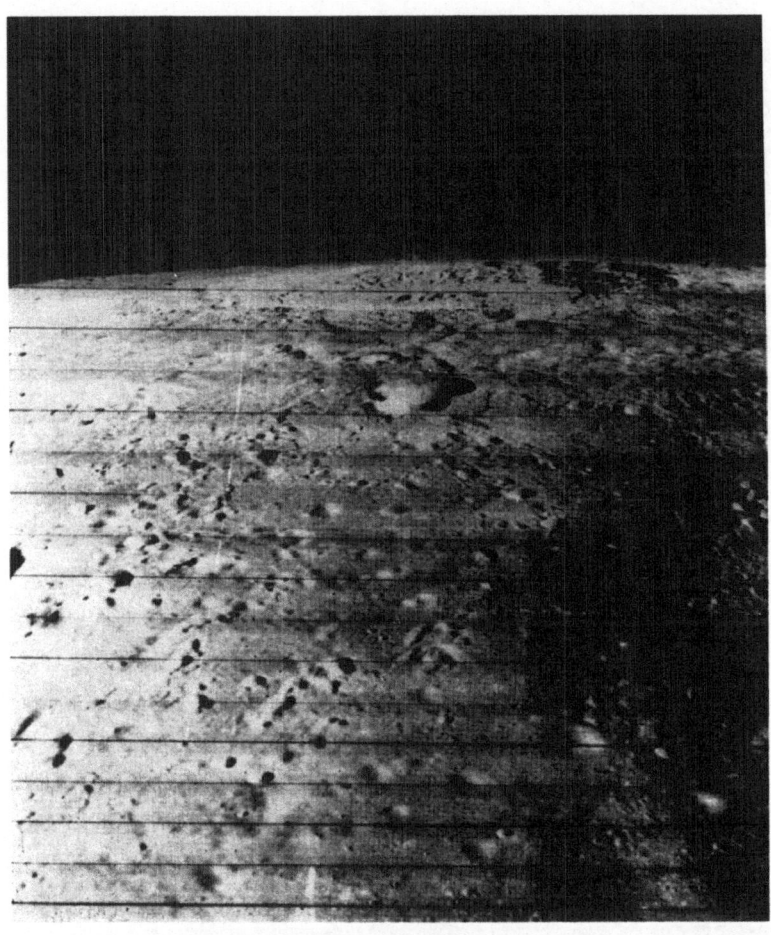

Fig. 14.

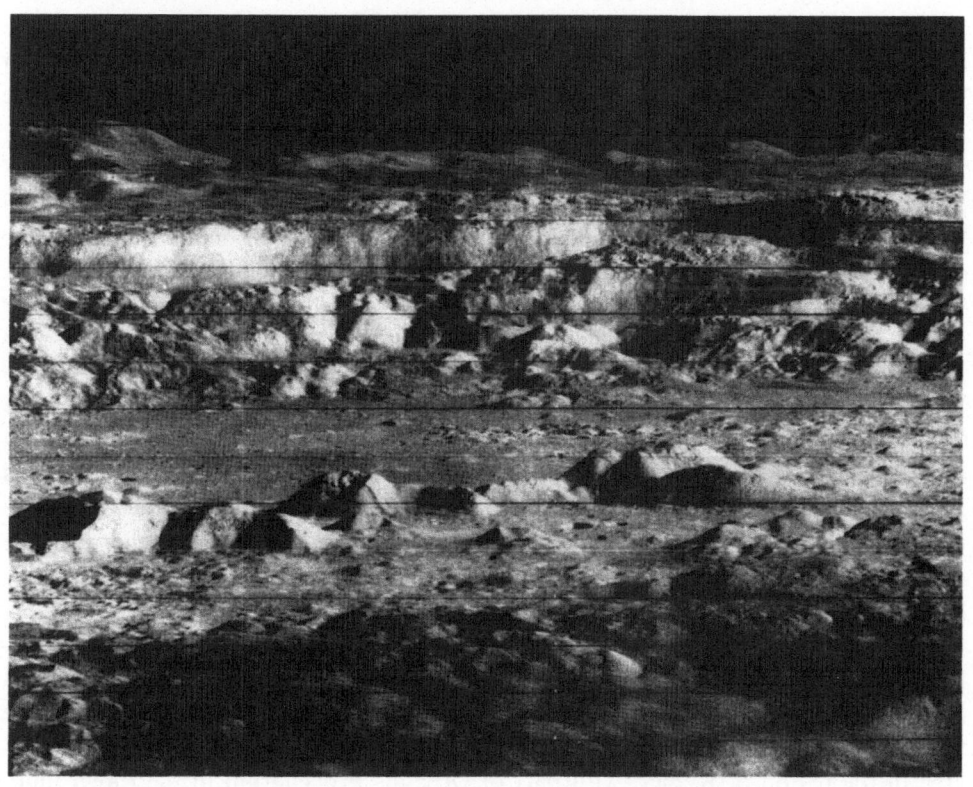

Fig. 15.

moderate-resolution oblique photograph of the region of the Marius Hills is shown in Figure 16. The crater Marius is seen near the limb slightly to the right of the center. A portion of Oceanus Procellarium near the crater Reiner Gamma is shown in Figure 17. Light-colored ray material can be seen just above center.

The coverage of the lunar far side was extended during the Orbiter-II mission. Oblique photographs were taken looking both north, as seen in Figure 18, and south. Figure 19 is a moderate-resolution vertical photograph of a region of the far side near the eastern limb and centered approximately on the equator. The mare region in the upper left-hand corner is Mare Smythii, which can be observed from the Earth.

The orbit of the Lunar Orbiter-III mission differed from that of the two earlier missions by a change in orbital inclination from 12° to 21°. This change permitted high-resolution photography of additional regions of scientific interest. Several examples of the vertical and oblique photography are shown in Figures 20–31. A moderate-resolution photograph of a prospective Apollo site in southeastern Mare Tranquillitatis is shown in Figure 20. The area covered is approximately 46 by 39 km. A 42 by 50 km section of the floor of the crater Hevelius on the western edge of Oceanus Procellarum is shown in Figure 21. Several interesting geological formations are shown in fine detail in Figure 22, which covers an area of approximately 23 km² in Oceanus Procellarum. An area of approximately 1 km² near Mare Tranquillitatis is shown in Figure 23. Numerous rocks can be seen on the sloping crater wall. The unusual appearing crater seen in Figure 24 is located in Oceanus Procellarum. The double-walled appearance of the 155-m diameter crater is believed due to a continuous landslide around its circumference.

A familiar feature on the lunar surface is shown in Figure 25. This is a moderate-resolution oblique photograph of the Hyginus Rille with the 11-km diameter Hyginus crater at the center. An area known as the Schneckenberg Uplands appears in the upper left corner of the picture. Figure 26 is a moderate-resolution photograph of an area in the region of Kepler, with part of Oceanus Procellarum in the foreground. Kepler is the large crater near the center and Kepler A is seen to the right. The 100-km diameter crater Theophilus is seen in Figure 27. A moderate-resolution photograph of the region near the crater Damoiseau is shown in Figure 28. The general area is in southwestern Oceanus Procellarum. Of considerable interest is the sharply defined contact between the mare floor and the upland areas. Damoiseau, seen to the left of center, is 36 km in diameter and has an interesting concentric ring structure. Figure 29 shows an area near the crater Galilei, the largest crater seen near the limb, in the northern portion of Oceanus Procellarum. The Cavalerius Hills are shown in the foreground. Located in the area included in this photograph is the landing site of the U.S.S.R. Luna-IX soft-landing spacecraft.

Only one frame of the lunar far side was taken during the Lunar Orbiter-III mission. It is shown in Figure 30 and includes a large portion of the southern portion of the far side. The dark floored crater Tsiolkovsky shows prominently in this photo-

Fig. 16.

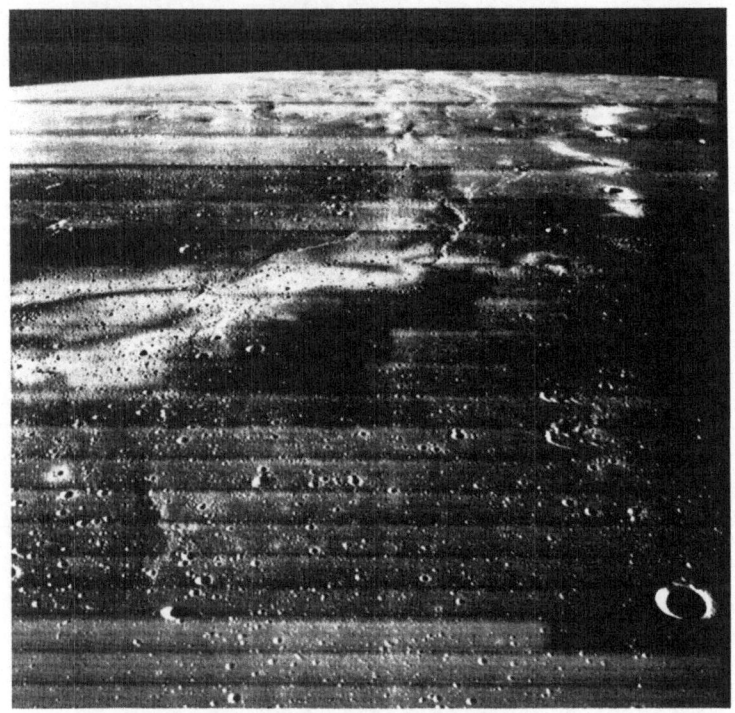

Fig. 17.

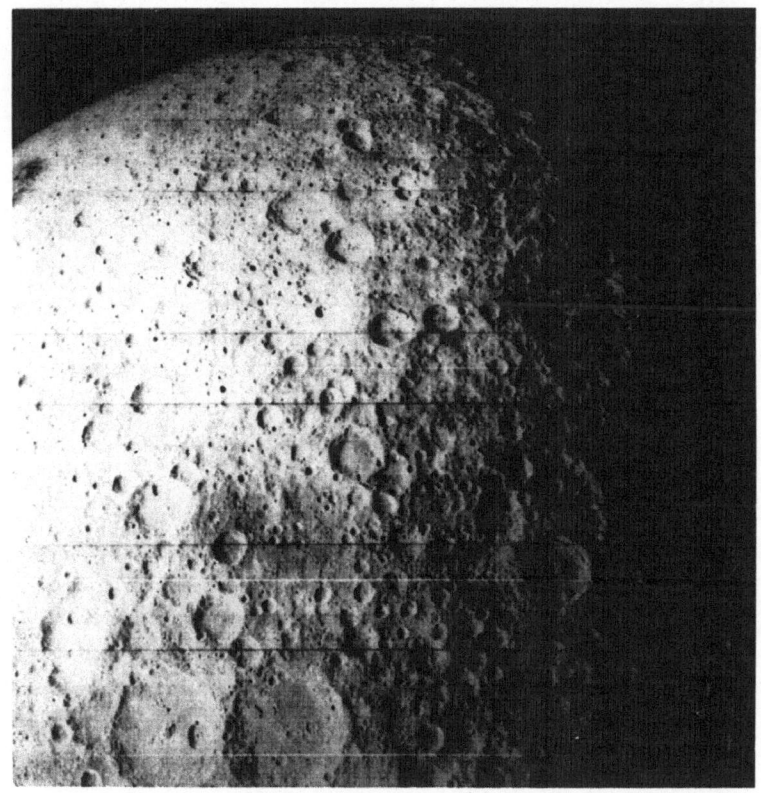

FIG. 18.

FIG. 19.

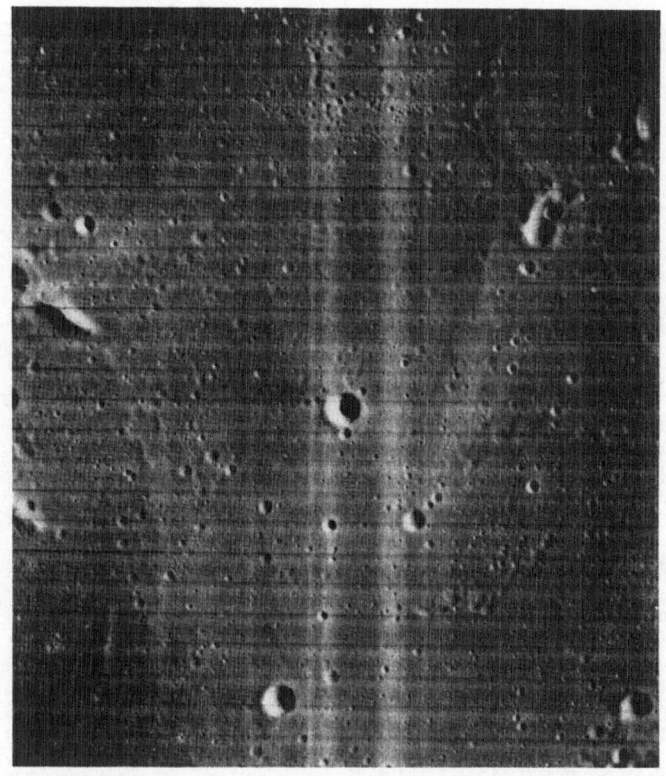

FIG. 20.

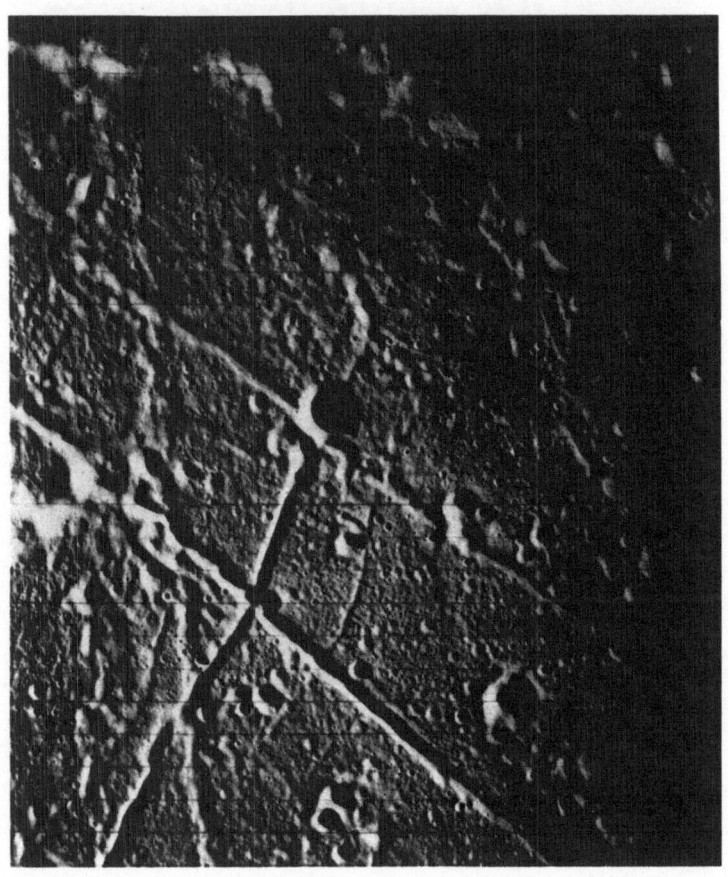

Fig. 21.

FIG. 22.

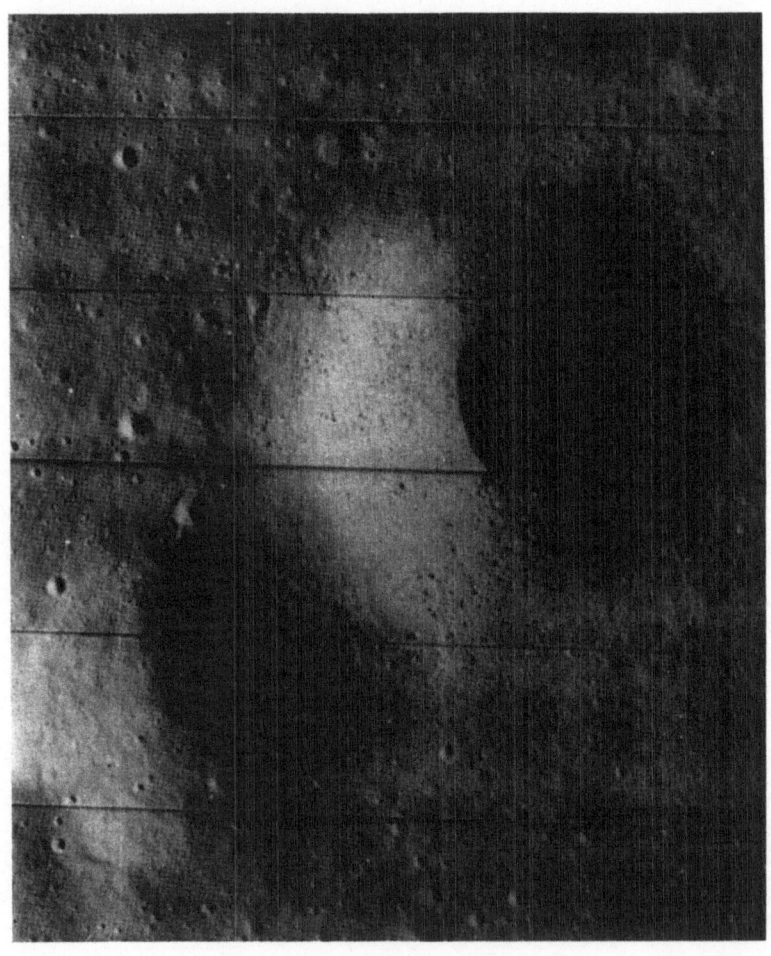

FIG. 23.

FIG. 24.

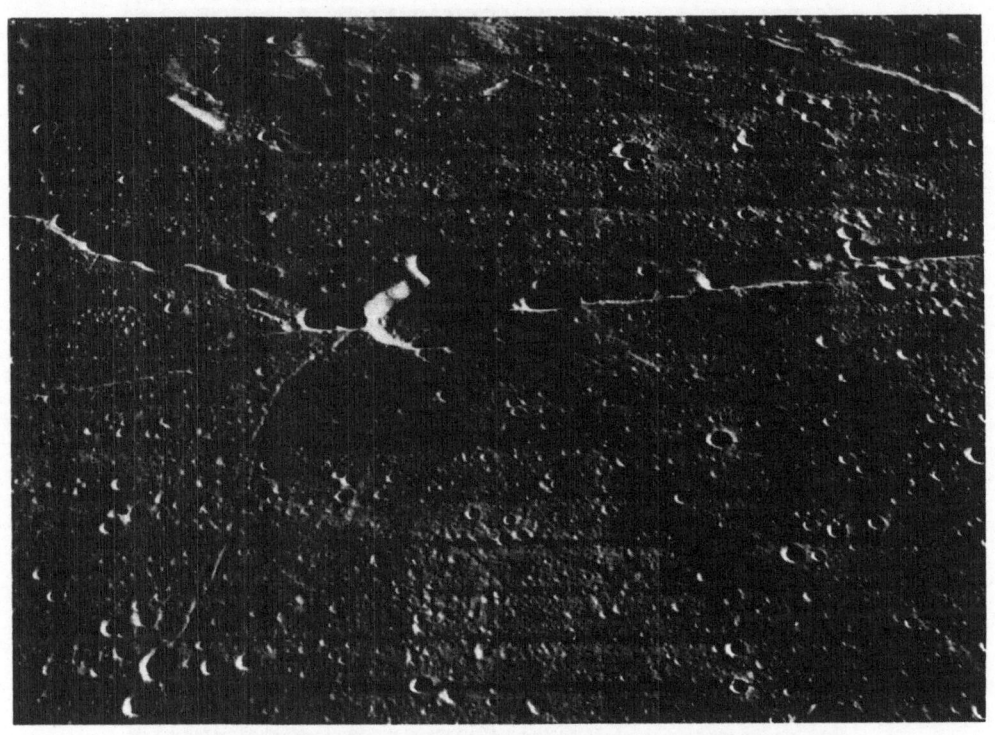

Fig. 25.

Fig. 26.

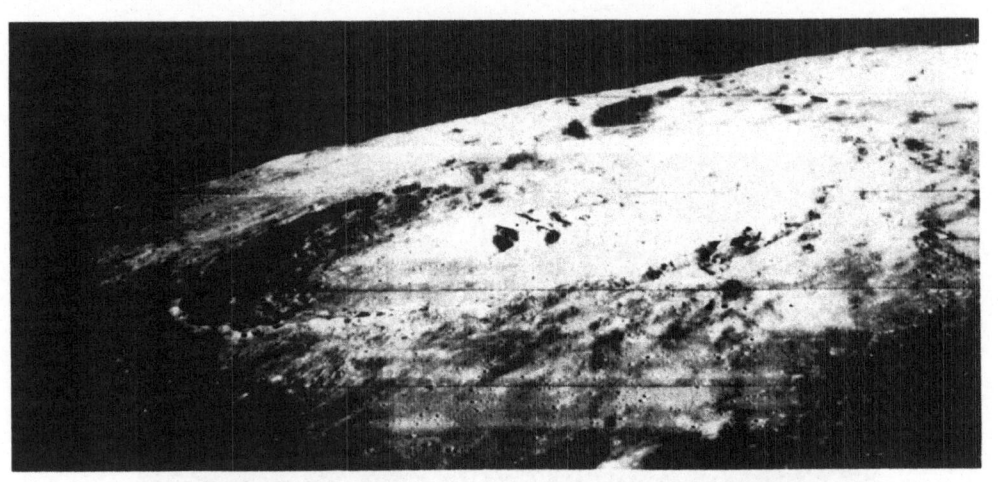

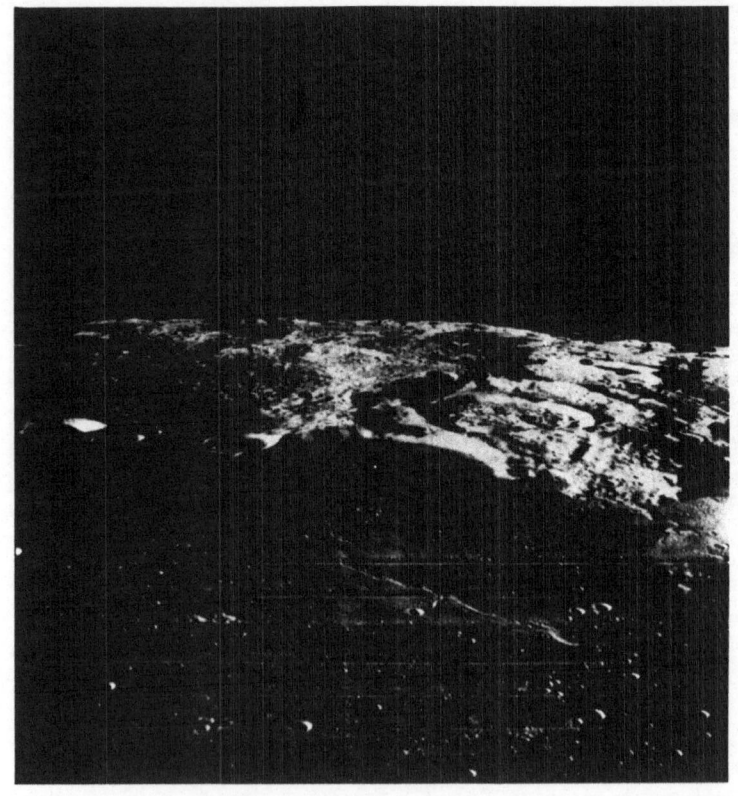

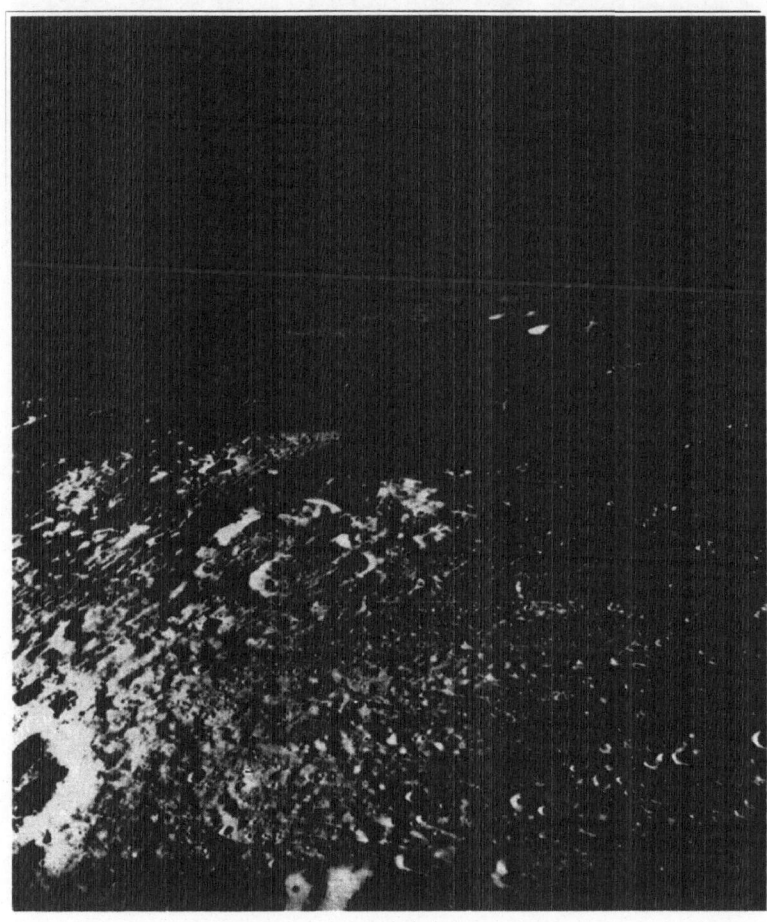

FIG. 29.

FIG. 30.

graph. A section of the high-resolution photograph of this region is shown in Figure 31. Details of Tsiolkovsky and the surrounding region are clearly seen.

The front-side coverage of Lunar Orbiters-I – III is diagrammed in Figure 32. The closed rectangles represent vertical photographs, while the open-ended outlines represent oblique photographs.

The prime mission of Lunar Orbiter-IV was to obtain maximum photographic coverage of the lunar front side at a resolution of 100 km or better. To accomplish this, the orbital inclination was changed to 85° and the perilune and apolune altitudes were changed to 2700 and 6100 km, respectively. The high-resolution coverage of the front side that was obtained is shown in Figure 33. The line drawing on the right should be considered an overlay of the lunar reference mosaic on the left. Each outlined area represents a high-resolution photograph. The gaps in the coverage near the Eastern limb were to be filled in with Lunar Orbiter-V photography. Each of the high-resolution photographs was a vertical photograph, that is, the center of each region was directly under the spacecraft at the time the exposure was made.

Moderate-resolution photographs from Lunar Orbiter-IV are of great interest as

FIG. 31.

FIG. 32.

FIG. 33.

Fig. 34.

they show the lunar disk as seen from different directions. Several of these will be shown in Figures 34–38 and, in each case, the coordinates of the center of the observed lunar disk will be given. The southeastern portion of the far side of the Moon is shown in Figure 34, coordinates 93 °E, 72 °S. The long scars are newly discovered features located near the south polar region. The northern portion of the eastern limb, coordinates 98 °E, 42 °N, is seen in Figure 35. Mare Smythii is visible near the bottom of the picture. In the next two figures, the Moon is observed from a longitude near zero. In Figure 36, the northern region of the Moon is seen, coordinates 13 °W, 72 °N, with Mare Imbrium at the bottom of the picture. A view from a southern latitude, coordinates 6 °W, 72 °S, is seen in Figure 37 with Mare Nubium at the top of the picture. This series is concluded with one view of the western limb. Figure 38, coordinates 89 °W, 14 °S, shows a spectacular view of Mare Orientale. This outstanding feature is partially observable on Earth-based photographs as seen in Figure 39. It is not difficult to recognize the corresponding features in the two photographs. Much of the area surrounding Mare Orientale shows surface structure not observed over other large areas of the lunar surface. Two examples of this unusual surface structure are shown in Figures 40 and 41. The appearance is very similar to that of a mud flow.

The next group of pictures show sections of some of the high-resolution photographs containing well-known lunar features. The Alpine Valley and a portion of the lunar Alps are shown in Figure 42. A sinuous rille is clearly seen running the length of the Alpine Valley. The crater Eratosthenes and the mare region surrounding it are seen in Figure 43. The line of white dots in the lower right is a flaw produced during the processing of the film in the spacecraft. The crater Archimedes and surroundings are shown in Figure 44, while Figure 45 shows the crater Tycho and the neighboring highland region. The crater Aristarchus, of great interest because of reported gaseous emissions emanating from it, is seen in Figure 46.

Lunar Orbiter-IV photography included a limited coverage of the lunar far side. Added to the earlier coverage were details of the eastern limb and the polar regions. The combined coverage of the far side by the first four orbiters is shown in Figure 47. Approximately 70% of the hemisphere was covered at a resolution greater than would be obtainable from Earth if this side were ever turned in that direction. The Lunar Orbiter-V spacecraft obtained sufficient additional far-side coverage to fill in most of the blank regions. There is presently greater than 95% coverage of the lunar far side but at varying resolution.

Based on the Lunar Orbiter photography, the Aeronautical Chart and Information Center has prepared a chart of the lunar far-side at the request of NASA. The chart consists of a Mercator projection for the latitude range from 48 °N to 48 °S and Gnomonic projections of the polar regions down to a latitude of 48°. The chart was printed in two sizes, 1:5 000 000 and 1:10 000 000 which are identical except for scale. A photograph of the 1:10 000 000 scale chart is shown in Figure 48. Details in the region from 105° to approximately 130 °W longitude were based on U.S.S.R. Zond-III

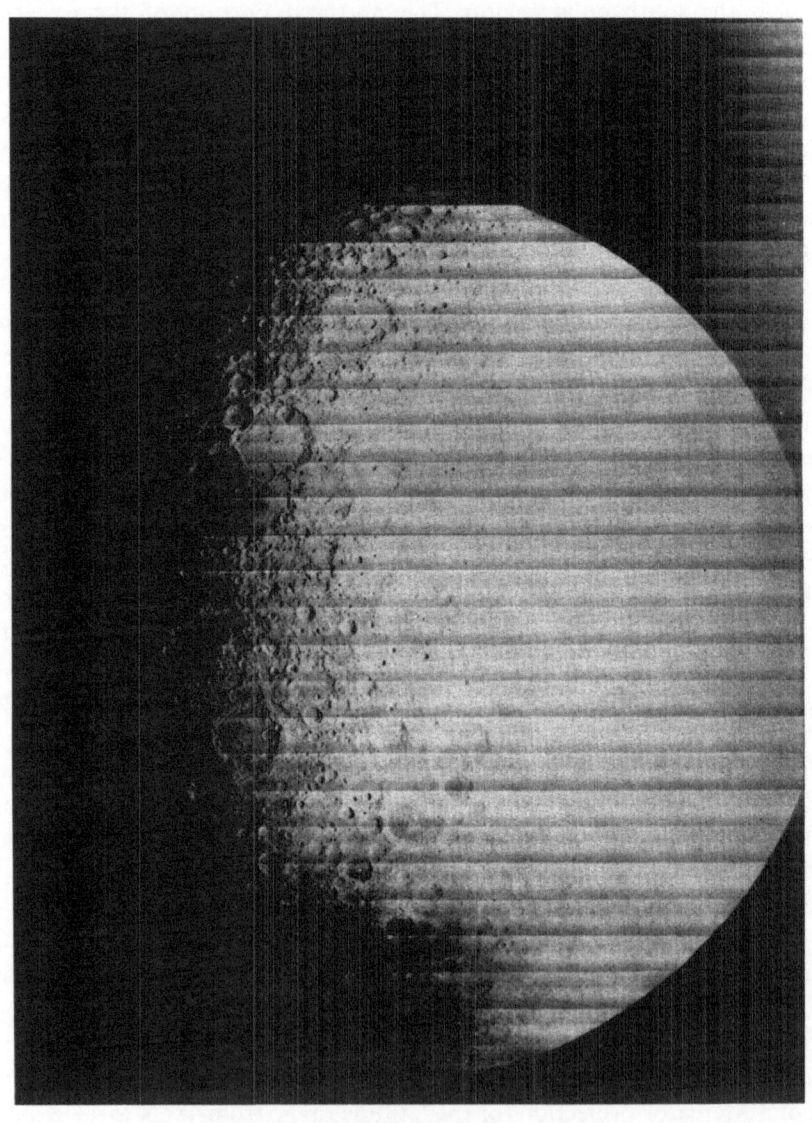

Fig. 35.

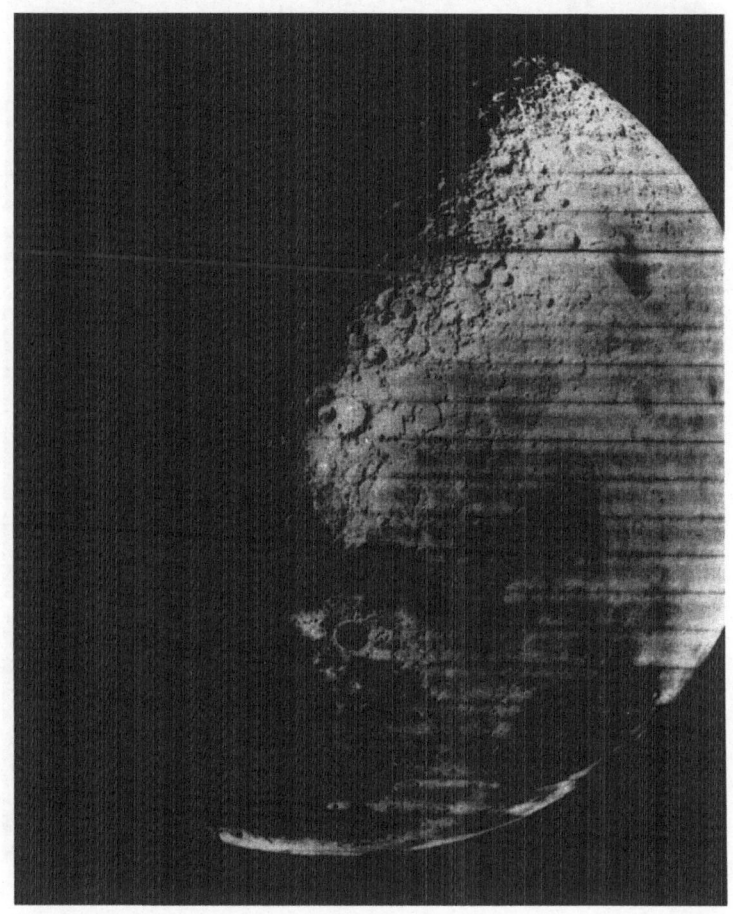

FIG. 36.

FIG. 37.

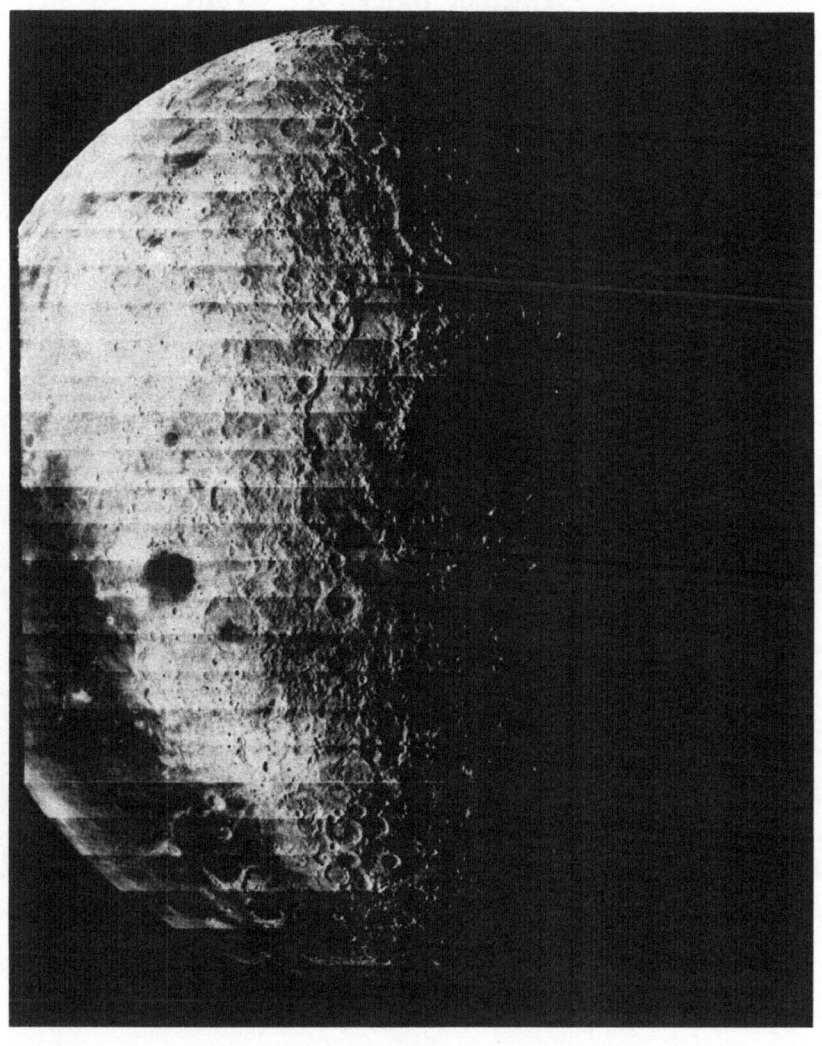

Fig. 38.

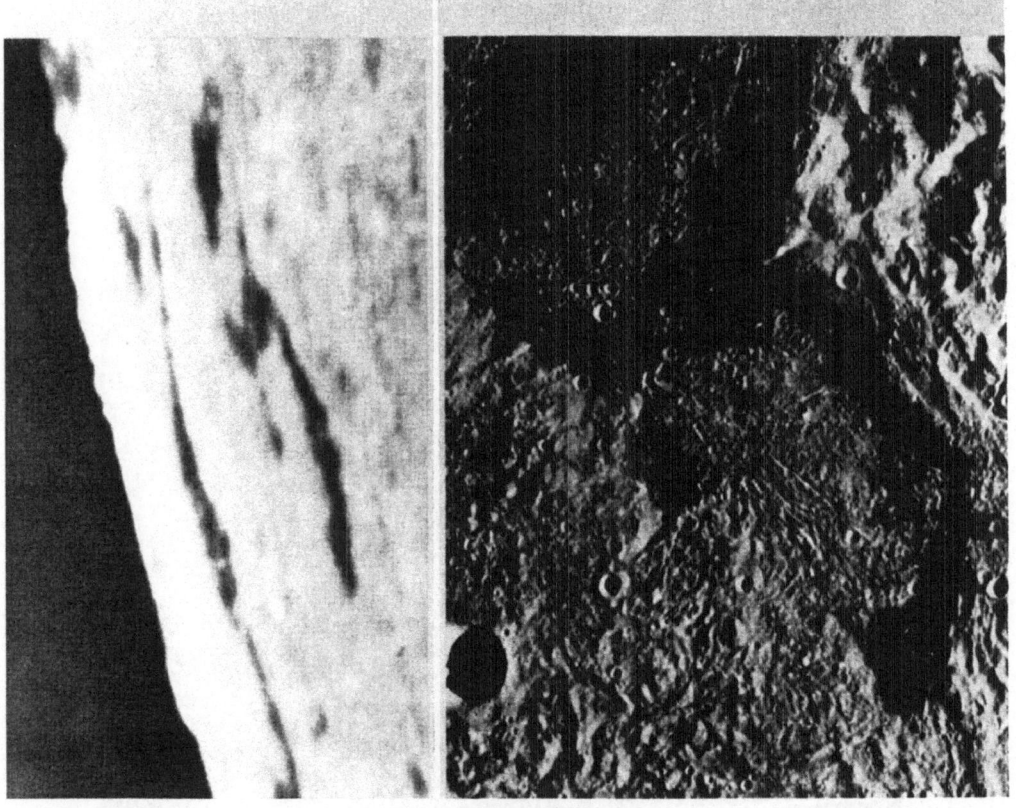

FIG. 39.

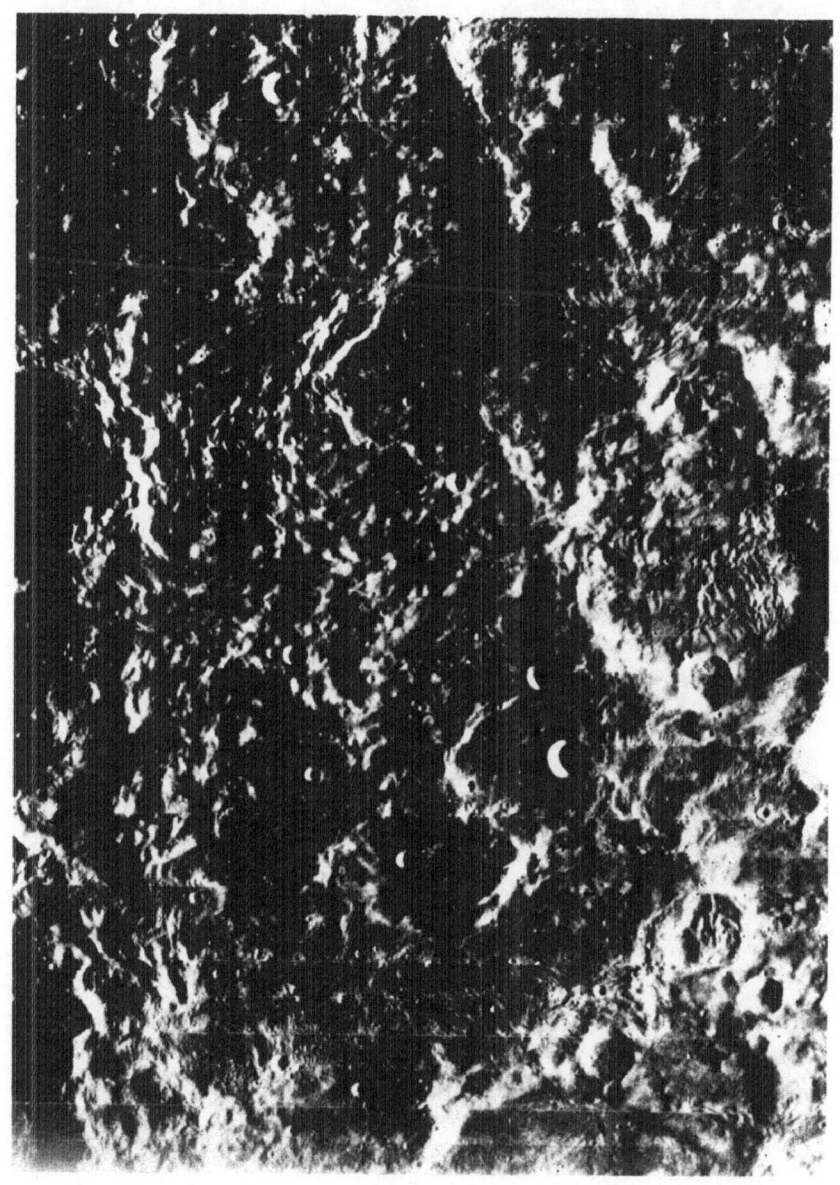

FIG. 40.

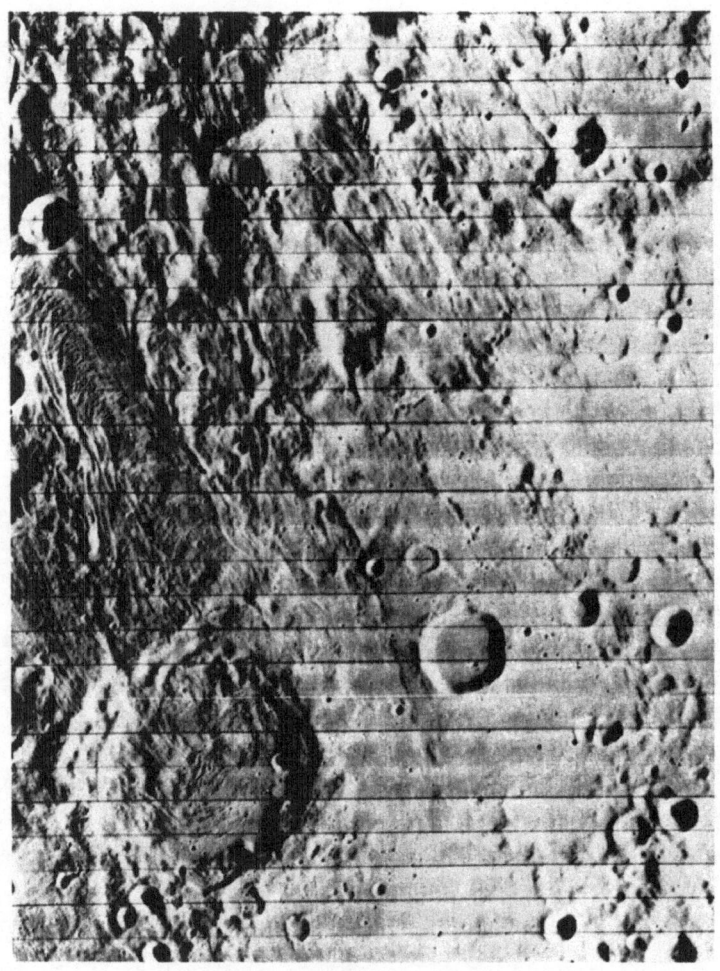

Fig. 41.

Fig. 42.

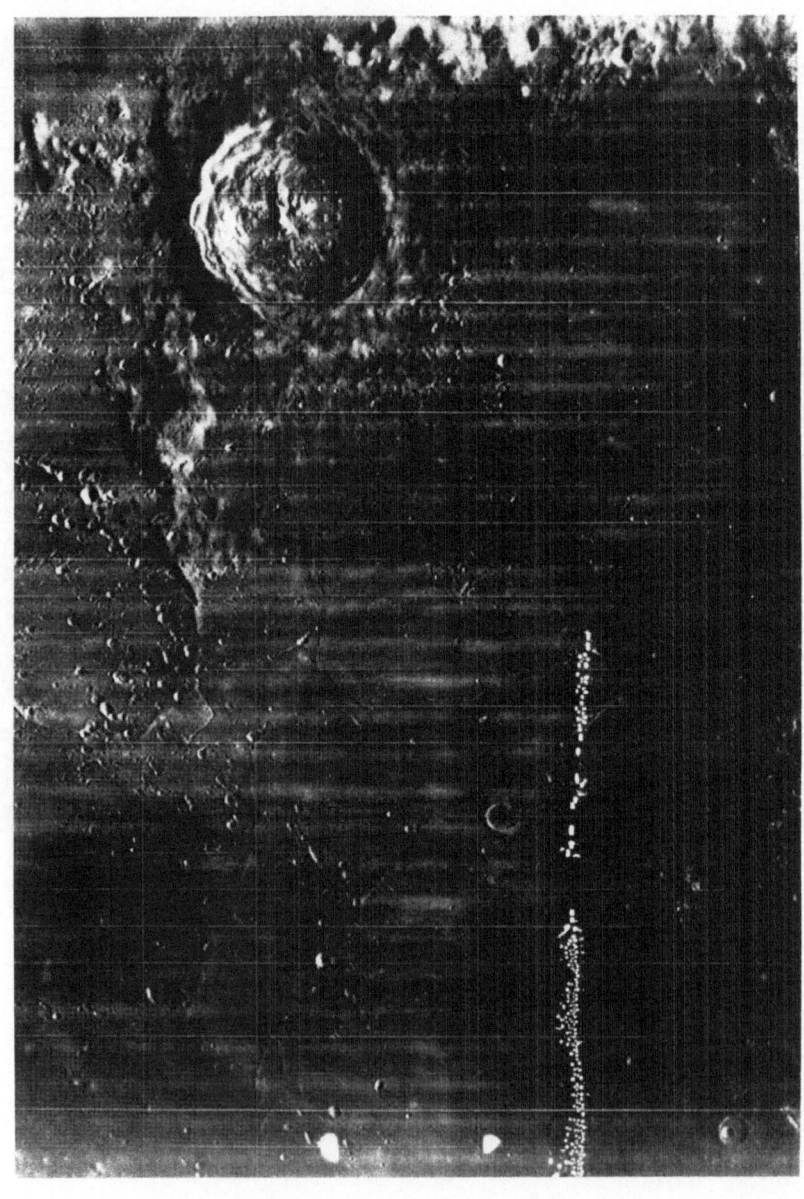

FIG. 43.

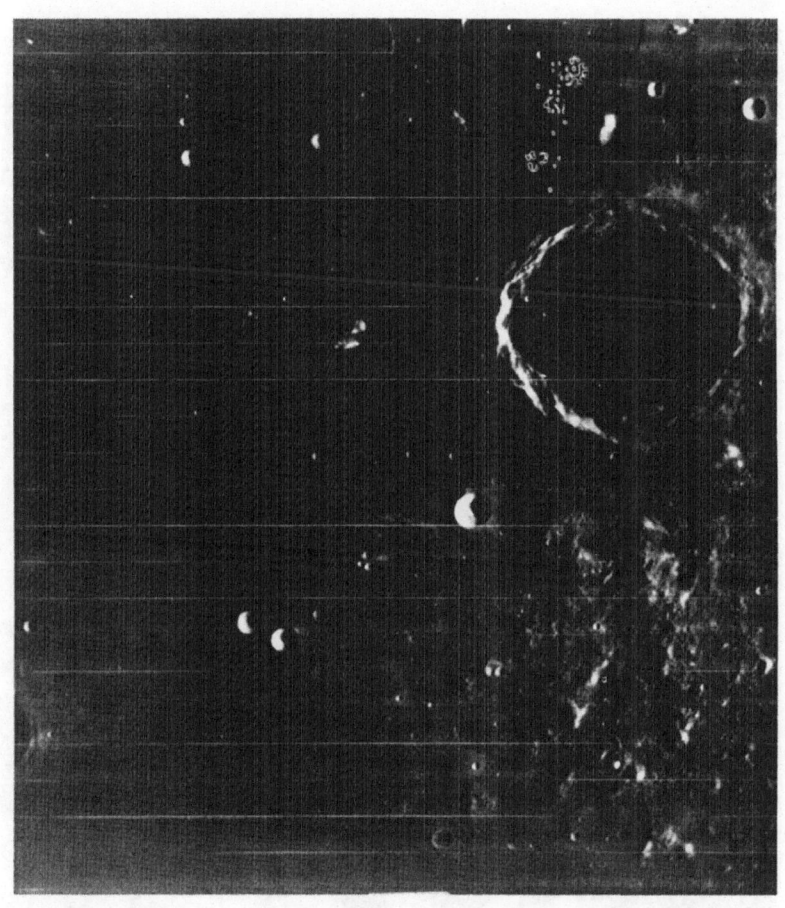

Fig. 44.

FIG. 45.

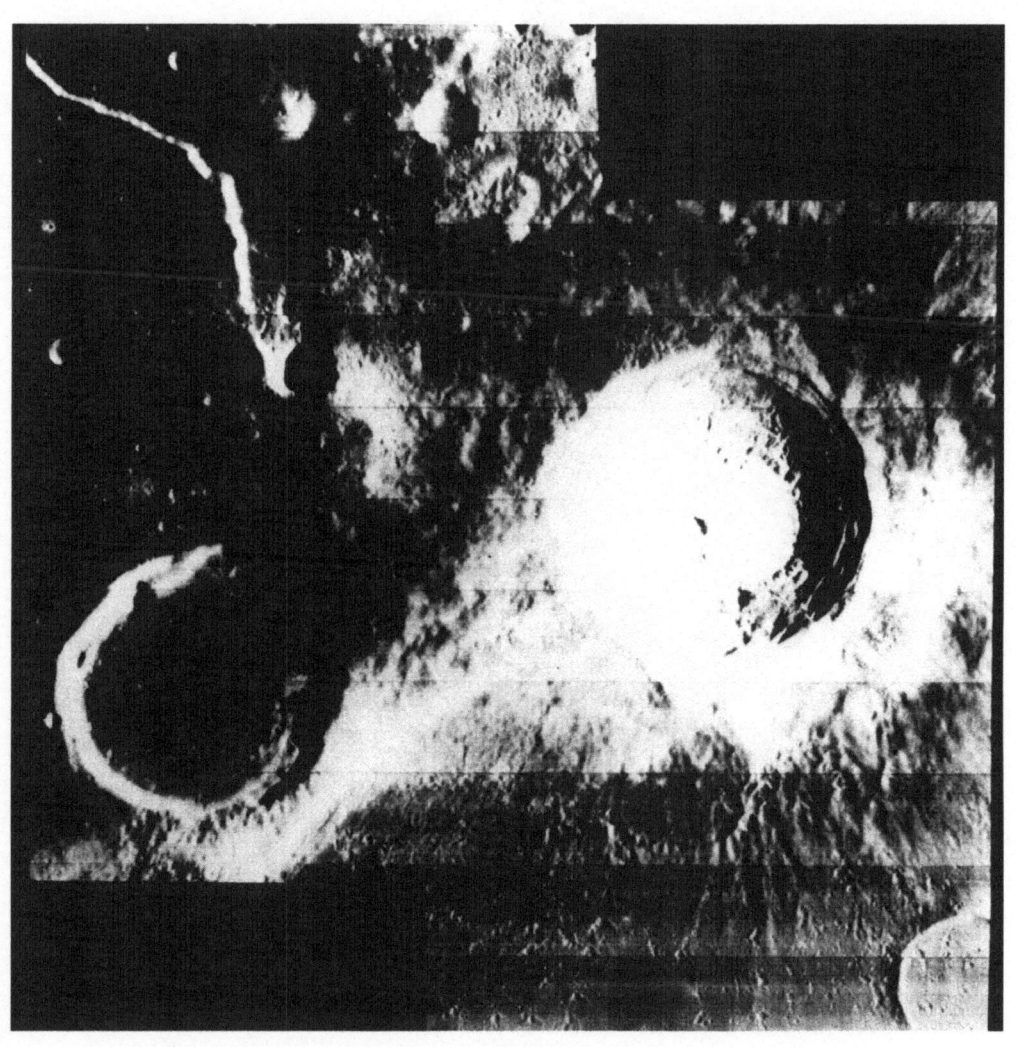

FIG. 46.

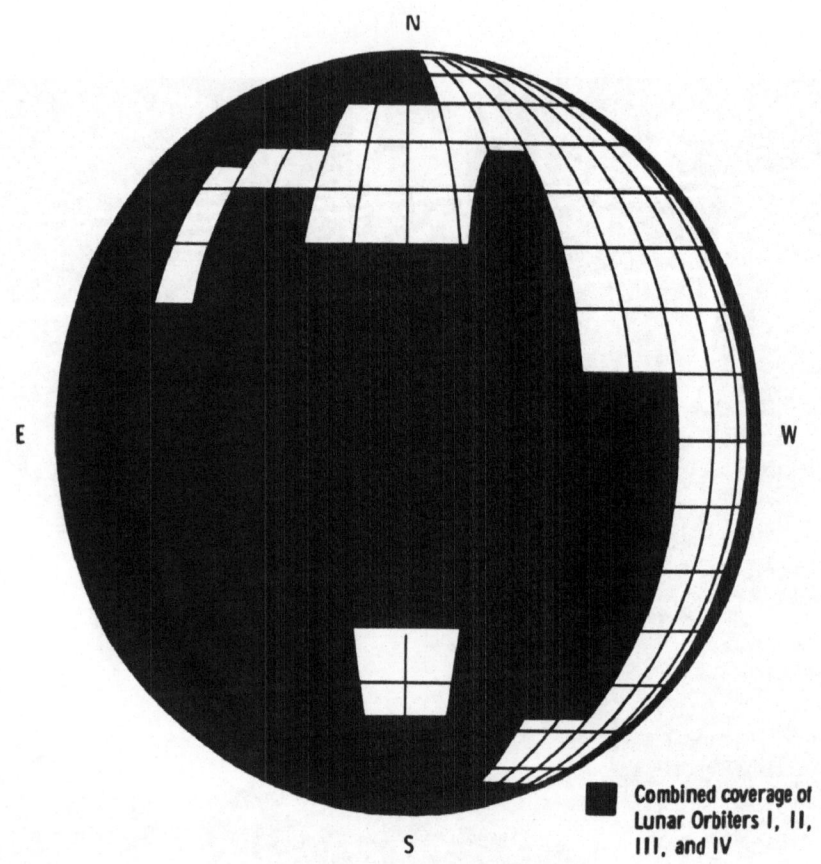

Combined coverage of
Lunar Orbiters I, II,
III, and IV

FIG. 47.

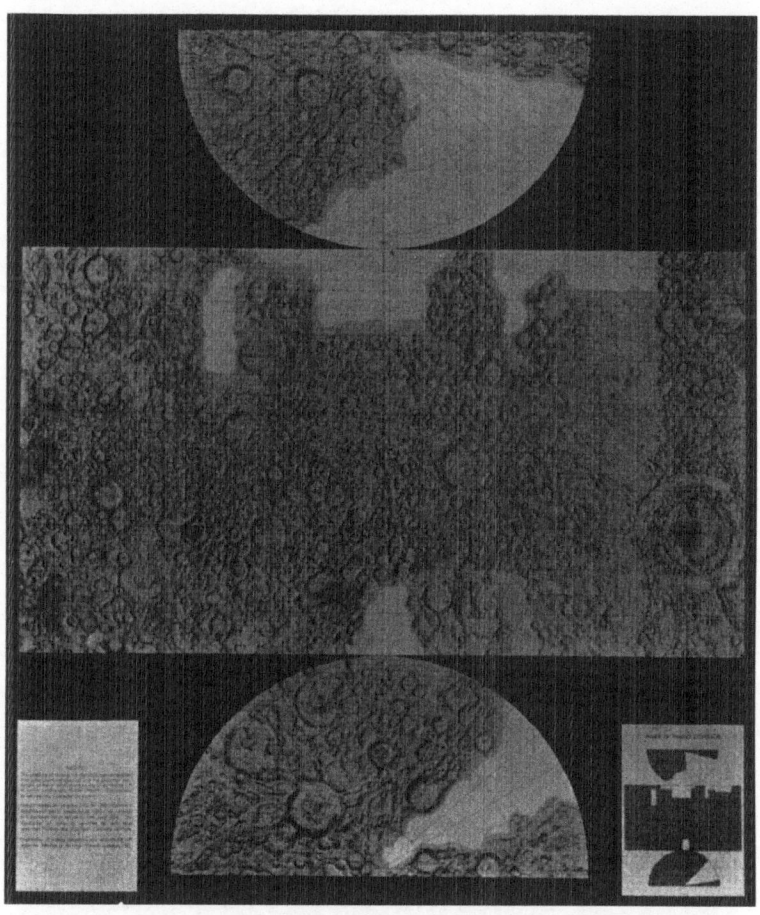

FIG. 48.

Fig. 49.

FIG. 50.

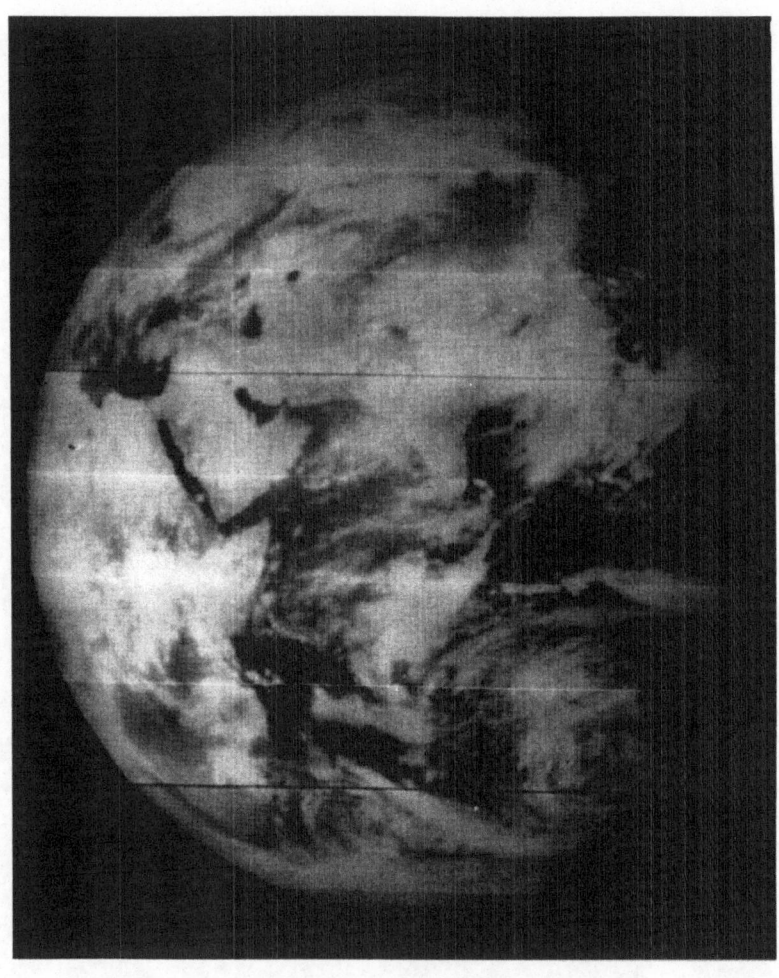

FIG. 51.

photography. The absence of detail in this region was due to the lower resolution available from the Zond-III results.

Lunar Orbiter-V completed its photographic mission by the end of August 1967. One of its missions was to complete the photographic coverage of the far side. One of the first moderate-resolution frames of the far side from that mission is shown in Figure 49. Figure 50 shows $\frac{1}{3}$ of the high-resolution photograph taken at the same time. The last figure, Figure 51, does not show the Moon at all. It is a section of a high-resolution photograph of the Earth taken by Lunar Orbiter-V on August 8, 1967. Clearly seen between the scattered clouds are the outlines of the Mediterranean and Red Sea, the east coast of Africa and the Indian Peninsula.

The photographs presented in this paper have been limited to a small selection of the total number that have been obtained. There is sufficient photographic material available from the Lunar Orbiter program to provide lunar specialists with research material for a long time into the future.

COMMISSIONS 10 AND 44
B. COORDINATION OF SOLAR OBSERVATIONS MADE AT GROUND-BASED OBSERVATORIES AND WITH SPACE VEHICLES

INTRODUCTION

The newly established Inter-Union Commission on Solar-Terrestrial Physics (IUCSTP) has inquired of selected experts about actions useful in the period of the next solar maximum, which need an organization and coordination on international basis. These inquiries have shown that one of the problems, in which a broad international cooperation might lead to substantial improvement of scientific results, is the coordination of ground-base and space-vehicle observations of the Sun and its active phenomena.

Therefore, in order to help the IUCSTP in its preparatory work, the chairmen of IAU Commissions 10 (Solar Activity) and 44 (Observations from outside the Atmosphere) agreed to hold a joint meeting of these two Commissions during the IAU Assembly in Prague, on 'Coordination of solar observations made at ground-base Observatories and with space vehicles'. It was anticipated that other IAU members interested in this problem, particularly members of Commissions 12 and 40, would also take part in the discussion.

The joint session held on August 24, was organized and chaired by C. de Jager, of the Space Research Laboratory in Utrecht. At this meeting, the following papers were presented:

1. K. O. Kiepenheuer: The Need of Ground-Correlated High-Resolution Space Observations in the Visible.
2. A. B. Severny: Space Observations Needed for Improving Our Knowledge of Solar Flares.
3. Y. Öhman: Space Observations of Flare Sprays and Related Surge Phenomena.
4. E. M. Reeves: The Needs and Requirements from the Standpoint of the Ultraviolet Solar Observations.
5. J. R. Winckler: Hard Solar X-Rays in the deka-keV Range.
6. C. de Jager: Hard Solar X-ray Bursts.
7. G. G. Fazio: Solar Gamma Rays and their Correlation with Space and Ground-Based Observations.
8. C. de Jager: The Coordination of Ground-Based and Space Observations: Summary.

Seven of these papers, which are of general interest and have not been published elsewhere, have been selected for publication in this volume. The contribution by Winckler appears in *Proceedings of the IAU Symposium*, **35**.

<table>
<tr><td>L. GOLDBERG,</td><td>Z. ŠVESTKA,</td></tr>
<tr><td>*President, IAU Commission 44*</td><td>*President, IAU Commission 10*</td></tr>
</table>

THE NEED OF GROUND-CORRELATED
HIGH-RESOLUTION SPACE OBSERVATIONS
IN THE VISIBLE*

K. O. KIEPENHEUER

(Fraunhofer Institute, Freiburg i. Br., Germany)

Solar Research during the last decade has definitely shown that almost all primary processes which build up the chromosphere and the inner corona, as well as the observable basic events leading to the optical manifestation of solar activity occur practically all in the subtelescopic range, as far as ground-based observation is concerned. I think here mainly of the great variety of phenomena occurring in the intergranular space (width $<0''.3$) and intersupergranular space (width $<1''$). Because of this situation we have only very few reliable information about the structural transition between photosphere and chromosphere, how the granulation goes over into the complex chromospheric structures as bright and dark mottles, spicules, threads, loops, or fibrils. In this region theory can still get along without being disturbed too much by observation. The situation is somewhat like looking into a human face with a resolving power of 1 or 2 cm!

For the magnetic field – nobody doubts its crucial importance – the situation is still much worse. Every scientist who has *not* been brought up – like most of us here – with the implied agreement to get along with our atmosphere without complaint, would first try to create the conditions which would enable him to resolve $0''.1$ or $0''.2$ and then go on with solar research. For the ground-based solar observer $1''$ or $0''.5$ are already something extraordinary (at least for monochromatic images or high-resolution spectrograms). There is no hope for him to reach $0''.2$ or $0''.1$. And it is astonishing to see with how much patience this sad fact is being accepted.

One way out is balloon astronomy, another one observing in space outside the atmosphere. It must be made quite clear here that the gain of angular resolution by observing from the stratosphere or from space is by no means less important than the extension of the spectrum in the EUV- or X-region. Both will bring us an unpredictable wealth of information.

The balloon technique – first brought to success by Schwarzschild and his group – has the advantage that the instrument can be flown several times; instrumental improvements can be brought in this way. However, balloons with such heavy payloads

---

* Mitteilungen aus dem Fraunhofer Institut Nr. 77.

Perek (ed.), Highlights of Astronomy, 527–529. © *I.A.U.*

cannot be launched at short notice. Waiting times of the order of weeks might arrive because of weather situations at the launching site or along the balloon's trajectory. The observing time from a balloon can be 8 hours. Angular resolutions better than 0″.3 or 0″.2 in the presence of sunlight might meet with serious difficulties because of the remaining air around the instrument.

In contrast to the balloon projects, the manned ATM project – and only about this I will speak here – will offer an observing period of ~ 50 days. This implies a significant probability that

(1) special events like larger flares, formation of sunspots etc., will occur during the mission;

(2) the astronauts have enough time to get adapted to their TV monitor and to a fruitful communication with ground-based astronomers; and

(3) the necessary adjustments or changes of program or necessary repairs can be made during flight.

Without ignoring the great value of the existing high-resolution projects on the ground or in the stratosphere – they will always be indispensable and form the intermediate steps to obtaining better optical high-resolution results in space – I would like to plead here strongly to use even the ATM facilities for getting the highest possible angular resolution in the *visible* part of the spectrum. By combining a prime mirror of 100 cm aperture with a high-resolution spectrograph and a Lyot filter (for better locating the position of the spectrographic slit on the Sun) an angular resolution of 0″.1 can be reached, which will be equivalent to an increase of flux of information (per unit of the Sun's surface) by a factor 100 as compared to the best conditions on ground. The guiding of such an instrument will certainly not be an easy problem. However, the experiences which we have collected with our balloon-borne spectro-stratoscope, which is locked in solar granules, look quite promising.

A film size of 60×60 mm will correspond for this high resolution to a field of $1' \times 1'$ on the Sun. In other words, this instrument will be a kind of 'space microscope' in the hands of the cooperating ground-based astronomer, handled through his space fellow astronaut.

The amount of pairs of spectrograms and pictures to be obtained during a 50-days mission could be of the order of 400 000 (24 000 m of film) and could be supplied after return to the ground to a number of observatories for examination and evaluation.

The possibility of investigating flares with such a 'space microscope' needs special mentioning. Flare theories have become an exciting battle field of advanced theoreticians, while unfortunately, flare observation is still in its very infancy. Most of the observers are convinced that the initial and basic process of a flare, as well as its close environment has not yet been observed with adequate angular resolution; we are at least off by one order of magnitude. The probability to succeed with a balloon-borne instrument during an 8 hours' flight is very small. The coincidence of a flare with excellent seeing on the ground is still more improbable. During the 50-day ATM observing period it should be possible (even with the small field available) by continuous contact between the operating astronaut and a competent group on the ground, to follow a flare from its very beginning (structure, spectrum and magnetic field) down to dimensions of $\lesssim 100$ km. Although the magnetic-field pattern probably cannot be resolved down to $0''.1$, the fine structure (as seen in certain lines or in integrated light) together with somewhat less resolved magnetograms will give us an idea of the magnetic fine structure.

There is no need to mention that apart from getting the highest possible angular resolution (which again for many reasons should be first practised in the visible) additional information in the EUV will be of greatest importance, especially if correlated to better than $1''$ to the visible features.

I think that the community of solar physicists all over the world would appreciate enormously, if NASA could make available such a 'space microscope' to supplement ground and balloon efforts.

SPACE OBSERVATIONS NEEDED FOR IMPROVING
OUR KNOWLEDGE OF SOLAR FLARES

A. B. SEVERNY

(Crimean Astrophysical Observatory, U.S.S.R.)

1. Up to the present time the technical possibilities of space research were not quite adequate for obtaining the ultraviolet spectrum of flares that occupy only a small portion (less than 10″ of arc) of the solar disk. Up to now all observations have been made in integrated sunlight. The importance of the knowledge of UV spectrum of flares is hardly necessary to emphasize. For example we just could mention that if ultraviolet spectra were available for a flare say from 850 Å up to 2000 Å, we would be able to estimate such extremely important parameters as the number of hydrogen atoms N_1 in the first quantum state (by using L-α, L-β, etc.), electronic temperature and density (from Lyman continuum) and other physical parameters of a flare which at the present time we try to derive by different indirect and inadequate methods. Several resonance (ultimate) lines are concentrated in this spectral interval and their careful examination in flares can bring additional important information about conditions prevailing in flares and in the underlying chromosphere. The same applies of course to the whole UV spectrum below 800 Å, and in particular to the resonance lines of He I (λ 512 Å) and He II (λ 304 Å), as well as to a number of lines of highly ionized atoms.

The purpose we are discussing should be to put forward before space-research technique a requirement to improve the pointing and guiding of spacecraft to secure the pointing with an accuracy of several seconds of arc during the lifetime of a flare. Probably this requirement cannot be solved without manned space flights for final adjustment and correction of the position of a flare on the slit of an UV spectrograph. Until we are able to get an UV spectrum of any interesting small area of the solar disk with the aid of spacecraft, we are far from any solution of the most important problems relating to the physics of solar activity. The task under consideration (accurate pointing) is not hopeless, especially for such a bright object as our Sun, and we hope that the accuracy of even unmanned pointing of $\sim 1''$ could be considered as feasible in the nearest future.

However, these space observations (probably with small dispersion ~ 3–5 Å/mm as a first step) should be accompanied by simultaneous ground-based observations including (as a minimum program) the spectrograms of flare in the visible spectrum in the course of the whole flare development, as well as Hα and K direct photography of the process.

Perek (ed.), Highlights of Astronomy, 530–532. © *I.A.U.*

2. The information on flares in integrated sunlight which we can obtain by using spacecraft (rockets and satellites) is of course still important, despite the fact that the contribution to this emission of a flare, by area, is no more than 0·1%. We think that it is important to know the time variation of the L-α and λ 304 emissions during important flares, provided we can evaluate adequately the energy excess due to a flare from these data. Also valuable would be the information on the variations of the Lyman continuum during the flare process and other active processes, if these variations could be detected at all. In this connection I would like to mention the recent successful experiment realized by Ing. Bruns of the Crimean Observatory, who found some variations in the profile of Lyman continuum as recorded photoelectrically during periods of enhanced solar activity in July this year. The photoelectric scanning spectrometer was working for ∼20 days during the flight of Sputnik Cosmos-166. The analysis of the telemetric records is now in progress, and we hope to get some important information about the possible changes of the electron temperature of the chromosphere in connection with the varying solar activity.

3. The space observations of X-ray emission indicate that sometimes Hα flares are not accompanied by X-rays, and some X-ray bursts do not seem to have counterparts in the optical region. (These indications have been obtained mainly during quiet periods on the Sun.) In principle the processes leading to X-rays take place in the corona and this is the reason why they may remain unnoticed by Hα patrol of solar flares. But we should not exclude the possibility that such violent and impulsive processes as 'moustaches' or bombs could also be a source of X-rays. Moustaches are extremely transient phenomena, their lifetime varies (in the mean) form several minutes to 20 min. At the same time short-lived bursts of X-rays of similar duration are frequently observed during the period of enhanced activity and, in particular, they have been recorded by Dr. Kreplin during the period preceding the proton flare of July 7, when moustaches appeared. Therefore, we think that the on- and off-band observations of moustaches combined with spacecraft records of X-rays would be desirable, because they could shed some light on the origin of small bursts of X-rays. At the usual on-band solar Hα-patrol these processes (moustaches) are likely to remain unnoticed.

If X-ray emission is partly due to bremsstrahlung we will probably have some continuation of this emission in the visible part of spectrum, and we found several years ago that observed continuous visible spectrum of moustaches could be considered as a continuation of the bremsstrahlung spectrum of electrons. In any case, one of the most important problems of X-ray space investigations of flares would be to derive the course of continuous emission of flare in the X-ray and UV region. We still do not have adequate observations, although there are some indications on the emission of thermal origin at the temperatures $\simeq 10^7 \,°K$. Still there are some very short impulsive bursts of X-rays, the origin of which is difficult to attribute to some thermal source.

4. For our knowledge of how a flare could be initiated, it would be important to have very accurate time marking for both the space and the ground-based observations (in particular Hα-solar patrol). At the present time the keeping of time for solar observations deserves improvement at most of the observatories. The adequate comparison of the time of onset of Hα flare and X-ray bursts could be done only if we have precise time markings. In principle it is not excluded that the process in the X-ray region can precede the process observed in the visible spectrum, especially for flares with a rapid, impulsive growth of Hα intensity. There are some indications that small X-ray bursts precede the optical Hα flare, as pointed out by Mandelštam. If accelerated particles appear during these bursts, we should conclude that the optical flare phenomenon is related to the supply of high-energy protons and other particles, which produce a high degree of ionization and excitation of atoms.

5. The X-ray space observations show that the area responsible for X-ray emission is small, not larger than 0.1% of solar disk ($< 30''$). I do not think we would be surprised if further observations show that this area is as small as the fine structure of an active region, or that the X-ray emission as well as the emission in optical range is concentrated mainly in small grains and filaments with characteristic size $\leqslant 1''$. The mottled structure of the Sun in λ 304 Å and fine structure of L-α ($\sim 2''$–$3''$) are probably first indications to that effect. Moreover, for the understanding of the flare process it would be important to know whether the X-rays emitting area coincides with the area responsible for the visible spectrum. All these problems could probably be solved if we can secure a resolution of the order of $1''$, for direct image X-ray photographs of the Sun. As far as we know such resolution is about to be realized in the space observations in the nearest future.

Without being complete, these are just some comments on possible observations relating to solar flares.

SPACE OBSERVATIONS OF FLARE SPRAYS
AND RELATED SURGE PHENOMENA

YNGVE ÖHMAN
(Stockholm Observatory, Saltsjöbaden, Sweden)

In a publication of the Dunsink Observatory in 1960 (Ellison *et al.*, 1960) the late Dr. Ellison and two of his collaborators presented light curves of 30 solar flares in relation to sudden ionospheric disturbances. Among these flares a solar-limb event of April 22, 1959 was of a particular interest, as it indicated strong X-ray emission from an object situated well outside the limb. This limb flare consisted of a small globule, not particularly bright, which rose 18000 km in 4 min and had a total duration of only 5 min. Yet it produced a fine 'sudden enhancement of atmospherics' which persisted for 66 min after the flare had faded to invisibility.

In my opinion Ellison's fine observation is a good illustration of the fact that our present means for optical recording at ground-based observatories are not always quite satisfactory. A coronagraph with wide band filter may have given a more complete optical recording in this case. And simultaneous X-ray photographs and X-ray spectra would no doubt have been of extremely great interest for a comparison with the optical recording.

In 1960 Kleczek and Křivský (1960) reported some similar observations of sudden enhancement of atmospherics for solar-limb surges. Six such objects not coinciding in time with other Hα events on the Sun gave very clear S.E.A.-effects, interpreted by Kleczek and Křivský as X-ray radiation emitted by the adjacent region of the solar corona.

With the introduction of direct X-ray recording from satellites the indirectly revealed effects could be controlled and studied in detail. In a paper published in 1962 by Kreplin *et al.*, I quote from the abstract the following clear statement: "Measurements made from the SR–1 satellite confirm the hypothesis that solar X-rays are the cause of flare SID-events. Results also indicate that active prominence regions, bright surges on the limb and certain limb flares have the same X-ray characteristics as major disk flares etc."

Among the active prominences listed in this first catalogue of X-ray events was the spectacular object of July 24, 1960. Very good photographs of this prominence were secured by Kerstin Fredga at the Swedish Solar Observatory in Anacapri. Simultaneous photographs were obtained at the German Solar Observatory on the island. These photographs show a twisted structure indicating the presence of magnetic fields.

There have been some contradictory opinions as to the origin of the strong X-ray

Perek (ed.), Highlights of Astronomy, 533–537. © *I.A.U.*

YNGVE ÖHMAN

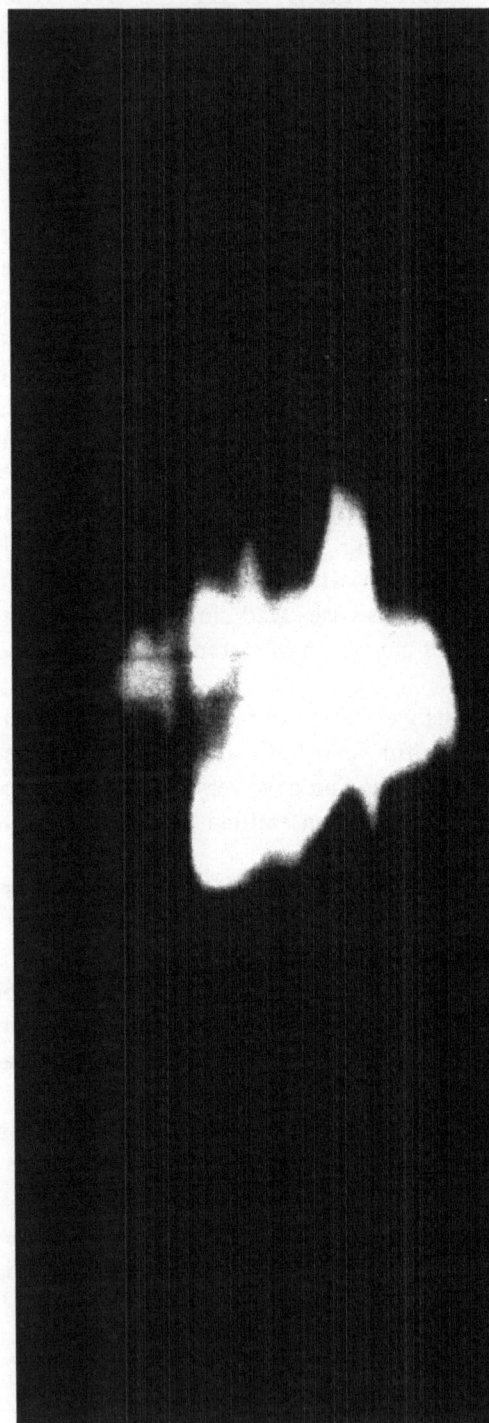

FIG. 1. *Hα spectrum of the flare spray of July 11, 1966 secured at the Swedish Solar Observatory in Anacapri with a Babcock grating.*

emission below 8 Å accompanying this event. In our opinion the visible surface activity on that occasion could not in itself produce the strong increase in the X-ray flux which was observed. Several similar events have in fact been recorded later on.

A very spectacular one appeared on July 11, 1966 (Öhman *et al.*, 1967). This was a typical flare-spray event. During the brilliant phase the intensity of the object, when measured in our Hα filter in Anacapri, was 1·6 times that of the solar surface near the centre of the disk. Few minutes later the ejected clouds formed a more or less circular or spiral configuration. Because of considerable Doppler velocities the image secured with the 0·7 Å Hα filter gave a rather incomplete recording of the spray.

This is evident when comparing these images with our spectrographic recordings made with a Babcock grating (Figure 1). They show not only a very broad Hα line but also displacements corresponding to velocities of up to about 500 km per second. A coronagraphic picture obtained with a 5 Å Hα filter 72 min after the start of the spray still shows a very spectacular prominence, and in spite of the fact that the narrow pass-band filter did not show any clouds at all at this late moment. This shows again the need for a coronagraphic limb-spray patrol, and for improved lineshifters in the solar disk patrol.

The spray of July 11, 1966 appeared suddenly, and perhaps in consequence of a flare situated on the non-visible surface beyond the limb. Sometimes the appearance of such a flare can be inferred from the appearance of bright surges. Very interesting effects of activation have been noted sometimes when such surges penetrate already existing quiescent prominences on the limb. Such an activation was observed in Anacapri on March 25, 1967 (Gimse and Hosinsky, 1967).

Sometimes the sudden activation of a prominence can be followed on the disk not only as a metamorphosis of the filament but as a rapid motion of the whole object as well, and, in fact, in such a way that the filament is transformed to a spectacular ascending prominence far outside the limb. A very fine example of this kind of sudden disappearance was observed by us on June 23, 1967 (Figure 2).

Hα spectra of this object showed great Doppler shifts, but at the same time a fairly narrow line, indicating a slow disintegration process. A very remarkable circularly shaped Hα-line picture resulted from some of the structure elements of this object.

We have found similar circular Hα spectra in quiescent prominences too. Some of them show a 'limb brightening', which partly may be due to self-absorption but which may perhaps indicate a rapid rotatory mass motion of filamentary elements as well and, if so, with the axis of rotation more or less parallel to the slit.

Another type of activation of an already existing prominence is characterized not only by Doppler-shifted lines but also by very broad lines suggesting a rapid disintegration. In our opinion this type of activation is more likely to be accompanied by X-ray emission than the one characterized by narrow spectral lines.

In connection with the tentative suggestion that rotatory mass motions may appear in prominences I want to take this opportunity to mention that somewhat similar

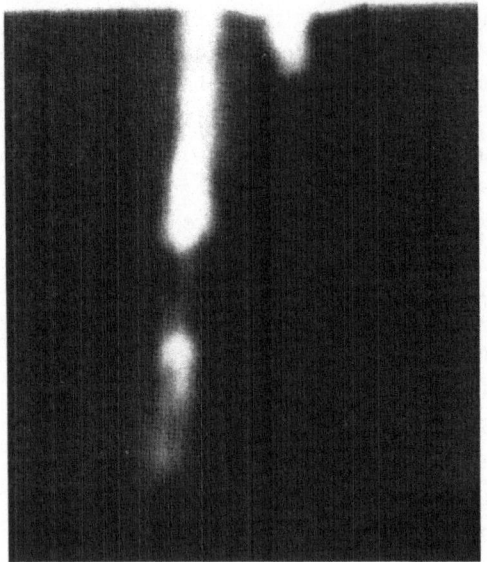 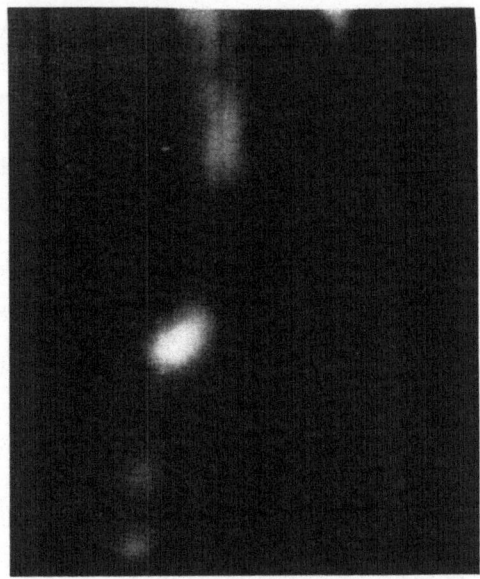

FIG. 2.   *Hα spectra of a sudden disappearance observed in Anacapri on June 23, 1967 at UT 13:16 (left) and UT 13:22 (right). Note the narrow width of the Hα line in spite of the great average violet shift.*

phenomena have been found by us in solar flares too. In this case the rotation of the gases is indicated by a slightly inclined Hα streak. The axis of rotation would in such a case be more or less perpendicular to the slit. In private conversations with Professor Alfvén, I have been informed that a rotation may well be expected according to his and Carlqvist's recent theory of solar flares (1967). Reference should also be given to an interesting paper by Maria Cristina Ballario, who has discussed the possibility that complex Doppler shifts may appear in flares if these have a loop structure and with rapid mass motion (Ballario, 1963).

But can such a rotation of structure elements in say flares, surges, spicules or pro-minences be of any interest to space research and X-ray studies in particular? In my opinion this may well be the case. In a recent paper Kuperus and Tandberg-Hanssen (1967) have examined in detail how the filamentary structures of prominences may be thermally insulated from the hot corona. It seems to me that a rotation is likely to produce a greater leakage of gas. Even if such a phenomenon may produce only a minor thermal increase in X-ray emission and no bursts of the type appearing in connection with explosive events such as surges and sprays we may have an effect worth studying.

The extremely good X-ray pictures secured on November 12, 1966 by Underwood and Mune have shown to us that fairly small details can already be studied on X-ray solar photographs (Underwood and Mune, 1967). It is important therefore that the methods in optical observation are developed so as to take advantage of this rapid development of X-ray solar research. Similar requirements may be made from micro-wave solar astronomy, particularly because of the intimate connection between the burst phenomena observed in X-rays and microwaves.

We have considered here mainly one branch of space research, the X-ray astronomy. Needless to say such research as ultraviolet spectroscopy etc. would certainly be of a great interest in connection with flare sprays too. With the good progress made recently in the study of the solar limb by Wilson and others (Burton *et al.*, 1967) such spectroscopic studies seem well possible.

Even such a difficult task as observing say the Lyman continuum in such objects seems possible. If so, a monochromator developed by us recently by the use of sub-tractive dispersion (Öhman, 1967) may perhaps find applications. With two gratings the spectrum can be made to turn, and at the turning-point a sharp image of say a spray flare can be produced in any selected wavelength range of its continuous spectrum.

References

Alfvén, H., Carlqvist, P. (1967)   *Solar Phys.*, **1**, 220.
Ballario, M.C. (1963)   *Osservatorio Arcetri, Contr.*, **78**.
Burton, W.M., Ridgeley, A., Wilson, R. (1967)   *Mon. Not. R. astr. Soc.*, **135**, 207.
Ellison, M.A., McKenna, Susan, M.P., Reid, J.H. (1960)   *Cape Lyot Heliograph Results No. 1*.
Gimse, O., Hosinsky, G. (1967)   *Solar Phys.*, **2**, 192.
Kleczek, J., Křivský, L. (1960)   *Nature*, **186**, 1035.
Kreplin, R.W., Chubb, T.A., Friedman, H. (1962)   *J. geophys. Res.*, **67**, 2231.
Kuperus, M., Tandberg-Hanssen, E. (1967)   *Solar Phys.*, **2**, 39.
Öhman, Y. (1967)   *Nature, 215*, 606.
Öhman, Y., Stiber, G., Kusoffsky, U. (1967)   *Solar Phys.*, **1**, 60.
Underwood, J.H., Muney, W.S. (1967)   *Solar Phys.*, **1**, 129.

THE NEEDS AND REQUIREMENTS FROM
THE STANDPOINT OF THE ULTRAVIOLET
SOLAR OBSERVATIONS

E. M. REEVES

(Harvard College Observatory)

Extending from the present to the early part of 1969 there are three Orbiting Solar Observatories to be launched, and these will all be capable of constructing spectro-heliograms of the Sun in solar emission lines of the EUV and X-ray region. The recently launched and highly successful OSO-III has obtained EUV and X-ray spectra with high-time resolution, but without spatial resolution on the solar disk. The later OSO satellites will provide spatial resolution of 1' of arc to 30" of arc, and will provide the basis for the extension to even higher spatial resolution in the future.

The comparatively short periods covered by these satellites, coupled with a real probability of only partial success, make it particularly important to obtain the fullest possible use of the data by implementing a complementary and simultaneous series of ground-based observations.

The data from solar satellite experiments are frequently capable of providing information in several fields other than that for which they were initially designed. Thus the correlation of data from such satellites can be useful in the fuller interpretation of the experiments on the interplanetary medium, and the structure of the Earth's atmosphere, as well as for solar experimentation.

At the present time the Interplanetary Monitoring Probe (IMP-F) is successfully recording data on the hydrogen and helium ion concentrations in a highly eccentric orbit, which extends at certain times into the region of the undisturbed solar wind. The OSO-D satellite which we currently expect to be launched in October, will probably be in orbit at the same time with the capability of producing spectroheliograms in H$_I$, He$_I$ and He$_{II}$ lines, as well as other solar emission lines with 1' of arc spatial resolution. The experimenters on the two satellites have agreed to program the observations of the two satellites so that enhancements of H and He ions detected in the solar wind can be correlated with the spatial distribution of enhancements of optical radiation in the same elements in various sectors of the solar disk several days previously.

There seems to be a good probability that two OSO satellites, namely OSO-III and OSO-IV, will also be in orbit together in the next few months, and it will then be possible to compare the interesting whole Sun enhancements in the shorter wavelengths and X-ray regions which have been reported in Commission 44 with spatially

Perek (ed.), Highlights of Astronomy, 538–540. © *I.A.U.*

resolved enhancements in the OSO–IV experiments. Simultaneous comparisons of absolute fluxes as well as time variations can be made. The instrument packages on each satellite generally cover a fairly wide range of wavelengths, and here again inter-comparison of data among experimenters is highly desirable and profitable.

The observation of emissions of the EUV and X-ray fluxes from the Sun can be interpreted to yield information on the structure and variation of the Earth's atmos-phere and hence may be correlated with effects in the terrestrial atmosphere. Both the present and the next OSO have the capability to observe the absorption of solar ultra-violet lines as a function of height in the Earth's atmosphere during sunrise and sunset portions of the orbit. These observations have been referred to by Hinteregger during the present meetings of Commission 44 and will be continued by our own experiment on OSO–IV.

The third area of coordination is that of simultaneous solar observations from the ground and from space, and is of most immediate interest to the present discussion. A general program of ground-based observations such as fairly continuous H I and Ca II spectroheliograms from stations around the world, and at least daily coronagraph observations with good spatial resolution to be used for both flare prediction and data correlation, will all be required for the optimum use of the satellite ultraviolet data as will communication with the daily solar activity and flare warning systems presently in operation. Coordination is now being planned for the OSO–IV satellite, and a number of observatories have already been contacted. There is a particular need to extend the coverage of the ground-based spectroheliograms to include more lines, and especially to increase the number of observing sites to extend the amount of coverage as well as to compensate for seasonal and weather variabilities at existing facilities.

The OSO–IV satellite will be able to construct spectroheliograms in lines of O I–VI, N II–V, Si III and XII, C II and III, He I and II, Ne VIII, Mg X, Si XII and the lines of the hydrogen Lyman series as well as the Lyman continuum.

A number of special programs are being arranged in order to obtain specific ground-based spectroheliograph observations simultaneously with OSO–IV observations. Spectroheliograms in lines of C, N, O, He, H, etc., will be made on the ground at the same time that observations in various stages of ionization of the same element are being observed from the spacecraft. This is being done in order to evaluate the extent to which the observations from solar satellites indicate unique variations or structures and to assess the extent to which the ground-based observations can be used for those extended periods when satellite observations will not be available. We solicit the cooperation of any observatories in this venture and welcome any suggestions from other astronomers on how our satellite observations can be scheduled to support spe-cific researches which will be in progress this autumn and winter.

A manned solar observatory, referred to as ATM (Apollo Telescope Mount), is being prepared at the present time for several launches in the period 1969–71. This manned satellite will use an astronaut observer to program the solar observations to take maximum advantage of the activity on the disk. Coronagraphs, UV spectrographs and spectrometers, photographic and photoelectric spectroheliograms and X-ray and Hα telescopes are all included in the payload. Most of the experiments employ recoverable photographic records although several of the experiments, including our own, are photoelectric. This mission will be limited to 4–8 weeks and the maximum possible cordination with ground-based observations will be required to make optimum use of these observations. Primarily required are extended temporal coverage by coronagraphs, magnetographs and spectroheliographs in the lines I have already mentioned, and the establishment of an effective means of communications between the participating observatories and the satellite experimenters. The spatial resolution from the ATM experiments will be approximately 5″ of arc, which begins to become comparable with the resolution which can be obtained from the ground at much longer wavelengths. The intercomparison can then be made at about the same resolution and will be much more meaningful.

The prediction of regions on the solar disk which are likely to exhibit larger flare activity as well as estimates of the onset time for such events will be necessary if the ATM experiments are to have any hope of catching a moderate to large flare, particularly in the important initial phases of the flare. Magnetograph observations with good accuracy and wider global coverage seem to offer some chance of observing such an event, if adequate coordination and communication can be established.

The relatively short lifetime of the ATM satellite experiments underlines the need for a careful pre-arrangement of the observing program of ground-based observatories to take maximum advantage of the data from the space experiments.

HARD SOLAR X-RAY BURSTS

C. DE JAGER

(University Observatory 'Sonnenborgh' and Space Research Laboratory, Utrecht, The Netherlands)

1. Hard and Soft X-ray Bursts

A tentative spectrum of the Sun in the X-ray region is shown in Figure 1 (De Jager, 1967). The quiet Sun emits a measurable spectrum at photon energies below about 3 keV. During the occurrence of solar flares an enhancement of the emitted energy, and a hardening of the spectrum is clearly visible. The soft X-ray bursts apparently show a maximum radiation flux at wavelengths of about 10 Å.

In the spectral region above about 10 keV hardly any quiet solar radiation is observable. During the occurrence of solar flares hard X-ray bursts are occasionally observed. From the point of view of observational techniques these bursts may be divided in the so-called deka–keV bursts, covering the range 10–200 keV, and the deci–MeV bursts in the range between 0·2 and 1 MeV. Many deka–keV bursts have been observed during the years 1966–67 by Winckler *et al.*, by means of the OGO–I and OGO–III satellites (see Arnoldy *et al.*, 1967). The existence of deci–MeV bursts has been doubted various times (see e.g. Chubb *et al.*, 1966). Its reality seems now to be proved (see also De Jager, 1967). However, they may be much rarer than the deka–keV bursts, although it is not yet completely sure that the selection is not observational.

2. Characteristics of Hard X-ray Bursts

A sharp distinction between the properties of soft, deka–keV and deci–MeV X-ray bursts can certainly not be made. There is rather a continuous transition between them. Roughly the duration of the bursts seems to decrease with increasing photon energies. It also seems that the number of electrons involved is larger for the sources of soft X-bursts than for those of deci–MeV bursts (see Table 1).

As to their interpretation it is generally assumed that the soft X-bursts are of a quasi-thermal nature; this means that the radiation can be described with the normal radiation laws of a plasma not in local thermal equilibrium. It also seems quite probable that the deci–MeV bursts are of a non-thermal origin. The deka–keV bursts may be in an intermediate state. Part of them may be due to quasi-thermal radiation, others perhaps to non-thermal emission of the plasma. In one case a deka–keV burst

Perek (ed.), Highlights of Astronomy, 541–543. © *I.A.U.*

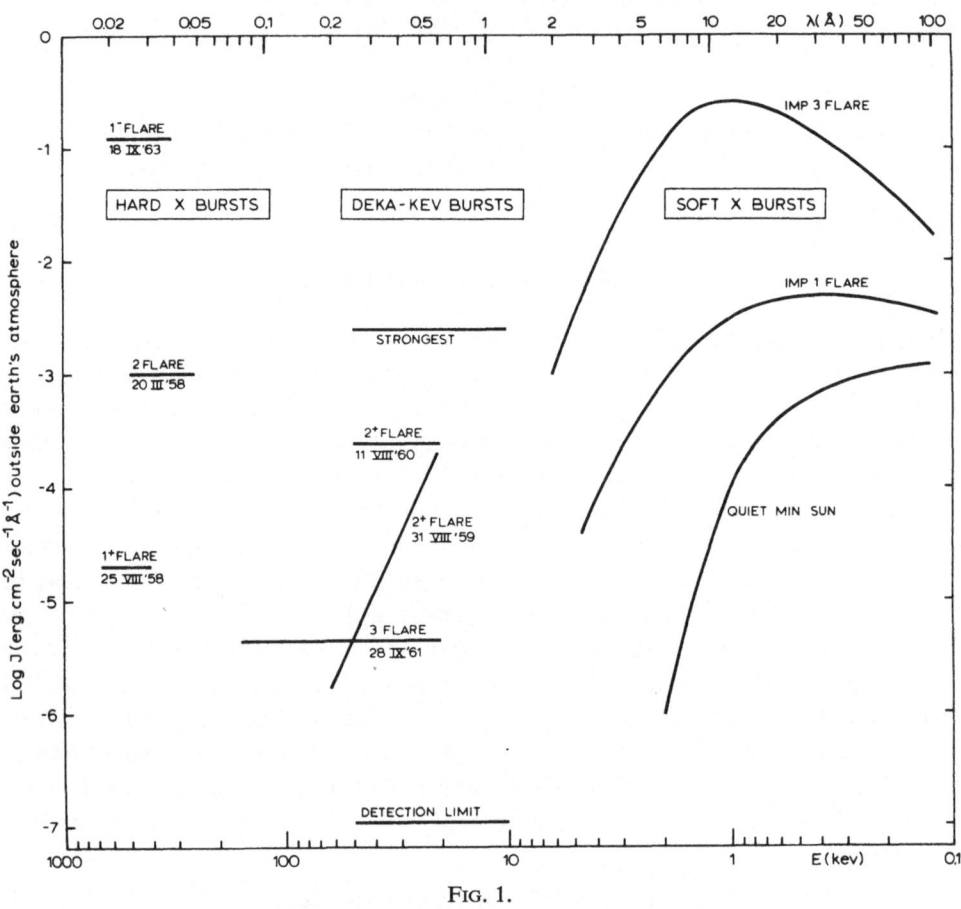

Fig. 1.

Table 1

Properties of hard X-ray bursts and the associated high-energy flare plasma

		duration	$\int Ne^2 \, dV$	$\int Ne \, dV$	Origin
soft X bursts ($E < 10$ keV)		as flare	10^{47}–10^{49}	10^{37}–10^{39}*	quasi-thermal
	deka–keV (10–200 keV)	5–10$^{\min}$		10^{33}–10^{37}	quasi-thermal and/or non-thermal
hard X bursts					
	deci–MeV (0·2–1 MeV)	few min		10^{32}–10^{35}	non-thermal

* Assuming $Ne = 10^{10}$.

had been assumed to consist of a quasi-thermal basis radiation component with a non-thermal burst superimposed on it.

Clearly, the interpretation of these features is still in its infancy, and many more observations should be obtained.

3. Coordination of Ground and Space Observations in the Investigation of the High-Energy Flare

We assume that the hard X-ray bursts are due to processes occurring in what we call the *high-energy flare plasma*: a highly excited transient plasma occurring in connection and together with solar flares. The number of electrons involved in this high-energy plasma ranges between 10^{32} (for the weakest deci–MeV bursts) to 10^{39} (for the strongest soft X-ray bursts). So, the volume involved may be very small for the high-energy phenomena and seems to be rather large, of the order of that of the sporadic coronal condensations for the plasmas producing the soft X-bursts. Assuming as example an electron density of 10^{10} and a thickness of 10^4 km, one would obtain a projected area, which, if circular, had a diameter ranging between 10^4 and 10 km. A model that is often assumed for the high-energy flare plasma is the one of Takakura and Kai (1966), in which electrons are supposed to be trapped in a magnetic field in an active region.

Important problems to be investigated in order to improve our understanding of the high-energy flare may be the following:

(a) *Space observations*: complete spectral observations should be obtained of the X-ray bursts in the whole energy range between 100 and 0·01 Å (0·1 and 1000 keV). It is not sufficient to have observations in only a small part of the spectrum. In order to understand the acceleration mechanism of the high-energy flare, to define its structure in relation to the other solar features, to investigate the time history of the high-energy flare, it is necessary to have the spectrum of the bursts in this whole energy range, and its variation with time.

(b) *Ground-based observations*: it seems necessary to obtain radioheliograms with an angular resolution, sufficient to enable the observer to establish the detailed structure of the high-energy radio flare, also in connection to the other solar features and in connection to the high-energy X-ray flare plasma. Spectral observations should be obtained in the whole microwave region, down to wavelengths of about 1 cm, or preferably lower.

References

Arnoldy, R.L., Kane, S.R., Winckler, J.R. (1967)   *Solar Phys.*, **2**, 171.
Chubb, T.A., Kreplin, R.W., Friedman, H. (1966)   *J. geophys. Res.*, **71**, 3611.
De Jager, C. (1967)   *Solar Phys.*, **2**, 347.
Takakura, T., Kai, K. (1966)   *Publ. Astron. Soc. Japan*, **18**, 57.

SOLAR GAMMA RAYS AND THEIR CORRELATION
WITH SPACE AND GROUND-BASED OBSERVATIONS

G. G. FAZIO

*(Smithsonian Astrophysical Observatory and Harvard College Observatory,
Cambridge, Mass., U.S.A.)*

Thus far, only two experiments have detected solar γ-radiation with energy significantly greater than 200 keV. In both events the γ-ray emission occurred during a solar
flare. The first observation was in 1958 by Peterson and Winckler (1959), who recorded
a burst of radiation that occurred in less than 18 sec from a class-2 solar flare. The
radiation spectrum peaked in the 200- to 500-keV region. Recently, Cline *et al.* (1967)
recorded in the OGO–3 satellite three rapid γ-ray bursts in the 80-keV to 1-MeV
energy range and measured the integral energy spectrum. The measurements were
made on July 7, 1966, during the first high-intensity flare (importance 3) of the new
solar cycle. Many attempts have been made to measure higher energy γ-radiation
from the quiet Sun and from solar flares, but no flux has been detected; this is primarily due to the fact that no high-energy γ-ray detectors have viewed a major solar flare
during the maximum of the optical or microwave burst. However, theoretical estimates of the flux of solar γ-rays, based on a simple flare model, indicate a readily
detectable flux from a major flare even to photon energies of 100 MeV. It is therefore
important that experiments be performed during the coming maximum of the solar
cycle to investigate this region of the electromagnetic spectrum.

To understand better how solar γ-ray measurements can improve our knowledge of
flares and to specify what other simultaneous observations should be performed, let
us briefly review the origin of γ-rays in a flare region. Gamma radiation from the Sun
results primarily from the interactions of high-energy electrons, protons, and nuclei
in the solar atmosphere. The intensity of the radiation is proportional to the product
of the high-energy particle intensity and the solar gas density in the volume where the
particles interact. The duration of the prompt γ-ray burst is related to the length of
time the particles are trapped. The important source mechanisms are:

(1) Electron bremsstrahlung.
(2) Compton scattering (inverse Compton effect).
(3) Nuclear deexcitation.
(4) Positron-electron annihilation.
(5) Decay of π^0 mesons.

From theoretical calculations the most important source mechanism in the 100-keV
to 1-MeV energy region is bremsstrahlung by high-energy electrons. The photon ener-

Perek (ed.), Highlights of Astronomy, 544–546. © *I.A.U.*

gy spectrum is continuous. Line emission occurs at 0·51 MeV because of positron annihilation, at 2·23 MeV because of deuteron formation, and throughout the spectrum from 0·5 to 10 MeV because of nuclear deexcitation. The most important lines in the last energy band are at 4·43 MeV owing to C^{14} and at 6·14 MeV owing to O^{16}. All the above sources of line emission indicate the occurrence of nuclear reactions during a flare. The intensity of the 2·23-MeV line is a measure of the neutron density in the flare region. The γ-ray flux above 20 MeV is primarily caused by π^0-meson decay, which produces a continuous energy spectrum peaking at about 70 MeV. The π^0 mesons are produced in high-energy proton and nucleon collisions with the solar gas.

The most important γ-ray measurements that can be performed are the following:

(1) The flux and energy spectrum as a function of time of the electron bremsstrahlung continuum. Microwave bursts that accompany X-ray radiation and γ-radiation may be caused by the synchrotron radiation of the same electrons.

(2) Time dependence of the flux of γ-ray line emission at 4·43 MeV and 6·14 MeV. This radiation is prompt and accurately reflects the time dependence of the product of the high-energy proton intensity and gas density in the flare volume. The 0·51-MeV and 2·23-MeV lines are delayed from the primary interactions and are therefore less useful. In addition, the 0·51-MeV line may be completely masked by the bremsstrahlung radiation.

(3) The flux and energy spectrum of the radiation above 20 MeV. This radiation is also prompt and accurately reflects the time dependence of the product of the high-energy proton intensity and gas density in the flare volume. In addition, the intensity of the radiation is very sensitive to the energy of the accelerated protons.

From the nature of the source mechanisms it is evident that knowledge of the time dependence of the solar γ-ray energy spectrum and flux can yield direct and important new information on the manner in which high-energy electrons, protons, and nuclei in a flare are accelerated. Gamma radiation is not affected by solar and interplanetary magnetic fields and is not readily absorbed; therefore, flux and energy measurements are directly related to the source mechanism. However, the value of γ-ray experiments in interpreting acceleration mechanisms is dependent on other simultaneous measurements from both satellites and ground-based observatories.

Probably the most important supporting observations are in the X-ray and radio regions of the spectrum. Time correlations of the fine structure of radio, X-ray, and γ-ray solar bursts are essential for the understanding of the motions of high-energy electrons. In the radio spectrum, microwave bursts appear most directly related to high-energy X-ray and γ-ray emission. Location of the fine space structure of X-ray radiation and radio radiation on the solar disk is extremely desirable. Heliographs of high spatial resolution (10″) at X-ray and radio wavelengths have already been proposed by De Jager (1964).

The production of secondary nuclear particles such as neutrons, deutrons, tritons, and He^3 is proportional to the product of the mean gas density and the length of time the high-energy particles are trapped. It is therefore important to measure the flux and energy of these particles that have escaped into the interplanetary medium, and to combine these data with γ-ray measurements to understand the details of the particle-acceleration mechanism. The high-energy neutron measurements are of particular value since these particles are not affected by magnetic fields.

The flux and energy spectrum of high-energy electrons, protons, and α particles injected into the interplanetary medium during a solar flare can also be correlated with γ-ray measurements to understand the trapping and modulation of these particles.

Finally, knowledge of the magnetic-field structure in the solar-flare region, particularly before and after a flare, is necessary for the understanding of the particular acceleration process.

It may be premature to discuss what observations should be correlated with solar γ-ray bursts when so few bursts have been detected, but the importance of these observations cannot be overemphasized. More experiments should be performed during the coming maximum of the solar cycle to measure solar γ-rays during a flare. At energies above 1 MeV, very few experiments are currently planned for satellites.

References

Cline, T. L., Holt, S. S., Hones, E. W., Jr. (1967)   *NASA Preprint X*-611-67-348.
De Jager, C. (1964)   *Research in Geophysics*, Vol. I, The Massachusetts Institute of Technology Press, Cambridge, Mass., p. 1.
Peterson, L. E., Winckler, J. R. (1959)   *J. geophys. Res.*, **64**, 697.

THE COORDINATION OF GROUND-BASED
AND SPACE OBSERVATIONS: SUMMARY

C. DE JAGER

*(University Observatory 'Sonnenborgh' and Space Research Laboratory,
Utrecht, The Netherlands)*

We briefly summarize the main contents of the papers presented by the contributors to the present Joint Discussion. Partly the remarks refer to the spacecraft payload, partly they aim at improving the ground observing facilities.

1. Space Observations

Both for the quiet Sun and for the active region it is clear that space observations are limited in their angular resolution. Facilities like those presented by the Apollo Telescope Mount should be used for observations of solar fine structures both in the visible and in the ultraviolet spectral regions: the observation of granules, their origin and evolution; the observations of flares and the correlation with ground-made magnetograms; the observations of the various structures visible in Hα filtergrams..., all these structures should eventually be observed with the highest possible resolution from space observatories (Kiepenheuer, Severny).

The need for simultaneous observations of X-ray heliograms during the occurrence of solar flares, surges and sprays has been stressed by various observers. These X-ray heliograms should be obtained with the highest resolution possible. Because of the limited photon flux this resolution can perhaps in no case be comparable to that obtainable in the visible and near ultraviolet region (Öhman, Fazio, De Jager).

Our understanding of solar flares could be greatly improved when flare spectra could be obtained in the Lyman series and in the Lyman continuum (Severny).

Also the study of detailed line profiles in the region between 1 and 10 Å may shed light on the question: what is the state of the high-energy flare plasma, is the observed broadening of the lines near 1–3 Å due to non-thermal effects and to an important amount of turbulent motions? (Neupert).

The necessity of obtaining complete spectral observations of flare-associated emissions in the whole spectral region between 0·1 and 1000 keV is also manifest for a detailed study of the mechanisms of electron acceleration in connection with flares, and the subsequent decay of the high-energy flare plasma (De Jager).

Perek (ed.), Highlights of Astronomy, 547–548. © I.A.U.

2. Ground-Based Observations

Ground-based observatories should look for the various high-energy phenomena like those called moustaches, points or Ellerman bombs in their relation to solar X-ray bursts. Is the source for the X-rays to be found in the fine structure of the active region, or is it the same area that is responsible for the emission of the optical flare? (Severny).

Plans were described for extensive ground-based observations coordinated with OSO observations: spectroheliograms should be obtained in Hα, the K line of Ca II, and in other spectral lines; means should be found to communicate between the participating observatories and the satellite (Reeves).

The relation between loop prominences and solar proton events was stressed (Jefferies, Fazio).

Most contributors stressed the importance of high-resolution magnetic-field observations and their variation with time.

Others also mentioned the need for high-resolution microwave spectroheliograms, as well as for detailed radiospectra of flares in the microwave region (De Jager).